Applied Mathematical Sciences
Volume 73

Applied Mathematical Sciences

(continued following index)

Stephen Wiggins

Global Bifurcations and Chaos
Analytical Methods

With 200 Illustrations

Springer Science+Business Media, LLC

Stephen Wiggins
Applied Mechanics 104-44
California Institute of Technology
Pasadena, CA 91125
USA

Editors

F. John
Courant Institute of
  Mathematical Sciences
New York University
New York, NY 10012
USA

J.E. Marsden
Department of
  Mathematics
University of California
Berkeley, CA 94720
USA

L. Sirovich
Division of
  Applied Mathematics
Brown University
Providence, RI 02912
USA

Mathematics Subject Classification (1980): 34xx, 58xx, 70

Library of Congress Cataloging-in-Publication Data
Wiggins, Stephen.
    Global bifurcations and chaos : analytical methods / Stephen Wiggins.
      p. cm.—(Applied mathematical sciences ; v. 73)
    Bibliography: p.
    Includes index.
    ISBN 978-1-4612-1041-2   ISBN 978-1-4612-1042-9 (eBook)
    DOI 10.1007/978-1-4612-1042-9
  3. Differential equations—Numerical solutions. I. Title.
  II. Series: Applied mathematical sciences (Springer Science+Business Media,
LLC) ; v. 73.
  QA1.A647 vol. 73
  [QA372]
  510 s—dc19
  [514'.74]   88-19959

Camera-ready copy provided by the author.

9 8 7 6 5 4 3 2 1

ISBN 978-1-4612-1041-2

To Meredith

PREFACE

The study of chaotic phenomena in deterministic nonlinear dynamical systems has attracted much attention over the last fifteen years. For the applied scientist, this study poses three fundamental questions. First, and most simply, what is meant by the term "chaos"? Second, by what mechanisms does chaos occur, and third, how can one predict when it will occur in a specific dynamical system? This book begins the development of a program that will answer these questions.

I have attempted to make the book as self-contained as possible, and thus have included some introductory material in Chapter One. The reader will find much new material in the remaining chapters. In particular, in Chapter Two, the techniques of Conley and Moser (Moser [1973]) and Afraimovich, Bykov, and Silnikov [1983] for proving that an invertible map has a hyperbolic, chaotic invariant Cantor set are generalized to arbitrary (finite) dimensions and to subshifts of finite type. Similar techniques are developed for the nonhyperbolic case. These nonhyperbolic techniques allow one to demonstrate the existence of a chaotic invariant set having the structure of the Cartesian product of a Cantor set with a surface or a "Cantor set of surfaces". In Chapter Three the nonhyperbolic techniques are applied to the study of the orbit structure near orbits homoclinic to normally hyperbolic invariant tori.

In Chapter Four, I develop a class of global perturbation techniques that enable one to detect orbits homoclinic or heteroclinic to hyperbolic fixed points, hyperbolic periodic orbits, and normally hyperbolic invariant tori in a large class of systems. The methods developed in Chapter Four are similar in spirit to a technique originally developed by Melnikov [1963] for periodically forced, two-dimensional systems;

however, they are much more general in that they are applicable to arbitrary (but finite) dimensional systems and allow for slowly varying parameters and quasiperiodic excitation. This general theory will hopefully be of interest to the applied scientist, since it allows one to give a criterion for chaotic dynamics in terms of the system parameters. Moreover, the methods apply in arbitrary dimensions, where much work remains to be done in chaos and nonlinear dynamics.

In this book I do not deal with the question of the existence of strange attractors. Indeed, this remains a major outstanding problem in the subject. However, this book does provide useful techniques for studying strange attractors, in that the first step in proving that a system possesses a chaotic attracting set is to prove that it possesses chaotic dynamics and then to show that the dynamics are contained in an attracting set that has no stable "regular" motions. One cannot deny that chaotic Cantor sets can radically influence the dynamics of a system; however, the extent and nature of this influence needs to be studied. This will require the development of new ideas and techniques.

Over the past two years many people have offered much encouragement and help in this project, and I take great pleasure in thanking them now.

Phil Holmes and Jerry Marsden gave me the initial encouragement to get started and criticized several early versions of the manuscript.

Steve Schecter provided extremely detailed criticisms of early versions of the manuscript which prevented many errors.

Steve Shaw read and commented on all of the manuscript.

Pat Sethna listened patiently to my explanations of various parts of the book and helped me considerably in clarifying my thoughts and presentation style.

John Allen and Roger Samelson called my attention to a crucial error in some earlier work.

Darryl Holm, Daniel David, and Mike Tratnik listened to several lengthy explanations of material in Chapters Three and Four and pointed out several errors in the manuscript.

Much of the material in Chapters Three and Four was first tried out in graduate applied math courses at Caltech. I am grateful to the students in those courses for enduring many obscure lectures and offering useful suggestions.

During the past two years Donna Gabai and Jan Patterson worked tirelessly on the layout and typing of this manuscript. They unselfishly gave of their time (often evenings and weekends) so that various deadlines could be met. Their skill and help made the completion of this book immensely easier.

I would also like to acknowledge the artists who drew the figures for this book and pleasantly tolerated my many requests for revisions. The figures for Chapter One were done by Betty Wood, and those for Chapter Four by Cecilia Lin. Peggy Firth, Pat Marble, and Bob Turring of the Caltech Graphic Arts Facilities and Joe Pierro, Haydee Pierro, Melissa Loftis, Gary Hatt, Marcos Prado, Bill Contado, Abe Won, and Stacy Quinet of Imperial Drafting Inc. drew the figures for Chapters Two and Three.

Finally, Meredith Allen gave indispensable advice and editorial assistance throughout this project.

CONTENTS

CHAPTER 1
Introduction: Background for Ordinary Differential Equations and Dynamical Systems

The purpose of this first chapter is to review and develop the necessary concepts from the theory of ordinary differential equations and dynamical systems which we will need for the remainder of the book. We will begin with some results from classical ordinary differential equations theory such as existence and uniqueness of solutions, dependence of solutions on initial conditions and parameters, and various concepts of stability. We will then discuss more modern ideas such as genericity, structural stability, bifurcations, and Poincaré maps. Standard references for the theory of ordinary differential equations are Coddington and Levinson [1955], Hale [1980], and Hartman [1964]. We will take a more global, geometric point of view of the theory; some references which share this viewpoint are Arnold [1973], Guckenheimer and Holmes [1983], Hirsch and Smale [1974], and Palis and deMelo [1982].

1.1. The Structure of Solutions of Ordinary Differential Equations

In this book we will regard an ordinary differential equation as a system of equations having the following form

$$\dot{x} = f(x,t) , \qquad (x,t) \in \mathbb{R}^n \times \mathbb{R}^1 \qquad (1.1.1)$$

where $f : U \to \mathbb{R}^n$ with U an open set in $\mathbb{R}^n \times \mathbb{R}^1$ and $\dot{x} \equiv dx/dt$. The space of dependent variables is often referred to as the *phase* or *state space* of the system (1.1.1). By a solution of (1.1.1) we will mean a map

$$\phi : I \to \mathbb{R}^n \qquad (1.1.2)$$

where I is some interval in $\mathbb{R}$ such that

$$\dot{\phi}(t) = f\left(\phi(t), t\right) . \qquad (1.1.3)$$

Thus, geometrically (1.1.1) can be viewed as defining a vector at every point in U, and a solution of (1.1.1) is a curve in $\mathbb{R}^n$ whose tangent or velocity vector at each point is given by $f(x, t)$ evaluated at the specific point. For this reason (1.1.1) is often referred to as a *vector field*.

Now, the existence of solutions of (1.1.1) is certainly not obvious and evidently must rely in some way on the properties of f, so now we want to give some classical results concerning existence of solutions of (1.1.1) and their properties.

1.1a. Existence and Uniqueness of Solutions

Suppose that f is C^r in U (note: by C^r, $r \geq 1$, we mean that f has r derivatives which are continuous at each point of U; C^0 means that f is continuous at each point of U) and for some $\epsilon_1, \epsilon_2 > 0$ let $I_1 = \{ t \in \mathbb{R} \mid t_0 - \epsilon_1 < t < t_0 + \epsilon_1 \}$ and $I_2 = \{ t \in \mathbb{R} \mid t_0 - \epsilon_2 < t < t_0 + \epsilon_2 \}$; then we have the following theorem.

Theorem 1.1.1. *Let (x_0, t_0) be a point in U. Then for ϵ_1 sufficiently small there exists a solution of (1.1.1), $\phi_1 : I_1 \to \mathbb{R}^n$, satisfying $\phi_1(t_0) = x_0$. Moreover, if f is C^r in U, $r \geq 1$, and $\phi_2 : I_2 \to \mathbb{R}^n$ is also a solution of (1.1.1) satisfying $\phi_2(t_0) = x_0$, then $\phi_1(t) = \phi_2(t)$ for all $t \in I_3 = \{ t \in \mathbb{R} \mid t_0 - \epsilon_3 < t < t_0 + \epsilon_3 \}$ where $\epsilon_3 = \min\{\epsilon_1, \epsilon_2\}$.*

PROOF: See Arnold [1973] or Hale [1980].   □

We make the following remarks concerning Theorem 1.1.1.

1) For a solution of (1.1.1) to exist, only continuity of f is required; however, in this case the solution passing through a given point in U may not be unique (see Hale [1980] for an example). If f is at least C^1 in U, then there is only one solution passing through any given point of U (note: for uniqueness of solutions one actually only needs f to be Lipschitz in the x variable uniformly in t, see Hale [1980] for the proof). The degree of differentiability of the vector field will not be a major concern to us in this book since all of the examples we consider will be infinitely differentiable.

2) The differentiability of solutions with respect to t was not explicitly considered in the theorem, although evidently they must be at least C^r since f is C^r. This result will be stated shortly.

3) Notation: In denoting the solutions of (1.1.1) it may be useful to note the dependence on initial conditions explicitly. For ϕ, a solution of (1.1.1) passing through the point $x = x_0$ at $t = t_0$, the notation would be

$$\phi(t, t_0, x_0) \quad \text{with} \quad \phi(t_0, t_0, x_0) = x_0 \,. \tag{1.1.4}$$

In some cases, the initial time is always understood to be a specific value (often $t_0 = 0$); in this case, the explicit dependence on the initial time is omitted and the solution is written as

$$\phi(t, x_0) \quad \text{with} \quad \phi(t_0, x_0) = x_0 \,. \tag{1.1.5}$$

1.1b. Dependence on Initial Conditions and Parameters

In the computation of stability properties of solutions and in the construction of Poincaré maps (see Section 1.6) the differentiability of solutions with respect to initial conditions is very important.

Theorem 1.1.2. *If $f(x,t)$ is C^r in U, then the solution of 1.1.1, $\phi(t, t_0, x_0)$ $(x_0, t_0) \in U$, is a C^r function of t, t_0 and x_0.*

PROOF: See Arnold [1973] or Hale [1980]. □

Theorem 1.1.2 justifies the procedure of computing the Taylor series expansion of a solution of (1.1.1) about a given initial condition. This enables one to determine the nature of solutions near a particular solution. Often the linear term in such an expansion is sufficient for determining many of the local properties near a particular solution (e.g., stability). The following theorem gives an equation which the first derivative of the solution with respect to x_0 must obey.

Theorem 1.1.3. *Suppose $f(x,t)$ is C^r, $r \geq 1$, in U and let $\phi(t, t_0, x_0)$, $(x_0, t_0) \in U$, be a solution of (1.1.1). Then the $n \times n$ matrix $D_{x_0}\phi$ is the solution of the following linear ordinary differential equation*

$$\dot{Z} = D_x f(\phi(t), t) Z, \qquad Z(t_0) = \mathrm{id}, \tag{1.1.6}$$

where Z is an $n \times n$ matrix and id denotes the $n \times n$ identity matrix.

PROOF: See Arnold [1973], Hale [1980], or Irwin [1980].   □

Equation (1.1.6) is often referred to as the *first variational equation*. We remark that it is possible to find linear ordinary differential equations which the higher order derivatives of solutions with respect to the initial conditions must obey; however, we will not need these in this book.

Now suppose that equation (1.1.1) depends on parameters

$$\dot{x} = f(x, t; \epsilon), \qquad (x, t, \epsilon) \in \mathbf{R}^n \times \mathbf{R}^1 \times \mathbf{R}^p \qquad (1.1.7)$$

where $f : U \to \mathbf{R}^n$ with U an open set in $\mathbf{R}^n \times \mathbf{R}^1 \times \mathbf{R}^p$. We have the following theorem.

Theorem 1.1.4. *Suppose $f(x, t; \epsilon)$ is C^r in U. Then the solution of (1.1.7), $\phi(t, t_0, x_0, \epsilon)$ $(x_0, t_0, \epsilon) \in U$, is a C^r function of ϵ.*

PROOF: See Arnold [1973] or Hale [1980].   □

In many applications it is useful to seek Taylor series expansions in ϵ of solutions of (1.1.7) (e.g., in perturbation theory and bifurcation theory). Analogous to Theorem 1.1.3, the following theorem gives an ordinary differential equation which the first derivative of a solution of (1.1.7) with respect to ϵ must obey.

Theorem 1.1.5. *Suppose $f(x, t, \epsilon)$ is C^r, $r \geq 1$, in U and let $\phi(t, t_0, x_0, \epsilon)$, $(x_0, t_0, \epsilon) \in U$, be a solution of (1.1.7). Then the $n \times p$ matrix $D_\epsilon \phi$ satisfies the following linear ordinary differential equation*

$$\dot{Z} = D_x f\left(\phi(t), t; \epsilon\right) Z + D_\epsilon f\left(\phi(t), t; \epsilon\right), \qquad z(t_0) = 0, \qquad (1.1.8)$$

where Z is a $n \times p$ matrix and 0 represents the $n \times p$ matrix of zeros.

PROOF: See Hale [1980].   □

1.1c. Continuation of Solutions

Theorem 1.1.1 gave sufficient conditions for the existence of solutions of (1.1.1) but only on a sufficiently small time interval. We will now give a theorem which justifies the extension of this time interval, but first we need the following definition.

Definition 1.1.1. Let ϕ_1 be a solution of (1.1.1) defined on the interval I_1, and let ϕ_2 be a solution of (1.1.1) defined on the interval I_2. We say that ϕ_2 is a *continuation* of ϕ_1 if $I_1 \subset I_2$ and $\phi_1 = \phi_2$ on I_1. A solution is *noncontinuable* if no such continuation exists; in this case, I_1 is called the *maximal interval of existence* of ϕ_1.

We now state the following theorem concerning continuation of solutions.

Theorem 1.1.6. *Suppose* $f(x,t)$ *is* C^r *in* U *and* $\phi(t, t_0, x_0)$, $(x_0, t_0) \in U$, *is a solution of (1.1.1), then there is a continuation of* ϕ *to a maximal interval of existence. Furthermore, if* (t_1, t_2) *is a maximal interval of existence for* ϕ, *then* $(\phi(t), t)$ *tends to the boundary of* U *as* $t \to t_1$ *and* $t \to t_2$.

PROOF: See Hale [1980]. $\qquad\qquad\qquad\qquad\qquad\qquad\qquad\qquad\qquad$ $\square$

Terminology

At this point we want to introduce some common terminology that applies to solutions of ordinary differential equations. Recall that a solution of (1.1.1) is a map $\phi: I \to \mathbf{R}^n$ where I is some interval in $\mathbf{R}$. Geometrically, the image of I under ϕ is a curve in $\mathbf{R}^n$, and this geometrical picture gives rise to the following terminology.

1) A solution $\phi(t, t_0, x_0)$ of (1.1.1) may also be called the *trajectory, phase curve or motion* through the point x_0.

2) The graph of the solution $\phi(t, t_0, x_0)$, i.e.,

$$\left\{ (x, t) \in \mathbf{R}^n \times \mathbf{R}^1 \mid x = \phi(t, t_0, x_0),\ t \in I \right\}$$

is called an *integral curve*.

3) Suppose we have a solution $\phi(t, t_0, x_0)$; then the set of points in $\mathbf{R}^n$ through which this solution passes as t varies through I is called the *orbit through* x_0, denoted $O(x_0)$ and written as follows.

$$O(x_0) = \{ x \in \mathbf{R}^n \mid x = \phi(t, t_0, x_0),\quad t \in I \}\ .$$

We remark that it follows from this definition that, for any $T \in I$,

$$O\left(\phi(T, t_0, x_0)\right) = O(x_0)\ .$$

The following example should serve to illustrate the terminology.

EXAMPLE 1.1.1. Consider the following equation

$$\ddot{x} + x = 0 \,. \tag{1.1.9}$$

This is just the equation for a simple harmonic oscillator having frequency one. Writing (1.1.9) as a system we obtain

$$\dot{x} = y \,,$$
$$\dot{y} = -x \,. \tag{1.1.10}$$

Equation (1.1.10) has the form of equation (1.1.1) with phase space $\mathbf{R}^2$. The solution of (1.1.10) passing through the point (1,0) at $t = 0$ is given by $\phi(t) = (\cos t, -\sin t)$.

1) The *trajectory, phase curve or motion* through the point (1,0) is illustrated in Figure 1.1.1.

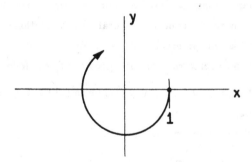

Figure 1.1.1. Trajectory through the Point (1,0).

2) The *integral curve* of the solution $\phi(t) = (\cos t, -\sin t)$ is illustrated in Figure 1.1.2.

3) The orbit through the point (1,0) is given by $\left\{ (x,y) \in \mathbf{R}^2 \mid x^2 + y^2 = 1 \right\}$ and is illustrated in Figure 1.1.3.

We remark that, although the solution through (1,0) passes through the same set of points in $\mathbf{R}^2$ as the orbit through (1,0), and thus both appear to be the same object when viewed as a locus of points in $\mathbf{R}^2$, we stress that they are indeed different objects. A solution must pass through a specific point at a specified time and an

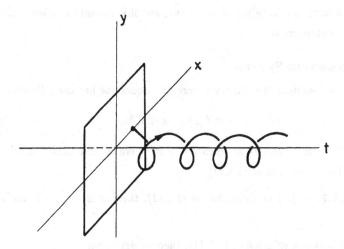

Figure 1.1.2. Integral Curve of $\phi(t) = (\cos t, -\sin t)$.

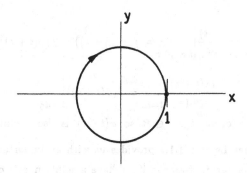

Figure 1.1.3. Orbit through (1,0).

orbit can be thought of as a one parameter family of solutions corresponding to a curve of possible initial conditions for different solutions at a specific time. In the qualitative theory of ordinary differential equations it is not unusual to use the terms orbit and solution interchangeably and, usually, no harm comes from this.

There is a difference in the nature of solutions depending upon whether or not the vector field depends explicitly on the independent variable (note: we will henceforth always refer to the independent variable as time). If the vector field is independent of time it is called *autonomous*, and if it depends explicitly on time it is called *nonautonomous*.

1.1d. Autonomous Systems

An autonomous system of ordinary differential equations has the following form

$$\dot{x} = f(x), \quad x \in \mathbf{R}^n \tag{1.1.11}$$

where $f : U \to \mathbf{R}^n$ with U an open set in $\mathbf{R}^n$. We assume that f is C^r, $r \geq 1$, and let $\phi(t)$ be a solution of (1.1.11).

Lemma 1.1.7. *If $\phi(t)$ is a solution of (1.1.11), then so is $\phi(t + \tau)$ for any real number τ.*

PROOF: If $\phi(t)$ is a solution of (1.1.11), then by definition

$$\frac{d\phi(t)}{dt} = f\big(\phi(t)\big). \tag{1.1.12}$$

So we have

$$\frac{d\phi(t + \tau)}{dt}\bigg|_{t=t_0} = \frac{d\phi(t)}{dt}\bigg|_{t=t_0+\tau} = f\big(\phi(t_0 + \tau)\big) = f\big(\phi(t + \tau)\big)\bigg|_{t=t_0} \tag{1.1.13}$$

or

$$\frac{d\phi(t + \tau)}{dt}\bigg|_{t=t_0} = f\big(\phi(t + \tau)\big)\bigg|_{t=t_0}. \tag{1.1.14}$$

Now (1.1.14) is true for any t_0, $\tau \in \mathbf{R}$, so $\phi(t + \tau)$ is also a solution of (1.1.11). $\square$

We remark that Lemma 1.1.7 provides us with an important fact which will prove useful in Chapter 4. Namely, if we have a solution $\phi(t)$ of an autonomous equation, then we immediately have a parametric representation of the orbit of this solution of the form $\phi(t + \tau)$, where $\tau \in I$ is regarded as the parameter. Thus, we can view t as fixed, and varying τ carries us through the orbit of $\phi(t)$.

Two important properties of solutions of autonomous equations are given in the following lemmas.

Lemma 1.1.8. *Suppose that f is C^r in U, $r \geq 1$, and $\phi_1(t)$, $\phi_2(t)$ are solutions of (1.1.11) defined on I_1 and I_2, respectively, with $\phi_1(t_1) = \phi_2(t_2) = p$. Then $\phi_1(t - (t_2 - t_1)) = \phi_2(t)$ on their common interval of definition.*

PROOF: Let $\gamma(t) = \phi_1\big(t - (t_2 - t_1)\big)$; then by Lemma 1.1.7 $\gamma(t)$ is also a solution of (1.1.11). Now notice that $\gamma(t_2) = \phi_1(t_1) = p = \phi_2(t_2)$. Thus, $\gamma(t)$ and $\phi_2(t)$

are solutions of (1.1.11) which satisfy the same initial condition (taking the initial time as $t = t_2$); thus, by uniqueness of solutions (Theorem 1.1.1), $\gamma(t)$ (and hence $\phi_1(t - (t_2 - t_1))$) and $\phi_2(t)$ must coincide on their common interval of definition. $\square$

Lemma 1.1.9. *Suppose that f is C^r in U, $r \geq 1$, and $\phi(t)$ is a solution of (1.1.11) defined on I. Suppose there exist two points $t_1, t_2 \in I$, $t_1 < t_2$, such that $\phi(t_1) = \phi(t_2)$. Then $\phi(t)$ exists for all $t \in \mathbb{R}$ and is periodic in t with period $T = t_2 - t_1$, i.e., $\phi(t) = \phi(t + T)$ for all $t \in \mathbb{R}$.*

PROOF: Let $\psi(t) = \phi(t + t_1)$, by Lemma 1.1.7 $\psi(t)$ is a solution of (1.1.11). Then we have

$$\psi(t + T) = \phi(t_1 + t + T) = \phi(t + t_2) . \tag{1.1.15}$$

Now, since $\phi(t_1) = \phi(t_2)$, we must have $\phi(t + t_1) = \phi(t + t_2)$ by uniqueness of solutions (Theorem 1.1.1). Therefore,

$$\psi(t + T) = \phi(t + t_2) = \phi(t + t_1) = \psi(t) . \tag{1.1.16}$$

Therefore, $\psi(t)$ is periodic in time with period T and $\phi(t)$ is likewise periodic in time with period T, and since every $t \in \mathbb{R}$ can be written in the form $t = nT + \tau$, $0 \leq \tau < T$, $\phi(t)$ exists for all time. $\square$

Lemmas 1.1.8 and 1.1.9 tell us that solutions (and hence all orbits) of autonomous equations cannot intersect themselves or each other in isolated points without coinciding on their common intervals of definition. These facts can be extremely useful in determining certain global properties of the orbit structure of an ordinary differential equation (e.g., this fact is largely responsible for the Poincaré-Bendixson theorem, see Hale [1980] or Palis and deMelo [1982]).

1.1e. Nonautonomous Systems

A nonautonomous system of ordinary differential equations has the following form

$$\dot{x} = f(x, t) , \quad (x, t) \in \mathbb{R}^n \times \mathbb{R}^1 \tag{1.1.17}$$

where $f : U \to \mathbb{R}^n$ with U an open set in $\mathbb{R}^n \times \mathbb{R}^1$. Lemma 1.1.7 does not follow for nonautonomous systems. Consider the following example.

EXAMPLE 1.1.2. Consider the following nonautonomous ordinary differential equation

$$\dot{x} = e^t .$$ (1.1.18)

The solution of equation (1.1.18) is obviously $\phi(t) = e^t$, and it is clear that $\phi(t+\tau) = e^{t+\tau}$ is not a solution of (1.1.18) for $\tau \neq 0$.

Example 1.1.2 shows that time translations of solutions of nonautonomous equations are not likewise solutions of the equations. This was the crucial property which led to the proofs of Lemmas 1.1.8 and 1.1.9 so we conclude that it is possible for solutions of nonautonomous ordinary differential equations to intersect themselves and each other. This can lead to a very complicated geometrical structure of the solutions of nonautonomous ordinary differential equations.

Often the geometry of the solution structure of nonautonomous ordinary differential equations is clarified by enlarging the phase space by redefining time as a new *dependent* variable. This is done as follows: by writing (1.1.17) as

$$\frac{dx}{dt} = \frac{f(x,t)}{1}$$ (1.1.19)

and using the chain rule, we can introduce a new independent variable, s, so that (1.1.19) becomes

$$\frac{dx}{ds} \equiv x' = f(x,t) ,$$
$$\frac{dt}{ds} \equiv t' = 1 .$$ (1.1.20)

If we define $y = (x,t)$ and $g(y) = (f(y),1)$, we see that (1.1.20) has the form of an autonomous ordinary differential equation with phase space $\mathbf{R}^n \times \mathbf{R}^1$.

$$y' = g(y) , \qquad y \in \mathbf{R}^n \times \mathbf{R}^1 .$$ (1.1.21)

Of course, knowledge of the solutions of (1.1.21) implies knowledge of the solutions of (1.1.17) and vice versa. For example, if $\phi(t)$ is a solution of (1.1.17) passing through $x = x_0$ at $t = t_0$, i.e., $\phi(t_0) = x_0$, then $\psi(s) = (\phi(s+t_0), t(s) = s+t_0)$ is a solution of (1.1.17) passing through $y = y_0 \equiv (x_0,t_0)$ at $s = 0$. This apparently trivial trick is a great aid in the construction of Poincaré maps, as we shall see in Section 1.6, and it also justifies the consideration of autonomous systems exclusively. Henceforth we will state concepts only for autonomous ordinary differential equations. For the most part we will consider autonomous ordinary differential equations in this book; the nonautonomous equations which we consider will have either periodic or quasiperiodic time dependence, and in each case we will reduce the study of such systems to the study of an associated Poincaré map (see Section 1.6).

1.1f. Phase Flows

Consider the following autonomous ordinary differential equation

$$\dot{x} = f(x), \qquad x \in \mathbb{R}^n \qquad\qquad (1.1.22)$$

where f is C^r, $r \geq 1$, on some open set $U \subset \mathbb{R}^n$. Let $\phi(t, t_0, x_0)$ be a solution of (1.1.22) defined on the interval I. We will henceforth take $t_0 = 0$ and drop the explicit dependence on t_0 from the solution of (1.1.22); i.e., we have $\phi(t, x_0)$ with $\phi(0, x_0) = x_0$.

Lemma 1.1.10. *i)* $\quad \phi(t, x_0)$ *is* C^r.

$\qquad\qquad$ *ii)* $\quad \phi(0, x_0) = x_0$.

$\qquad\qquad$ *iii)* $\quad \phi(t + s, x_0) = \phi(t, \phi(s, x_0)), t + s \in I$.

PROOF: *i)* follows from Theorem 1.1.2 and *ii)* is by definition. The proof of *iii)* goes as follows: let $\gamma(t) = \phi(t + s, x_0)$, then $\gamma(t)$ solves (1.1.22) with $\gamma(0) = \phi(s, x_0)$. Also, we have $\phi(t, \phi(s, x_0))$ satisfies (1.1.22) with $\phi(0, \phi(s, x_0)) = \phi(s, x_0)$. So $\gamma(t) \equiv \phi(t + s, x_0)$ and $\phi(t, \phi(s, x_0))$ are both solutions of (1.1.22) satisfying the same initial condition at $t = 0$; hence by Lemma 1.1.8 $\phi(t + s, x_0) = \phi(t, \phi(s, x_0))$ on their common interval of definition. $\qquad\square$

Since $\phi(t, x_0)$ is C^r, viewing t as fixed we see that $\phi(t, x_0) \equiv \phi_t(x_0)$ defines a C^r map of U into $\mathbb{R}^n$. Thus $\phi_t(x_0)$ is a one parameter family of maps of $U \to \mathbb{R}^n$. By property *iii)* of Lemma 1.1.10 we see that this C^r one parameter family of maps is invertible with C^r inverse. A C^r invertible map having a C^r inverse is called a C^r *diffeomorphism* (if $r = 0$, the term *homeomorphism* is used). So we see that the solutions of an autonomous ordinary differential equation generate a one parameter family of diffeomorphisms of the phase space onto itself. This one parameter family of diffeomorphisms is called a *phase flow*.

1.1g. Phase Space

As mentioned earlier, the phase space of a system is the space of dependent variables which was taken to be some open set in $\mathbf{R}^n$. However, in some applications the space of dependent variables naturally arises as a surface such as a cylinder or torus or, more generally, a differentiable manifold (see Section 1.3 for the definition of a differentiable manifold). We consider several common examples.

Circle: Consider the ordinary differential equation

$$\dot{\theta} = \omega , \qquad \theta \in (0, 2\pi]$$

where $\omega > 0$ is constant. The phase space of this equation is the interval $(0, 2\pi]$ with 0 and 2π identified. Thus, the phase space has the structure of a circle having length 2π which we denote as S^1 (note: the superscript one refers to the dimension of the phase space).

Cylinder: Mathematically the cylinder is denoted by $\mathbf{R}^1 \times S^1$. Consider the following equation which describes the dynamics of a free undamped pendulum

$$\dot{\theta} = v$$
$$\dot{v} = - \sin \theta .$$

(1.1.23)

The angular velocity, v, can take on any value in $\mathbf{R}$ but, since the motion is rotational, the position, θ, is periodic with period 2π. Hence, the phase space of the pendulum is the cylinder $\mathbf{R}^1 \times S^1$. Figure 1.1.4a shows the orbits of the pendulum on $\mathbf{R}^1 \times S^1$ and Figure 1.1.4b gives an alternate representation of the cylinder.

Torus: Heuristically, we think of a torus as the surface of a donut; if we consider the surface plus its interior, we speak of the solid torus. Mathematically, the two dimensional torus is denoted by $T^2 = S^1 \times S^1$, i.e., the Cartesian product of two circles.

Consider the following ordinary differential equation

$$\dot{\theta}_1 = \omega_1$$
$$\dot{\theta}_2 = \omega_2$$
$$\theta_1, \theta_2 \in (0, 2\pi]$$

(1.1.24)

where ω_1 and ω_2 are positive constants. Since θ_1 and θ_2 are angular variables, the phase space of (1.1.24) is $S^1 \times S^1 = T^2$. If we draw the torus as the surface of

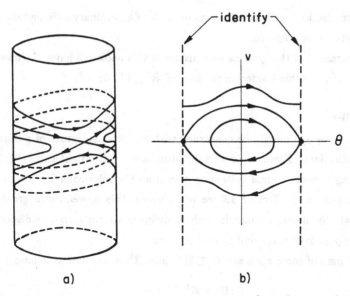

Figure 1.1.4. Phase Space of the Pendulum on a) $\mathbf{R}^1 \times S^1$. b) $\mathbf{R}^1 \times \mathbf{R}^1$.

a donut in $\mathbf{R}^3$ the orbits of (1.1.24) spiral around the surface and close (i.e., they are periodic) when ω_1/ω_2 is a rational number; alternatively, they densely fill the surface when ω_1/ω_2 is an irrational number, see Arnold [1973] for a detailed proof of these statements.

Another useful way of representing the two torus is to first cut the torus along $\theta_1 = 0$ (resulting in a tube), next cut along $\theta_2 = 0$ (resulting in a sheet), and finally flatten out the sheet into a square. Thus, we can make a torus from a square by identifying the two vertical sides of the square and the two horizontal sides of the square. Mathematically, this representation of the torus is written $\mathbf{R}^2/\mathbf{Z}^2$ (read "r two mod z two") which means that, given any two points $x, y \in \mathbf{R}^2$, we consider x and y to be the same point if $x = y + 2\pi n$ where n is some integer two vector.

These ideas and notations also go through in n dimensions. The n torus is written as $T^n = \underbrace{S^1 \times \cdots \times S^1}_{n \text{ factors}}$ (note: $T^1 \equiv S^1$) or, equivalently, T^n can be thought of as $\mathbf{R}^n/\mathbf{Z}^n$, i.e., as a n dimensional cube with opposite sides identified.

Sphere: The n sphere of radius R is denoted by S^n and is defined as

$$S^n = \left\{ x \in \mathbf{R}^{n+1} \big| \, |x| = R \right\}. \tag{1.1.25}$$

See the introduction to Chapter 3 for an example of an ordinary differential equation whose orbits lie on a sphere.

We remark that the systems we consider in this book will have as phase spaces $\mathbf{R}^n$, T^n, S^n, or some Cartesian product of $\mathbf{R}^n$, T^n, or S^n.

1.1h. Maps

In this book we will mostly be concerned with the orbit structure of ordinary differential equations. However, in certain situations much insight can be gained by constructing a discrete time system or *map* from the solutions of an ordinary differential equation. In Section 1.6 we will consider this procedure in great detail; however, at this point, we merely wish to define the term map and discuss some various properties of maps and their dynamics.

A C^r map of some open set $U \subset \mathbf{R}^n$ into $\mathbf{R}^n$ is denoted as follows

$$\begin{aligned} f : U &\to \mathbf{R}^n \\ x &\mapsto f(x), \qquad x \in U \end{aligned} \tag{1.1.26}$$

with f C^r in U. We will be interested in the dynamics of f. By this we mean the nature of the iterates of points in U under f. For a point $x \in U$, equivalent notations for the nth iterate of x under f are

$$\underbrace{f\left(f\left(\cdots\left(f\left(x\right)\right)\right)\cdots\right)}_{n \text{ times}} \equiv \underbrace{f \circ f \circ \cdots \circ f}_{n \text{ times}}(x) \equiv f^n(x). \tag{1.1.27}$$

By the orbit of x under f we will mean the following bi-infinite sequence if f is invertible

$$\left\{\ldots, f^{-n}(x), \ldots, f^{-1}(x), x, f(x), \ldots, f^n(x), \ldots\right\}, \tag{1.1.28}$$

and the following infinite sequence if f is noninvertible.

$$\left\{x, f(x), \ldots, f^n(x), \ldots\right\}. \tag{1.1.29}$$

This brings up an important difference between orbits of ordinary differential equations and orbits of maps. Namely, orbits of ordinary differential equations are curves and orbits of maps are discrete sets of points. In Chapter 3 we will see that this difference is significant.

1.1i. Special Solutions

At this time we want to consider various special solutions and orbits which are often important in applications.

1) *Fixed point, equilibrium point, stationary point, rest point, singular point, or critical point*. These are all synonyms for a point p in the phase space of an ordinary differential equation that is also a solution of the equation, i.e., for the equation $\dot{x} = f(x)$ we have

$$0 = f(p) \qquad (1.1.30)$$

or a point in the phase space of a map $x \mapsto f(x)$ such that

$$p = f(p) . \qquad (1.1.31)$$

For a map, p may also be called a period one point. In this book we will exclusively use the term fixed point when referring to such solutions.

2) *Periodic Motions*. A *periodic solution*, $\phi(t)$, of an ordinary differential equation is a solution which is periodic in time, i.e., $\phi(t) = \phi(t + T)$ for some fixed positive constant T. T is called the period of $\phi(t)$. A *periodic orbit* of an ordinary differential equation is the orbit of any point through which a periodic solution passes.

For maps, a *period k point*, p, is a point such that $f^k(p) = p$. The orbit of a period k point is a sequence of k distinct points

$$\left\{ p, f(p), \ldots, f^{k-1}(p) \right\} \qquad (1.1.32)$$

and the orbit is called a *periodic orbit of period k*.

3) *Quasiperiodic Motions*

Definition 1.1.2. A function

$$h : \quad \mathbf{R}^1 \to \mathbf{R}^m$$
$$t \mapsto h(t)$$

is called *quasiperiodic* if it can be represented in the form

$$h(t) = H(\omega_1 t, \ldots, \omega_n t) \qquad (1.1.33)$$

where $H(x_1, \ldots, x_n)$ is a function of period 2π in $x_1, \ldots, x_n$. The real numbers $\omega_1, \ldots, \omega_n$ are called the *basic frequencies*. We shall denote by $C^r(\omega_1, \ldots, \omega_n)$ the class of $h(t)$ for which $H(x_1, \ldots, x_n)$ is r times continuously differentiable.

EXAMPLE 1.1.3. $h(t) = \gamma_1 \cos \omega_1 t + \gamma_2 \sin \omega_2 t$ is a quasiperiodic function.

(Note: There exists a more general class of functions called *almost periodic functions* which can be viewed as quasiperiodic functions having an infinite number of basic frequencies. These will not be considered in this book, see Hale [1980] for a discussion and rigorous definitions.)

A *quasiperiodic solution* $\phi(t)$, of an ordinary differential equation is a solution which is quasiperiodic in time. A *quasiperiodic orbit* is the orbit of any point through which $\phi(t)$ passes. A quasiperiodic orbit may be interpreted geometrically as lying on an n dimensional torus. This can be seen as follows. Consider the equation

$$y = H(x_1, \ldots, x_n).$$

Then, if $m \geq n$ and $D_x H$ has rank n for all $x = (x_1, \ldots, x_n)$, then this equation can be viewed as an embedding of an n-torus in m space with $x_1, \ldots, x_n$ serving as coordinates on the torus. Now, viewing $h(t)$ as a solution of an ordinary differential equation, since $x_i = \omega_i t$, $i = 1, \ldots, n$, $h(t)$ can be viewed as tracing a curve on the n-torus as t varies.

In recent years quasiperiodic orbits of maps have received much attention, mainly in the context of maps of the circle and annulus. These will not be studied in this book but see Katok [1983] for an overview and recent references.

4) *Homoclinic and Heteroclinic Motions.* These will be defined and studied in great detail in Chapter 3.

1.1j. Stability

The general theory of stability is a very large subject to which many books have been devoted. However, in this section we will only consider those aspects of the theory which have particular relevance to the subjects covered in this book, namely, the stability of specific solutions of ordinary differential equations and its determination and the stability of periodic orbits of maps and its determination. We refer the reader to Rouche, Habets, and Laloy [1977], Yoshizawa [1966], LaSalle [1976],

Abraham and Marsden [1978], and references therein for a more complete discussion of stability.

Consider the ordinary differential equation

$$\dot{x} = f(x), \qquad x \in \mathbb{R}^n \tag{1.1.34}$$

where $f : U \to \mathbb{R}^n$ with U an open set in $\mathbb{R}^n$ and f is C^r, $r \geq 1$. Let $\phi(t)$ be a solution of (1.1.34).

Definition 1.1.3. $\phi(t)$ is said to be Liapunov stable, or stable if given $\epsilon > 0$ we can find a $\delta = \delta(\epsilon) > 0$ such that for any other solution $\psi(t)$ of (1.1.34) with $|\psi(t_0) - \phi(t_0)| < \delta$ then we have $|\psi(t) - \phi(t)| < \epsilon$ for $t \in [t_0, \infty)$.

If $\phi(t)$ is not stable then it is said to be *unstable*.

Definition 1.1.4. $\phi(t)$ is said to be asymptotically stable if it is Liapunov stable and there exists $\bar{\delta} > 0$ such that if $|\phi(t_0) - \psi(t_0)| < \bar{\delta}$, then $\lim\limits_{t \to \infty} |\phi(t) - \psi(t)| = 0$.

We remark that, for autonomous systems, δ and $\bar{\delta}$ are independent of t_0, see Hale [1980].

Heuristically, these definitions say that solutions starting near a Liapunov stable solution remain nearby thereafter, and solutions starting near an asymptotically stable solution approach the solution as $t \to \infty$, see Figure 1.1.5.

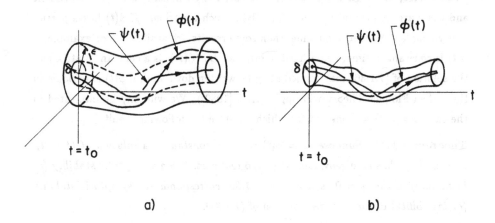

Figure 1.1.5. a) Liapunov Stability.   b) Asymptotic Stability.

Now that we have defined stability of solutions we need to address its determination for specific problems. One method for determining the stability of a specific solution is the direct method of Liapunov, and for this we refer the reader to the references given at the beginning of this section. Another method for determining stability is *linearization*, which we will discuss in some detail.

Let us make the coordinate transformation $x = y + \phi(t)$ for (1.1.34) and Taylor expand $f(y + \phi(t))$ about $y = 0$. Then we get the equation

$$\dot{y} = Df(\phi(t))y + O\left(|y|^2\right) . \tag{1.1.35}$$

Now the $y = 0$ solution of (1.1.35) corresponds to the $x = \phi(t)$ solution of (1.1.34). So, if the $y = 0$ solution of (1.1.35) is stable, then the $x = \phi(t)$ solution of (1.1.34) will likewise be stable. Now, (1.1.35) is no less difficult to solve than (1.1.34), so for y small we assume that the $O\left(|y|^2\right)$ terms can be neglected, and we arrive at the linear equation

$$\dot{y} = Df(\phi(t))y . \tag{1.1.36}$$

Now we would like to do two things: 1) determine the stability of the $y = 0$ solution of (1.1.36), and 2) conclude that stability (or instability) of the $y = 0$ solution of (1.1.36) corresponds to stability (or instability) for the $x = \phi(t)$ solution of (1.1.34). In general, the determination of the stability of the $y = 0$ solution of (1.1.36) is a formidable problem (e.g., see the discussion of Hill's equation in Hale [1980]) since, although the equation is linear, the coefficients are time dependent and there are no general methods for solving such equations. If $\phi(t)$ has a particularly simple dependence on time, then some results are available. For example, if $\phi(t)$ is constant in time, i.e., a fixed point, the $Df(\phi(t))$ is a constant matrix and the solution of (1.1.36) may immediately be written down and, if $\phi(t)$ is periodic in time, then Floquet Theory will apply (Hale [1980]). We will only be interested in the case $x = \phi(t) =$ constant for which we state the following result.

Theorem 1.1.11. *Suppose* $x = \phi(t) = x_0 =$ *constant is a solution of (1.1.34), and* $Df(x_0)$ *has no eigenvalues with zero real part. Then asymptotic stability (or instability) of the* $y = 0$ *solution of (1.1.36) corresponds to asymptotic stability (or instability) of the* $x = x_0$ *solution of (1.1.34).*

PROOF: This follows from the Hartman-Grobman theorem, see Hartman [1964].
□

Next we want to state some similar results for maps. Let

$$x \mapsto f(x), \qquad x \in \mathbb{R}^n \tag{1.1.37}$$

be a C^r map, $r \geq 1$ with f defined on some open set $U \subset \mathbb{R}^n$. Given an orbit of f, we leave as as exercise for the reader the task of writing down discrete versions of Definitions 1.1.3 and 1.1.4. Here we will concentrate on the stability of periodic orbits of (1.1.37).

Let p be a period k point of f, i.e., the orbit of p under f is given by

$$O(p) = \left\{ p,\, p_1 \equiv f(p),\, p_2 \equiv f^2(p),\, \ldots,\, p_k \equiv f^k(p) = p \right\}. \tag{1.1.38}$$

We ask whether or not $O(p)$ is stable. Notice that $p_1,\ldots,p_k$ are each fixed points for $f^k(x)$ and that, by the chain rule, $Df^k(x) = Df\big(f^{k-1}(x)\big) Df\big(f^{k-2}(x)\big) \cdots Df(x)$. Therefore, stability of $O(p)$ is reduced to the question of the stability of a fixed point p_j for any $j = 1,\ldots,k$ of $f^k(x)$. The question of stability for fixed points of maps has an answer analogous to that given for fixed points of ordinary differential equations described in Theorem 1.1.11. Consider the map

$$x \mapsto f^k(x), \qquad x \in \mathbb{R}^n \tag{1.1.39}$$

which has fixed points of p_j, $j = 1,\ldots,k$. Following an argument similar to that given for ordinary differential equations consider the associated linear map

$$y \mapsto Df^k(p_j)y, \qquad y \in \mathbb{R}^n, \quad \text{for any } j = 1,\ldots,k \tag{1.1.40}$$

which has a fixed point at $y = 0$. We have the following result.

Theorem 1.1.12. *Suppose p is a period k point for (1.1.37) and $Df^k(p)$ has no eigenvalues of modulus one. Then asymptotic stability (or instability) of the fixed point $y = 0$ of (1.1.40) corresponds to asymptotic stability (or instability) of $O(p)$.*

PROOF: This is a consequence of the discrete version of the Hartman-Grobman Theorem, see Hartman [1964]. □

We remark that, in general, any theorem pertaining to fixed points of maps has an analogous statement for periodic orbits of maps which can be obtained by replacing the map by its kth iterate where k is the period of the orbit. For more information on stability of maps see Bernoussou [1977] or Guckenheimer and Holmes [1983].

1.1k. Asymptotic Behavior

In this section we want to develop some concepts necessary for describing the asymptotic or observable behavior of dynamical systems. We will do this simultaneously for ordinary differential equations and maps.

Consider the following ordinary differential equation

$$\dot{x} = f(x), \qquad x \in U \tag{1.1.41}$$

and map

$$x \mapsto g(x), \qquad x \in U \tag{1.1.42}$$

where in each case $f : U \to \mathbb{R}^n$ and $g : U \to \mathbb{R}^n$ are C^r diffeomorphisms, $r \geq 1$, on some open set $U \subset \mathbb{R}^n$. We assume that (1.1.41) generates a flow for all time, and we denote this flow by $\phi_t(\cdot)$.

Definition 1.1.5. A set $S \subset U$ is said to be invariant under $\phi_t(\cdot)$ (resp. g) if

$$\phi_t(S) \subset S \qquad (\text{ resp. } g^n(S) \subset S) \quad \text{for all } t \in \mathbb{R} \quad (\text{ resp. } n \in \mathbb{Z}).$$

S is called an *invariant set*. If the above statement is true for all $t \in \mathbb{R}^+$ (resp. $n \in \mathbb{Z}^+$), then S is called a *positive invariant set*, and if true for all $t \in \mathbb{R}^-$ (resp. $n \in \mathbb{Z}^-$), then S is called a *negative invariant set*.

Recurrent behavior is contained in the nonwandering set of a flow or map.

Definition 1.1.6. A point $p \in U$ is said to be a *nonwandering point* for $\phi_t(\cdot)$ (resp. $g(\cdot)$) if *for any* neighborhood V of p there exists some nonzero $T \in \mathbb{R}$ (resp. $N \in \mathbb{Z}$) such that $\phi_T(V) \cap V \neq \emptyset$ (resp. $g^N(V) \cap V \neq \emptyset$). The collection of all nonwandering points for $\phi_t(\cdot)$ (resp. $g(\cdot)$) is called the *nonwandering set* for $\phi_t(\cdot)$ (resp. $g(\cdot)$).

EXAMPLE 1.1.4. Fixed points as well as all the points on periodic orbits are nonwandering points for both flows and maps.

EXAMPLE 1.1.5. Consider the equation

$$\begin{aligned} \dot{\theta}_1 &= \omega_1 \\ \dot{\theta}_2 &= \omega_2 \end{aligned} \qquad (\theta_1, \theta_2) \in S^1 \times S^1 \equiv T^2. \tag{1.1.43}$$

The flow generated by this equation is

$$\phi(t) = (\theta_1(t), \theta_2(t)) = (\omega_1 t + \theta_{10}, \omega_2 t + \theta_{20}) \ . \qquad (1.1.44)$$

It is easy to see that if ω_1/ω_2 is a rational number all points on T^2 lie on periodic orbits, and if ω_1/ω_2 is an irrational number then all points lie on orbits that never close but densely cover the surface of T^2. Hence, in both cases, all points of T^2 are nonwandering points.

Attracting sets are thought of as the "observable" states of dynamical systems.

Definition 1.1.7. A closed invariant set $A \subset U$ is called an *attracting set* if there exists some neighborhood V of A such that for all $x \in V$ $\phi_t(x) \in V$ (resp. $g^n(x) \in V$) for all $t \geq 0$ (resp. $n \geq 0$) and $\phi_t(x) \to A$ (resp. $g^n(x) \to A$) as $t \to \infty$ (resp. $n \to \infty$).

Definition 1.1.8. The basin or domain of attraction of A, denoted $\mathcal{D}_A$, is defined as follows.

$$\mathcal{D}_A = \bigcup_{t \leq 0} \phi_t(V) \qquad \left(\text{resp.} \ \bigcup_{n \leq 0} g^n(V) \right).$$

EXAMPLE 1.1.6. Consider the following equation

$$\begin{aligned} \dot{x} &= y \\ \dot{y} &= x - x^3 - \delta y \end{aligned} \qquad (x,y) \in \mathbb{R}^1 \times \mathbb{R}^1, \qquad (1.1.45)$$

with $\delta > 0$. The phase space of (1.1.45) is shown in Figure 1.1.6.

Equation 1.1.45 has three fixed points, an unstable (saddle) fixed point at the origin and two stable (sinks) fixed points at $(\pm 1, 0)$. The stable fixed points are attractors, and the domains of attraction of the two sinks are as indicated in Figure 1.1.6. Notice the two pairs of curves which issue from the saddle point at the origin. One pair of curves consists of points which recede from the origin in positive time and the other pair of curves consists of points which approach the origin in positive time; these curves are called the unstable and stable manifolds of the origin, respectively (see Section 1.3 for a discussion of invariant manifolds), and are examples of invariant sets. Note that the unstable manifold of the origin serves to separate the domains of attraction of the two sinks.

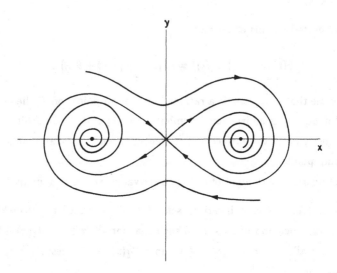

Figure 1.1.6. Phase Space of (1.1.45).

1.2. Conjugacies

The importance of coordinate transformations in the study of dynamical systems cannot be overestimated. For example, in the study of systems of linear constant coefficient ordinary differential equations, coordinate transformations allow one to decouple the system and hence reduce the system to a set of decoupled linear first order equations which are easily solved. In the study of completely integrable Hamiltonian systems, the transformation to action-angle coordinates results in a trivially solvable system (see Arnold [1978]), and these coordinates are also useful in the study of near integrable systems. If we consider general properties of dynamical systems, coordinate transformations provide us with a way of classifying dynamical systems according to properties which remain unchanged after a coordinate transformation. In Section 1.4 we will see that the notion of structural stability is based on such a classification scheme. In this section we want to discuss coordinate transformations or, to use the more general mathematical term, *conjugacies* in general, giving some results which describe properties which must be retained by a map or vector field after a coordinate transformation of a specific differentiability class. We will discuss conjugacies for both maps and vector fields separately, beginning with maps.

Let us consider two C^r diffeomorphisms $f: \mathbf{R}^n \to \mathbf{R}^n$, $g: \mathbf{R}^n \to \mathbf{R}^n$, and a C^k diffeomorphism $h: \mathbf{R}^n \to \mathbf{R}^n$.

Definition 1.2.1. f and g are said to be C^k *conjugate* $(k \leq r)$ if there exists a C^k diffeomorphism $h: \mathbf{R}^n \to \mathbf{R}^n$ such that $g \circ h = h \circ f$. If $k = 0$, f and g are said to be *topologically conjugate*.

The conjugacy of two diffeomorphisms is often represented by the following diagram.

$$
\begin{array}{ccc}
\mathbf{R}^n & \xrightarrow{\ f\ } & \mathbf{R}^n \\
h \downarrow & & \downarrow h \\
\mathbf{R}^n & \xrightarrow{\ g\ } & \mathbf{R}^n
\end{array}
\tag{1.2.1}
$$

The diagram is said to *commute* if the relation $g \circ h = h \circ f$ holds, meaning that you can start at a point in the upper left hand corner of the diagram and reach the same point in the lower right hand corner of the diagram by either of the two possible routes. We note that h need not be defined on all of $\mathbf{R}^n$, but possibly only locally about a given point. In such cases, f and g are said to be *locally C^k conjugate*.

If f and g are C^k conjugate then we have the following results.

Proposition 1.2.1. *If f and g are C^k conjugate, then orbits of f map to orbits of g under h.*

PROOF: Let $x_0 \in \mathbf{R}^n$; then the orbit of x_0 under f is given by

$$
O_f(x_0) = \left\{ \ldots, f^{-n}(x_0), \ldots, f^{-1}(x_0), x_0, f(x_0), \ldots, f^n(x_0), \ldots \right\}.
\tag{1.2.2}
$$

From Definition 1.2.1, $f = h^{-1} \circ g \circ h$, so for a given $n > 0$ we have

$$
f^n(x_0) = \underbrace{\left(h^{-1} \circ g \circ h\right) \circ \left(h^{-1} \circ g \circ h\right) \circ \cdots \circ \left(h^{-1} \circ g \circ h\right)}_{n \text{ factors}}(x_0)
\tag{1.2.3}
$$

$$
= h^{-1} \circ g^n \circ h(x_0)
\tag{1.2.4}
$$

or

$$
h \circ f^n(x_0) = g^n \circ h(x_0).
\tag{1.2.5}
$$

Also from Definition 1.2.1, $f^{-1} = h^{-1} \circ g^{-1} \circ h$ so, by the same argument for $n > 0$ we obtain

$$
h \circ f^{-n}(x_0) = g^{-n} \circ h(x_0).
\tag{1.2.6}
$$

Therefore, from (1.2.5) and (1.2.6) we see that the orbit of x_0 under f is mapped by h to the orbit of $h(x_0)$ under g.   □

Proposition 1.2.2. *If f and g are C^k conjugate, $k \geq 1$, and x_0 is a fixed point of f. Then the eigenvalues of $Df(x_0)$ are equal to the eigenvalues of $Dg(h(x_0))$.*

PROOF: From Definition 1.2.1 $f(x) = h^{-1} \circ g \circ h(x)$. Note that since x_0 is a fixed point then $g(h(x_0)) = x_0$. Since h is differentiable we have

$$Df\Big|_{x_0} = Dh^{-1}\Big|_{x_0} Dg\Big|_{h(x_0)} Dh\Big|_{x_0} , \qquad (1.2.7)$$

so recalling that similar matrices have equal eigenvalues gives the result.   □

Next we turn to flows. Let f and g be C^r vector fields on $\mathbf{R}^n$.

Definition 1.2.2. *f and g are said to be C^k-equivalent if there exists a C^k diffeomorphism h, which takes orbits of the flow generated by f, $\phi(t,x)$, to orbits of the flow generated by g, $\psi(t,y)$, preserving orientation but not necessarily parametrization by time. If h does preserve parametrization by time, then f and g are said to be C^k-conjugate.*

We remark that, as for maps, the conjugacies do not need to be defined on all of $\mathbf{R}^n$.

Now we examine some of the consequences of Definition 1.2.2.

Proposition 1.2.3. *Suppose f and g are C^k-conjugate. Then*

   a) fixed points of f are mapped to fixed points of g,

   b) T-periodic orbits of f map to T-periodic orbits of g.

PROOF: f, g C^k conjugate under h implies the following.

$$h \circ \phi(t,x) = \psi(t, h(x)) \qquad (1.2.8)$$

$$Dh\dot{\phi} = \dot{\psi} \qquad (1.2.9)$$

The proof of a) follows from (1.2.9) and the proof of b) follows from (1.2.8).   □

Proposition 1.2.4. *Suppose f and g are C^k-conjugate ($k \geq 1$) and $f(x_0) = 0$. Then $Df(x_0)$ has the same eigenvalues as $Dg(h(x_0))$.*

PROOF: We have the two vector fields, $\dot{x} = f(x)$, $\dot{y} = g(y)$. By differentiating (1.2.8) with respect to t we have

$$Dh\Big|_x f(x) = g(h(x)) . \qquad (1.2.10)$$

Differentiating (1.2.10) gives

$$D^2h\Big|_x f(x) + Dh\Big|_x Df\Big|_x = Dg\Big|_{h(x)} Dh\Big|_x . \tag{1.2.11}$$

Evaluating (1.2.11) at x_0 gives,

$$Dh\Big|_{x_0} Df\Big|_{x_0} = Dg\Big|_{h(x_0)} Dh\Big|_{x_0} \tag{1.2.12}$$

or

$$Df\Big|_{x_0} = Dh^{-1}\Big|_{x_0} Dg\Big|_{h(x_0)} Dh\Big|_{x_0} \tag{1.2.13}$$

and, since similar matrices have equal eigenvalues, the proof is complete. □

The previous two propositions dealt with C^k-*conjugacies*. We next examine the consequences of C^k-*equivalence* under the assumption that the change in parametrization by time along orbits is C^1.

Proposition 1.2.5. *Suppose f and g are C^k-equivalent; then*

a) fixed points of f are mapped to fixed points of g,

b) periodic orbits of f are mapped to periodic orbits of g, but the periods need not be equal.

PROOF: If f and g are C^k-equivalent then

$$h \circ \phi(t, x) = \psi(\alpha(x, t), h(x)) \tag{1.2.14}$$

where α is an increasing function of time along orbits (note: α must be increasing in order to preserve orientations of orbits).

Differentiating (1.2.14) gives:

$$Dh\,\dot{\phi} = \frac{\partial \alpha}{\partial t}\frac{\partial \psi}{\partial \alpha} . \tag{1.2.15}$$

So (1.2.15) implies a). Also, b) follows automatically since C^k-diffeomorphisms map closed curves to closed curves. (If this were not true then the inverse would not be continuous.) □

Proposition 1.2.6. *Suppose f and g are C^k-equivalent ($k \geq 1$) and $f(x_0) = 0$; then the eigenvalues of $Df(x_0)$ and the eigenvalues of $Dg(h(x_0))$ differ by a positive multiplicative constant.*

PROOF: Proceeding as in the proof of Proposition 1.2.4 we have

$$Dh\Big|_x f(x) = \frac{\partial \alpha}{\partial t}\, g(h(x)) . \tag{1.2.16}$$

Differentiating (1.2.16) gives

$$D^2h\bigg|_x f(x) + Dh\bigg|_x Df\bigg|_x = \frac{\partial\alpha}{\partial t} Dg\bigg|_{h(x)} Dh\bigg|_x + \frac{\partial^2\alpha}{\partial x \partial t}\bigg|_x g\big(h(x)\big) . \qquad (1.2.17)$$

Evaluating at x_0 gives

$$Dh\bigg|_{x_0} Df\bigg|_{x_0} = \frac{\partial\alpha}{\partial t} Dg\bigg|_{h(x_0)} Dh\bigg|_{x_0} \qquad (1.2.18)$$

so $Df\bigg|_{x_0}$ and $Dg\bigg|_{h(x_0)}$ are similar, up to the multiplicative constant $\partial\alpha/\partial t$ which is positive since α increases on orbits. $\qquad\qquad\square$

1.3. Invariant Manifolds

In this section we want to describe some aspects of the theory of invariant manifolds which we will use repeatedly in Chapters 3 and 4. Roughly speaking, an invariant manifold is a surface contained in the phase space of a dynamical system which has the property that orbits starting on the surface remain on the surface throughout the course of their dynamical evolution: i.e., an invariant manifold is a collection of orbits which form a surface (note: later we will relax this requirement by introducing the idea of locally invariant manifolds). Additionally, the set of orbits which approach or recede from an invariant manifold M asymptotically in time under certain conditions are also invariant manifolds which are called the stable and unstable manifolds, respectively, of M. Knowledge of the invariant manifolds of a dynamical system as well as the interactions of their respective stable and unstable manifolds is absolutely crucial in order to obtain a complete understanding of the global dynamics. In Chapters 3 and 4 this statement will become evident. Also, under certain general conditions, invariant manifolds often possess the property of persistence under perturbations. This property is used in Chapter 4 to develop global perturbation methods for systems where we have a global knowledge of the invariant manifold structure.

The amount of work on the subject of invariant manifolds in the past fifty years has been prodigious and is still continuing today at a rapid pace. It is, therefore, not possible to give a full account of the various aspects of the theory or to even give an adequate historical survey in this section. However, we will give a chronology of some of the major results as well as some references where further information can be obtained. (Note: results concerning invariant manifolds may be expressed in terms of continuous time systems (vector fields) or discrete time systems (maps). In either case, it usually poses little difficulty to translate results for one type of system into corresponding results for the other type, see Fenichel [1971], [1974], [1977], Hirsch, Pugh and Shub [1977] or Palis and deMelo [1982] for a discussion of some examples.)

The first rigorous results concerning invariant manifolds are due to Hadamard [1901] and Perron [1928], [1929], [1930]. They proved the existence of stable and unstable manifolds of fixed points of maps and ordinary differential equations using different techniques. Levinson [1950] constructed invariant two-tori in his studies of coupled oscillators. This work was extended and generalized by Diliberto [1960], [1961] with additional contributions by Kyner [1956], Hufford [1956], Marcus [1956], Hale [1961], Kurzweil [1968], McCarthy [1955] and Kelley [1967]; at the same time, similar work was being carried out independently by the Russian school led by Bogoliubov and Mitropolsky [1961]. The existence of stable and unstable manifolds and their persistence under perturbation for an arbitrary invariant manifold was first proved by Sacker [1964]. This work was later extended and generalized by Fenichel [1971], [1974], [1977] with similar work and even more extensions being done independently by Hirsch, Pugh and Shub [1977]. Some more recent results include the work of Sacker and Sell [1978], [1974] and Sell [1978] which was used by Sell [1979] in the study of bifurcations of n-tori (note: Sell's work represents the first rigorous results dealing with the Ruelle-Takens-Newhouse scenario for the transition to turbulence, see Sell [1981], [1982]), and the work of Pesin [1976], [1977] dealing with the existence of invariant manifolds under nonuniform hyperbolicity assumptions. We have not mentioned any results relating to center manifolds (see Carr [1981] or Sijbrand [1985]) or invariant manifolds in infinite dimensional systems (see Hale, Magalhães and Oliva [1984] and Henry [1981]) since we will not use those ideas or results in this book.

The results from invariant manifold theory which we will describe will be taken from Fenichel [1971] since they are most closely suited for the perturbation techniques which we will develop in Chapter 4. However, first we will begin with a motivational example.

EXAMPLE 1.3.1. We consider a nonlinear, autonomous ordinary differential equation defined on $\mathbf{R}^n$,

$$\dot{x} = f(x) , \qquad x(0) = x_0 , \qquad x \in \mathbf{R}^n \tag{1.3.1}$$

where $f : \mathbf{R}^n \to \mathbf{R}^n$ is at least C^1. We make the following assumptions on (1.3.1).

A1) (1.3.1) has a fixed point at $x = 0$, i.e., $f(0) = 0$.

A2) $Df(0)$ has $n - k$ eigenvalues having positive real parts and k eigenvalues having negative real parts.

Thus, (1.3.1) possesses a particularly trivial type of invariant manifold, namely the fixed point at $x = 0$. Let us now study the nature of the linear system obtained by linearizing (1.3.1) about the fixed point $x = 0$. We denote the linearized system by

$$\dot{\xi} = Df(0) \, \xi , \qquad \xi \in \mathbf{R}^n \tag{1.3.2}$$

and note that the linearized system possesses a fixed point at the origin. Let $v^1, \ldots, v^{n-k}$ denote the generalized eigenvectors corresponding to the eigenvalues having positive real parts, and $v^{n-k+1}, \ldots, v^n$ denote the generalized eigenvectors corresponding to the eigenvalues having negative real parts. Then the linear subspaces of $\mathbf{R}^n$ defined as

$$\begin{aligned} E^u &= \text{ span } \{v^1, \ldots, v^{n-k}\} \\ E^s &= \text{ span } \{v^{n-k+1}, \ldots, v^n\} \end{aligned} \tag{1.3.3}$$

are invariant manifolds for the linear system (1.3.2) which are known as the unstable and stable subspaces, respectively. E^u is the set of points such that orbits of (1.3.2) through these points approach the origin asymptotically in negative time, and E^s represents the set of points such that orbits of (1.3.2) through these points approach the origin asymptotically in positive time (note: these statements are not hard to prove, and we refer the reader to Arnold [1973] or Hirsch and Smale [1974] for a thorough discussion of linear, constant coefficient systems). We represent this situation geometrically in Figure 1.3.1.

The question we now ask is what is the behavior of the nonlinear system (1.3.1) near the fixed point $x = 0$? We might expect that the linearized system should give us some indication of the nature of the orbit structure near the fixed point of the nonlinear system, since the fact that none of the eigenvalues of $Df(0)$ have

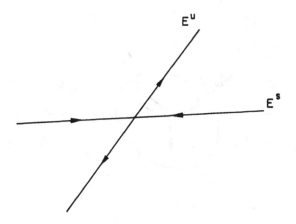

Figure 1.3.1.

zero real part implies that near $x = 0$ the flow of (1.3.1) is dominated by the flow of (1.3.2). (Note: fixed points of vector fields which have the property that the eigenvalues of the matrix associated with the linearization of the vector field about the fixed point have nonzero real parts are called *hyperbolic* fixed points.) Indeed, the stable manifold theorem for fixed points (see Palis and deMelo [1982]) tells us that in a neighborhood N of the fixed point $x = 0$ for (1.3.1), there exists a differentiable (as differentiable as the vector field (1.3.1)) $n - k$ dimensional surface, $W^u_{loc}(0)$, tangent to E^u at $x = 0$ and a differentiable k dimensional surface, $W^s_{loc}(0)$, tangent to E^s at $x = 0$ with the properties that orbits of points on $W^u_{loc}(0)$ approach $x = 0$ asymptotically in negative time (i.e., as $t \to -\infty$) and orbits of points on $W^s_{loc}(0)$ approach $x = 0$ asymptotically in positive time (i.e., as $t \to +\infty$). $W^u_{loc}(0)$ and $W^s_{loc}(0)$ are known as the local unstable and stable manifolds, respectively, of $x = 0$. We represent this situation geometrically in Figure 1.3.2.

Let us denote the flow generated by (1.3.1) as $\phi_t(\cdot)$; then we can define global stable and unstable manifolds of $x = 0$ by using points on the local manifolds as initial conditions.

$$W^u(0) = \bigcup_{t \geq 0} \phi_t \left(W^u_{loc}(0) \right)$$

$$W^s(0) = \bigcup_{t \leq 0} \phi_t \left(W^s_{loc}(0) \right) \tag{1.3.4}$$

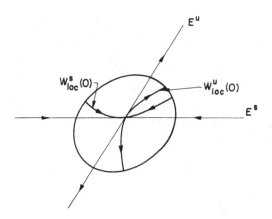

Figure 1.3.2. Phase Space of (1.3.1) near $x = 0$.

$W^u(0)$ and $W^s(0)$ are called the unstable and stable manifolds, respectively, of $x = 0$. We represent the situation geometrically in Figure 1.3.3.

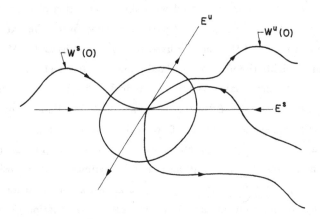

Figure 1.3.3. Global Stable and Unstable Manifolds of $x = 0$.

Now suppose we add a small autonomous perturbation, $\epsilon g(x)$, to (1.3.1) where $g(x)$ is as differentiable as $f(x)$ and $\epsilon \in I \subset \mathbb{R}$ where $I = \{\epsilon \in \mathbb{R} \mid 0 < \epsilon < \epsilon_0\}$, we denote the perturbed system by

$$\dot{x} = f(x) + \epsilon g(x), \qquad x(0) = x_0, \qquad x \in \mathbb{R}^n. \tag{1.3.5}$$

The question we now ask is how much of the structure of (1.3.1) is preserved in the perturbed system (1.3.5). Specifically, we will be concerned with what happens to the fixed point at the origin and its stable and unstable manifolds.

The fate of the fixed point is easy to determine by a simple application of the implicit function theorem (note: recall that a fixed point of (1.3.5) is a solution of $f(x) + \epsilon g(x) = 0$). We will set up the problem for application of the implicit function theorem. Let us consider the function

$$G : \mathbb{R}^n \times I \to \mathbb{R}^n \tag{1.3.6}$$
$$(x, \epsilon) \mapsto f(x) + \epsilon g(x) .$$

It is clear that $G(0,0) = 0$, and we wish to determine if there exists a solution of $G(x, \epsilon) = 0$ for (x, ϵ) close to (0,0). Now the derivative of G with respect to x evaluated at $(x, \epsilon) = (0,0)$ is given by

$$D_x G(0,0) = D_x f(0) . \tag{1.3.7}$$

By our assumption on the eigenvalues of $Df(0)$ (specifically, there are no zero eigenvalues) it is clear that $\det[D_x G(0,0)] = \det D_x f(0) \neq 0$; thus, by the implicit function theorem there exists a function of ϵ, $\bar{x}(\epsilon)$ (with $\bar{x}(\epsilon)$ as differentiable as $G(x, \epsilon)$), such that

$$G(\bar{x}(\epsilon), \epsilon) = 0 \tag{1.3.8}$$

for ϵ sufficiently small contained in I. Thus, the fixed point is preserved in the perturbed system, although it may move slightly.

The fate of the unstable and stable manifolds of $x = 0$ follows from the persistence theory for stable and unstable manifolds (see Fenichel [1971] or Hirsch, Pugh and Shub [1977]) which we will describe in some detail later on. However, for now we will state the consequence of this theory, which tells us that in some neighborhood $\tilde{N}$ containing $x = 0$ and $x = \bar{x}(\epsilon)$ there exist differentiable manifolds $\tilde{W}_{\text{loc}}^u(\bar{x}(\epsilon))$ and $\tilde{W}_{\text{loc}}^s(\bar{x}(\epsilon))$ passing through $\bar{x}(\epsilon)$ with the properties that orbits of points in $\tilde{W}_{\text{loc}}^u(\bar{x}(\epsilon))$ under the perturbed flow approach $x = \bar{x}(\epsilon)$ asymptotically in negative time and orbits of points in $\tilde{W}_{\text{loc}}^s(\bar{x}(\epsilon))$ under the perturbed flow approach $x = \bar{x}(\epsilon)$ asymptotically in positive time. $\tilde{W}_{\text{loc}}^u(\bar{x}(\epsilon))$ and $\tilde{W}_{\text{loc}}^s(\bar{x}(\epsilon))$ have the same dimensions and differentiability as $W_{\text{loc}}^u(0)$ and $W_{\text{loc}}^s(0)$, respectively. Utilizing the flow generated by the perturbed system (1.3.5) and $\tilde{W}_{\text{loc}}^u(\bar{x}(\epsilon))$ and

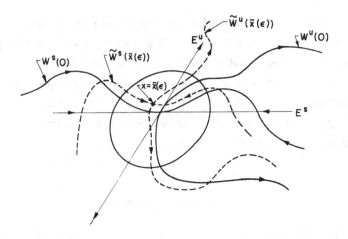

Figure 1.3.4. Perturbed and Unperturbed Structure.

$\tilde{W}^s_{\text{loc}}(\bar{x}(\epsilon))$ as initial conditions, we can define global unstable and stable manifolds of $x = \bar{x}(\epsilon)$ in exactly the same manner as we defined them for the unperturbed system. See Figure 1.3.4 for a geometrical interpretation.

This simple example illustrates several points that arise in invariant manifold theory which we now want to emphasize.

1) For the unperturbed equation it is first necessary to locate the invariant manifold. In our simple example the invariant manifold is a fixed point which can be found by solving for the zeros of a system of coupled nonlinear algebraic relations. Locating more general types of invariant manifolds may involve having quite a detailed knowledge of the orbit structure of a nonlinear ordinary differential equation, which in general is a formidable task.

2) Once the invariant manifold of the unperturbed system is obtained, it is then necessary to study the linear system obtained by linearizing the unperturbed system about the invariant manifold. This procedure, where the invariant manifold is a fixed point, periodic orbit, or quasiperiodic orbit, is quite familiar; if the unperturbed system is of the form

$$\dot{x} = f(x), \quad x \in \mathbf{R}^n, \quad f \in C^1 \tag{1.3.9}$$

with an invariant manifold $\phi(t)$ being a fixed point, periodic orbit, or quasiperiodic orbit, then letting $x(t) = \phi(t) + \xi(t)$, we obtain

$$\dot{x} = \dot{\phi} + \dot{\xi} = f(\phi + \xi) = f(\phi) + Df(\phi)\,\xi + O\left(|\xi|^2\right)$$
$$\text{or,} \quad \dot{\xi} = Df(\phi)\,\xi + O\left(|\xi|^2\right)$$

(1.3.10)

since $\dot{\phi} = f(\phi)$ (i.e., ϕ is a solution of (1.3.9)). If we retain only terms linear in ξ we obtain the associated linearized system

$$\dot{\xi} = Df(\phi)\,\xi.$$

(1.3.11)

Now if the invariant manifold is more general, such as a surface containing many different orbits of (1.3.9), then linearizing about the invariant manifold is not a straightforward procedure, especially if the invariant manifold is not globally representable as a graph of a function. In this case one obtains a collection of linear equations representing the linearized vector field in different "coordinate charts" on the invariant manifold. The techniques for describing the vector field near a general invariant manifold are obtained from the theory of differentiable manifolds which we will describe shortly.

3) Once the linearized system is obtained, it is then necessary to study its stability. This information will allow us to determine the dimension of the stable and unstable manifolds of the invariant manifold as well as the persistence and smoothness properties of the structure under perturbations. In general, this is a formidable task, since the coefficients of the linear system may have a complicated time dependence. There are two approaches to the problem which are essentially equivalent, one involves the computation of Lyapunov type numbers or exponents (this is the approach we shall take) and the other a consideration of exponential dichotomies (see Coppel [1978] and Sacker and Sell [1974]).

Before discussing the general theory of invariant manifolds, we need to give some background material from differential geometry. More specifically, we will need to understand the definition of a differentiable manifold, the tangent space at a point, the tangent bundle, and the derivatives of maps defined on differentiable manifolds. We will not develop these concepts in the most abstract or mathematically crisp manner, but rather along the lines where they occur most frequently in applications. In applications involving the modelling of the dynamics of some physical system, we typically choose certain quantities describing various aspects of the system and write down equations describing the time evolution of these quantities. These quantities constitute the phase space of the system with invariant manifolds arising as surfaces in the phase space. Consequently, we choose to develop the concept of a differentiable manifold as a surface embedded in $\mathbb{R}^n$ (loosely following the exposition of Milnor [1965] and Guillemin and Pollack [1974]) and refer the reader to any differential geometry textbook for the abstract development of the theory of differentiable manifolds (e.g., a standard and very thorough textbook is Spivak [1979]). Our approach will allow us to bypass certain set theoretic and topological technicalities, since our manifolds will inherit much structure from $\mathbb{R}^n$, whose topology is relatively familiar. Additionally, it is hoped that this approach will appeal to the intuition of the reader who has little or no experience with the subject of differential geometry.

We begin by defining the derivative of a map defined on an arbitrary subset of $\mathbb{R}^n$.

Definition 1.3.1. Consider a map $f: X \to \mathbb{R}^m$ where X is an arbitrary subset of $\mathbb{R}^n$. f is said to be C^r on X if for every point $x \in X$ there exists an open set $U \subset \mathbb{R}^n$ containing x and a C^r map $F: U \to \mathbb{R}^m$ such that $f = F$ on $U \cap X$.

Definition 1.3.2. A map $f: X \to Y$ of subsets of two Euclidean spaces is called a C^r *diffeomorphism* if it is one to one and onto and if the inverse map $f^{-1}: Y \to X$ is also C^r.

We are now in a position to give the definition of a differentiable manifold.

Definition 1.3.3. A subset $M \subset \mathbb{R}^n$ is called a C^r *manifold of dimension* m if it possesses the following two structural characteristics.

1) There exists a countable collection of open sets $V^\alpha \subset \mathbb{R}^n$, $\alpha \in A$ where A is some countable index set, with $U^\alpha \equiv V^\alpha \cap M$ such that $M = \bigcup_{\alpha \in A} U^\alpha$.

2) There exists a C^r diffeomorphism x^α defined on each U^α which maps U^α onto some open set in $\mathbb{R}^m$.

We make the following remarks regarding Definition 1.3.3.

1. A standard terminology is that the pair $(U^\alpha; x^\alpha)$ is called a *chart* for M and the union of all charts, i.e., $\bigcup_{\alpha \in A} (U^\alpha; x^\alpha)$, is called an *atlas* for M.

2. The sets U^α are often called relatively open sets, i.e., open with respect to M.

3. From 2) of Definition 1.3.3 we see that the degree of differentiability of a manifold is the same as the degree of differentiability of the x^α. This implies a
   certain compatibility condition which must be satisfied on overlapping charts.
   More specifically, let $(U^\alpha; x^\alpha)$ and $(U^\beta; x^\beta)$, $\alpha, \beta \in A$, be two charts such that
   $U^\alpha \cap U^\beta \neq \emptyset$, see Figure 1.3.5.

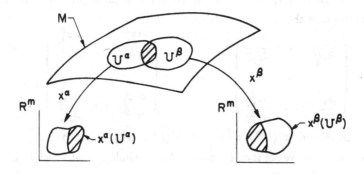

Figure 1.3.5. Coordinate Charts on a Manifold.

Then the region $U^\alpha \cap U^\beta$ can be described by two different coordinatizations,
namely $(U^\alpha \cap U^\beta; x^\alpha)$ and $(U^\alpha \cap U^\beta; x^\beta)$. We denote

$$
\begin{aligned}
x^\alpha &: p \in U^\alpha \cap U^\beta \longrightarrow (x_1^\alpha, \ldots, x_m^\alpha) \in \mathbf{R}^m \\
x^\beta &: p \in U^\alpha \cap U^\beta \longrightarrow (x_1^\beta, \ldots, x_m^\beta) \in \mathbf{R}^m
\end{aligned}
\tag{1.3.12}
$$

where $(x_1^\alpha, \ldots, x_m^\alpha)$ and $(x_1^\beta, \ldots, x_m^\beta)$ represent points in the Euclidean space
$\mathbf{R}^m$. Now the maps

$$
\begin{aligned}
x^\beta \circ (x^\alpha)^{-1} &: x^\alpha(U^\alpha \cap U^\beta) \to x^\beta(U^\alpha \cap U^\beta) \\
(x_1^\alpha, \ldots, x_m^\alpha) &\mapsto (x_1^\beta(x_1^\alpha, \ldots, x_m^\alpha), \ldots, x_m^\beta(x_1^\alpha, \ldots, x_m^\alpha)) \\
x^\alpha \circ (x^\beta)^{-1} &: x^\beta(U^\alpha \cap U^\beta) \to x^\alpha(U^\alpha \cap U^\beta) \\
(x_1^\beta, \ldots, x_m^\beta) &\mapsto (x_1^\alpha(x_1^\beta, \ldots, x_m^\beta), \ldots, x_m^\alpha(x_1^\beta, \ldots, x_m^\beta))
\end{aligned}
\tag{1.3.13}
$$

represent the change of coordinates from x^β to x^α coordinates and from x^α to
x^β coordinates, respectively, and the fact that x^α and x^β are C^r diffeomorphisms

implies that the maps describing the change of coordinates must likewise be C^r diffeomorphisms. (Note: in the change of coordinate maps in formula (1.3.13) we should more correctly write $\left(x_1^\beta(p(x_1^\alpha,\ldots,x_m^\alpha)),\ldots,x_m^\beta(p(x_1^\alpha,\ldots,x_m^\alpha))\right)$ for the image of $(x_1^\alpha,\ldots,x_m^\alpha)$ under $x^\beta \circ (x^\alpha)^{-1}$ (and similarly for the map $x^\alpha \circ (x^\beta)^{-1}$); however, it is standard and somewhat intuitive to identify points in the manifold with their images in a coordinate chart, especially when the manifold is a surface in $\mathbb{R}^n$.) In particular, for $r \geq 1$ we get the familiar requirement on changes of coordinates that the jacobian matrices

$$
\begin{pmatrix}
\dfrac{\partial x_1^\beta}{\partial x_1^\alpha} & \cdots & \dfrac{\partial x_1^\beta}{\partial x_m^\alpha} \\
\vdots & & \vdots \\
\dfrac{\partial x_m^\beta}{\partial x_1^\alpha} & \cdots & \dfrac{\partial x_m^\beta}{\partial x_m^\alpha}
\end{pmatrix}
\quad \text{and} \quad
\begin{pmatrix}
\dfrac{\partial x_1^\alpha}{\partial x_1^\beta} & \cdots & \dfrac{\partial x_1^\alpha}{\partial x_m^\beta} \\
\vdots & & \vdots \\
\dfrac{\partial x_m^\alpha}{\partial x_1^\beta} & \cdots & \dfrac{\partial x_m^\alpha}{\partial x_m^\beta}
\end{pmatrix}
\tag{1.3.14}
$$

be nonsingular on $x^\alpha(U^\alpha \cap U^\beta)$ and $x^\beta(U^\alpha \cap U^\beta)$, respectively.

Heuristically, we see that a differentiable manifold is a set which locally has the structure of ordinary Euclidean space. We now give several examples of manifolds.

EXAMPLE 1.3.2. The Euclidean space $\mathbb{R}^n$ is a trivial example of a C^∞ manifold. We take as the single coordinate chart $(i; \mathbb{R}^n)$ where i is the identity map identifying each "point" in $\mathbb{R}^n$ with its coordinates; it should be clear that i is infinitely differentiable and hence $\mathbb{R}^n$ is a C^∞ manifold.

EXAMPLE 1.3.3. Let $f : I \to \mathbb{R}$ be a C^r function where $I \subset \mathbb{R}$ is some open connected set. Then the graph of f is defined as follows:

$$
\text{graph } f = \{\, (s,t) \in \mathbb{R}^2 \mid t = f(s), \quad s \in I \,\}.
\tag{1.3.15}
$$

Geometrically, graph f might appear as in Figure 1.3.6.

We claim that graph f is a C^r one dimensional manifold. In order to verify this we must show that the two requirements of Definition 1.3.3 can be satisfied.

1) Let $U = \mathbb{R}^2 \cap$ graph f; then, by definition, $U =$ graph f.

2) We define a coordinate chart on U in the following manner

$$
\begin{aligned}
x &: U \to \mathbb{R}^1 \\
&\bigl(s, f(s)\bigr) \mapsto s
\end{aligned}
\tag{1.3.16}
$$

with the inverse defined in the obvious way,

$$
\begin{aligned}
x^{-1} &: \mathbb{R}^1 \to U \\
&s \mapsto \bigl(s, f(s)\bigr).
\end{aligned}
\tag{1.3.17}
$$

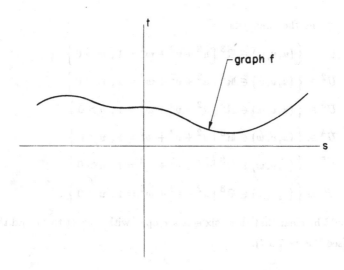

Figure 1.3.6. Graph f.

It is clear that x and x^{-1} are C^r since f is C^r. Thus, graph f is a C^r one dimensional manifold described by a single coordinate chart. We remark that this example should remind the reader of some of the heuristic aspects of elementary calculus, where it is common to visualize scalar functions as curves in the plane and to identify points on the curve with the corresponding points in the domain of the function.

EXAMPLE 1.3.4. Consider the following set of points contained in $\mathbb{R}^3$

$$M = \left\{ (u, v, w) \in \mathbb{R}^3 \mid u^2 + v^2 + w^2 = 1 \right\} . \tag{1.3.18}$$

This is just the two dimensional sphere of unit radius. We want to show that M is a C^∞ two dimensional manifold.

Let us define the open sets

$$U^1 \equiv \left\{ (u,v,w) \in \mathbb{R}^3 \mid u^2 + v^2 + w^2 = 1 \, , \, w > 0 \right\}$$
$$U^2 \equiv \left\{ (u,v,w) \in \mathbb{R}^3 \mid u^2 + v^2 + w^2 = 1 \, , \, w < 0 \right\}$$
$$U^3 \equiv \left\{ (u,v,w) \in \mathbb{R}^3 \mid u^2 + v^2 + w^2 = 1 \, , \, v > 0 \right\}$$
$$U^4 \equiv \left\{ (u,v,w) \in \mathbb{R}^3 \mid u^2 + v^2 + w^2 = 1 \, , \, v < 0 \right\} \tag{1.3.19}$$
$$U^5 \equiv \left\{ (u,v,w) \in \mathbb{R}^3 \mid u^2 + v^2 + w^2 = 1 \, , \, u > 0 \right\}$$
$$U^6 \equiv \left\{ (u,v,w) \in \mathbb{R}^3 \mid u^2 + v^2 + w^2 = 1 \, , \, u < 0 \right\} .$$

It should be clear that these six sets are open with respect to M and that they cover M (see Figure 1.3.7).

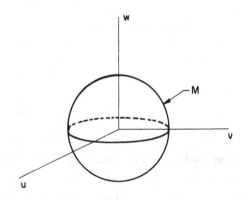

Figure 1.3.7. $M = U^1 \cup U^2 \cup U^3 \cup U^4 \cup U^5 \cup U^6$.

In these six sets, points of M can be represented as follows

$$U^1 : \left(u, v, \sqrt{1 - u^2 - v^2} \right)$$
$$U^2 : \left(u, v, -\sqrt{1 - u^2 - v^2} \right)$$
$$U^3 : \left(u, \sqrt{1 - u^2 - w^2}, w \right)$$
$$U^4 : \left(u, -\sqrt{1 - u^2 - w^2}, w \right) \tag{1.3.20}$$
$$U^5 : \left(\sqrt{1 - v^2 - w^2}, v, w \right)$$
$$U^6 : \left(-\sqrt{1 - v^2 - w^2}, v, w \right) .$$

We define maps of the U^α, $\alpha = 1, 2, 3, 4, 5, 6$, into $\mathbb{R}^2$ as follows

$$x^1 : U^1 \to \mathbb{R}^2$$
$$\left(u, v, \sqrt{1 - u^2 - v^2}\right) \mapsto (u, v)$$
$$x^2 : U^2 \to \mathbb{R}^2$$
$$\left(u, v, -\sqrt{1 - u^2 - v^2}\right) \mapsto (u, v)$$
$$x^3 : U^3 \to \mathbb{R}^2$$
$$\left(u, \sqrt{1 - u^2 - w^2}, w\right) \mapsto (u, w)$$
$$x^4 : U^4 \to \mathbb{R}^2 \qquad\qquad (1.3.21)$$
$$\left(u, -\sqrt{1 - u^2 - w^2}, w\right) \mapsto (u, w)$$
$$x^5 : U^5 \to \mathbb{R}^2$$
$$\left(\sqrt{1 - v^2 - w^2}, v, w\right) \mapsto (v, w)$$
$$x^6 : U^6 \to \mathbb{R}^2$$
$$\left(-\sqrt{1 - v^2 - w^2}, v, w\right) \mapsto (v, w)$$

with the inverse maps defined in the obvious manner (see Example 1.3.3). It should be clear that x^α and $(x^\alpha)^{-1}$, $\alpha = 1, 2, 3, 4, 5, 6$, are C^∞.

Let us now demonstrate the compatibility of the coordinatizations on overlapping regions for a particular example. The open set in M

$$U^1 \cap U^4 = \left\{ (u, v, w) \mid u^2 + v^2 + w^2 = 1, \; w > 0, \; v < 0 \right\} \qquad (1.3.22)$$

may be given coordinates by either x^1 or x^4. The formulas for the coordinate changes are given as follows:

$$x^4 \circ \left(x^1\right)^{-1} : x^1(U^1 \cap U^4) \to x^4(U^1 \cap U^4)$$
$$(u, v) \mapsto \left(u, \sqrt{1 - u^2 - v^2}\right) \equiv (u, w)$$
$$\qquad\qquad (1.3.23)$$
$$x^1 \circ \left(x^4\right)^{-1} : x^4(U^1 \cap U^4) \to x^1(U^1 \cap U^4)$$
$$(u, w) \mapsto \left(u, -\sqrt{1 - u^2 - w^2}\right) \equiv (u, v) .$$

It is easy to see that these two coordinate change maps are mutual inverses and that they are C^∞.

The reader should note the similarities between this example and Example 1.3.3. In the present example we were not able to represent the manifold globally as the graph of a function; however, we divided up the manifold into regions where we could represent it as a graph and, in these regions, the construction of the coordinate maps is exactly the same as in Example 1.3.3. Notice that the choice of (relatively) open sets to cover M is certainly not unique, but this does not result in any practical difficulties (see Spivak [1979] for a discussion of "maximal" atlases).

Although in Definition 1.3.1 we defined the derivative of a map defined on a manifold, there is a geometric object associated with a manifold called the tangent space which plays an important role in the concept of the derivative of a function defined on a manifold. We want to motivate its construction by first recalling the definition of differentiability of a map defined on Euclidean space. We consider a map

$$f : U \to V \tag{1.3.24}$$

where $U \subset \mathbb{R}^l$ and $V \subset \mathbb{R}^k$ are open sets. The map f is said to be differentiable at a point $x_0 \in U$ if there exists a linear map

$$L : \mathbb{R}^l \to \mathbb{R}^k \tag{1.3.25}$$

such that

$$|f(x_0 + h) - f(x_0) - Lh| = O\left(|h|^2\right) \tag{1.3.26}$$

where $|\cdot|$ is any norm on Euclidean space. The linear map L is called the derivative of f at x_0 and consists of the $l \times k$ matrix of partial derivatives of f. The linear map L acts on elements $h \in \mathbb{R}^l$ which can be viewed as vectors emanating from the point $x_0 \in U$. This previous sentence is quite important. The derivative is a linear map, but linearity of a map depends crucially on the linear structure of the space on which it operates. If we want to define the derivative of a map intrinsic to the manifold on which it is defined, we must somehow associate a linear space on which the derivative can operate in a way that is "natural" for the manifold. This linear space will be the tangent space at a point of the manifold at which the derivative is computed. We begin with two preliminary definitions.

Definition 1.3.4. Let $I = \{ t \in \mathbb{R} \mid -\epsilon < t < \epsilon \}$ for some fixed $\epsilon > 0$. Then a C^r curve in M is a C^r map from I into M.

Definition 1.3.5. Let $C : I \to M$ be a C^r curve such that $C(0) = p$. Then the vector tangent to C at p is

$$\frac{d}{dt}C(t)\Big|_{t=0} \equiv \dot{C}(0).$$

See Figure 1.3.8 for an illustration of the geometry.

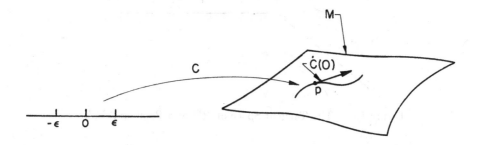

Figure 1.3.8. A Curve and Its Tangent Vector at a Point.

Before discussing the case of the tangent space at a point for a general manifold, let us first discuss the case where the manifold is $\mathbf{R}^m$ (see Example 1.3.2 following Definition 1.3.3).

Let x be a point in $\mathbf{R}^m$; then the tangent space to $\mathbf{R}^m$ at x, $T_x\mathbf{R}^m$, is defined to be $\mathbf{R}^m$. A more geometrical, but equivalent, definition would be that $T_x\mathbf{R}^m$ is the collection of all vectors tangent to curves passing through x at the point x. It is easily seen that this set is equal to $\mathbf{R}^m$, since every point in $\mathbf{R}^m$ can be viewed as the tangent vector to some differentiable curve, e.g., take as the curve $C(t) = x + t\xi$, $\xi \in \mathbf{R}^m$, then $\frac{dC(t)}{dt}\Big|_{t=0} = \xi$, see Figure 1.3.9.

Recalling our brief discussion of the differentiation of maps defined on $\mathbf{R}^m$, it should now be clear what role the tangent space at a point plays in the definition of the derivative at a point. Namely, $T_x\mathbf{R}^m$ is the domain of the derivative $Df(x)$, and locally it reflects the structure of the manifold $\mathbf{R}^m$, thus allowing the linear map $Df(x)$ to locally reflect the character of $f(x)$. In the case where the manifold has the structure of a linear vector space we usually do not bother with formalizing the notion of the tangent space at a point, since the tangent space at a point is the space itself. However, in the case where the domain of the map has no linear structure, then in order to discuss a local linear approximation to the map at a

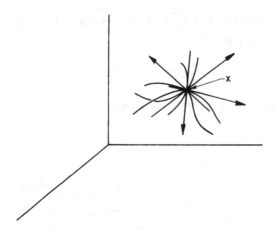

Figure 1.3.9. The Tangent Space of $\mathbb{R}^m$ at a Point $x \in \mathbb{R}^m$.

point, i.e., the derivative of the map, it is necessary to introduce the structure of a linear vector space for the domain of the derivative, since the definition of linearity of a map depends crucially on the fact that the domain of the map is linear.

Now we will define the tangent space at a point for an arbitrary differentiable manifold. Let (U^α, x^α) be a chart containing the point p in the m dimensional differentiable manifold M. Then $x^\alpha(U^\alpha)$ is an open set in $\mathbb{R}^m$ containing the point $x^\alpha(p)$. From the previous discussion, the tangent space at $x^\alpha(p)$ in $\mathbb{R}^m$, $T_{x^\alpha(p)}\mathbb{R}^m$, is just $\mathbb{R}^m$. To construct T_pM we carry $T_{x^\alpha(p)}\mathbb{R}^m$ to p in M by using x^α. Since $x^\alpha : U^\alpha \to \mathbb{R}^m$ is a diffeomorphism, then $(x^\alpha)^{-1} : x^\alpha(U^\alpha) \to U^\alpha$ is also a diffeomorphism from $\mathbb{R}^m$ into $\mathbb{R}^m$. Therefore, we can compute $D\left[(x^\alpha)^{-1}\right]$ which is a linear isomorphism mapping $\mathbb{R}^m$ into $\mathbb{R}^m$. The tangent space at $p \in M$, T_pM is then defined to be $D\left[(x^\alpha)^{-1}\right]\Big|_{x^\alpha(p)} \cdot \mathbb{R}^m$.

Definition 1.3.6. Let $(U^\alpha; x^\alpha)$ be a chart on M with $p \in U^\alpha$. Then the tangent space to M at the point p, denoted T_pM, is defined to be $D\left[(x^\alpha)^{-1}\right]\Big|_{x^\alpha(p)} \cdot \mathbb{R}^m$.

The tangent space at p in M has the same geometrical interpretation as the tangent space at a point at $\mathbb{R}^m$; namely, it can be thought of as the collection of vectors tangent to curves which pass through p at p. This can be seen as follows: let $(U^\alpha; x^\alpha)$ be a chart containing the point $p \in M$. Then, as previously discussed, $T_{x^\alpha(p)}\mathbb{R}^m$ consists of the collection of vectors tangent to differentiable curves passing through $x^\alpha(p)$ at $x^\alpha(p)$. Let $\gamma(t)$ be such a curve with $\gamma(0) = x^\alpha(p)$;

then $\dot{\gamma}(0)$ is a vector tangent to $\gamma(t)$ at $x^\alpha(p)$. Using the chart, since x^α is a diffeomorphism of U^α onto $x^\alpha(U^\alpha)$, then $(x^\alpha)^{-1}(\gamma(t)) \equiv C(t)$ is a differentiable curve satisfying $C(0) = p$. Using the chain rule, the tangent vector to $C(t)$ at p is given by $D\left[(x^\alpha)^{-1}\right]\Big|_{x^\alpha(p)} \cdot \dot{\gamma}(0) \equiv \dot{C}(0)$. Now $\dot{\gamma}(0)$ is a vector in $\mathbf{R}^m$; thus $\dot{C}(0)$ is an element of T_pM. So we see that the elements of T_pM consist of the vectors tangent to differentiable curves passing through p at p.

Before leaving the tangent space at a point, there is one last detail to be considered; namely, in our construction of the tangent space at a point of a manifold we utilized a specific chart, but the tangent space is a geometrical object which should be intrinsic to the manifold being representative of the manifold's local structure. Therefore, the tangent space should be independent of the specific chart used in its construction.

Proposition 1.3.1. *The construction of T_pM is independent of the specific chart.*

PROOF: Let $(U^\alpha; x^\alpha)$, $(U^\beta; x^\beta)$ be two charts with $U^\alpha \cap U^\beta \neq \emptyset$ and $p \in U^\alpha \cap U^\beta$. Then by Definition 1.3.5 T_pM can be constructed as either $D\left[(x^\alpha)^{-1}\right]\Big|_{x^\alpha(p)} \cdot \mathbf{R}^m$ or $D\left[(x^\beta)^{-1}\right]\Big|_{x^\beta(p)} \cdot \mathbf{R}^m$. We must show that

$$D\left[(x^\alpha)^{-1}\right]\Big|_{x^\alpha(p)} \cdot \mathbf{R}^m = D\left[(x^\beta)^{-1}\right]\Big|_{x^\beta(p)} \cdot \mathbf{R}^m \,.$$

This can be established by the following argument. Consider Figure 1.3.5, on $U^\alpha \cap U^\beta$ we have the relationship

$$(x^\alpha)^{-1} = (x^\beta)^{-1} \circ \left[x^\beta \circ (x^\alpha)^{-1}\right] \,. \tag{1.3.27}$$

Differentiating (1.3.27) we get

$$D\left[(x^\alpha)^{-1}\right]\Big|_{x^\alpha(p)} = D\left[(x^\beta)^{-1}\right]\Big|_{x^\beta(p)} D\left[x^\beta \circ (x^\alpha)^{-1}\right]\Big|_{x^\alpha(p)} \tag{1.3.28}$$

but $D\left[x^\beta \circ (x^\alpha)^{-1}\right]$ is an isomorphism of $\mathbf{R}^m$ so $D\left[x^\beta \circ (x^\alpha)^{-1}\right]\Big|_{x^\alpha(p)} \cdot \mathbf{R}^m = \mathbf{R}^m$. Therefore, we get

$$D\left[(x^\alpha)^{-1}\right]\Big|_{x^\alpha(p)} \cdot \mathbf{R}^m = D\left[(x^\beta)^{-1}\right]\Big|_{x^\beta(p)} D\left[x^\beta \circ (x^\alpha)^{-1}\right]\Big|_{x^\alpha(p)} \cdot \mathbf{R}^m$$

$$= D\left[(x^\beta)^{-1}\right]\Big|_{x^\beta(p)} \cdot \mathbf{R}^m \,.$$

$$\tag{1.3.29}$$

So we see that the tangent space at a point is independent of the particular chart chosen in a neighborhood of that point.   □

Now that we have defined the tangent space at a point of a manifold, we want to demonstrate the role it plays in defining the derivative of maps between manifolds. Let $f : M^m \to N^s$ be a C^r map where $M^m \subseteq \mathbf{R}^n$, $m \leq n$, is an m dimensional manifold and $N^s \subset \mathbf{R}^q$, $s \leq q$, is an s dimensional manifold. From Definition 1.3.1, f being C^r means that, for every point $p \in M$, there is an open set $V \subset \mathbf{R}^n$ (depending on p) and a C^r map $F : V \to \mathbf{R}^q$ with $F = f$ on $V \cap M^m$.

Proposition 1.3.2. *Let $f : M^m \to N^s$ be as defined above; then*

$$DF|_p : T_p M^m \to T_{f(p)} N^s . \tag{1.3.30}$$

PROOF: Let $(x^\alpha; U^\alpha)$ be a coordinate chart on M^m containing p and let $(y^\beta; W^\beta)$ be a coordinate chart on N^s containing $f(p)$. Let $V \subset \mathbf{R}^n$ be the open set around p in $\mathbf{R}^n$ and, if necessary, shrink U^α so that $U^\alpha \subset V$. Then we have

$$F(p) = (y^\beta)^{-1} \circ \left[y^\beta \circ f \circ (x^\alpha)^{-1} \right] \circ x^\alpha(p) . \tag{1.3.31}$$

Now we must show that $DF(p) \cdot D\left[(x^\alpha)^{-1} \right]\big|_{x^\alpha(p)} \cdot \mathbf{R}^m$ is contained in

$$D\left[(y^\beta)^{-1} \right]\big|_{y^\beta(f(p))} \cdot \mathbf{R}^n \equiv T_{f(p)} N^s .$$

Differentiating (1.3.31) we obtain

$$DF|_p = D\left[(y^\beta)^{-1} \right]\big|_{y^\beta(f(p))} D\left[y^\beta \circ f \circ (x^\alpha)^{-1} \right]\big|_{x^\alpha(p)} Dx^\alpha\big|_p \tag{1.3.32}$$

or equivalently,

$$DF|_p D\left[(x^\alpha)^{-1} \right]\big|_{x^\alpha(p)} = D\left[(y^\beta)^{-1} \right]\big|_{y^\beta(f(p))} D\left[y^\beta \circ f \circ (x^\alpha)^{-1} \right]\big|_{x^\alpha(p)} . \tag{1.3.33}$$

So we get

$$DF|_p D\left[(x^\alpha)^{-1} \right]\big|_{x^\alpha(p)} \cdot \mathbf{R}^m = D\left[(y^\beta)^{-1} \right]\big|_{y^\beta(f(p))} D\left[y^\beta \circ f \circ (x^\alpha)^{-1} \right]\big|_{x^\alpha(p)} \cdot \mathbf{R}^m \tag{1.3.34}$$

but

$$D\left[y^\beta \circ f \circ (x^\alpha)^{-1} \right]\big|_{x^\alpha(p)} \cdot \mathbf{R}^m \subset \mathbf{R}^s \tag{1.3.35}$$

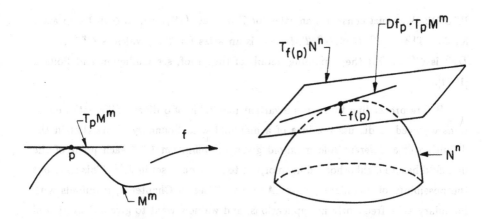

Figure 1.3.10. $T_p M^m$ and Its Image under $Df|_p$.

and

$$D\left[(y^\beta)^{-1}\right]\Big|_{y^\beta(f(p))} \cdot \mathbf{R}^n = T_{f(p)} N^s. \tag{1.3.36}$$

So we see that $DF|_p \cdot T_p M^m \subset T_{f(p)} N^s$. From equation (1.3.34) it should be clear that this result is independent of the particular extension F of f to some open set in $\mathbf{R}^n$.   □

Geometrically, Proposition 1.3.2 may be visualized as in Figure 1.3.10.

When we study manifolds which are invariant under the flow generated by a vector field it will be important to have information concerning the tangent spaces at different points on the manifold. In this regard, it is useful to consider the geometric object formed by the disjoint union of all the tangent spaces at all possible points of the manifold. This is called the *tangent bundle.*

Definition 1.3.7. The tangent bundle of a C^r manifold $M \subset \mathbf{R}^n$, denoted TM, is defined as

$$TM = \{ (p, v) \in M \times \mathbf{R}^m \mid v \in T_p M \}. \tag{1.3.37}$$

So the tangent bundle is the set of all possible tangent vectors to M, and TM itself has the structure of a $2m$ dimensional C^{r-1} manifold, as we show next.

Proposition 1.3.3. Let $M \subset \mathbf{R}^n$ be a C^r manifold of dimension m; then the tangent bundle of M, $TM \subset \mathbf{R}^{2n}$, is a C^{r-1} manifold of dimension $2m$.

PROOF: We must construct an atlas for TM. Let $(x^\alpha; U^\alpha)$, $\alpha \in A$, be an atlas for M. Then $(x^\alpha, Dx^\alpha; U^\alpha, TM|_{U^\alpha})$ is an atlas for TM, which is C^{r-1}, since Dx^α is C^{r-1}. For the remaining details of the proof, see Guillemin and Pollack [1974]. $\qquad\qquad\qquad\qquad\qquad\qquad\qquad\qquad\qquad\qquad\qquad\qquad\qquad\qquad$ □

Before proceeding to discuss invariant manifolds of ordinary differential equations we need to discuss the idea of a manifold with boundary. Note that in the definition of a differentiable manifold given in Definition 1.3.3 each point of the manifold has a neighborhood diffeomorphic to some open set in $\mathbf{R}^m$. This rules out the possibility of boundary points. As we shall see in Chapter 4, manifolds with boundary arise frequently in applications, and we now want to give a definition of a C^r manifold with boundary. We begin with a preliminary definition.

Definition 1.3.8. The *closed half space*, $\mathbf{R}^m_- \subset \mathbf{R}^m$, is defined as follows

$$\mathbf{R}^m_- = \{ (x_1, x_2, \ldots, x_m) \in \mathbf{R}^m \mid x_1 \leq 0 \}. \tag{1.3.38}$$

The *boundary* of $\mathbf{R}^m_-$, denoted $\partial \mathbf{R}^m_-$, is $\mathbf{R}^{m-1}$.

We now give the definition of a differentiable manifold with boundary.

Definition 1.3.9. A subset $M \subset \mathbf{R}^n$ is called a C^r *manifold of dimension m with boundary* if it possesses the following two structural characteristics:

1) There exists a countable collection of open sets $V^\alpha \subset \mathbf{R}^n$, $\alpha \in A$ where A is some countable index set, with $U^\alpha \equiv V^\alpha \cap M$ such that $M = \bigcup_{\alpha \in A} U^\alpha$.

2) There exists a C^r diffeomorphism x^α defined on each U^α which maps U^α onto some set $W \cap \mathbf{R}^m_-$ where W is some open set in $\mathbf{R}^n$.

We make the following remarks concerning Definition 1.3.9.

1) The *boundary of M*, denoted ∂M, is defined to be the set of points in M which are mapped to $\partial \mathbf{R}^m_-$ under x^α. It is necessary to show that this set is independent of the particular chart that is chosen, see Guillemin and Pollack [1974] for the details.

2) The boundary of M is a C^r manifold of dimension $m-1$, and $M - \partial M$ is a C^r manifold of dimension m.

3) The tangent space of M at a point is defined just as in Definition 1.3.6 even if the point is a boundary point.

We are now at the point where we can state some general results on invariant manifolds of ordinary differential equations. As mentioned earlier, we will follow Fenichel's development of the theory, since he explicitly treats the case of invariant manifolds with boundary which we will encounter in our applications.

We consider a general autonomous ordinary differential equation defined on $\mathbb{R}^n$

$$\dot{x} = f(x), \qquad x \in \mathbb{R}^n \tag{1.3.39}$$

where f is a C^r function of x. Let us denote the flow generated by (1.3.39) by $\phi_t(p)$, i.e., $\phi_t(p)$ denotes the solution of (1.3.39) passing through the point $p \in \mathbb{R}^n$ at $t = 0$. We remark that $\phi_t(p)$ need not be defined for all $t \in \mathbb{R}$ or all $p \in \mathbb{R}^n$. Let $\bar{M} \equiv M \cup \partial M$ be a compact, connected C^r manifold with boundary contained in $\mathbb{R}^n$.

Definition 1.3.10. a) $\bar{M} \equiv M \cup \partial M$ is said to be *overflowing invariant* under (1.3.39) if for every $p \in \bar{M}$, $\phi_t(p) \in \bar{M}$ for all $t \leq 0$ and the vector field (1.3.39) is pointing strictly outward and is nonzero on ∂M.   b) $\bar{M} \equiv M \cup \partial M$ is said to be *inflowing invariant* under (1.3.39) if for every $p \in \bar{M}$, $\phi_t(p) \in \bar{M}$ for all $t \geq 0$ and the vector field (1.3.39) is pointing strictly inward and is nonzero on ∂M.   c) $\bar{M} \equiv M \cup \partial M$ is said to be *invariant* under (1.3.39) if for every $p \in \bar{M}$, $\phi_t(p) \in \bar{M}$ for all $t \in \mathbb{R}$.

We make the following remarks concerning this definition.

1) The phrase "the vector field (1.3.39) is pointing strictly outward and is nonzero on ∂M" means that for every $p \in \partial M$, $\phi_t(p) \notin \bar{M}$ for all $t > 0$. A similar definition is obtained for the "... pointing strictly inward ..." by reversing time.

2) Overflowing invariant manifolds become inflowing invariant under time reversal and vice versa.

3) Since $\bar{M}$ is compact, $\phi_t(\cdot)|_{\bar{M}}$ exists for all $t \leq 0$ if $\bar{M}$ is overflowing invariant, for all $t \geq 0$ if $\bar{M}$ is inflowing invariant, and for all $t \in \mathbb{R}$ if $\bar{M}$ is invariant.

4) $\bar{M}$ can be an invariant manifold only if the vector field (1.3.39) is identically zero on ∂M, if $\partial M = \emptyset$, or if the vector field (1.3.39) is parallel to ∂M.

The following definition will also be useful.

Definition 1.3.11. Let $M \subset \mathbf{R}^n$ be a compact, connected C^r manifold in $\mathbf{R}^n$. We say that M is *locally invariant* under (1.3.39) if for each $p \in M$ there exists a time interval $I_p = \{t \in \mathbf{R} \,|\, t_1 < t < t_2$ where $t_1 < 0, t_2 > 0\}$ such that $\phi_t(p) \in M$ for all $t \in I_p$.

We remark that the overflowing and inflowing invariant manifolds of Definition 1.3.10 are examples of locally invariant manifolds.

Next we want to describe the stability characteristics of the invariant manifolds. This will be done by describing the asymptotic behavior of vectors tangent and normal to the invariant manifold under the action of the linearized flow.

Let $\bar{M} \equiv M \cup \partial M$ be a C^r overflowing invariant manifold contained in $\mathbf{R}^n$. Let $T\mathbf{R}^n|_{\bar{M}}$ denote the tangent bundle of $\mathbf{R}^n$ restricted to $\bar{M}$, i.e., $T\mathbf{R}^n|_{\bar{M}} \equiv \{(p,v) \in \mathbf{R}^n \times \mathbf{R}^n | p \in \bar{M}, \ v \in T_p\mathbf{R}^n\}$. Then, by Definition 1.3.7, $T\bar{M} \subset T\mathbf{R}^n|_{\bar{M}}$ and, by Proposition 1.3.2, $T\bar{M}$ is invariant under $D\phi_t(p)$, $p \in \bar{M}$, for all $t \leq 0$. $T\bar{M}$ is referred to as a *negatively invariant subbundle*. At each point $p \in \bar{M}$ we can use the standard metric on $\mathbf{R}^n$ to choose a complementary subspace of $\mathbf{R}^n$, N_p, such that $T_p\mathbf{R}^n = T_p\bar{M} + N_p$, where "+" denotes the usual direct sum of vector spaces. If we form the union of all such decompositions of $T_p\mathbf{R}^n$ over all points $p \in \bar{M}$, we obtain a decomposition or *splitting* of $T\mathbf{R}^n|_{\bar{M}}$, i.e., $T\mathbf{R}^n|_{\bar{M}} = T\bar{M} \oplus N \equiv \bigcup_{p \in M} (T_p\bar{M} + N_p)$. The sum of two subbundles is denoted by $\oplus$ and is referred to as the Whitney sum (see Spivak [1979]).

We denote the projection onto N corresponding to the splitting $T\mathbf{R}^n|_{\bar{M}} = T\bar{M} \oplus N$ by Π^N, and the projection onto $T\bar{M}$ by Π^T. For the C^r perturbation theorem for overflowing invariant manifolds, we will require the manifold to be stable, in the sense that vectors complementary to $T\bar{M}$ grow in length as $t \to -\infty$ under the action of the linearized flow, i.e., $w_0 \in N_p$ implies $\left|\Pi^N D\phi_t(p)w_0\right| \to \infty$ as $t \to -\infty$, where $|\cdot|$ denotes the norm associated with the chosen metric on $\mathbf{R}^n$, and that neighborhoods of the invariant manifold "flatten out" as $t \to -\infty$ under the action of the linearized flow. This "flattening out" property is expressed as $|D\phi_t(p)v_0| / \left|\Pi^N D\phi_t(p)w_0\right|^s \to 0$ as $t \to -\infty$ for every $v_0 \in T_p\bar{M}$, $w_0 \in N_p$. The real number s is a measure of the degree of the flattening of the neighborhoods of $\bar{M}$. This situation might be visualized geometrically as in Figure 1.3.11.

For computations and proving theorems the stability properties of $\bar{M}$ are more conveniently phrased in terms of rates of growth of vectors under the linearized

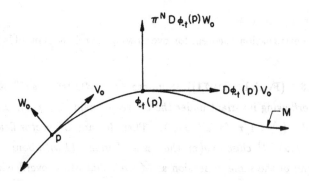

Figure 1.3.11. Action of Tangent Vectors under $D\phi_t(p)$.

flow. We define the following quantities

$$\gamma(p) = \inf\{\, a \in \mathbf{R}^+ \mid a^t / \left|\Pi^N D\phi_t(p) w_0\right| \to 0 \quad \text{as } t \to -\infty \quad \text{for all } w_0 \in N_p \,\}$$

$$\text{and if } \gamma(p) < 1 \tag{1.3.40}$$

$$\sigma(p) = \inf\{\, s \in \mathbf{R} \mid \left|D\phi_t(p) v_0\right| / \left|\Pi^N D\phi_t(p) w_0\right|^s \to 0 \quad \text{as } t \to -\infty$$

$$\text{for all } v_0 \in T_p\bar{M},\ w_0 \in N_p \,\}.$$

The functions $\gamma(p)$ and $\sigma(p)$ are called *generalized Lyapunov type numbers* (Fenichel [1971]) and have several properties which we now state.

Proposition 1.3.4. *The functions $\gamma(p)$ and $\sigma(p)$ have the following properties:*

i) *$\gamma(p)$ and $\sigma(p)$ are constant on orbits, i.e., $\gamma(p) = \gamma(\phi_t(p))$, $\sigma(p) = \sigma(\phi_t(p))$,*
   $t \le 0$.

ii) *$\gamma(p)$ and $\sigma(p)$ are bounded and achieve their suprema on $\bar{M}$ (although in general*
   they are neither continuous nor semicontinuous).

iii) *$\gamma(p)$ and $\sigma(p)$ are independent of the choice of metric and of the choice of N.*

PROOF: See Fenichel [1971]. □

We remark that a more computable form for the generalized Lyapunov type numbers which can be derived from (1.3.40) is the following:

$$\gamma(p) = \varlimsup_{t \to -\infty} \left\|\Pi^N D\phi_t(p)\right\|^{1/t}$$

$$\sigma(p) = \varlimsup_{t \to -\infty} \frac{\log\left\|D\phi_t(p)\Pi^T\right\|}{\log\left\|\Pi^N D\phi_t(p)\right\|} \tag{1.3.41}$$

where $\|\cdot\|$ is a matrix norm.

The C^r perturbation theorem for overflowing invariant manifolds is stated as follows:

Theorem 1.3.5 (Fenichel [1971]). *Suppose* $\bar{M} = M \cup \partial M$ *is a* C^r *manifold with boundary overflowing invariant under the* C^r *vector field* $\dot{x} = f(x)$, $x \in \mathbb{R}^n$ *with* $\gamma(p) < 1$ *and* $\sigma(p) < 1/r$ *for all* $p \in \bar{M}$. *Then, for any* C^r *vector field* $\dot{x} = g(x)$, $x \in \mathbb{R}^n$ *with* $f(x)$ C^1*-close to* $g(x)$, *there is a* C^r *manifold with boundary* $\bar{M}_g$, C^r*-close to* $\bar{M}$ *and of the same dimension as* $\bar{M}$ *such that* $\bar{M}_g$ *is overflowing invariant under* $\dot{x} = g(x)$, $x \in \mathbb{R}^n$.

PROOF: See Fenichel [1971]. $\qquad\qquad\qquad\qquad\qquad\qquad\qquad\qquad\qquad$ □

This theorem can also be applied to inflowing invariant manifolds. In that case the generalized Lyapunov type numbers are computed using the time reversed flow and taking the limits as $t \to +\infty$. Theorem 1.3.5 will then read exactly the same except that the word overflowing will be replaced by inflowing.

We illustrate Theorem 1.3.5 with the following simple example.

EXAMPLE 1.3.5. Consider the following C^∞ planar vector field

$$\begin{aligned} \dot{x} &= ax \\ \dot{y} &= -by \end{aligned} \qquad , a, b > 0 . \qquad (1.3.42)$$

It should be clear that (0,0) is a fixed point of (1.3.42) with the x axis being the unstable manifold of (0,0) and the y axis being the stable manifold of (0,0). Consider the set

$$\bar{M} = \{ (x,y) \in \mathbb{R}^2 \mid -\delta \leq x \leq \delta , \qquad y = 0 , \qquad \text{for some } \delta > 0 \} . \qquad (1.3.43)$$

It is easy to verify that $\bar{M}$ is an overflowing invariant manifold under the flow generated by (1.3.42). We now show that $\bar{M}$ satisfies the hypotheses of Theorem 1.3.3. We have

$$\begin{aligned} T\bar{M} &= \bar{M} \times (\mathbb{R}^1, 0) \\ N &= \bar{M} \times (0, \mathbb{R}^1) \\ \Pi^N &= \begin{pmatrix} 0 & 0 \\ 0 & 1 \end{pmatrix} \\ \Pi^T &= \begin{pmatrix} 1 & 0 \\ 0 & 0 \end{pmatrix} \\ D\phi_t &= \begin{pmatrix} e^{at} & 0 \\ 0 & e^{-bt} \end{pmatrix} , \end{aligned} \qquad (1.3.44)$$

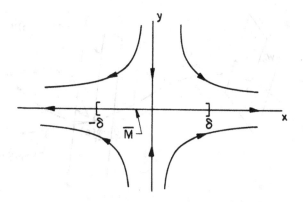

Figure 1.3.12. The Geometry of (1.3.42)

see Figure 1.3.12.

The generalized Lyapunov type numbers can now be computed, and we find that

$$\gamma(p) = \gamma = \varlimsup_{t \to -\infty} \left\| e^{-bt} \right\|^{1/t} = e^{-b} < 1$$

$$\sigma(p) = \sigma = \varlimsup_{t \to -\infty} \frac{\log \left\| e^{at} \right\|}{\log \left\| e^{-bt} \right\|} = -a/b < 0 \,.$$

(1.3.45)

Thus, $\bar{M}$ satisfies the hypotheses of Theorem 1.3.3 so that any C^r vector field ($r \geq 1$) which is C^1-close to (1.3.42) has an overflowing invariant manifold C^r-close to $\bar{M}$. We have therefore established the local unstable manifold theorem for a fixed point of a planar nonlinear vector field whose linear part is given by (1.3.42). Local stable manifolds can be shown to exist by considering inflowing invariant manifolds.

Now for overflowing invariant manifolds it makes sense to consider the unstable manifold of the overflowing invariant manifold, and we have an existence and perturbation theorem for unstable manifolds of overflowing invariant manifolds. The set-up is as follows: Let $\bar{M} = M \cup \partial M$ be overflowing invariant under (1.3.39), and let $N^u \subset T\mathbb{R}^n|_{\bar{M}}$ be a subbundle which contains $T\bar{M}$ and is negatively invariant under the linearized flow generated by (1.3.39). Let $I \subset N^u$ be any subbundle complementary to $T\bar{M}$, and let $J \subset T\mathbb{R}^n|_{\bar{M}}$ be any subbundle complementary

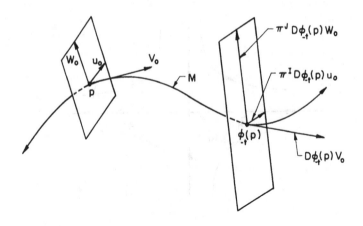

Figure 1.3.13.

to N^u. Then we have the splitting $T\mathbb{R}^n\big|_{\bar{M}} = T\bar{M} \oplus I \oplus J$. Let Π^I, Π^J, and Π^T be the projections onto I, J, and $T\bar{M}$, respectively, corresponding to this splitting. We define generalized Lyapunov type numbers as follows:

For any $p \in \bar{M}$

$$\lambda(p) = \inf\{\, b \in \mathbb{R}^+ \big|\, \big|\Pi^I D\phi_t(p)u_0\big|b^t \to 0 \quad \text{as } t \to -\infty \quad \text{for all } u_0 \in I_p \,\}$$

$$\gamma(p) = \inf\{\, a \in \mathbb{R}^+ \big|\, a^t \big/ \big|\Pi^J D\phi_t(p)w_0\big| \to 0 \quad \text{as } t \to -\infty \quad \text{for all } w_0 \in J_p \,\}$$

and if $\gamma(p) < 1$ define $\hspace{4cm}$ (1.3.46)

$$\sigma(p) = \inf\{\, s \in \mathbb{R} \big|\, \big|D\phi_t(p)v_0\big| \big/ \big|\Pi^J D\phi_t(p)w_0\big|^s \to 0 \quad \text{as } t \to -\infty$$

$$\text{for all } v_0 \in T_p\bar{M},\ w_0 \in J_p \,\}.$$

Conclusions identical to those in Proposition 1.3.4 follow for $\lambda(p)$, $\gamma(p)$ and $\sigma(p)$ defined above. More computable expressions for the generalized Lyapunov type numbers can be derived from (1.3.46) and have the following form

$$\lambda(p) = \varlimsup_{t \to -\infty} \left\| \Pi^I D\phi_t(p) \right\|^{-1/t}$$

$$\gamma(p) = \varlimsup_{t \to -\infty} \left\| \Pi^J D\phi_t(p) \right\|^{1/t} \hspace{3cm} (1.3.47)$$

$$\sigma(p) = \varlimsup_{t \to -\infty} \frac{\log \left\| D\phi_t(p)\Pi^T \right\|}{\log \left\| \Pi^J D\phi_t(p) \right\|}$$

where $\|\cdot\|$ is some matrix norm. See Figure 1.3.13 for an illustration of the geometry.

We have the following theorem.

Theorem 1.3.6 (Fenichel [1971], [1979]). *Suppose $\bar{M} = M \cup \partial M$ is a C^r manifold with boundary overflowing invariant under the C^r vector field $\dot{x} = f(x)$, $x \in \mathbb{R}^n$, with $N^u \subset T\mathbb{R}^n|_{\bar{M}}$ a subbundle containing $\bar{M}$ negatively invariant under the linearized flow generated by $\dot{x} = f(x)$, $x \in \mathbb{R}^n$. Then if $\lambda(p) < 1$, $\gamma(p) < 1$, and $\sigma(p) < 1/r$ for all $p \in \bar{M}$, the following conclusions hold:*

i) *(Existence) There exists a C^r manifold W^u overflowing invariant under $\dot{x} = f(x)$, $x \in \mathbb{R}^n$, such that W^u contains $\bar{M}$ and is tangent to N^u along $\bar{M}$.*

ii) *(Persistence) Suppose $\dot{x} = g(x)$, $x \in \mathbb{R}^n$, is a C^r vector field C^1 close to $\dot{x} = f(x)$, $x \in \mathbb{R}^n$. Then there exists a C^r manifold W^u_g overflowing invariant under $\dot{x} = g(x)$, $x \in \mathbb{R}^n$ which is C^r close to W^u and has the same dimension as W^u.*

PROOF: See Fenichel [1971]. □

We remark that Theorem 1.3.6 can be applied to inflowing invariant manifolds. In that case, the generalized Lyapunov type numbers are computed using the time reversed flow with the limits taken as $t \to +\infty$, and the phrase "overflowing invariant" in Theorem 1.3.6 is replaced by "inflowing invariant." Also, N^u and W^u are replaced by N^s and W^s, with N^s taken to be a positively invariant subbundle under the linearized flow generated by $\dot{x} = f(x)$, $x \in \mathbb{R}^n$ (i.e., N^s is negatively invariant under the time reversed linearized flow).

We illustrate this theorem with the following example.

EXAMPLE 1.3.6. Consider the vector field (1.3.42). We will regard the fixed point $(0,0)$ as the overflowing invariant manifold $\bar{M}$. Then we have $N^u \equiv (0,0) \times (\mathbb{R}^1, 0)$ is a negatively invariant subbundle. We now show that $\bar{M}$ and N^u satisfy the hypotheses of Theorem 1.3.6. We have

$$\bar{M} = (0,0)$$

$$T\bar{M} = \emptyset$$

$$I \equiv (0,0) \times (\mathbb{R}^1, 0)$$

$$J \equiv (0,0) \times (0, \mathbb{R}^1)$$

$$\Pi^I = \begin{pmatrix} 1 & 0 \\ 0 & 0 \end{pmatrix}$$

$$\Pi^J = \begin{pmatrix} 0 & 0 \\ 0 & 1 \end{pmatrix}$$

$$D\phi_t = \begin{pmatrix} e^{at} & 0 \\ 0 & e^{-bt} \end{pmatrix} .$$

(1.3.48)

The generalized Lyapunov type numbers are given by

$$\lambda(0) = \varlimsup_{t \to -\infty} \left\| e^{bt} \right\|^{-1/t} = e^{-b} < 1$$

$$\gamma(0) = \varlimsup_{t \to -\infty} \left\| e^{-at} \right\|^{-1/t} = e^{-a} < 1$$

$$\sigma(0) = 0 \qquad \text{since } T\bar{M} = \emptyset .$$

(1.3.49)

Thus, $\bar{M}$ and N^u satisfy the hypotheses of Theorem 1.3.6 so any C^r vector field C^1-close to (1.3.42) contains an overflowing invariant manifold.

We remark that in Examples 1.3.5 and 1.3.6 the same conclusions can be made concerning the vector field (1.3.42); however, different conditions which are necessary in order to arrive at these conclusions are computed in each case. In Example 1.3.5 generalized Lyapunov type numbers are computed on a manifold, and in Example 1.3.6 generalized Lyapunov type numbers are computed at a point. In fact, Theorem 1.3.6 is proven by showing that N^u under the hypotheses given in the theorem is an overflowing invariant manifold satisfying the hypotheses of Theorem 1.3.5.

Let us now give the usual theorem (Sacker [1964], Hirsch, Pugh and Shub [1977]) for compact, boundaryless manifolds invariant under (1.3.39).

Theorem 1.3.7. *Let M be a compact, boundaryless C^r manifold invariant under $\dot{x} = f(x)$, $x \in \mathbb{R}^n$. Let N^s and N^u be subbundles of $T\mathbb{R}^n|_M$ such that $N^s \oplus N^u = T\mathbb{R}^n|_M$ and $N^s \cap N^u = TM$. Suppose N^u satisfies the hypotheses of Theorem*

1.3.6, and N^s satisfies the hypotheses of Theorem 1.3.6 for the time reversed flow. Then the following conclusions hold.

   i) (Existence) There exist C^r manifolds W^u, W^s tangent to N^u, N^s along M with W^u overflowing invariant under $\dot{x} = f(x)$, $x \in R^n$, and W^s overflowing invariant under $\dot{x} = -f(x)$, $x \in R^n$, moreover $M = W^u \cap W^s$.

  ii) (Persistence) Suppose $\dot{x} = g(x)$, $x \in R^n$ is a C^r vector field with $g(x)$ C^1 close to $f(x)$. Then there are C^r manifolds W^u_g, W^s_g C^r close to W^u, W^s, respectively, and having the same respective dimensions with W^u_g overflowing invariant under $\dot{x} = g(x)$, $x \in R^n$ and W^s_g overflowing invariant under $\dot{x} = -g(x)$, $x \in R^n$; moreover, $W^u_g \cap W^s_g \equiv M_g$ is a C^r manifold invariant under $\dot{x} = g(x)$, $x \in R^n$ and C^r close to M.

We remark that compact boundaryless invariant manifolds satisfying the hypotheses of Theorem 1.3.7 are said to be *normally hyperbolic*.

EXAMPLE 1.3.1 - continued. Let us now return to our original example concerning invariant manifolds in order to demonstrate the calculations necessary for the verification of the hypotheses of the theorems. Recall that we considered the equation

$$\dot{x} = f(x), \qquad x \in \mathbb{R}^n \tag{1.3.50}$$

under the following assumptions.

  A1) (1.3.50) has a fixed point at $x = 0$, i.e., $f(0) = 0$.

  A2) $Df(0)$ has $n - k$ eigenvalues having positive real parts and k eigenvalues having negative real parts.

In this simple case, the invariant manifold M is just the fixed point $x = 0$ and $T\mathbb{R}^n|_M = \mathbb{R}^n$ with the unstable and stable subspaces of the linear problem, E^u and E^s, corresponding to the invariant subbundles N^u and N^s where $T\mathbb{R}^n|_M = \mathbb{R}^n = E^u + E^s$. We must show that E^u satisfies the hypotheses of Theorem 1.3.6, and E^s satisfies the hypotheses of Theorem 1.3.6 under the time reversed flow.

Step 1: Show that E^u satisfies the hypotheses of Theorem 1.3.6.

Let $I = E^u$ and $J = E^s$; then, in coordinates given by the unstable and stable subspaces, the linearized vector field written as

$$\dot{\xi} = Df(0)\,\xi, \qquad \xi \in \mathbb{R}^n \tag{1.3.51}$$

with

$$Df(0) = \begin{pmatrix} A_1 & 0 \\ 0 & A_2 \end{pmatrix} \qquad (1.3.52)$$

where A_1 is a $n - k \times n - k$ matrix with all eigenvalues having positive real parts and A_2 is a $k \times k$ matrix with all eigenvalues having negative real parts. Then we have

$$\Pi^I e^{Df(0)t} = e^{A_1 t}$$
$$\Pi^J e^{Df(0)t} = e^{A_2 t}. \qquad (1.3.53)$$

Let λ_1 be the real part of the eigenvalue of A_1 having the smallest real part, and λ_2 the real part of eigenvalue of A_2 having the largest real part. Then it is easy to see that

$$\lambda(0) = e^{-\lambda_1} < 1$$
$$\gamma(0) = e^{\lambda_2} < 1 \qquad (1.3.54)$$
$$\sigma(0) = 0 \qquad \text{since } TM = \emptyset.$$

Step 2: Show that E^s satisfies the hypotheses of Theorem 1.3.6 for the time reversed vector field.

Under the time reversed vector field E^u and E^s are interchanged, so now, letting $I = E^s$, $J = E^u$ and using (1.3.52), we get

$$\lambda(0) = e^{\lambda_2} < 1$$
$$\gamma(0) = e^{-\lambda_1} < 1 \qquad (1.3.55)$$
$$\sigma(0) = 0 \qquad \text{since } TM = \emptyset.$$

Thus, we can conclude that there exist manifolds W^u, W^s as differentiable as f and tangent to E^u and E^s at $x = 0$. W^u is overflowing invariant under (1.3.50) having the same dimension as E^u, and W^s is overflowing invariant under (1.3.50), with time reversed, having the same dimension as E^s; moreover, for any other vector field C^1 close to (1.3.50) this structure persists.

1.4. Transversality, Structural Stability, and Genericity

The concepts of transversality, structural stability, and genericity have played an important role in the development of dynamical systems theory, and in this section we want to present a brief discussion of these ideas.

Transversality is a geometric notion which deals with the intersection of surfaces or manifolds (see Section 1.3). Let M and N be differentiable (at least C^1) manifolds in $\mathbf{R}^n$.

Definition 1.4.1. Let p be a point in $\mathbf{R}^n$; then M and N are said to be *transversal at p* if $p \notin M \cap N$ or, if $p \in M \cap N$, then $T_pM + T_pN = \mathbf{R}^n$ where T_pM and T_pN denote the tangent spaces of M and N, respectively, at the point p. M and N are said to be *transversal* if they are transversal at every point $p \in \mathbf{R}^n$, see Figure 1.4.1.

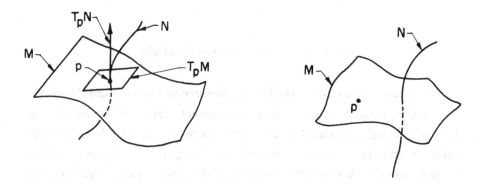

Figure 1.4.1. M and N Transverse at p.

Note that transversality of two manifolds at a point requires more than just the two manifolds geometrically piercing each other at the point. Consider the following example.

EXAMPLE 1.4.1. Let M be the x axis in $\mathbf{R}^2$, and let N be the graph of the function $f(x) = x^3$, see Figure 1.4.2. Then M and N intersect at the origin in $\mathbf{R}^2$, but they are not transversal at the origin, since the tangent space of M is just the x axis and the tangent space of N is the span of the vector $(1,0)$; thus, $T_{(0,0)}N = T_{(0,0)}M$ and, therefore, $T_{(0,0)}N + T_{(0,0)}M \neq \mathbf{R}^2$.

The most important characteristic of transversality is that it persists under sufficiently small perturbations. This fact will play a useful role in many of our geometric arguments in Chapters 3 and 4. Finally, we remark that a term often used synonymously for transversal is *general position*, i.e., two or more manifolds which are transversal are said to be in general position.

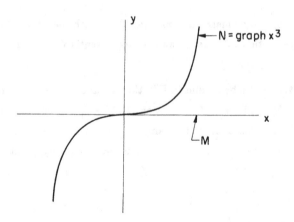

Figure 1.4.2. Nontransversal Manifolds.

The concept of structural stability was introduced by Andronov and Pontryagin [1931] and has played a central role in the development of dynamical systems theory. Roughly speaking, a dynamical system (vector field or map) is said to be structurally stable if nearby systems have qualitatively the same dynamics. This sounds like a simple enough idea; however, first one must provide a recipe for determining when two systems are "close" and then one must specify what is meant by saying that, qualitatively, two systems have the same dynamics. We will discuss each question separately.

Let $C^r(\mathbb{R}^n, \mathbb{R}^n)$ denote the space of C^r maps of $\mathbb{R}^n$ into $\mathbb{R}^n$. In terms of dynamical systems, we can think of the elements of $C^r(\mathbb{R}^n, \mathbb{R}^n)$ as being vector fields. We denote the subset of $C^r(\mathbb{R}^n, \mathbb{R}^n)$ consisting of the C^r diffeomorphisms by $\mathrm{Diff}^r(\mathbb{R}^n, \mathbb{R}^n)$.

Two elements of $C^r(\mathbb{R}^n, \mathbb{R}^n)$ are said to be C^k ϵ-close $(k \leq r)$, or just C^k-close, if they, along with their first k derivatives, are within ϵ as measured in some norm. There is a problem with this definition; namely, $\mathbb{R}^n$ is unbounded, and the behavior at infinity needs to be brought under control (note: this explains why most of dynamical systems theory has been developed using compact phase spaces; however, in applications this is not sufficient, and appropriate modifications must be made).

There are several ways of handling this difficulty. For the purpose of our discussion we will choose the usual way and assume that our maps act on compact, boundaryless n dimensional differentiable manifolds, M, rather than all of $\mathbf{R}^n$. The topology induced on $C^r(M, M)$ by this measure of distance between two elements of $C^r(M, M)$ is called the C^k *topology*, and we refer the reader to Palis and deMelo [1982] or Hirsch [1976] for a more thorough discussion.

The question of what is meant by saying that two dynamical systems have qualitatively the same dynamics is usually answered in terms of conjugacies (see Section 1.2). Specifically, C^0 conjugate maps and C^0 equivalent vector fields have qualitatively the same orbit structures in the sense of the propositions given in Section 1.2. It should also be clear from Section 1.2 why we do not use differentiable conjugacies, e.g., from Proposition 1.2.2, two maps having a fixed point cannot be C^k ($k \geq 1$) conjugate unless the eigenvalues associated with the linearized maps about the respective fixed points are equal. This is a much too strong requirement if we are only interested in distinguishing qualitative differences between the dynamics of different dynamical systems. We remark that in recent years, as our knowledge of the global dynamics of nonlinear systems has increased, it is beginning to appear possible that even C^0 conjugacies may be too strong for distinguishing the important dynamical features in different dynamical systems. This is evidenced by the great difficulties encountered in trying to ascertain generic properties and structural stability criteria for higher dimensional systems (i.e., n dimensional maps, $n \geq 2$ and n dimensional vector fields, $n \geq 3$) using the classical concepts (i.e., C^0 conjugacies). This could be caused by the fact that C^0 conjugacies are relations between specific orbits, and much of the complicated and chaotic phenomena that occur in higher dimensional dynamical systems arise via interactions amongst families of orbits (note: we will see many examples of this in Chapters 3 and 4).

We are now at the point where we can formally define structural stability.

Definition 1.4.2. Consider a map $f \in \text{Diff}^r(M, M)$ (resp. a C^r vector field in $C^r(M, M)$); then f is said to be *structurally stable* if there exists a neighborhood, $\mathcal{N}$, of f in the C^k-topology such that f is C^0 conjugate (resp. C^0 equivalent) to every map (resp. vector field) in $\mathcal{N}$.

Now that we have defined structural stability, it would be nice if we could determine the characteristics of a specific system which result in the system being structurally stable. From the point of view of the applied scientist, this would be useful, since one might presume that dynamical systems modelling phenomena occurring in nature should possess the property of structural stability. Unfortunately, such a characterization does not exist, although some partial results are known which we will describe shortly. One approach to the characterization of structural stability has been through the identification of typical or generic properties of dynamical systems, and we now discuss this idea.

Naively, one might expect a typical or generic property of a dynamical system to be one that is common to a dense set of dynamical systems in $C^r(M, M)$. This is not quite adequate, since it is possible for a set and its complement to both be dense. For example, the set of rational numbers is dense in the real line, and so is its complement, the set of irrational numbers; however, there are many more irrational numbers than rational numbers, and one might expect the irrationals to be more typical than the rationals in some sense. The proper sense in which this is true is captured by the idea of a *residual set*.

Definition 1.4.3. Let X be a topological space and let U be a subset of X. U is called a *residual set* if it is the intersection of a countable number of sets each of which are open and dense in X. If a residual set in X is itself dense in X then X is called a *Baire space*.

We remark that $C^r(M, M)$ equipped with the C^k topology $(k \leq r)$ is a Baire space (see Palis and deMelo [1982]). We now give the definition of a generic property.

Definition 1.4.4. A property of a map (resp. vector field) is said to be C^k *generic* if the set of maps (resp. vector fields) possessing that property contains a residual subset in the C^k topology.

An example of some important generic properties are listed in the following theorem due to Kupka and Smale.

Theorem 1.4.1. *Let $\mathcal{N}$ be the set of diffeomorphisms of M, where M has dimension ≥ 2, such that all fixed points and periodic orbits of elements of $\mathcal{N}$ are hyperbolic and the stable and unstable manifolds of each fixed point and periodic orbit intersect transversely. Then $\mathcal{N}$ is a residual set.*

For a proof of the Kupka-Smale theorem, see Palis and deMelo [1982].

In utilizing the idea of a generic property to characterize the structurally stable systems, one first identifies some generic property. Then, since a structurally stable system is C^0 conjugate (resp. equivalent for vector fields) to all nearby systems, the structurally stable systems must have this property if the property is one that is preserved under C^0 conjugacy (resp. equivalence for vector fields). Now, one would like to go the other way with this argument; namely, it would be nice to show that structurally stable systems are generic. For two dimensional vector fields on compact manifolds we have the following result due to Peixoto [1962].

Theorem 1.4.2. *A C^r vector field on a compact boundaryless two dimensional manifold M is structurally stable if and only if*

 1) *The number of fixed points and periodic orbits is finite and each is hyperbolic.*

 2) *There are no orbits connecting saddle points.*

 3) *The nonwandering set consists of fixed points and periodic orbits.*

Moreover, if M is orientable, then the set of such vector fields is open and dense in $C^r(M)$ (note: this is stronger than generic).

This theorem is useful because it spells out precise conditions on the dynamics of a vector field on a compact boundaryless two manifold under which it is structurally stable. Unfortunately, we do not have a similar theorem in higher dimensions. This is in part due to the presence of complicated recurrent motions (e.g., the Smale horseshoe, see Section 2.1) which are not possible for two dimensional vector fields. Even more disappointing is the fact that structural stability is not a generic property for n dimensional diffeomorphisms ($n \geq 2$) or n dimensional vector fields ($n \geq 3$). This fact was first demonstrated by Smale [1966].

At this point we will conclude our brief discussion of the ideas of transversality, structural stability, and genericity. For more information, we refer the reader to Chillingworth [1976], Hirsch [1976], Arnold [1982], Nitecki [1971], Smale [1967], and Shub [1987]. However, before ending this section, we want to make some brief comments concerning the relevance of these ideas to the applied scientist, i.e., someone who has a specific dynamical system and must somehow discover what types of dynamics are present in that system.

Genericity and structural stability as defined above have been guiding forces behind much of the development of dynamical systems theory. The approach often taken has been to postulate some "reasonable" form of dynamics for a certain class of dynamical systems and then to prove that this form of dynamics is structurally stable and/or generic within this class. If one is persistent and occasionally successful in this approach, eventually a significant catalogue of generic and structurally stable dynamical properties is obtained. This catalogue is useful to the applied scientist in that it gives him or her some idea of what dynamics to expect in a specific dynamical system. However, this is hardly adequate. Given a specific dynamical system, is it structurally stable and/or generic? If this question could be answered, then very general and powerful theorems such as the Kupka-Smale theorem could be invoked, resulting in far-reaching conclusions concerning the dynamics of the system in question. So we would like to give computable conditions under which a specific dynamical system is structurally stable and/or generic. For certain special types of motions such as periodic orbits and fixed points this can be done in terms of the eigenvalues of the linearized system. However, for more general global motions such as homoclinic orbits and quasiperiodic orbits, this cannot be done so easily, since the nearby orbit structure may be exceedingly complicated and defy any local description (see Chapter 3). What this boils down to is that, in order to determine whether or not a specific dynamical system is structurally stable, one needs a fairly complete understanding of its orbit structure or, to put it more cynically, one needs to know the answer before asking the question. It might therefore seem that these ideas are of little use to the applied scientist; however, this is not exactly true, since the theorems describing structural stability and generic properties do give one a good idea of what to *expect*, although they cannot tell one what is precisely happening in a specific system.

1.5. Bifurcations

The term bifurcation is broadly used to describe significant qualitative changes that occur in the orbit structure of a dynamical system as the parameters on which the dynamical system depends are varied. In this section, we want to describe some of the ideas behind bifurcation theory, beginning with the general framework of the theory and then addressing various special situations.

Let us consider the infinite dimensional space of dynamical systems, either vector fields or diffeomorphisms. The set of structurally stable dynamical systems forms an open set S in this infinite dimensional space. The complement of S, denoted S^c, is defined to be the *bifurcation set*. We would like to describe the structure of the bifurcation set S^c; to begin with, we would like to show that S^c is a codimension one submanifold or, more generally, a stratified subvariety (see Arnold

[1983]) in the infinite dimensional space of dynamical systems. In order to motivate this we must first make a slight digression and explain the term "codimension."

Let M be an m dimensional manifold and let N be an n dimensional submanifold contained in M. Then the codimension of N is defined to be $m - n$. Thus, the codimension of a submanifold is a measure of the avoidability of the submanifold as one moves about the ambient space; in particular, the codimension of a submanifold N is equal to the minimum dimension of a submanifold $P \subset M$ that intersects N such that the intersection is transversal. This defines codimension in a finite dimensional setting and allows one to gain some intuition. Now we move to the infinite dimensional setting. Let M be an infinite dimensional manifold and let N be a submanifold contained in M. (Note: for the definition of an infinite dimensional manifold see Hirsch [1976]. Roughly speaking, an infinite dimensional manifold is a set which is locally diffeomorphic to an infinite dimensional Banach space. Since we only mention infinite dimensional manifolds in this section, and mainly in a heuristic fashion, we refer the reader to the literature for the proper definitions.) We say that N is of codimension k if every point of N is contained in some open set in M which is diffeomorphic to $U \times \mathbf{R}^k$ where U is an open set in N. This implies that k is the smallest dimension of a submanifold $P \subset M$ that intersects N such that the intersection is transversal. Thus, the definition of codimension in the infinite dimensional case has the same geometrical connotations as in the finite dimensional case. Now we return to our main discussion.

Suppose S^c is a codimension one submanifold or, more generally, a stratified subvariety. We might think of S^c as a surface dividing the infinite dimensional space of dynamical systems as depicted in Figure 1.5.1. Bifurcations (i.e., topologically distinct orbit structures) occur as one passes through S^c. Thus, in the infinite dimensional space of dynamical systems, one might define a *bifurcation point* as being any dynamical system which is structurally unstable.

Now in this setting one might initially conclude that bifurcations seldom occur and are unimportant, since any point p on S^c may be perturbed to S by (most) arbitrarily small perturbations. Also, from a practical point of view, dynamical systems contained in S^c are probably not very good models for physical systems, since any model is only an approximation to reality and, therefore, we should require our model to be structurally stable. However, suppose we have a curve γ of dynamical systems transverse to S^c, i.e., a one parameter family of dynamical systems. Then

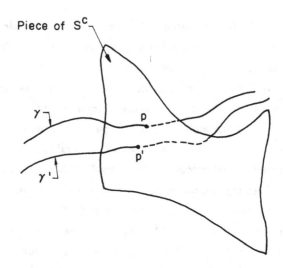

Figure 1.5.1. The Bifurcation Surface S^c Contained in S.

any sufficiently small perturbation of this curve γ of systems still results in a curve γ' transverse to S^c. So even though any particular point on S^c may be removed from S^c by (most) arbitrarily small perturbations, a curve transverse to S^c remains transverse to S^c under perturbation. Thus, bifurcation may be unavoidable in a parametrized family of dynamical systems.

Although it may be possible to show that S^c is a codimension one submanifold or stratified subvariety, the detailed structure of S^c may be quite complicated, for it may be divided up into submanifolds of higher codimension corresponding to more degenerate forms of bifurcations. Then a particular type of codimension k bifurcation in S^c would be persistent in a k parameter family of dynamical systems transverse to the codimension k submanifold.

This is essentially the program for bifurcation theory originally outlined by Poincaré. In order to utilize it in practice one would proceed as follows:

1) Given a specific dynamical system, determine whether or not it is structurally stable.

2) If it is not structurally stable, compute the codimension of the bifurcation.

3) Embed the system in a parametrized family of systems transverse to the bifurcation surface with the number of parameters equal to the codimension of the

bifurcation. These parametrized systems are called *unfoldings* or *deformations* and, if they contain all possible qualitative dynamics that can occur near the bifurcation, they are called *universal unfoldings* or *versal deformations*, see Arnold [1982].

4) Study the dynamics of the parametrized systems.

In this way one obtains structurally stable *families* of systems. Moreover, this provides a method for gaining a complete understanding of the qualitative dynamics of the space of dynamical systems with as little work as possible. Namely, one uses the degenerate bifurcation points as "organizing centers" around which one studies the dynamics. Since elsewhere the dynamical systems are structurally stable, one need not worry about the details of their dynamics, since qualitatively they will be topologically conjugate to the structurally stable dynamical systems in a neighborhood of the bifurcation point.

Now this program for the development of bifurcation theory is far from complete, and the obstacles preventing its completion are exactly those discussed at the end of Section 1.4; namely, in the space of dynamical systems, one must first identify S and S^c, and this involves a detailed knowledge of the orbit structure of each element in the space of dynamical systems. Although the situation appears hopeless, some progress has been made along two fronts:

1) Local Bifurcations.
2) Global Bifurcations of Specific Orbits.

We will discuss each of these situations separately.

Local bifurcation theory is concerned with the bifurcation of fixed points of vector fields and maps, or in situations where the problem can be cast into this form, such as in the study of bifurcations of periodic motions; for vector fields one can construct a local Poincaré map (see Section 1.6) near the periodic orbit, thus reducing the problem to one of studying the bifurcation of a fixed point of a map, and for maps with a k periodic orbit one can consider the kth iterate of the map thus reducing the problem to one of studying the bifurcation of a fixed point of the kth iterate of the map (see Section 1.1h). Utilizing a procedure such as the center manifold theorem (see Carr [1981] or Guckenheimer and Holmes [1983]) or the Lyapunov-Schmidt reduction (see Chow and Hale [1982]), one can usually reduce the problem to that of studying an equation of the form

$$f(x, \lambda) = 0 \tag{1.5.1}$$

where $x \in \mathbb{R}^n$, $\lambda \in \mathbb{R}^p$ are the system parameters, and $f \colon \mathbb{R}^n \times \mathbb{R}^p \to \mathbb{R}^n$ is assumed to be sufficiently smooth. The goal is to study the nature of the solutions of (1.5.1) as λ varies. In particular, it would be interesting to know for what parameter values solutions disappear or are created. These particular parameters are called *bifurcation values*. Now there exists an extensive mathematical machinery called *singularity theory* (see Golubitsky and Guillemin [1973]) which deals with such questions. Singularity theory is concerned with the local properties of smooth functions near a zero of the function. It provides a classification of the various cases based on codimension in a spirit similar to that described in the beginning of this section. The reason this is possible is that the codimension k submanifolds in the space of all smooth functions having zeros can be described algebraically by imposing conditions on derivatives of the functions. This gives us a way of classifying the various possible bifurcations and of computing the proper unfoldings. From this one might be led to believe that local bifurcation theory is a well-understood subject; however, this is not the case. The problem arises in the study of degenerate local bifurcations, specifically, in codimension k ($k \geq 2$) bifurcations of vector fields. Fundamental work of Takens [1974], Langford [1979], and Guckenheimer [1981] has shown that arbitrarily near these degenerate bifurcation points complicated global dynamical phenomena such as invariant tori and Smale horseshoes may arise. These phenomena cannot be described or detected via singularity theory techniques. We refer the reader to Chapter 7 of Guckenheimer and Holmes [1983] for a more thorough discussion of these issues.

Global bifurcations will be defined to be bifurcations which are not local in the sense described above, i.e., a qualitative change in the orbit structure of an extended region of phase space. Typical examples are homoclinic and heteroclinic bifurcations. In both of these examples the complete story is far from known, mainly because techniques for the global analysis of the orbit structure of dynamical systems are just now beginning to be developed. In Chapter 3 we will present a great deal of what is known at this point regarding homoclinic and heteroclinic bifurcations and comment also on the large gaps in our knowledge, and in Chapter 4 we will develop a variety of analytical techniques suitable for dealing with these situations.

1.6. Poincaré Maps

The idea of reducing the study of continuous time systems (flows) to the study of an associated discrete time system (map) is due to Poincaré [1899], who first utilized it in his studies of the three body problem in celestial mechanics. Nowadays virtually any discrete time system which is associated to an ordinary differential equation is referred to as a *Poincaré map*. This technique offers several advantages in the study of ordinary differential equations, three of which are the following:

1) *Dimensional Reduction.* Construction of the Poincaré map involves the elimination of *at least* one of the variables of the problem resulting in a lower dimensional problem to be studied.

2) *Global Dynamics.* In lower dimensional problems (say dimension ≤ 4) numerically computed Poincaré maps provide an insightful and striking display of the global dynamics of a system, see Guckenheimer and Holmes [1983] and Lichtenberg and Lieberman [1982] for examples of numerically computed Poincaré maps.

3) *Conceptual Clarity.* Many concepts that are somewhat cumbersome to state for ordinary differential equations may often be succinctly stated for the associated Poincaré map. An example would be the notion of orbital stability of a periodic orbit of an ordinary differential equation (see Hale [1980]). In terms of the Poincaré map, this problem would reduce to the problem of the stability of a fixed point of the map which is simply characterized in terms of the eigenvalues of the map linearized about the fixed point (see Case 1 to follow in this section).

It would be useful to give methods for constructing the Poincaré map associated with an ordinary differential equation. Unfortunately, there exist no general methods applicable to arbitrary ordinary differential equations, since construction of the Poincaré map of an ordinary differential equation requires some knowledge of the geometrical structure of the phase space of the ordinary differential equation. So constructing a Poincaré map requires ingenuity specific to the problem at hand; however, in four cases which come up frequently, the construction of a specific type of Poincaré map can in some sense be said to be canonical. The four cases are:

1) In the study of the orbit structure near a periodic orbit of an ordinary differential equation.

2) In the case where the phase space of an ordinary differential equation is periodic such as in periodically forced oscillators.

3) In the case where the phase space of an ordinary differential equation is quasiperiodic such as in quasiperiodically forced oscillators.

4) In the study of the orbit structure near a homoclinic or heteroclinic orbit.

We will discuss Cases 1, 2 and 3 now; all of Chapter 3 is devoted to Case 4.

Case 1. Consider the following ordinary differential equation

$$\dot{x} = f(x), \qquad x \in \mathbb{R}^n \tag{1.6.1}$$

where $f: U \to \mathbb{R}^n$ is C^r on some open set $U \subset \mathbb{R}^n$. Let $\phi(t, \cdot)$ denote the flow generated by (1.6.1). Suppose that (1.6.1) has a periodic solution of period T which we denote by $\phi(t, x_0)$, where $x_0 \in \mathbb{R}^n$ is any point through which this periodic solution passes (i.e., $\phi(t + T, x_0) = \phi(t, x_0)$). Let Σ be an $n - 1$ dimensional surface transverse to the vector field at x_0 (note: "transverse" means that $f(x) \cdot n(x_0) \neq 0$ where "$\cdot$" denotes the vector dot product and $n(x_0)$ is the normal to Σ at x_0); we refer to Σ as a cross-section to the vector field (1.6.1). Now in Theorem 1.1.2 we proved that $\phi(t, x)$ is C^r if $f(x)$ is C^r; thus, we can find an open set $V \subset \Sigma$ such that trajectories starting in V return to Σ in a time close to T. The map which associates points in V with their points of first return to Σ is called the *Poincaré map* which we denote by P. To be more precise,

$$P: V \to \Sigma \tag{1.6.2}$$
$$x \mapsto \phi(\tau(x), x)$$

where $\tau(x)$ is the time of first return of the point x to Σ. Note that by construction we have $\tau(x_0) = T$ and $P(x_0) = x_0$.

So a fixed point of P corresponds to a periodic orbit of (1.6.1) and a period k point of P (i.e., a point $x \in V$ such that $P^k(x) = x$ provided $P^i(x) \in V$, $i = 1, \ldots, k$) corresponds to a periodic orbit of (1.6.1) which pierces Σ k times before closing, see Figure 1.6.1.

A question that arises is how does the Poincaré map change if the cross-section Σ is changed. Let x_0 and x_1 be two points on the periodic solution of (1.6.1), and

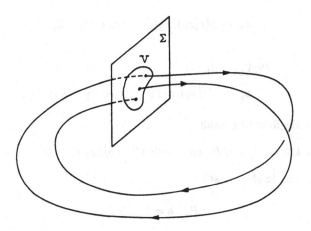

Figure 1.6.1. The Geometry of the Poincaré Map.

let Σ_0 and Σ_1 be two $n-1$ dimensional surfaces at x_0 and x_1, respectively, which are transverse to the vector field, and suppose that Σ_1 is chosen such that it is the image of Σ_0 under the flow generated by (1.6.1) see Figure 1.6.2. This defines a C^r diffeomorphism

$$h: \Sigma_0 \to \Sigma_1 . \tag{1.6.3}$$

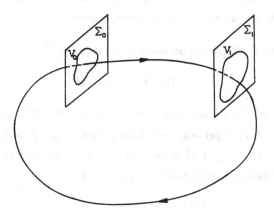

Figure 1.6.2. The Cross-sections Σ_0 and Σ_1.

We define Poincaré maps P_0 and P_1 as in the previous construction.

$$P_0: V_0 \to \Sigma_0$$

$$x_0 \mapsto \phi\big(\tau(x_0), x_0\big), \qquad x_0 \in V_0 \subset \Sigma_0 \tag{1.6.4}$$

$$P_1 : V_1 \to \Sigma_1$$
$$x_1 \mapsto \phi\big(\tau(x_1), x_1\big), \qquad x_1 \in V_1 \subset \Sigma_1. \tag{1.6.5}$$

Then we have the following result.

Proposition 1.6.1. P_0 and P_1 are locally C^r conjugate.

PROOF: We need to show that

$$P_1 \circ h = h \circ P_0$$

from which the result is immediate since h is a C^r diffeomorphism. However, we need to worry a bit about the domains of the maps. We have

$$h(\Sigma_0) = \Sigma_1$$
$$P_0(V_0) \subset \Sigma_0 \tag{1.6.6}$$
$$P_1(V_1) \subset \Sigma_1.$$

So $h \circ P_0 : V_0 \to \Sigma_1$ is well-defined; but $P_1 \circ h$ need not be defined since P_1 is not defined on all of Σ_1; however, this problem is solved if we choose Σ_1 such that $\Sigma_1 = h(V_0)$ rather than $h(\Sigma_0)$ and the result follows. $\qquad\square$

Case 2. Consider the following ordinary differential equation

$$\dot{x} = f(x, t), \qquad x \in \mathbf{R}^n \tag{1.6.7}$$

where $f : U \to \mathbf{R}^n$ is C^r on some open set $U \subset \mathbf{R}^n \times \mathbf{R}^1$. Suppose the time dependence of (1.6.7) is periodic with fixed period $T = \frac{2\pi}{\omega} > 0$, i.e., $f(x, t) = f(x, t + T)$. We rewrite (1.6.7) in the form of an autonomous equation in $n + 1$ dimensions (see Section 1.1e) by defining the function

$$\theta : \mathbf{R}^1 \to S^1,$$
$$t \mapsto \theta(t) = \omega t, \qquad \mathrm{mod}\, 2\pi. \tag{1.6.8}$$

Using (1.6.8) the equation (1.6.7) becomes

$$\dot{x} = f(x, \theta)$$
$$\dot{\theta} = \omega \qquad (x, \theta) \in \mathbf{R}^n \times S^1. \tag{1.6.9}$$

We denote the flow generated by (1.6.9) by $\phi(t) = \big(x(t), \theta(t) = \omega t + \theta_0 \pmod{2\pi}\big)$. We define a cross-section $\Sigma^{\bar{\theta}_0}$ to the vector field (1.6.9) by

$$\Sigma^{\bar{\theta}_0} = \Big\{ (x, \theta) \in \mathbf{R}^n \times S^1 \mid \theta = \bar{\theta}_0 \in (0, 2\pi] \Big\} . \tag{1.6.10}$$

The unit normal to $\Sigma^{\bar{\theta}_0}$ in $\mathbf{R}^n \times S^1$ is given by the vector $(0,1)$, and it is clear that $\Sigma^{\bar{\theta}_0}$ is transverse to the vector field (1.6.9) for all $x \in \mathbf{R}^n$, since $\big(f(x,\theta), \omega\big) \cdot (0,1) = \omega \neq 0$. In this case $\Sigma^{\bar{\theta}_0}$ is called a *global cross-section*.

We define the Poincaré map of $\Sigma^{\bar{\theta}_0}$ into itself as follows:

$$P_{\bar{\theta}_0} \colon \Sigma^{\bar{\theta}_0} \to \Sigma^{\bar{\theta}_0} \tag{1.6.11}$$

$$\Big(x\Big(\frac{\bar{\theta}_0 - \theta_0}{\omega}\Big), \bar{\theta}_0 \Big) \mapsto \Big(x\Big(\frac{\bar{\theta}_0 - \theta_0 + 2\pi}{\omega}\Big), \bar{\theta}_0 + 2\pi \equiv \bar{\theta}_0 \Big)$$

or
$$x\Big(\frac{\bar{\theta}_0 - \theta_0}{\omega}\Big) \mapsto x\Big(\frac{\bar{\theta}_0 - \theta_0 + 2\pi}{\omega}\Big) .$$

Thus, the Poincaré map merely tracks initial conditions in x at a fixed phase after successive periods of the vector field.

It should be clear that fixed points of $P_{\bar{\theta}_0}$ correspond to $2\pi/\omega$-periodic orbits of (1.6.9) and k-periodic points of $P_{\bar{\theta}_0}$ correspond to periodic orbits of (1.6.9) which pierce $\Sigma^{\bar{\theta}_0}$ k times before closing.

As in Case 1, suppose we construct a different Poincaré map $P_{\bar{\theta}_1}$ as above but with cross-section

$$\Sigma^{\bar{\theta}_1} = \Big\{ (x, \theta) \in \mathbf{R}^n \times S^1 \mid \theta = \bar{\theta}_1 \in (0, 2\pi] \Big\} . \tag{1.6.12}$$

Then we have the following result.

Proposition 1.6.2. $P_{\bar{\theta}_0}$ and $P_{\bar{\theta}_1}$ are C^r conjugate.

PROOF: The proof follows a construction similar to that given in Proposition 1.6.1. We construct a C^r diffeomorphism, h of $\Sigma^{\bar{\theta}_0}$ into $\Sigma^{\bar{\theta}_1}$ by mapping points on $\Sigma^{\bar{\theta}_0}$ into $\Sigma^{\bar{\theta}_1}$ under the action of the flow generated by (1.6.9). Points starting on $\Sigma^{\bar{\theta}_0}$ have initial time $t_0 = (\bar{\theta}_0 - \theta_0)/\omega$, and they reach $\Sigma^{\bar{\theta}_1}$ after time

$$t = \frac{\bar{\theta}_1 - \bar{\theta}_0}{\omega} ;$$

thus we have

$$h \colon \Sigma^{\bar{\theta}_0} \to \Sigma^{\bar{\theta}_1} \tag{1.6.13}$$

$$\Big(x\Big(\frac{\bar{\theta}_0 - \theta_0}{\omega}\Big), \bar{\theta}_0 \Big) \mapsto \Big(x\Big(\frac{\bar{\theta}_1 - \theta_0}{\omega}\Big), \bar{\theta}_1 \Big) .$$

Using (1.6.13) and the expressions for the Poincaré maps defined on the different cross-sections we obtain

$$h \circ P_{\bar{\theta}_0} : \Sigma^{\bar{\theta}_0} \to \Sigma^{\bar{\theta}_1} \tag{1.6.14}$$

$$\left(x\left(\frac{\bar{\theta}_0 - \theta_0}{\omega} \right), \bar{\theta}_0 \right) \mapsto \left(x\left(\frac{\bar{\theta}_1 - \theta_0 + 2\pi}{\omega} \right), \bar{\theta}_1 + 2\pi \equiv \bar{\theta}_1 \right)$$

and

$$P_{\bar{\theta}_1} \circ h : \Sigma^{\bar{\theta}_0} \to \Sigma^{\bar{\theta}_1} \tag{1.6.15}$$

$$\left(x\left(\frac{\bar{\theta}_0 - \theta_0}{\omega} \right), \bar{\theta}_0 \right) \mapsto \left(x\left(\frac{\bar{\theta}_1 - \theta_0 + 2\pi}{\omega} \right), \bar{\theta}_1 + 2\pi \equiv \bar{\theta}_1 \right).$$

Thus from (1.6.14) and (1.6.15) we have that

$$h \circ P_{\bar{\theta}_0} = P_{\bar{\theta}_1} \circ h. \tag{1.6.16}$$

$\square$

Case 3. Consider the following ordinary differential equation

$$\dot{x} = f(x,t), x \in \mathbf{R}^n \tag{1.6.17}$$

where $f : U \to \mathbf{R}^n$ is C^r on some open set $U \subset \mathbf{R}^n \times \mathbf{R}^1$. We assume that for fixed x, $f(x,t)$ is a quasiperiodic function of time. Recalling the definition of a quasiperiodic function given in Section 1.1i, (1.6.17) can be written as

$$\begin{aligned} \dot{x} &= f(x, \theta_1, \ldots, \theta_m) \\ \dot{\theta}_1 &= \omega_1 \\ &\vdots \qquad \vdots \\ \dot{\theta}_m &= \omega_m \end{aligned} \qquad (x, \theta_1, \ldots, \theta_m) \in \mathbf{R}^n \times \underbrace{S^1 \times \cdots \times S^1}_{m \text{ factors}}. \tag{1.6.18}$$

We denote the flow generated by (1.6.18) by $\phi(t) = \left(x(t), \omega_1 t + \theta_{10}, \cdots, \omega_m t + \theta_{m0} \right)$.

In analogy with Case 2, we construct a cross-section to the vector field (1.6.18) by fixing the phase of *one* of the angular variables. To be more precise, the global cross-section $\Sigma^{\bar{\theta}_{j0}}$ is defined as

$$\Sigma^{\bar{\theta}_{j0}} \equiv \left\{ (x, \theta_1, \ldots, \theta_m) \in \mathbf{R}^n \times S^1 \times \cdots \times S^1 \mid \theta_j = \bar{\theta}_{j0} \in (0, 2\pi] \right\}. \tag{1.6.19}$$

(Note: the fact that (1.6.19) is a global cross-section to (1.6.18) follows by an argument similar to that given in Case 2.) Using the flow generated by (1.6.18), the Poincaré map, $P_{\bar{\theta}_j 0}$, is constructed by choosing an initial time $t_0 = (\bar{\theta}_j 0 - \theta_j 0)/\omega_j$ such that solutions of (1.6.14) start with $\theta_j = \bar{\theta}_j 0$ and evolve for time $t = t_0 + \frac{2\pi}{\omega_j}$, thus returning to $\Sigma^{\bar{\theta}_j 0}$. To be more precise, we have

$$P_{\bar{\theta}_j 0} : \Sigma^{\bar{\theta}_j 0} \to \Sigma^{\bar{\theta}_j 0} \tag{1.6.20}$$

$$\left(x\left(\frac{\bar{\theta}_j 0 - \theta_j 0}{\omega_j} \right), \frac{\omega_1}{\omega_j} \left(\bar{\theta}_j 0 - \theta_j 0 \right) + \theta_1 0, \cdots, \bar{\theta}_j 0, \cdots, \frac{\omega_m}{\omega_j} \left(\bar{\theta}_j 0 - \theta_j 0 \right) + \theta_m 0 \right)$$

$$\mapsto \left(x\left(\frac{\bar{\theta}_j 0 - \theta_j 0 + 2\pi}{\omega_j} \right), \frac{\omega_1}{\omega_j} \left(\bar{\theta}_j 0 - \theta_j 0 + 2\pi \right) + \theta_1 0, \cdots, \bar{\theta}_j 0 + 2\pi \equiv \bar{\theta}_j 0, \cdots \right.$$

$$\left. \cdots, \frac{\omega_m}{\omega_j} \left(\bar{\theta}_j 0 - \theta_j 0 + 2\pi \right) + \theta_m 0 \right).$$

We remark that changing the cross-section by changing the angle corresponding to a *fixed frequency* (i.e., $\Sigma^{\bar{\theta}_j 0}$ and $\Sigma^{\bar{\bar{\theta}}_j 0}$) results in two Poincaré maps defined on the respective cross-sections which are C^r conjugate. However, changing the cross-section by changing angles which correspond to different frequencies (i.e., $\Sigma^{\bar{\theta}_j 0}$ and $\Sigma^{\bar{\theta}_k 0}$) results in two Poincaré maps defined on the respective cross-sections which are C^r conjugate only if the frequencies ω_j and ω_k are commensurate.

Before concluding our discussion on Poincaré maps we want to address an important issue of a more general nature. In Cases 1, 2, and 3 the Poincaré maps were all constructed by considering a portion of the phase space and allowing it to evolve in time under the action of the flow generated by the vector field. The region of the phase space and the "time of flight" were not chosen arbitrarily but in such a manner that the dynamics of the resulting Poincaré map could be directly related to the dynamics of the flow. In these three cases, this was accomplished by choosing a portion of the phase space which was mapped back onto itself (or at least near itself) after a certain amount of time. The ability to make this choice depended on our knowledge of certain recurrent properties in the dynamics of the vector field (e.g., a periodic orbit, periodic or quasiperiodic phase space, or, as we shall see, a homoclinic or heteroclinic orbit). There is a property which is common to all such "flow maps" which is quite useful in making certain global arguments; namely,

Poincaré maps constructed as discrete time flow maps from the flow generated by an ordinary differential equation are *orientation preserving* (note: recall that a map $f: U \to V$, U, V open sets in $\mathbb{R}^n$, is said to be orientation preserving if the determinant of Df (denoted $\det Df$) is positive in U). We now want to give a proof of this fact. The set up is as follows.

Consider the ordinary differential equation

$$\dot{x} = f(x), \quad x \in \mathbb{R}^n \tag{1.6.21}$$

where $f: U \to \mathbb{R}^n$ is C^r on some open set $U \subset \mathbb{R}^n$. Let $\phi(t, x)$ denote the flow generated by (1.6.21), and we assume that it exists for a sufficiently long time on some set $V \subset U \subset \mathbb{R}^n$. Consider the map

$$P: V \to \mathbb{R}^n \tag{1.6.22}$$

$$x \to \phi(\tau, x)$$

where τ is some fixed real number which may depend on x. Then we have the following result.

Proposition 1.6.3. *The determinant of $DP \equiv D_x\phi(\tau, x)$ is positive on V; hence P is orientation preserving on V.*

PROOF: We have $\phi(0, x) = x$ and, therefore, $D_x\phi(0, x) = $ id where id denotes the $n \times n$ identity matrix. It follows that $\det D_x\phi(0, x) = 1$.

Now from Theorem 1.1.3 $D_x\phi(t, x)$ is a solution of the linear matrix equation

$$\dot{z} = D_x f(\phi(t, x))z \tag{1.6.23}$$

where we regard the x in the argument of ϕ as fixed. Utilizing the formula for the determinant of the fundamental solution matrix of a linear system based on the knowledge of the determinant at a fixed time (see Hale [1980], Chapter 3, Lemma 1.5) we see that

$$\det D_x\phi(\tau, x) = \det D_x\phi(0, x) \exp\left[\int_0^\tau tr\, D_x f(\phi(t, x))dt\right]$$

$$= \exp\left[\int_0^\tau tr\, D_x f(\phi(t, x))dt\right]. \tag{1.6.24}$$

Since (1.6.24) holds for each $x \in V$, it follows that $\det DP \equiv \det D_x\phi(\tau, x) > 0$ for each $x \in V$. $\qquad\square$

CHAPTER 2
Chaos: Its Descriptions
and Conditions for Existence

In this chapter we will discuss and derive sufficient conditions for a dynamical system to exhibit complicated dynamical, or chaotic, behavior. We will also discuss a characterization of this behavior in terms of symbolic dynamics.

We begin in Section 2.1 with a discussion of the two dimensional, piecewise linear, Smale horseshoe map, which is the prototypical chaotic dynamical system. We will use this specific example to introduce many techniques and concepts such as symbolic dynamics, sensitive dependence on initial conditions, and chaos that will appear in a broader context later on. We believe that a complete understanding of this example is absolutely essential in order to understand the meaning of the term "chaos" as it applies to deterministic dynamical systems.

In Section 2.2 we will discuss separately the subject of symbolic dynamics, since it will play a crucial role in most of our examples in Chapter 3. We will derive many of the topological properties of the space of symbol sequences as well as discuss the dynamics of the shift map and the subshift of finite type which acts on the space of symbol sequences.

In Section 2.3 we will give sufficient conditions in order for a map to possess an invariant set of points on which the dynamics can be described via the techniques of symbolic dynamics. Among other things, these conditions will require that there be a uniform splitting of the domain of the map into strongly expanding and contracting directions; this results in the invariant set being an invariant set of points.

In Section 2.4 we will weaken the conditions of Section 2.3 in order to allow the map to possess directions which exhibit neutral growth. This will result in the invariant set not being an invariant set of points, but rather an invariant set of surfaces or, more precisely, the Cartesian product of a Cantor set and a surface. The dynamics of the map on this invariant set will still admit a description via the techniques of symbolic dynamics.

2.1. The Smale Horseshoe

In this section we will describe a two dimensional map possessing an invariant set having a delightfully complicated structure. Our map is a simplified version of a map first studied by Smale [1963], [1980] and, due to the shape of the image of the domain of the map, is called a *Smale horseshoe*. At this stage of our understanding of chaotic dynamics it is safe to call the Smale horseshoe the prototypical map possessing a chaotic invariant set (note: the phrase "chaotic invariant set" will be precisely defined later on in the discussion). Therefore, we feel that a thorough understanding of the Smale horseshoe is absolutely essential for understanding what is meant by the term "chaos" as it is applied to the dynamics of specific physical systems. For this reason we will first endeavor to define as simple a two dimensional map as possible that contains the necessary ingredients for possessing a complicated and chaotic dynamical structure so that the reader may get a feel for what is going on in the map with a minimum of distractions. As a result, our construction may not appeal to those interested in applications, since it may appear rather artificial. However, following our discussion of the simplified Smale horseshoe map, we will give sufficient conditions for the existence of Smale horseshoe-like dynamics in n-dimensional maps which are of a very general nature. We will begin by defining the map and then proceed to a geometrical construction of the invariant set of the map. We will utilize the nature of the geometrical construction in such a way as to motivate a description of the dynamics of the map on its invariant set by symbolic dynamics, following which we will make precise the idea of chaotic dynamics.

2.1a. Definition of the Smale horseshoe map

We will give a combination geometrical-analytical definition of the map. Consider a map, f, from the square having sides of unit length into R^2

$$f : D \longrightarrow R^2, \qquad D = \left\{ (x,y) \in R^2 \mid 0 \leq x \leq 1, \quad 0 \leq y \leq 1 \right\} \qquad (2.1.1)$$

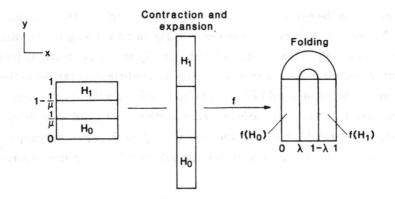

Figure 2.1.1. The Action of f on D.

which contracts the x-direction, expands the y-direction, and twists D around, laying it back on itself as shown in Figure 2.1.1.

We will assume that f acts affinely on the "horizontal" rectangles

$$H_0 = \{(x,y) \in R^2 \mid 0 \le x \le 1, 0 \le y \le 1/\mu\} \tag{2.1.2a}$$

and

$$H_1 = \{(x,y) \in R^2 \mid 0 \le x \le 1, 1 - 1/\mu \le y \le 1\} \tag{2.1.2b}$$

taking them to the "vertical" rectangles

$$f(H_0) \equiv V_0 = \{(x,y) \in R^2 \mid 0 \le x \le \lambda, 0 \le y \le 1\} \tag{2.1.3}$$

and

$$f(H_1) \equiv V_1 = \{(x,y) \in R^2 \mid 1 - \lambda \le x \le 1, 0 \le y \le 1\} \tag{2.1.4}$$

with the form of f on H_0 and H_1 given by

$$H_0 : \begin{pmatrix} x \\ y \end{pmatrix} \mapsto \begin{pmatrix} \lambda & 0 \\ 0 & \mu \end{pmatrix} \begin{pmatrix} x \\ y \end{pmatrix}$$

$$H_1 : \begin{pmatrix} x \\ y \end{pmatrix} \mapsto \begin{pmatrix} -\lambda & 0 \\ 0 & -\mu \end{pmatrix} \begin{pmatrix} x \\ y \end{pmatrix} + \begin{pmatrix} 1 \\ \mu \end{pmatrix} \tag{2.1.5}$$

with $0 < \lambda < 1/2$, $\mu > 2$ (note: the fact that, on H_1, the matrix elements are negative means that, in addition to being contracted in the x-direction by a factor λ

and expanded in the y-direction by a factor μ, H_1 is also rotated 180°). Additionally, it follows that f^{-1} acts on D as shown in Figure 2.1.2, taking the "vertical" rectangles V_0 and V_1 to the "horizontal" rectangles H_0 and H_1, respectively (note: by "vertical rectangle" we will mean a rectangle in D whose sides parallel to the y axis each have length one, and by "horizontal rectangle" we will mean a rectangle in D whose sides parallel to the x axis each have length one). This serves to define f; however, before proceeding to study the dynamics of f on D, there is a consequence of the definition of f which we want to single out, since it will be very important later.

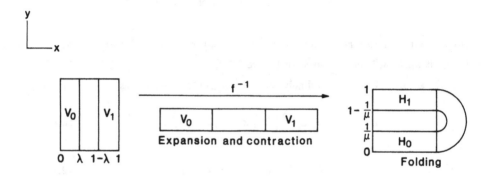

Figure 2.1.2. The Action of f^{-1} on D.

Lemma 2.1.1. a) *Suppose V is a vertical rectangle; then $f(V) \cap D$ consists of precisely two vertical rectangles, one in V_0 and one in V_1, with their widths each being equal to a factor of λ times the width of V.* b) *Suppose H is a horizontal rectangle; then $f^{-1}(H) \cap D$ consists of precisely two horizontal rectangles, one in H_0 and one in H_1, with their widths being a factor of $1/\mu$ times the width of H.*

PROOF: We will prove case a). Note that from the definition of f the horizontal and vertical boundaries of H_0 and H_1 are mapped to the horizontal and vertical boundaries of V_0 and V_1. So let V be a vertical rectangle. Then V intersects the horizontal boundaries of H_0 and H_1; hence, $f(V) \cap D$ consists of two vertical rectangles, one in H_0 and one in H_1. The contraction of the width follows from the form of f on H_0 and H_1, which indicates that the x-direction is contracted uniformly by a factor λ on H_0 and H_1. Case b) is proved similarly. See Figure 2.1.3. $\square$

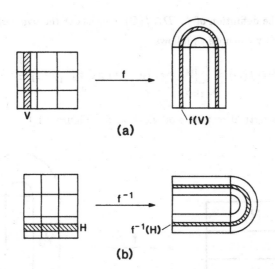

Figure 2.1.3. (a) Geometry of Lemma 2.1.1a. (b) Geometry of Lemma 2.1.1b.

2.1b. Construction of the Invariant Set

We now will geometrically construct the set of points, Λ, which remain in D under all possible iterations by f; thus Λ is defined as

$$\cdots \cap f^{-n}(D) \cap \cdots \cap f^{-1}(D) \cap D \cap f(D) \cap \cdots \cap f^n(D) \cap \cdots$$

$$\text{or} \quad \bigcap_{n=-\infty}^{\infty} f^n(D). \qquad (2.1.6)$$

We will construct this set inductively, and it will be convenient to construct separately the "halves" of Λ corresponding to the positive iterates and the negative iterates, and then take their intersections to obtain Λ. Before proceeding with the construction, we need some notation in order to keep track of the iterates of f at each step of the inductive process. Let $S = \{0, 1\}$ be an index set, and let s_i denote one of the two elements of S, i.e., $s_i \in S$, $i = 0, \pm 1, \pm 2, \ldots$ (note: the reason for this notation will become apparent later on).

We will construct $\bigcap\limits_{n=0}^{\infty} f^n(D)$ by constructing $\bigcap\limits_{n=0}^{n=k} f^n(D)$ and then determining the nature of the limit as $k \to \infty$.

$\underline{D \cap f(D)}$: By the definition of f, $D \cap f(D)$ consists of the two vertical rectangles V_0 and V_1, which we denote as follows:

$$D \cap f(D) = \bigcup_{s_{-1} \in S} V_{s_{-1}} = \{p \in D \mid p \in V_{s_{-1}}, s_{-1} \in S\} \qquad (2.1.7)$$

where $V_{s_{-1}}$ is a vertical rectangle of width λ. See Figure 2.1.4.

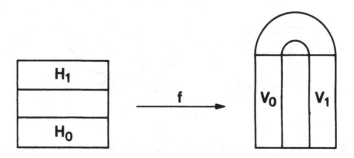

Figure 2.1.4. $D \cap f(D)$.

$\underline{D \cap f(D) \cap f^2(D)}$: It is easy to see that this set is obtained by acting on $D \cap f(D)$ with f and taking the intersection with D, since $D \cap f(D \cap f(D)) = D \cap f(D) \cap f^2(D)$. Thus, by Lemma 2.1.1, since $D \cap f(D)$ consists of the vertical rectangles V_0 and V_1 with each intersecting H_0 and H_1 and their respective horizontal boundaries in two components, then $D \cap f(D) \cap f^2(D)$ corresponds to four vertical rectangles, two each in V_0 and V_1, with each of width λ^2. We label this set as follows:

$$D \cap f(D) \cap f^2(D) = \bigcup_{\substack{s_{-i} \in S \\ i=1,2}} (f(V_{s_{-2}}) \cap V_{s_{-1}}) \equiv \bigcup_{\substack{s_{-i} \in S \\ i=1,2}} V_{s_{-1} s_{-2}} \qquad (2.1.8)$$

$$= \{p \in D \mid p \in V_{s_{-1}}, f^{-1}(p) \in V_{s_{-2}}, s_{-i} \in S, i = 1,2\}.$$

Pictorially, this set is described in Figure 2.1.5.

$\underline{D \cap f(D) \cap f^2(D) \cap f^3(D)}$: Using the same reasoning as in the previous steps, this set consists of eight vertical rectangles, each having width λ^3, which we denote as

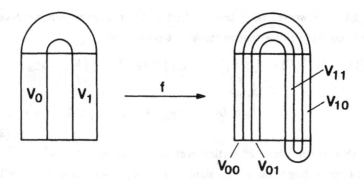

Figure 2.1.5. $D \cap f(D) \cap f^2(D)$.

follows:

$$D \cap f(D) \cap f^2(D) \cap f^3(D) = \bigcup_{\substack{s_{-i} \in S \\ i=1,2,3}} (f(V_{s_{-2}s_{-3}}) \cap V_{s_{-1}}) \equiv \bigcup_{\substack{s_{-i} \in S \\ i=1,2,3}} V_{s_{-1}s_{-2}s_{-3}} =$$

$$\left\{ p \in D \mid p \in V_{s_{-1}}, \ f^{-1}(p) \in V_{s_{-2}}, \ f^{-2}(p) \in V_{s_{-3}}, \ s_{-i} \in S, \ i = 1, 2, 3 \right\},$$

$$(2.1.9)$$

and is represented pictorially in Figure 2.1.6.

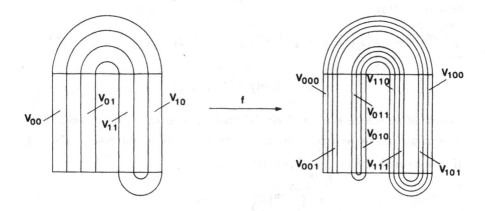

Figure 2.1.6. $D \cap f(D) \cap f^2(D) \cap f^3(D)$.

If we continually repeat this procedure, we almost immediately encounter extreme difficulty in trying to represent this process pictorially as in Figures 2.1.4

through 2.1.6. However, using Lemma 2.1.1 and our labelling scheme developed above, it is not hard to see that at the kth step we obtain

$$D \cap f(D) \cap \cdots \cap f^k(D) = \bigcup_{\substack{s_{-i} \in S \\ i=1,2,\ldots,k}} (f\left(V_{s_{-2}\cdots s_{-k}}\right) \cap V_{s_{-1}}) \equiv \bigcup_{\substack{s_{-i} \in S \\ i=1,2,\ldots,k}} V_{s_{-1}\cdots s_{-k}}$$

$$= \left\{ p \in D | f^{-i+1}(p) \in V_{s_{-i}}, \quad s_{-i} \in S, \quad i = 1, \ldots, k \right\}$$

$$(2.1.10)$$

and that this set consists of 2^k vertical rectangles, each of width λ^k.

Before proceeding to discuss the limit as $k \to \infty$, we want to make the following important observation concerning the nature of this construction process. Note that at the kth stage, we obtain 2^k vertical rectangles, and that each vertical rectangle can be labelled by a sequence of 0's and 1's of length k. The important point to realize is that there are 2^k possible distinct sequences of 0's and 1's having length k and that each of these is realized in our construction process; thus, the labelling of each vertical rectangle is unique at each step. This fact follows from the geometric definition of f and the fact that V_0 and V_1 are disjoint.

Letting $k \to \infty$, since a decreasing intersection of compact sets is nonempty, it is clear that we obtain an infinite number of vertical rectangles and that the width of each of these rectangles is zero, since $\lim_{k \to \infty} \lambda^k = 0$ for $0 < \lambda < 1/2$. Thus, we have shown that

$$\bigcap_{n=0}^{\infty} f^n(D) = \bigcup_{\substack{s_{-i} \in S \\ i=1,2,\ldots}} \left(f\left(V_{s_{-2}\cdots s_{-k}\cdots}\right) \cap V_{s_{-1}}\right) \equiv \bigcup_{\substack{s_{-i} \in S \\ i=1,2,\ldots}} V_{s_{-1}\cdots s_{-k}} \qquad (2.1.11)$$

$$= \left\{ p \in D | f^{-i+1}(p) \in V_{s_{-i}}, \quad s_i \in S, \quad i = 1,2,\ldots \right\}$$

consists of an infinite number of vertical lines and that each line can be labelled by a unique infinite sequence of 0's and 1's (note: we will give a more detailed set theoretic description of $\bigcap_{n=0}^{\infty} f^n(D)$ later on).

Next we will construct $\bigcap_{-\infty}^{n=0} f^n(D)$ inductively.

$\underline{D \cap f^{-1}(D)}$: From the definition of f, this set consists of the two horizontal rectangles H_0 and H_1 and is denoted as follows:

$$D \cap f^{-1}(D) = \bigcup_{s_0 \in S} H_{s_0} = \{p \in D | p \in H_{s_0}, \quad s_0 \in S\} . \qquad (2.1.12)$$

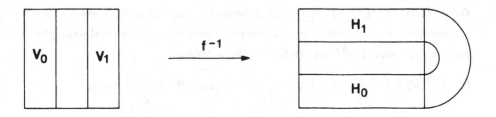

Figure 2.1.7. $D \cap f^{-1}(D)$.

See Figure 2.1.7.

$\underline{D \cap f^{-1}(D) \cap f^{-2}(D)}$: We obtain this set from the previously constructed set, $D \cap f^{-1}(D)$, by acting on $D \cap f^{-1}(D)$ with f^{-1} and taking the intersection with D, since $D \cap f^{-1}\left(D \cap f^{-1}(D)\right) = D \cap f^{-1}(D) \cap f^{-2}(D)$. Also, by Lemma 2.1.1, since H_0 intersects both vertical boundaries of V_0 and V_1 as does H_1, $D \cap f^{-1}(D) \cap f^{-2}(D)$ consists of four horizontal rectangles, each of width $1/\mu^2$, and we denote this set as follows:

$$D \cap f^{-1}(D) \cap f^{-2}(D) = \bigcup_{\substack{s_i \in S \\ i=0,1}} (f^{-1}(H_{s_1}) \cap H_{s_0}) \equiv \bigcup_{\substack{s_i \in S \\ i=0,1}} H_{s_0 s_1} \qquad (2.1.13)$$

$$= \{p \in D | p \in H_{s_0}, \quad f(p) \in H_{s_1}, \quad s_i \in S, \; i = 0, 1\}.$$

See Figure 2.1.8.

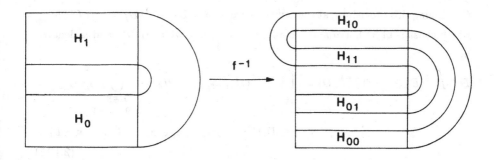

Figure 2.1.8. $D \cap f^{-1}(D) \cap f^{-2}(D)$.

$D \cap f^{-1}(D) \cap f^{-2}(D) \cap f^{-3}(D)$: Using the same arguments as those given in the previous steps, it is not hard to see that this set consists of eight horizontal rectangles each having width $1/\mu^3$ and can be denoted as follows:

$$D \cap f^{-1}(D) \cap f^{-2}(D) \cap f^{-3}(D) = \bigcup_{\substack{s_i \in S \\ i=0,1,2}} (f^{-1}(H_{s_1 s_2}) \cap H_{s_0}) \equiv \bigcup_{\substack{s_i \in S \\ i=0,1,2}} H_{s_0 s_1 s_2}$$

$$= \left\{ p \in D | p \in H_{s_0}, \, f(p) \in H_{s_1}, \, f^2(p) \in H_{s_2}, \quad s_i \in S, \, i = 0, 1, 2 \right\}.$$
$$(2.1.14)$$

See Figure 2.1.9.

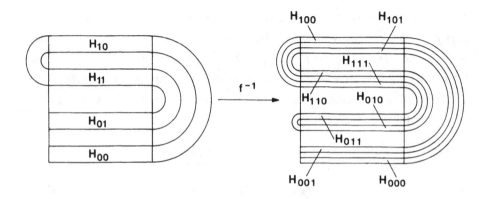

Figure 2.1.9. $D \cap f^{-1}(D) \cap f^{-2}(D) \cap f^{-3}(D)$.

Continuing this procedure, at the kth step we obtain $D \cap f^{-1}(D) \cap \cdots \cap f^{-k}(D)$, which consists of 2^k horizontal rectangles each having width $1/\mu^k$ and is denoted by

$$D \cap f^{-1}(D) \cap \cdots \cap f^{-k}(D) = \bigcup_{\substack{s_i \in S \\ i=0,\ldots,k-1}} (f^{-1}(H_{s_1 \cdots s_{k-1}}) \cap H_{s_0}) \equiv \bigcup_{\substack{s_i \in S \\ i=0,\ldots,k-1}} H_{s_0 \cdots s_{k-1}}$$

$$= \left\{ p \in D | f^i(p) \in H_{s_i}, \quad s_i \in S, \, i = 0, \ldots, k-1 \right\}.$$
$$(2.1.15)$$

As in the case of vertical rectangles, we note the important fact that at the kth step of the inductive process, each one of the 2^k can be labelled uniquely with a sequence of 0's and 1's of length k. Now, as we take the limit as $k \to \infty$, we arrive at $\bigcap\limits_{-\infty}^{n=0} f^n(D)$, which is an infinite set of horizontal lines, since a decreasing intersection of compact sets is nonempty and the width of each component of the intersection is given by $\lim\limits_{k\to\infty}(1/\mu^k) = 0$, $\mu > 2$. Each line is labelled by a unique infinite sequence of 0's and 1's as follows:

$$\bigcap_{n=-\infty}^{0} f^n(D) = \bigcup_{\substack{s_i \in S \\ i=0,1,\ldots}}(f(H_{s_1\cdots s_k\cdots}) \cap H_{s_0}) \equiv \bigcup_{\substack{s_i \in S \\ i=0,1,\ldots}} H_{s_0\cdots s_k\cdots} \tag{2.1.16}$$

$$= \Big\{ p \in D | f^i(p) \in H_{s_i}, \quad s_i \in S, \; i = 0,1,\ldots \Big\}.$$

Thus, we have

$$\Lambda = \bigcap_{n=-\infty}^{\infty} f^n(D) = \left[\bigcap_{n=-\infty}^{0} f^n(D) \right] \cap \left[\bigcap_{n=0}^{\infty} f^n(D) \right] \tag{2.1.17}$$

which consists of an infinite set of points, since each vertical line in $\bigcap\limits_{n=0}^{\infty} f^n(D)$ intersects each horizontal line in $\bigcap\limits_{n=-\infty}^{0} f^n(D)$ in a unique point. Furthermore, each point $p \in \Lambda$ can be labelled *uniquely* by a bi-infinite sequence of 0's and 1's which is obtained by concatenating the sequences associated with the respective vertical and horizontal lines which serve to define p. Stated more precisely, let $s_{-1} \cdots s_{-k} \cdots$ be a particular infinite sequence of 0's and 1's; then $V_{s_{-1}\cdots s_{-k}\cdots}$ corresponds to a unique vertical line. Let $s_0 \cdots s_k \cdots$ likewise be a particular infinite sequence of 0's and 1's; then $H_{s_0\cdots s_k\cdots}$ corresponds to a unique horizontal line. Now a horizontal line and vertical line intersect in a unique point p; thus, we have a well-defined map from points $p \in \Lambda$ to bi-infinite sequences of 0's and 1's which we call ϕ.

$$p \xmapsto{\phi} \cdots s_{-k} \cdots s_{-1}.s_0 \cdots s_k \cdots . \tag{2.1.18}$$

Notice that since

$$V_{s_{-1}\cdots s_{-k}\cdots} = \Big\{ p \in D | f^{-i+1}(p) \in V_{s_{-i}}, \quad i = 1,\ldots \Big\} \tag{2.1.19}$$

$$= \Big\{ p \in D | f^{-i}(p) \in H_{s_{-i}}, \quad i = 1,\ldots \Big\} \quad \text{since } f(H_{s_i}) = V_{s_i}$$

and

$$H_{s_0\cdots s_k\cdots} = \Big\{ p \in D | f^i(p) \in H_{s_i}, \quad i = 0,\ldots \Big\} \tag{2.1.20}$$

we have

$$p = V_{s_{-1}\cdots s_{-k}\cdots} \cap H_{s_0\cdots s_k\cdots} = \left\{ p \in D \,|\, f^i(p) \in H_{s_i}, \quad i = 0, \pm 1, \pm 2, \ldots \right\}.$$
$$(2.1.21)$$

Therefore, we see that the unique sequence of 0's and 1's which we have associated with p contains information concerning the behavior of p under iteration by f. In particular, the $s_{k\text{th}}$ element in the sequence associated with p indicates that $f^k(p) \in H_{s_k}$. Now, note that for the bi-infinite sequence of 0's and 1's associated with p, the decimal point separates the past iterates from the future iterates; thus, the sequence of 0's and 1's associated with $f^k(p)$ is obtained from the sequence associated with p merely by shifting the decimal point in the sequence associated with p k places to the right if k is positive or k places to the left if k is negative, until s_k is the symbol immediately to the right of the decimal point. We can define a map of bi-infinite sequences of 0's and 1's, called the shift map, σ, which takes a sequence and shifts the decimal point one place to the right. Therefore, if we consider a point $p \in \Lambda$ and its associated bi-infinite sequence of 0's and 1's, $\phi(p)$, we can take any iterate of p, $f^k(p)$, and we can immediately obtain its associated bi-infinite sequence of 0's and 1's given by $\sigma^k(\phi(p))$. So there is a direct relationship between iterating any point $p \in \Lambda$ under f and iterating the sequence of 0's and 1's associated with p under the shift map σ.

Now at this point it is not clear where we are going with this analogy between points in Λ and bi-infinite sequences of 0's and 1's since, although the sequence associated with a given point $p \in \Lambda$ contains information on the entire future and past as to whether or not it is in H_0 or H_1 for any given iterate, it is not hard to imagine different points, both contained in the same horizontal rectangle after any given iteration, whose orbits are completely different. The fact that this cannot happen for our map and that the dynamics of f on Λ are completely modeled by the dynamics of the shift map acting on sequences of 0's and 1's is an amazing fact which before we justify we must make a slight digression into symbolic dynamics.

2.1c. Symbolic Dynamics

Let $S = \{0, 1\}$ be the set of nonnegative integers consisting of 0 and 1. Let Σ be the collection of all bi-infinite sequences of elements of S, i.e., $s \in \Sigma$ implies

$$s = \{\cdots s_{-n} \cdots s_{-1}.s_0 \cdots s_n \cdots\}, \quad s_i \in S \quad \forall\, i. \qquad (2.1.22)$$

We will refer to Σ as the space of bi-infinite sequences of 2 symbols. We wish to introduce some structure on Σ in the form of a metric, $d(\,\cdot\,,\cdot\,)$, which we do as follows:

Consider $s = \{\cdots s_{-n} \cdots s_{-1}.s_0 \cdots s_n \cdots\}$, $\bar{s} = \{\cdots \bar{s}_{-n} \cdots \bar{s}_{-1}.\bar{s}_0 \cdots \bar{s}_n \cdots\} \in \Sigma$; we define the distance between s and $\bar{s}$

$$d(s, \bar{s}) = \sum_{i=-\infty}^{\infty} \frac{\delta_i}{2^{|i|}} \qquad \text{where } \delta_i = \begin{cases} 0 \text{ if } s_i = \bar{s}_i \\ 1 \text{ if } s_i \neq \bar{s}_i \end{cases}. \tag{2.1.23}$$

Thus, two sequences are "close" if they agree on a long central block. (Note: the reader should check that $d(\,\cdot\,,\cdot\,)$ does indeed satisfy the properties of a metric. See Devaney [1986] for a proof.)

We consider a map of Σ into itself, which we shall call the shift map, σ; it is defined as follows:

For $s = \{\cdots s_{-n} \cdots s_{-1}.s_0 s_1 \cdots s_n \cdots\} \in \Sigma$

$$\sigma(s) = \{\cdots s_{-n} \cdots s_{-1} s_0.s_1 \cdots s_n \cdots\} \tag{2.1.24}$$

or, $\sigma(s)_i = s_{i+1}$. Next, we want to consider the dynamics of σ on Σ (note: for our purposes the phrase "dynamics of σ on Σ" refers to the orbits of points in Σ under iteration by σ). It should be clear that σ has precisely two fixed points, namely, the sequence whose elements are all zeros and the sequence whose elements are all ones (notation: bi-infinite sequences which periodically repeat after some fixed length will be denoted by the finite length sequence with an overbar, e.g., $\{\ldots 101010.101010\ldots\}$ is denoted by $\{\overline{10.10}\}$).

In particular, it is easy to see that the orbits of sequences which periodically repeat are periodic under iteration by σ. For example, consider the sequence $\{\overline{10.10}\}$. We have

$$\sigma\{\overline{10.10}\} = \{\overline{01.01}\} \tag{2.1.25}$$

and

$$\sigma\{\overline{01.01}\} = \{\overline{10.10}\}. \tag{2.1.26}$$

Thus

$$\sigma^2\{\overline{10.10}\} = \{\overline{10.10}\}. \tag{2.1.27}$$

Therefore, the orbit of $\{\overline{10.10}\}$ is an orbit of period two for σ. So, from this particular example, it is easy to see that for any fixed k, the orbits of σ having period k correspond to the orbits of sequences made up of periodically repeating blocks of 0's and 1's with the blocks having length k. Thus, since for any fixed k the number of sequences having a periodically repeating block of length k is finite, we see that σ has a countable infinity of periodic orbits having all possible periods. We list the first few below.

$$
\begin{aligned}
&\text{Period 1 :} &&\{\overline{0.0}\}\,,\{\overline{1.1}\} \\
&\text{Period 2 :} &&\{\overline{01.01}\} \xrightarrow{\sigma} \{\overline{10.10}\} \xrightarrow{\sigma} \{\overline{01.01}\} \\
&\text{Period 3 :} &&\{\overline{001.001}\} \xrightarrow{\sigma} \{\overline{010.010}\} \xrightarrow{\sigma} \{\overline{100.100}\} \\
&\quad\ \ : &&\{\overline{110.110}\} \xrightarrow{\sigma} \{\overline{101.101}\} \xrightarrow{\sigma} \{\overline{011.011}\}
\end{aligned}
$$

$$(2.1.28)$$

$$\vdots$$

etc.

Also, σ has an uncountable number of nonperiodic orbits. To show this, we need only construct a non-periodic sequence and show that there are an uncountable number of such sequences. A proof of this fact goes as follows: we can easily associate an infinite sequence of 0's and 1's with a given bi-infinite sequence by the following rule:

$$\cdots s_{-n}\cdots s_{-1}.s_0\cdots s_n\cdots \quad \to \quad .s_0\,s_1\,s_{-1}\,s_2\,s_{-2}\cdots. \qquad (2.1.29)$$

Now, we will take it as a known fact that the irrational numbers in the closed unit interval $[0,1]$ constitute an uncountable set, and that every number in this interval can be expressed in base 2 as a binary expansion of 0's and 1's with the irrational numbers corresponding to non-repeating sequences. Thus, we have a one-to-one correspondence between an uncountable set of points and non-repeating sequences of 1's and 0's. As a result, the orbits of these sequences are the non-periodic orbits of σ, and there are an uncountable number of such orbits.

Another interesting fact concerning the dynamics of σ on Σ is that there exists an element, say $s \in \Sigma$, whose orbit is dense in Σ, i.e., for any given $s' \in \Sigma$ and $\epsilon > 0$, there exists some integer n such that $d\big(\sigma^n(s), s'\big) < \epsilon$. This is easiest to see by constructing s directly. We do this by first constructing all possible sequences of 0's and 1's having length 1, 2, 3, $\ldots$. This process is well defined in a

set theoretic sense, since there are only a finite number of possibilities at each step
(more specifically, there are 2^k distinct sequences of 0's and 1's of length k). The
first few of these sequences would be as follows:

length 1 : $\{0\}$, $\{1\}$

length 2 : $\{00\}$, $\{01\}$, $\{10\}$, $\{11\}$

length 3 : $\{000\}$, $\{001\}$, $\{010\}$, $\{011\}$, $\{100\}$, $\{101\}$, $\{110\}$, $\{111\}$

$\vdots$ $\vdots$

etc.

$$(2.1.30)$$

Now we can introduce an ordering on the collection of sequences of 0's and 1's in
order to keep track of the different sequences in the following way. Consider two
finite sequences of 0's and 1's

$$s = \{s_1 \cdots s_k\}, \qquad \bar{s} = \{\bar{s}_1 \cdots \bar{s}_{k'}\}. \qquad (2.1.31)$$

Then we say

$$s < \bar{s} \quad \text{if } k < k' \qquad (2.1.32)$$

and if $k = k'$

$$s < \bar{s} \quad \text{if } s_i < \bar{s}_i, \qquad (2.1.33)$$

where i is the *first* integer such that $s_i \neq \bar{s}_i$. For example, using this ordering we
have

$$\{0\} < \{1\},$$
$$\{0\} < \{00\}, \qquad (2.1.34)$$
$$\{00\} < \{01\}, \qquad \text{etc.}$$

This ordering gives us a systematic way of distinguishing different sequences that
have the same length. Thus, we will denote the sequences of 0's and 1's having
length k as follows:

$$s_1^k < \cdots < s_{2^k}^k \qquad (2.1.35)$$

where the superscript refers to the length of the sequence and the subscript refers to
a particular sequence of length k which is uniquely specified by the above ordering
scheme. This will give us a systematic way of writing down our candidate for a
dense orbit.

Now consider the following sequence.:

$$s = \{ \cdots s_8^3 \, s_6^3 \, s_4^3 \, s_2^3 \, s_4^2 \, s_2^2 \cdot s_1^1 \, s_1^2 \, s_3^2 \, s_1^3 \, s_3^3 \, s_5^3 \, s_7^3 \cdots \} \, . \qquad (2.1.36)$$

Thus, s contains all possible sequences of 0's and 1's of any fixed length. Now, in order to show that the orbit of s is dense in Σ, we argue as follows: let s' be an arbitrary point in Σ and let $\epsilon > 0$ be given. An ϵ-neighborhood of s' consists of all points $s'' \in \Sigma$ such that $d(s', s'') < \epsilon$, where d is the metric given in (2.1.23). Therefore, by definition of the metric on Σ, there must be some integer $N = N(\epsilon)$ such that $s_i' = s_i''$, $|i| \leq N$ (note: a proof of this statement can be found in Devaney [1986] or in Section 2.2). Now, by construction, the finite sequence $\{s_{-N}' \cdots s_{-1}' \, s_0' \cdots s_N'\}$ is contained somewhere in s; therefore, there must be some integer $\bar{N}$ such that $d\left(\sigma^{\bar{N}}(s), \, s'\right) < \epsilon$, so we can conclude that the orbit of s is dense in Σ.

We summarize these facts concerning the dynamics of σ on Σ in the following theorem.

Theorem 2.1.2. *The shift map σ acting on the space of bi-infinite sequences of 0's and 1's, Σ, has*

1) *a countable infinity of periodic orbits of arbitrarily high period;*

2) *an uncountable infinity of non-periodic orbits; and*

3) *a dense orbit.*

2.1d. The Dynamics on the Invariant Set

Now we want to relate the dynamics of σ on Σ, on which we have a great deal of information, to the dynamics of the Smale horseshoe f on its invariant set Λ, of which, at this point, we know little except for its complicated geometric structure. Recall that we have shown the existence of a well-defined map ϕ which associates to each point, $p \in \Lambda$, a bi-infinite sequence of 0's and 1's, $\phi(p)$. Furthermore, we noted that the sequence associated with any iterate of p, say $f^k(p)$, can be found merely by shifting the decimal point in the sequence associated with p k places to the right if k is positive or k places to the left if k is negative. In particular, the relation $\sigma \circ \phi(p) = \phi \circ f(p)$ holds for every $p \in \Lambda$. Now, if ϕ were invertible and continuous (continuity is necessary since f is continuous), the following relationship would hold:

$$\phi^{-1} \circ \sigma \circ \phi(p) = f(p), \qquad \forall \, p \in \Lambda \, . \qquad (2.1.37)$$

Thus, if the orbit of $p \in \Lambda$ under f is denoted by

$$\{\cdots f^n(p), \cdots, f^{-1}(p), p, f(p), \cdots, f^n(p), \cdots\} \tag{2.1.38}$$

since $\phi^{-1} \circ \sigma \circ \phi(p) = f(p)$, we see that

$$\begin{aligned} f^n(p) &= \left(\phi^{-1} \circ \sigma \circ \phi\right) \circ \left(\phi^{-1} \circ \sigma \circ \phi\right) \cdots \circ \left(\phi^{-1} \circ \sigma \circ \phi(p)\right) \\ &= \phi^{-1} \circ \sigma^n \circ \phi(p). \end{aligned} \tag{2.1.39}$$

Therefore, the orbit of $p \in \Lambda$ under f would correspond directly to the orbit of $\phi(p)$ under σ in Σ. In particular, the entire orbit structure of σ on Σ would be identical to the structure of f on Λ. So, in order to verify that this situation holds we need to show that ϕ is a homeomorphism of Λ and Σ.

Theorem 2.1.3. *The map* $\phi : \Lambda \to \Sigma$ *is a homeomorphism.*

PROOF: We need only show that ϕ is one-to-one, onto, and continuous, since continuity of the inverse will follow from the fact that one-to-one, onto, and continuous maps from compact sets into Hausdorff spaces are homeomorphisms (see Dugundji [1966]). We prove each condition separately.

ϕ *is one-to-one* : This means that given $p, p' \in \Lambda$, if $p \neq p'$, then $\phi(p) \neq \phi(p')$.

We give a proof by contradiction. Suppose

$$\phi(p) = \phi(p') = \{\cdots s_{-n} \cdots s_{-1}.s_0 \cdots s_n \cdots\}. \tag{2.1.40}$$

Then, by construction of Λ, p and p' lie in the intersection of the vertical line $V_{s_{-1} \cdots s_{-n} \cdots}$ and the horizontal line $H_{s_0 \cdots s_n \cdots}$. However, the intersection of a horizontal line and a vertical line consists of a unique point; therefore $p = p'$, contradicting our original assumption. This contradiction is due to the fact that we have assumed $\phi(p) = \phi(p')$; thus, for $p \neq p'$, $\phi(p) \neq \phi(p')$.

ϕ *is onto*: This means that given any bi-infinite sequence of 0's and 1's in Σ, say $\{\cdots s_{-n} \cdots s_{-1}.s_0 \cdots s_n \cdots\}$, there is a point $p \in \Lambda$ such that $\phi(p) = \{\cdots s_{-n} \cdots s_{-1}.s_0 \cdots s_n \cdots\}$.

The proof goes as follows: Recall the construction of $\bigcap\limits_{n=0}^{\infty} f^n(D)$ and $\bigcap\limits_{n=-\infty}^{0} f^n(D)$; given any infinite sequence of 0's and 1's, $\{.s_0 \cdots s_n \cdots\}$, there

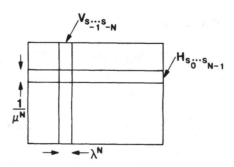

Figure 2.1.10. The Location of p and p'.

is a *unique* vertical line in $\bigcap\limits_{n=0}^{\infty} f^n(D)$ corresponding to this sequence. Similarly, given any infinite sequence of 0's and 1's, $\{\cdots s_{-n} \cdots s_{-1}.\}$, there is a unique horizontal line in $\bigcap\limits_{n=-\infty}^{0} f^n(D)$ corresponding to this sequence. Therefore, we see that for a given horizontal and vertical line, we can associate a unique bi-infinite sequence of 0's and 1's, $\{\cdots s_{-n} \cdots s_{-1}.s_0 \cdots s_n \cdots\}$ and since a horizontal and vertical line intersect in a unique point, p, to every bi-infinite sequence of 0's and 1's, there corresponds a unique point in Λ.

ϕ *is continuous*: This means that, given any point $p \in \Lambda$ and $\epsilon > 0$, we can find a $\delta = \delta(\epsilon, p)$ such that

$$|p - p'| < \delta \text{ implies } d\left(\phi(p), \phi(p')\right) < \epsilon \qquad (2.1.41)$$

where $|\cdot|$ is the usual distance measurement in R^2 and $d(\cdot, \cdot)$ is the metric on Σ introduced earlier.

Let $\epsilon > 0$ be given; then, if we are to have $d(\phi(p), \phi(p')) < \epsilon$, there must be some integer $N = N(\epsilon)$ such that if

$$\begin{aligned}
\phi(p) &= \{\cdots s_{-n} \cdots s_{-1}.s_0 \cdots s_n \cdots\} \\
\phi(p') &= \{\cdots s'_{-n} \cdots s'_{-1}.s'_0 \cdots s'_n \cdots\}
\end{aligned} \qquad (2.1.42)$$

then $s_i = s'_i$, $i = 0, \pm 1, \ldots, \pm N$. Thus, by construction of Λ, p and p' lie in the rectangle defined by $H_{s_0 \cdots s_N} \cap V_{s_{-1} \cdots s_{-N}}$, see Figure 2.1.10. Recall that the width and height of this rectangle are λ^N and $1/\mu^{N+1}$, respectively. Thus we have $|p - p'| \leq (\lambda^N + 1/\mu^{N+1})$. Therefore, if we take $\delta = \lambda^N + 1/\mu^{N+1}$, continuity is proved. $\qquad\square$

Remarks:

1) When ϕ is a homeomorphism recall from Section 1.2 that the dynamical systems f acting on Λ and σ acting on Σ are said to be *topologically conjugate* if $\phi \circ f(p) = \sigma \circ \phi(p)$. (Note: the equation $\phi \circ f(p) = \sigma \circ \phi(p)$ is also expressed by saying that the following diagram "commutes.")

$$
\begin{array}{ccc}
\Lambda & \xrightarrow{\ f\ } & \Lambda \\
\phi \downarrow & & \downarrow \phi \\
\Sigma & \xrightarrow{\ \sigma\ } & \Sigma
\end{array}
\qquad (2.1.43)
$$

2) The fact that Λ and ϕ are homeomorphic allows us to make several conclusions concerning the set theoretic nature of Λ. We have already shown that Σ is uncountable, and we state without proof that Σ is a closed, perfect (meaning every point is a limit point), totally disconnected set and that these properties carry over to Λ via the homeomorphism ϕ. A set having these properties is called a *Cantor Set*. We will give more detailed information concerning symbolic dynamics and Cantor sets in Section 2.2.

Now we can state a theorem regarding the dynamics of f on Λ that is almost precisely the same as Theorem 2.1.2, which describes the dynamics of σ on Σ.

Theorem 2.1.4. *The Smale horseshoe, f, has*

1) *a countable infinity of periodic orbits of arbitrarily high period. These periodic orbits are all of saddle type;*
2) *an uncountable infinity of non-periodic orbits; and*
3) *a dense orbit.*

PROOF: This is an immediate consequence of the topological conjugacy of f on Λ with σ on Σ, except for the stability result. The stability result follows from the form of f on H_0 and H_1 given in (2.1.5). $\qquad\square$

2.1e. Chaos

Now we can make precise the statement that the dynamics of f on Λ is chaotic.

Let $p \in \Lambda$ with corresponding symbol sequence

$$
\phi(p) = \{\cdots s_{-n} \cdots s_{-1} \cdot s_0 \cdots s_n \cdots\} . \qquad (2.1.44)
$$

We want to consider points close to p and how they behave under iteration by f as compared to p. Let $\epsilon > 0$ be given; then we consider an ϵ-neighborhood of p determined by the usual topology of the plane. Also, there exists an integer $N = N(\epsilon)$ such that the corresponding neighborhood of $\phi(p)$ includes the set of sequences $s' = \left\{ \cdots s'_{-n} \cdots s'_{-1} . s'_0 \cdots s'_n \cdots \right\} \in \Sigma$ such that $s_i = s'_i$, $|i| \leq N$. Now suppose the $N+1$ entry in the sequence corresponding to $\phi(p)$ is 0 and the $N+1$ sequence corresponding to s' is 1. Thus, after N iterations, no matter how small ϵ, the point p is in H_0 and the point, say p', corresponding to s' under ϕ^{-1} is in H_1 and they are at least a distance $1 - 2\lambda$ apart. Therefore, for any point $p \in \Lambda$, no matter how small a neighborhood of p we consider, there is at least one point in this neighborhood such that after a finite number of iterations, p and this point have separated by some fixed distance. A system displaying such behavior is said to exhibit *sensitive dependence on initial conditions*.

Now we want to end our discussion of this simplified version of the Smale horseshoe with some final observations.

1) If you consider carefully the main ingredients of f which led to Theorem 2.1.4, you will see that there are two key elements.

    a) The square is contracted, expanded, and folded in such a way that we can find disjoint regions which are mapped over themselves.

    b) There exists "strong" stretching and contraction in complementary directions.

2) From observation 1), the fact that the image of the square appears in the shape of a horseshoe is not important. Other possible scenarios are shown in Figure 2.1.11.

Notice that, in our study of the invariant set of f, we do not consider the question of the geometry of the points which escape from the square. We remark that this could be an interesting research topic, since this more global question may enable one to determine conditions under which the horseshoe becomes an attractor.

2.2. Symbolic Dynamics

In the previous section we saw an example of a two dimensional map which possessed an invariant Cantor set. The map, restricted to its invariant set, was shown to have a countable infinity of periodic orbits of all periods, an uncountable infinity

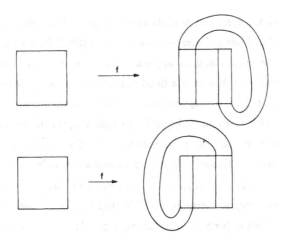

Figure 2.1.11. Two Other Possible Horseshoe Scenarios.

of nonperiodic orbits, and a dense orbit. Now, in general, the determination of
such detailed information concerning the orbit structure of a map is not possible.
However, in our example we were able to show that the map restricted to its invariant set behaved the same as the shift map acting on the space of bi-infinite
sequences of 0's and 1's (more precisely, these two dynamical systems were shown
to be topologically equivalent; thus their orbit structures are identical). The shift
map was no less complicated than our original map but, due to its structure, many
of the features concerning its dynamics (e.g., the nature and number of its periodic orbits) were more or less obvious. The technique of characterizing the orbit
structure of a dynamical system via infinite sequences of "symbols" (in our case 0's
and 1's) is known as *symbolic dynamics*. The technique is not new and appears
to have originally been applied by Hadamard [1898] in the study of geodesics on
surfaces of negative curvature and Birkhoff [1927], [1935] in his studies of dynamical
systems. The first exposition of symbolic dynamics as an independent subject was
given by Morse and Hedlund [1938]. Applications of this idea to differential equations can be found in Levinson's work on the forced Van der Pol equation (Levinson
[1949]), from which came Smale's inspiration for his construction of the horseshoe
map (Smale [1963] and [1980]), and also in the work of Alekseev [1968], [1969], who
gives a systematic account of the technique and applies it to problems arising from
celestial mechanics. These references by no means represent a complete account of

the history of symbolic dynamics or of its applications and we refer the reader to the bibliographies of the above listed references or Moser [1973] for a more complete list of references on the subject and its applications. In recent times (say from about 1965 to the present) there has been a flood of applications of the technique, and we will refer to many of these throughout the remainder of the book.

Symbolic dynamics will play a key role in explaining the dynamical phenomena which we encounter in the next two chapters. For this reason, we now want to describe some aspects of symbolic dynamics viewed as an independent subject.

We let $S = \{1, 2, 3, \ldots, N\}$, $N \geq 2$ be our collection of symbols. We will build our sequences from elements of S. Note that for the purpose of constructing sequences the elements of S could be anything, e.g., letters of the alphabet, Chinese characters, etc. We will use positive integers since they are familiar, easy to write down, and we have as many of them as we desire. At this time we will assume that S is finite (i.e., N is some fixed positive integer ≥ 2), since that is adequate for many purposes and will enable us to avoid certain technical questions in our exposition which at the moment we believe might be unnecessarily distracting. However, after discussing symbolic dynamics for a finite number of symbols, we will return to the case where N can be arbitrarily large and describe the technical modifications necessary to make our results go through in this case also.

2.2a. The Structure of the Space of Symbol Sequences

We now want to construct the space of all symbol sequences, which we will refer to as Σ^N, from elements of S and derive some properties of Σ^N. It will be convenient to construct Σ^N as a cartesian product of infinitely many copies of S. This construction will allow us to make some conclusions concerning the properties of Σ^N based only on our knowledge of S and the structure which we give to S. Also, this approach will make the generalization to an infinite number of symbols quite simple and straightforward later on.

Now we want to give some structure to S; specifically, we want to make S into a metric space. Since S is a finite set of points consisting of the first N positive integers, it is very natural to define the distance between two elements of S to be the absolute value of the difference of the two elements. We denote this as follows:

$$d(a, b) \equiv |a - b| , \qquad \forall\, a, b \in S . \tag{2.2.1}$$

Thus S is a discrete space (i.e., the open sets in S defined by the metric consist of the individual points which make up S so that all subsets of S are open) and hence it is totally disconnected. We summarize the properties of S in the following proposition.

Proposition 2.2.1. *The set S equipped with the metric (2.2.1) is a compact, totally disconnected metric space.*

We remark that compact metric spaces are automatically complete metric spaces (see Dugundji [1966]).

Now we will construct Σ^N as a bi-infinite cartesian product of copies of S

$$\Sigma^N \equiv \cdots \times S \times S \times S \times S \times \cdots \equiv \prod_{i=-\infty}^{\infty} S^i \qquad \text{where } S^i = S \quad \forall\, i. \qquad (2.2.2)$$

So a point in Σ^N is represented as a "bi-infinity-tuple" of elements of S

$$s \in \Sigma^N \Rightarrow s = \{\ldots, s_{-n}, \ldots, s_{-1}, s_0, s_1, \ldots, s_n, \ldots\} \qquad \text{where } s_i \in S \quad \forall\, i \qquad (2.2.3)$$

or, more succinctly, we will write s as

$$s = \{\cdots s_{-n} \cdots s_{-1}.s_0 s_1 \cdots s_n \cdots\}, \qquad \text{where } s_i \in S \quad \forall\, i. \qquad (2.2.4)$$

A word should be said about the "decimal point" that appears in each symbol sequence and has the effect of separating the symbol sequence into two parts with both parts being infinite (hence the reason for the phrase "bi-infinite sequence"). At present it does not play a major role in our discussion and could easily be left out with all of our results describing the structure of Σ^N going through just the same. In some sense, it serves as a starting point for constructing the sequences by giving us a natural way of subscripting each element of a sequence. This notation will prove convenient shortly when we define a metric on Σ^N. However, the real significance of the decimal point will become apparent when we define and discuss the shift map acting on Σ^N and its orbit structure.

In order to discuss limit processes in Σ^N it will be convenient to define a metric on Σ^N. Since S is a metric space it is also possible to define a metric on Σ^N. There are many possible choices for a metric on Σ^N; however, we will utilize the following:

for $s = \{\cdots s_{-n} \cdots s_{-1}.s_0 s_1 \cdots s_n \cdots\}$, $\bar{s} = \{\cdots \bar{s}_{-n} \cdots \bar{s}_{-1}.\bar{s}_0 \bar{s}_1 \cdots \bar{s}_n \cdots\} \in \Sigma^N$

$$\qquad (2.2.5)$$

the distance between s and $\bar{s}$ is defined as

$$d(s,\bar{s}) = \sum_{i=-\infty}^{\infty} \frac{1}{2^{|i|}} \frac{|s_i - \bar{s}_i|}{1 + |s_i - \bar{s}_i|}. \tag{2.2.6}$$

(Note: the reader should check that $d(\cdot,\cdot)$ satisfies the four properties which, by definition, a metric must possess.) Intuitively, this choice of metric implies that two symbol sequences are "close" if they agree on a long central block. The following lemma makes this precise.

Lemma 2.2.2. For $s,\bar{s} \in \Sigma^N$, 1) Suppose $d(s,\bar{s}) < 1/(2^{M+1})$; then $s_i = \bar{s}_i$ for all $|i| \leq M$. 2) Suppose $s_i = \bar{s}_i$ for $|i| \leq M$; then $d(s,\bar{s}) \leq 1/(2^{M-1})$.

PROOF: The proof of 1) is by contradiction. Suppose the hypothesis of 1) holds and there exists some j with $|j| \leq M$ such that $s_j \neq \bar{s}_j$. Then there exists a term in the sum defining $d(s,\bar{s})$ of the form

$$\frac{1}{2^{|j|}} \frac{|s_j - \bar{s}_j|}{1 + |s_j - \bar{s}_j|}.$$

But

$$\frac{|s_j - \bar{s}_j|}{1 + |s_j - \bar{s}_j|} \geq \frac{1}{2}$$

and each term in the sum defining $d(s,\bar{s})$ is positive, so we have

$$d(s,\bar{s}) \geq \frac{1}{2^{|j|}} \frac{|s_j - \bar{s}_j|}{1 + |s_j - \bar{s}_j|} \geq \frac{1}{2^{|j|+1}} \geq \frac{1}{2^{M+1}} \tag{2.2.7}$$

but this contradicts the hypothesis of 1).

We now prove 2). If $s_i = \bar{s}_i$ for $|i| \leq M$ we have

$$d(s,\bar{s}) = \sum_{-\infty}^{i=-(M+1)} \frac{1}{2^{|i|}} \frac{|s_i - \bar{s}_i|}{1 + |s_i - \bar{s}_i|} + \sum_{i=M+1}^{\infty} \frac{1}{2^{|i|}} \frac{|s_i - \bar{s}_i|}{1 + |s_i - \bar{s}_i|} \tag{2.2.8}$$

but $|s_i - \bar{s}_i| / (1 + |s_i - \bar{s}_i|) \leq 1$, so we get

$$d(s, \bar{s}) \leq 2 \sum_{i=M+1}^{\infty} \frac{1}{2^i} = \frac{1}{2^{M-1}}. \tag{2.2.9}$$

$\square$

Armed with our metric, we can define neighborhoods of points in Σ^N and describe limit processes. Suppose we are given a point

$$\bar{s} = \{ \cdots \bar{s}_{-n} \cdots \bar{s}_{-1}.\bar{s}_0 \bar{s}_1 \cdots \bar{s}_n \cdots \} \in \Sigma^N, \quad \bar{s}_i \in S \quad \forall i,$$

and a positive real number $\epsilon > 0$, and we wish to describe the "ϵ-neighborhood of $\bar{s}$", i.e., the set of $s \in \Sigma^N$ such that $d(s, \bar{s}) < \epsilon$. Then, by Lemma 2.2.2, given $\epsilon > 0$ we can find a positive integer $M = M(\epsilon)$ such that $d(s, \bar{s}) < \epsilon$ implies $s_i = \bar{s}_i \ \forall \ |i| \leq M$. Thus, our notation for an ϵ-neighborhood of an arbitrary $\bar{s} \in \Sigma^N$ will be as follows:

$$\mathcal{N}^{M(\epsilon)}(\bar{s}) = \{ s \in \Sigma^N \mid s_i = \bar{s}_i, \quad \forall \ |i| \leq M, \quad s_i, \bar{s}_i \in S \quad \forall i \}. \tag{2.2.10}$$

Before stating our theorem concerning the structure of Σ^N we need the following definition.

Definition 2.2.1. A set is called *perfect* if it is closed and every point in the set is a limit point of the set.

The following theorem of Cantor gives us information concerning the cardinality of perfect sets.

Theorem 2.2.3. *Every perfect set in a complete space has at least the cardinality of the continuum.*

PROOF: See Hausdorff [1957]. $\square$

We are now ready to state our main theorem concerning the structure of Σ^N.

Proposition 2.2.4. *The space Σ^N equipped with the metric (2.2.6) is*

  1) *compact,*

  2) *totally disconnected,*

      and

  3) *perfect.*

PROOF: 1) Since S is compact, Σ^N is compact by Tychonov's theorem (Dugundji [1966]). 2) By Proposition 2.2.1, S is totally disconnected, and therefore Σ^N is totally disconnected, since the product of totally disconnected spaces is likewise totally disconnected (Dugundji [1966]).   3) Σ^N is closed, since it is a compact metric space. Let $\bar{s} \in \Sigma^N$ be an arbitrary point in Σ^N; then, to show that $\bar{s}$ is a limit point of Σ^N, we need only show that every neighborhood of $\bar{s}$ contains a point $s \neq \bar{s}$ with $s \in \Sigma^N$. Let $\mathcal{N}^{M(\epsilon)}(\bar{s})$ be a neighborhood of $\bar{s}$ and let $\hat{s} = \bar{s}_{M(\epsilon)+1}+1$ if $\bar{s}_{M(\epsilon)+1} \neq N$, and $\hat{s} = \bar{s}_{M(\epsilon)+1}-1$ if $\bar{s}_{M(\epsilon)+1} = N$. Then the sequence $\{\cdots \bar{s}_{-M(\epsilon)-2} \, \bar{s}\bar{s}_{-M(\epsilon)} \cdots \bar{s}_{-1}.\bar{s}_0\bar{s}_1 \cdots \bar{s}_{M(\epsilon)} \, \hat{s} \, \bar{s}_{M(\epsilon)+2} \cdots\}$ is contained in $\mathcal{N}^{M(\epsilon)}(\bar{s})$ and is not equal to $\bar{s}$; thus Σ^N is perfect.   □

We remark that the three properties of Σ^N stated in Proposition 2.2.4 are often taken as the defining properties of a *Cantor set* of which the classical Cantor "middle-thirds" set is a prime example.

Next we want to make a remark which will be of interest later when we use Σ^N as a "model space" for the dynamics of maps defined on more "normal" domains than Σ^N (i.e., by "normal" domain we mean the type of domain which might arise as the phase space of a specific physical system). Recall that a map, $h : X \to Y$, of two topological spaces X and Y is called a *homeomorphism* if h is continuous, one-to-one, amd onto and h^{-1} is also continuous. Now there are certain properties of topological spaces which are invariant under homeomorphisms. Such properties are called *topological invariants*; compactness, connectedness, and perfectness are three examples of topological invariants (see Dugundji [1966] for a proof). We summarize this in the following proposition.

Proposition 2.2.5. *Let Y be a topological space and suppose that Σ^N and Y are homeomorphic, i.e., there exists a homeomorphism h taking Σ^N to Y. Then Y is compact, totally disconnected, and perfect.*

2.2b. The Shift Map

Now that we have established the structure of Σ^N, we want to define a map of Σ^N into itself, denoted by σ, as follows:

For $s = \{\cdots s_{-n} \cdots s_{-1}.s_0 s_1 \cdots s_n \cdots\} \in \Sigma^N$, we define

$$\sigma(s) \equiv \{\cdots s_{-n} \cdots s_{-1} s_0.s_1 \cdots s_n \cdots\} \tag{2.2.11}$$

or, $[\sigma(s)]_i \equiv s_{i+1}$.

The map, σ, is referred to as the *shift map*, and when the domain of σ is taken to be all of Σ^N, it is often referred to as a *full shift on N symbols*. We have the following proposition concerning some properties of σ.

Proposition 2.2.6. *1)* $\sigma(\Sigma^N) = \Sigma^N$. *2)* σ *is continuous.*

PROOF: 1) is obvious. To prove 2) we must show that, given $\epsilon > 0$, there exists a $\delta(\epsilon)$ such that $d(s, \bar{s}) < \delta$ implies $d(\sigma(s), \sigma(\bar{s})) < \epsilon$ for $s, \bar{s} \in \Sigma^N$. Suppose $\epsilon > 0$ is given; then choose M such that $1/(2^{M-2}) < \epsilon$. If we then let $\delta = 1/2^{M+1}$, we see by Lemma 2.2.2 that $d(s, \bar{s}) < \delta$ implies $s_i = \bar{s}_i$ for $|i| \leq M$; hence $[\sigma(s)]_i = [\sigma(\bar{s})]_i$, $|i| \leq M - 1$; then, also by Lemma 2.2.2, we have $d(\sigma(s), \sigma(\bar{s})) < 1/2^{M-2} < \epsilon$. $\square$

We now want to consider the orbit structure of σ acting on Σ^N. We have the following proposition.

Proposition 2.2.7. *The shift map σ has*

    1) a countable infinity of periodic orbits consisting of orbits of all periods;

    2) an uncountable infinity of nonperiodic orbits; and

    3) a dense orbit.

PROOF: 1) This is proved exactly the same way as the analogous result obtained in our discussion of the symbolic dynamics for the Smale horseshoe map in Section 2.1c. In particular, the orbits of the periodic symbol sequences are periodic, and there is a countable infinity of such sequences.   2) By Theorem 2.2.3 Σ^N is uncountable; thus, removing the countable infinity of periodic symbol sequences leaves an uncountable number of nonperiodic symbol sequences. Since the orbits of the nonperiodic sequences never repeat, this proves 2).   3) This is proved exactly the same way as the analogous result obtained in our discussion of the Smale horseshoe map in Section 2.1c; namely, we form a symbol sequence by stringing together all possible symbol sequences of any finite length. The orbit of this sequence is dense in Σ^N since, by construction, some iterate of this symbol sequence will be arbitrarily close to any given symbol sequence in Σ^N. $\square$

2.2c. The Subshift of Finite Type

In some applications which will arise in Chapter 3 it will be natural to restrict the domain of σ in such a way that it does not include all possible symbol sequences.

This will be accomplished by throwing out symbol sequences in which certain symbols appear as adjacent entries in the sequence. In order to describe this restriction of the domain of σ the following definition will be useful.

Definition 2.2.2. Let A be an $N \times N$ matrix of 0's and 1's constructed by the following rule: $(A)_{ij} = 1$ if the ordered pair of symbols ij may appear as adjacent entries in the symbol sequence, and $(A)_{ij} = 0$ if the ordered pair of symbols ij may not appear as adjacent entries in the symbol sequence. The matrix A is called the *transition matrix*. (Note: by the phrase "ordered pair of symbols ij" we mean that we are referring to the symbols i and j appearing in the pair with i immediately to the left of j.)

The collection of symbol sequences defined by a given transition matrix is denoted Σ_A^N and may be concisely written as

$$\Sigma_A^N = \left\{ s = \{\cdots s_{-n} \cdots s_{-1}.s_0 s_1 \cdots s_n \cdots\} \in \Sigma^N \mid (A)_{s_i s_{i+1}} = 1 \quad \forall i \right\}.$$
$$(2.2.12)$$

We remark that it should be clear that $\Sigma_A^N \subset \Sigma^N$ and that the metric (2.2.6) serves to define a topology on Σ_A^N with Lemma 2.2.2 also holding for Σ_A^N.

EXAMPLE 2.2.1. Let

$$A = \begin{pmatrix} 0 & 1 \\ 0 & 1 \end{pmatrix} \tag{2.2.13}$$

then Σ_A^2 consists of the collection of bi-infinite sequences of 1's and 2's where the combination of symbols 11 and 21 does not appear in any sequence.

We would like to characterize the structure of Σ_A^N similarly to the way in which Proposition 2.2.4 characterizes the structure of Σ^N. However, two different transition matrices may define two Σ_A^N's which have very different topological structures.

EXAMPLE 2.2.2. Let

$$A = \begin{pmatrix} 1 & 1 \\ 1 & 1 \end{pmatrix} \tag{2.2.14}$$

then Σ_A^2 consists of all possible bi-infinite sequences of 1's and 2's and, therefore, has the topological structure described in Proposition 2.2.4.

Let

$$A' = \begin{pmatrix} 1 & 0 \\ 0 & 1 \end{pmatrix} \tag{2.2.15}$$

then $\Sigma_{A'}^2$ consists of precisely two points, namely the sequences $\{\cdots 1111.1111 \cdots\}$ and $\{\cdots 2222.2222 \cdots\}$. (Notation: if all of the entries of A are one, then $\Sigma_A^N = \Sigma^N$, and we disregard the A in the notation for the space of symbol sequences.)

It is possible to impose restrictions on the transition matrix A in such a way that the properties of Σ^N described in Proposition 2.2.4 also hold for Σ_A^N. We now want to describe the nature of the necessary requirements on A for this to hold.

Definition 2.2.3. The transition matrix A is called *irreducible* if there is an integer $k > 0$ such that $(A^k)_{ij} \neq 0$ for all $1 \leq i,j \leq N$.

Recall from our previous discussion that many of our results concerning the structure of Σ^N and the orbit structure of σ involved constructions using symbol sequences of finite length. In this regard the following definition will be useful.

Definition 2.2.4. Let A be a transition matrix and let $s_1 \cdots s_k$, $s_i \in S$ $i = 1,\ldots,k$ be a finite string of symbols of length k. Then $s_1 \cdots s_k$ will be called an *admissible string of length k* if $(A)_{s_i s_{i+1}} = 1$, $i = 1,\ldots,k-1$.

Let $K > 0$ be the smallest integer such that $(A^K)_{ij} \neq 0$ for all $1 \leq i$, $j \leq N$. Then we have the following lemma.

Lemma 2.2.8. *Suppose A is an irreducible transition matrix, and let K be as described above; then, given any $i, j \in S$, there exists an admissible string of length $k \leq K-1$, $s_1 \cdots s_k$, such that $is_1 \cdots s_k j$ is an admissible string of length $k+2$.*

PROOF: The ij element of A is given by

$$\left(A^K\right)_{ij} = \sum_{s_1,\ldots,s_{K-1}}^{N} (A)_{is_1} (A)_{s_1 s_2} \cdots (A)_{s_{K-2} s_{K-1}} (A)_{s_{K-1} j} \qquad (2.2.16)$$

where each element of the sum is either 0 or 1. Then, since $(A^K)_{ij} \neq 0$, there must exist at least one sequence $\bar{s}_1 \cdots \bar{s}_{K-1}$ such that

$$(A)_{i\bar{s}_1} (A)_{\bar{s}_1 \bar{s}_2} \cdots (A)_{\bar{s}_{K-2} \bar{s}_{K-1}} (A)_{\bar{s}_{K-1} j} = 1.$$

Then each element of this product must also be 1; therefore, $i\bar{s}_1 \cdots \bar{s}_{K-1} j$ is an admissible string. $\qquad \square$

We make the following remarks regarding Lemma 2.2.8.

1) It follows from the proof of Lemma 2.2.8 that for any $i, j \in S$ there exists a maximal integer given by $K-1$ and an admissible string of length $K-1$, $s_1 \cdots s_{K-1}$, such that $is_1 \cdots s_{K-1} j$ is an admissible string of length $K+1$.

However, it is certainly possible that for any particular $i, j \in S$ j appears earlier in the sequence than the last place. This is the reason for stating the lemma in terms of an admissible string of length $k \le K - 1$. However, for some constructions the use of this maximal integer will play an important role (e.g., see Proposition 2.2.9).

2) For $i, j \in S$ regarded as fixed, we will often use the phrase "admissible string of length k connecting i and j," or, when the length of the string does not matter, "admissible string connecting i and j."

We are now in a position to prove the following result concerning the structure of Σ_A^N.

Proposition 2.2.9. *Suppose A is irreducible; then Σ_A^N equipped with the metric (2.2.6) is*

1) *compact,*

2) *totally disconnected,*

   and

3) *perfect.*

PROOF: 1) In order to show that Σ_A^N is compact, it suffices to show that Σ_A^N is closed since closed subsets of compact sets are compact (see Dugundji [1966]).

Let $\{s^i\}$ be a sequence of elements of Σ_A^N, i.e., a sequence of sequences, such that $\{s^i\}$ converges to $\bar{s}$. Σ_A^N is closed if $\bar{s} \in \Sigma_A^N$. The proof that $\bar{s} \in \Sigma_A^N$ is by contradiction. Suppose $\bar{s} \notin \Sigma_A^N$; then there must exist some integer M such that $(A)_{\bar{s}_M \bar{s}_{M+1}} = 0$. Now $\{s^i\}$ converges to $\bar{s}$, so there exists some integer $\bar{M}$ such that for $i \ge \bar{M}$ $d(s^i, \bar{s}) \le 1/2^{M+2}$. So by Lemma 2.2.2 for $i \ge \bar{M}$ we have $s_j^i = \bar{s}_j$ for all $|j| \le M + 1$ and, since $s^i \in \Sigma_A^N$, we have $(A)_{s_M^i s_{M+1}^i} = (A)_{\bar{s}_M \bar{s}_{M+1}} = 1$. This is a contradiction which arises from the fact that we have assumed $\bar{s} \notin \Sigma_A^N$.

2) This is obvious since the largest connected component of Σ^N is a point and the same must hold for any subset of Σ^N.

3) In 1) we showed that Σ_A^N is closed; hence, in order to show that Σ_A^N is perfect, it remains to show that each point of Σ_A^N is a limit point.

Let $\bar{s} \in \Sigma_A^N$; then to show that $\bar{s}$ is a limit point of Σ_A^N we need to show that every neighborhood of $\bar{s}$ contains a point $s \ne \bar{s}$ with $s \in \Sigma_A^N$. Let $\mathcal{N}^{M(\epsilon)}(\bar{s})$ be a neighborhood of $\bar{s} = \{\cdots \bar{s}_{-M} \cdots \bar{s}_{-1}.\bar{s}_0 \bar{s}_1 \cdots \bar{s}_M \cdots\}$. Now from the remark following the proof of Lemma 2.2.8, there exists an integer K such that for any $i, j \in S$ there exists an admissible string of length $K-1$ given by $s_1 \cdots s_{K-1}$ such that $is_1 \cdots s_{K-1}j$ is an admissible string of length $K + 1$. Now consider the $M + K$ entry in $\bar{s}$ given by $\bar{s}_{M+K}$. Let $\hat{s}_{M+K} = \bar{s}_{M+K} + 1$ if $\bar{s}_{M+K} \neq N$ or $\hat{s}_{M+K} = \bar{s}_{M+K} - 1$ if $\bar{s}_{M+K} = N$. Then consider the sequence

$$\hat{s} = \{ \cdots \bar{s}_{-M} \cdots \bar{s}_{-1} . \bar{s}_0 \bar{s}_1 \cdots \bar{s}_M s_1 \cdots s_{K-1} \hat{s}_{M+K} \tilde{s}_1 \cdots \tilde{s}_{K-1} \bar{s}_{M+K+1} \cdots \}$$

where $s_1 \cdots s_{K-1}$ is the admissible string connecting $\bar{s}_M$ and $\hat{s}_{M+K}$, $\tilde{s}_1 \cdots \tilde{s}_{K-1}$ is the admissible string connecting $\hat{s}_{M+K}$ and $\bar{s}_{M+K+1}$, and the "$\cdots$" before $\bar{s}_{-M}$ and after $\bar{s}_{M+K+1}$ indicates that the infinite tails of $\hat{s}$ are the same as $\bar{s}$. Now it should be clear by construction that $\hat{s} \in \Sigma_A^N$, $\hat{s} \in \mathcal{N}^{M(\epsilon)}(\bar{s})$, and $\hat{s} \neq s$. $\square$

Now we want to consider the orbit structure of the shift map σ restricted to Σ_A^N. In this case σ is called a *subshift of finite type* (note: the phrase "finite type" comes from the fact that we are considering only a finite number of symbols). We have the following proposition.

Proposition 2.2.10.   *1)* $\sigma \left(\Sigma_A^N \right) = \Sigma_A^N$.   *2)* σ is continuous.

PROOF: 1) is obvious and the proof of 2) follows from the same argument as that given for the proof of continuity of the full shift given in Proposition 2.2.6. $\square$

Now we give the main result concerning the orbit structure of σ restricted to Σ_A^N.

Proposition 2.2.11. *Suppose A is irreducible; then the shift map σ with domain Σ_A^N has:*

1) *a countable infinity of periodic orbits,*

2) *an uncountable infinity of nonperiodic orbits, and*

3) *a dense orbit.*

PROOF: 1) Recall the construction of the countable infinity of periodic orbits for the full shift. In that case, we merely wrote down the periodic symbol sequences of length $1, 2, 3, \ldots$ and the result immediately followed. Now for the case of subshifts of finite type it should be clear that this construction does not go through. However, a similar procedure will work.

Let $i, j \in S$; then by Lemma 2.2.8 there exists an admissible string $s_1 \cdots s_k$ such that $is_1 \cdots s_k j$ is also an admissible string. For a more compact notation we will denote the admissible string connecting i and j by $s_1 \cdots s_k \equiv s_{ij}$ (note: from the proof of Lemma 2.2.8 it should be clear that for a given $i, j \in S$ there may be more than one admissible string connecting i and j; henceforth, for each $i, j \in S$ we will choose one admissible string s_{ij} and consider it fixed). The construction of the countable infinity of periodic sequences proceeds as follows.

a) Write down all sequences of elements of S of length $2, 3, 4, 5, \ldots$ which begin and end with the same element. It should be clear that there is a countable infinity of such sequences.

b) Choose a particular sequence constructed in a). Now between each pair of entries ij in the sequence place the admissible string s_{ij} which connects these two elements of S. Repeating this procedure for each element constructed in a) yields a countable infinity of admissible strings. Since each admissible string begins and ends with the same element of S we can place copies of each admissible string back-to-back in order to create a bi-infinite periodic sequence. In this manner we construct a countable infinity of admissible periodic sequences. Repeating this procedure for all possible s_{ij} for each i,j should yield all admissible periodic sequences.

2) Since Σ_A^N is perfect, by Theorem 2.2.3 it has the cardinality of at least the continuum. Thus, subtracting the countable infinity of periodic sequences away from Σ_A^N leaves an uncountable infinity of nonperiodic sequences.

3) The construction of the sequence whose orbit under σ is dense in Σ_A^N is very similar to that in the case of the full shift. Write down all admissible strings of length $1, 2, 3, \ldots$; then connect them all together in one sequence by using connecting admissible strings provided by Lemma 2.2.8. The proof that the orbit of this sequence is dense in Σ_A^N is exactly the same as that for the analogous situation with the full shift, see Section 2.1c. $\square$

2.2d. The Case of $N = \infty$

We now want to consider the case of an infinite number of symbols. In particular, we want to determine how much of the previous results concerning the structure of the space of symbol sequences as well as the orbit structure of the shift map are still valid. We will only consider the case of the full shift but we will make some brief comments concerning the subshift at the end of our discussion.

We begin by discussing the structure of the space of symbol sequences, in this case denoted Σ^∞. Let $S = \{1, 2, 3, \ldots, N, \ldots\}$ be our collection of symbols. Then Σ^∞ is constructed as a Cartesian product of infinitely many copies of S. Therefore, it should be clear that the first casualty incurred by allowing an infinite number of symbols is the loss of compactness for Σ^∞ since S is now unbounded. In fact this is the only problem and is quite easily handled. We can compactify S by the usual one point compactification technique (see Dugundji [1966]) of adding a point at infinity. Thus, we have $\bar{S} = \{1, 2, \ldots, N, \ldots, \infty\}$ where ∞ is the point such that all other integers are less than infinity. We now construct $\bar{\Sigma}^\infty$ as an infinite Cartesian product of $\bar{S}$ as in Section 2.2a. Hence, by Tychonov's theorem (Dugundji [1966]), $\bar{\Sigma}^\infty$ is compact and clearly $\Sigma^\infty \subset \bar{\Sigma}^\infty$. $\bar{\Sigma}^\infty$ is called a compactification of Σ^∞. The metric (2.2.6) still works for $\bar{\Sigma}^\infty$ if we define

$$\frac{|N - \infty|}{1 + |N - \infty|} = 1, \qquad N \in S \tag{2.2.17}$$

and

$$\frac{|\infty - \infty|}{1 + |\infty - \infty|} = 0. \tag{2.2.18}$$

Thus, neighborhoods of points in $\bar{\Sigma}^\infty$ are defined exactly as in (2.2.10). Also, Lemma 2.2.2 and Proposition 2.2.4 apply for $\bar{\Sigma}^\infty$ (in particular, $\bar{\Sigma}^\infty$ is a Cantor set). Regarding the orbit structure of the shift map acting on $\bar{\Sigma}^\infty$, Propositions 2.2.6 and 2.2.7 still hold. Now, although the consideration of an infinite number of symbols does not greatly change our results concerning the structure of the space of symbol sequences or the orbit structure of the shift map, we will see in Section 2.3 and in Chapter 3 that the symbol ∞ does have a special meaning when we utilize symbolic dynamics to model the dynamics of maps.

Now let us make a few comments regarding subshifts on a space of symbol sequences having an infinite number of symbols. Recall from our discussion of subshifts of finite type that the transition matrix as well as powers of the transition matrix (specifically, the concept of irreducibility) played a central role in determining both the structure of Σ_A^N and the orbit structure of the shift acting on Σ_A^N. Thus, the immediate problems faced when dealing with *subshifts of infinite type* involve multiplication of infinite matrices as well as defining a concept of irreducibility for an infinite transition matrix. Since we will have no cause to utilize subshifts of infinite type in this book, we will leave the necessary generalizations as an exercise for the interested reader.

2.3. Criteria for Chaos: The Hyperbolic Case

In this section we will give verifiable conditions for a map to possess an invariant set on which it is topologically conjugate to the shift map acting on the space of bi-infinite sequences constructed from a countable set of symbols. Our plan follows that of Conley and Moser (see Moser [1973]); however, our criteria will be more general in the sense that they will apply to N dimensional invertible maps $(N \geq 2)$ and they will allow for subshifts as well as full shifts. The term "hyperbolic case" needs explanation. Roughly speaking, this term arises from the fact that at each point of the domain of the map there is a splitting of the domain into a part which is strongly contracting (the "horizontal" directions) and a part which is strongly expanding (the "vertical" directions). This results in the invariant set of the map being a set of discrete points; a more precise definition will be given later. The outline of this section is as follows:

1) We begin by developing introductory concepts, specifically the generalization of the horizontal and vertical rectangles described in our discussion of the Smale horseshoe in Section 2.1 and concepts for describing their behavior under maps.

2) We state and prove our main theorem.

3) We introduce and discuss the idea of a sector bundle, which provides an alternative and more easily verifiable hypothesis for our main theorem.

4) We give a second set of alternate conditions for verifying the hypotheses of our main theorem that are more convenient for applications near orbits homoclinic to fixed points.

5) We define and discuss the idea of a hyperbolic invariant set and show how it relates to our previous work.

2.3a. The Geometry of Chaos

In order to understand what the essential elements of such criteria might be, let us recall our discussion of the Smale horseshoe map given in Section 2.1. We saw that the horseshoe map contained an invariant Cantor set on which the dynamics was chaotic. Furthermore, we were able to obtain very detailed information concerning the orbit structure of the Smale horseshoe using symbolic dynamics. These results followed from two properties of the map:

1) The map possessed independent expanding and contracting directions.

2) We were able to locate two disjoint "horizontal" rectangles with horizontal and vertical sides parallel to the contracting and expanding directions, respectively, which were mapped onto two "vertical" rectangles with each vertical rectangle intersecting both horizontal rectangles and with the horizontal (resp. vertical) boundaries of the horizontal rectangles mapping onto the horizontal (resp. vertical) boundaries of the vertical rectangles.

These two properties were responsible for the existence of the invariant Cantor set on which the map was topologically conjugate to a full shift on two symbols. (We remark that the reason the map was topologically conjugate to a full shift was that the image of each horizontal rectangle intersected both horizontal rectangles.) So the fact that a map possesses a chaotic invariant set is largely due to geometric criteria and is not a consequence of the specific analytical form of the map (i.e., there is not a "horseshoe function" analogous to a trigonometric function or elliptic function, etc.). Now our goal will be to weaken the above two properties as much as possible and extend them to higher dimensional maps in order to establish criteria for a map to possess an invariant set on which it is topologically conjugate to a subshift.

We consider a map

$$f: D \to \mathbf{R}^n \times \mathbf{R}^m \qquad\qquad (2.3.1)$$

where D is a closed and bounded $n + m$ dimensional set contained in $\mathbf{R}^n \times \mathbf{R}^m$. We will discuss continuity and differentiability requirements on f when they are needed. We note that the domain of f is not required to be connected and, in several examples in Chapter 3, we will see that it is necessary to consider maps whose domains consist of several connected components. However, we will not be interested in the behavior of f on its entire domain, but rather in how it acts on a set of disjoint, specially defined "horizontal slabs". (Note: this situation is completely analogous to the Smale horseshoe discussed in Section 2.1. In that example the map was defined on the unit square, but all of the complicated dynamical consequences were derived from a knowledge of how the map acted on the two horizontal rectangles H_0 and H_1.)

We will now begin the development of the definition of horizontal and vertical slabs which will be the analogs of the H_i, V_i, $i = 0, 1$, in our discussion of the Smale horseshoe in Section 2.1. Following this, we will define the width of the slabs and various intersection properties of horizontal and vertical slabs. These will be of use in proving our main theorem, which will provide sufficient conditions in order for f to possess an invariant set on which it is topologically conjugate to a subshift of finite type.

We begin with some preliminary definitions. The following two sets will be useful for defining the domains of various maps:

$$D_x = \{x \in \mathbf{R}^n \text{ for which there exists a } y \in \mathbf{R}^m \text{ with } (x, y) \in D\}$$
$$D_y = \{y \in \mathbf{R}^m \text{ for which there exists an } x \in \mathbf{R}^n \text{ with } (x, y) \in D\}. \tag{2.3.2}$$

So D_x and D_y represent the projection of D onto $\mathbf{R}^n$ and $\mathbf{R}^m$, respectively, see Figure 2.3.1.

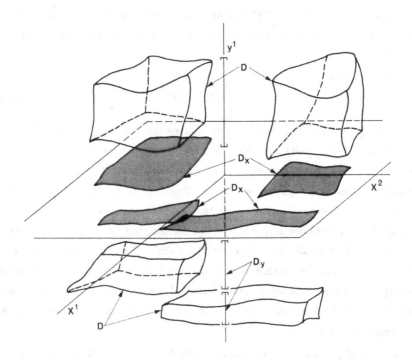

Figure 2.3.1. D, D_x, and D_y in $\mathbf{R}^{n+m}$; $n = 2$, $m = 1$.

Let I_x be a closed, simply connected n dimensional set contained in D_x, and let I_y be a closed, simply connected m dimensional set contained in D_y. We will need the following definitions.

Definition 2.3.1. A μ_h-*horizontal slice*, $\bar{H}$, is defined to be the graph of a function $h: I_x \to \mathbf{R}^m$ where h satisfies the following two conditions:

1) The set $\bar{H} = \{ (x, h(x)) \in \mathbf{R}^n \times \mathbf{R}^m \mid x \in I_x \}$ is contained in D.

2) For every $x_1, x_2 \in I_x$, we have

$$|h(x_1) - h(x_2)| \le \mu_h |x_1 - x_2| \qquad (2.3.3)$$

for some $0 \le \mu_h < \infty$.

Similarly, a μ_v-*vertical slice*, $\bar{V}$, is defined to be the graph of a function $v: I_y \to \mathbf{R}^n$ where v satisfies the following two conditions:

1) The set $\bar{V} = \{ (v(y), y) \in \mathbf{R}^n \times \mathbf{R}^m \mid y \in I_y \}$ is contained in D.

2) For every $y_1, y_2 \in I_y$, we have

$$|v(y_1) - v(y_2)| \le \mu_v |y_1 - y_2| \qquad (2.3.4)$$

for some $0 \le \mu_v < \infty$.

See Figure 2.3.2 for an illustration of Definition 2.3.1 (Note: hereafter in giving figures to describe the definitions of μ_h-horizontal slices, μ_h-horizontal slabs, widths of slabs, etc., we will not show the domain D of f in the figures so that they do not become too cluttered.)

Next we want to "fatten up" these horizontal and vertical slices into $n + m$ dimensional horizontal and vertical "slabs." We begin with the definition of μ_h-horizontal slabs.

Definition 2.3.2. Fix some μ_h, $0 \le \mu_h < \infty$. Let $\bar{H}$ be a μ_h-horizontal slice, and let $J^m \subset D$ be an m-dimensional topological disk intersecting $\bar{H}$ at any, but only one, point of $\bar{H}$. Let $\bar{H}^\alpha$, $\alpha \in I$, be the set of all μ_h-horizontal slices that intersect the boundary of J^m and have the same domain as $\bar{H}$ where I is some index set (note: it may be necessary to adjust the domain I_x of $\bar{H}$, or equivalently, adjust J^m in order for this situation to be obtained). Consider the following set in $\mathbf{R}^n \times \mathbf{R}^m$.

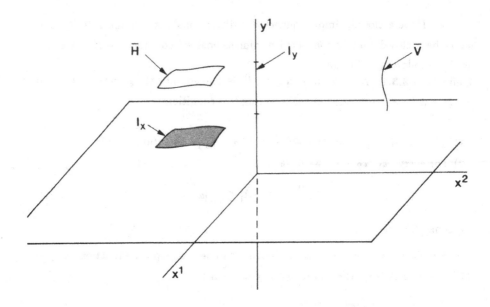

Figure 2.3.2. μ_h-Horizontal and μ_v-Vertical Slices in $\mathbb{R}^n \times \mathbb{R}^m$; $n = 2$, $m = 1$.

$S_H = \{\, (x,y) \in \mathbb{R}^n \times \mathbb{R}^m \mid x \in I_x$ and y has the property that, for each $x \in I_x$, given any line L through (x,y) with L parallel to the $x = 0$ plane, then L intersects the points $\big(x, h_\alpha(x)\big)$, $\big(x, h_\beta(x)\big)$ for some $\alpha, \beta \in I$ with (x,y) between these two points along L. $\}$

Then a μ_h-*horizontal slab* H is defined to be the closure of S_H.

When we discuss the behavior of μ_h-horizontal slabs under maps it will be useful to have the notion of horizontal and vertical boundaries.

Definition 2.3.3. The *vertical boundary* of a μ_h-horizontal slab H is denoted $\partial_v H$ and is defined as

$$\partial_v H \equiv \{\, (x,y) \in H \mid x \in \partial I_x \,\} . \tag{2.3.5}$$

The *horizontal boundary* of a μ_h-horizontal slab H is denoted $\partial_h H$ and is defined as

$$\partial_h H \equiv \partial H - \partial_v H . \tag{2.3.6}$$

We remark that it follows from this as well as Definition 2.3.2 that $\partial_v H$ is parallel to the $x = 0$ plane. See Figure 2.3.3 for an illustration of Definitions 2.3.2 and 2.3.3.

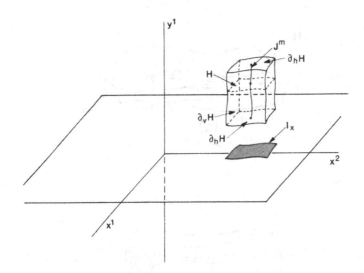

Figure 2.3.3. Horizontal Slab in $\mathbf{R}^n \times \mathbf{R}^m$; $n = 2$, $m = 1$.

Before proceeding further, let us give some motivation for Definition 2.3.2. We will see later on that the main properties we need for H are that it is an $n + m$ dimensional compact set such that any point on $\partial_h H$ lies on a μ_h-horizontal slice. In Definition 2.3.2 these properties are manifested as follows.

By construction $\partial_h H$ is made up of μ_h-horizontal slices. Therefore, any point lying on $\partial_h H$ lies on a μ_h-horizontal slice. We remark that it should be clear that I is an uncountable set.

The line L is used to fill out the space "between" the μ_h-horizontal slices that make up $\partial_h H$. By moving L through ∂I_x the vertical boundary of H is created. In this way one obtains an $n + m$ dimensional compact set.

We will be interested in the behavior of μ_h-horizontal slabs under maps. In particular, we will be interested in the situation where the image of a μ_h-horizontal slab intersects its preimage (note: the "preimage of the image" is just the slab itself). For describing this situation the following definition will be useful.

Definition 2.3.4. Let H and $\tilde{H}$ be μ_h-horizontal slabs. $\tilde{H}$ is said to intersect H *fully* if $\tilde{H} \subset H$ and $\partial_v \tilde{H} \subset \partial_v H$.

See Figure 2.3.4 for an illustration of Definition 2.3.4.

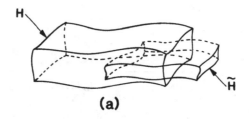

(a)

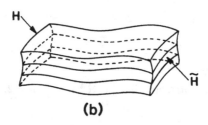

(b)

Figure 2.3.4. a) $\tilde{H}$ Does Not Intersect H Fully.   b) $\tilde{H}$ Intersects H Fully.

Next we will define μ_v-vertical slabs.

Definition 2.3.5. Fix some μ_v, $0 \leq \mu_v < \infty$. Let H be a μ_h-horizontal slab, and let $\bar{V}$ be a μ_v-vertical slice contained in H such that $\partial \bar{V} \subset \partial_h H$. Let $J^n \subset H$ be an n dimensional topological disk intersecting $\bar{V}$ at any, but only one, point of $\bar{V}$, and let $\bar{V}^\alpha$, $\alpha \in I$, be the set of all μ_v-vertical slices that intersect the boundary of J^n with $\partial \bar{V}^\alpha \subset \partial_h H$ where I is some index set. We denote the domain of the function $v_\alpha(y)$ whose graph is $\bar{V}^\alpha$ by I_y^α. Consider the following set in $\mathbf{R}^n \times \mathbf{R}^m$.

$$S_V = \{\, (x,y) \in \mathbf{R}^n \times \mathbf{R}^m \mid (x,y) \text{ is contained in the interior of the set bounded by } \bar{V}^\alpha, \ \alpha \in I, \text{ and } \partial_h H \,\}.$$

Then, a μ_v-*vertical slab* V is defined to be the closure of S_V.

We remark that the main properties we need for μ_v-vertical slabs are that they are $n + m$ dimensional compact sets such that any point on the vertical boundary lies on a μ_v-vertical slice (cf. the discussion following Definition 2.3.3).

The horizontal and vertical boundaries of μ_v-vertical slabs are defined as follows.

Definition 2.3.6. Let V be a μ_v-vertical slab. The *horizontal boundary* of V, denoted $\partial_h V$, is defined to be $V \cap \partial_h H$. The *vertical boundary* of V, denoted $\partial_v V$, is defined to be $\partial V - \partial_h V$.

See Figure 2.3.5 for an illustration of Definitions 2.3.5 and 2.3.6.

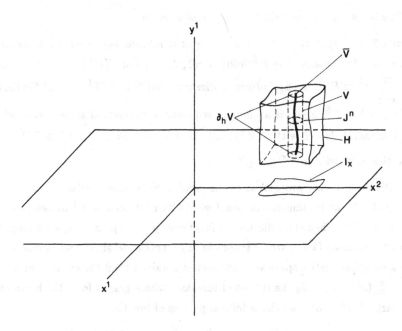

Figure 2.3.5. Vertical Slab in $\mathbf{R}^n \times \mathbf{R}^m$; $n = 2$, $m = 1$.

Notice in Figures 2.3.4 and 2.3.5 that we depict the μ_h-horizontal and μ_v-vertical slabs as slightly warped cubes or tubes. Definitions 2.3.2 and 2.3.5 certainly allow much more pathological behavior of the boundaries; however, for convenience we will continue to draw the slabs as in Figures 2.3.4 and 2.3.5.

Next we define the widths of μ_h-horizontal and μ_v-vertical slabs.

Definition 2.3.7. The width of a μ_h-horizontal slab H is denoted by $d(H)$ and is defined as follows:

$$d(H) = \sup_{\substack{x \in I_x \\ \alpha, \beta \in I}} \left| h_\alpha(x) - h_\beta(x) \right| . \qquad (2.3.7)$$

Similarly, the width of a μ_v-vertical slab V is denoted by $d(V)$ and is defined as follows:

$$d(V) = \sup_{\substack{y \in \tilde{I}_y \\ \alpha, \beta \in I}} \left| v_\alpha(y) - v_\beta(y) \right| \qquad (2.3.8)$$

where $\tilde{I}_y \equiv I_y^\alpha \cap I_y^\beta$.

The following lemmas will be very useful later on.

Lemma 2.3.1. a) If $H^1 \supset H^2 \supset H^3 \supset \cdots$ is an infinite sequence of μ_h-horizontal slabs with H^{k+1} intersecting H^k fully, $k = 1, 2, \ldots$, and $d(H^k) \to 0$ as $k \to \infty$, then $\bigcap_{k=1}^{\infty} H^k \equiv H^\infty$ is a μ_h-horizontal slice with $\partial H^\infty \subset \partial_v H^1$.   b) Similarly, if $V^1 \supset V^2 \supset V^3 \supset \cdots$ is an infinite sequence of μ_v-vertical slabs contained in a μ_h-horizontal slab H with $d(V^k) \to 0$ as $k \to \infty$, then $\bigcap_{k=1}^{\infty} V^k \equiv V^\infty$ is a μ_v-vertical slice with $\partial V^\infty \subset \partial_h H$.

PROOF: We will only prove a) since the proof of b) is quite similar.

Let J^m be an m-dimensional topological disk contained in H^1 as described in Definition 2.3.2. Then the collection of functions $y = h(x)$, $x \in I_x$, satisfying the Lipschitz condition (2.3.3) whose graphs form μ_h-horizontal slices that intersect J^m form a complete metric space with the metric obtained from the supremum norm. Let $\{h_{\alpha_k}^k(x)\}$, $\alpha_k \in I_k$ be the set of functions whose graphs form the horizontal boundary of H^k. Consider the infinite sequences of functions

$$\left\{ h_{\alpha_k}^k \right\}_{k=1}^{\infty}, \ \left\{ h_{\bar{\alpha}_k}^k \right\}_{k=1}^{\infty}, \ \left\{ h_{\alpha_k}^k, h_{\bar{\alpha}_k}^k \right\}_{k=1}^{\infty} \qquad (2.3.9)$$

where $\alpha_k, \bar{\alpha}_k \in I_k$ are regarded as fixed for each k. Now the condition

$$H^k \supset H^{k+1}, \ H^{k+1} \text{ intersecting } H^k \text{ fully for } k = 1, 2, \ldots$$

$$\text{with } d(H^k) \to 0, \ k \to \infty \qquad (2.3.10)$$

implies that the three sequences in (2.3.9) are Cauchy sequences. Since the elements of these Cauchy sequences lie in a complete metric space, it follows that they each converge to a limit $h^{\infty}(x)$. The limit functions for the three sequences in (2.3.9) must be identical since the first two sequences are subsequences of the third sequence. Moreover, the limit function must satisfy the condition

$$|h^{\infty}(x_1) - h^{\infty}(x_2)| \le \mu_h |x_1 - x_2|, \qquad x_1, x_2 \in I_x, \qquad 0 \le \mu_h < \infty. \quad (2.3.11)$$

Thus, we have shown that all the μ_h-horizontal slices that make up the boundary of H^k converge to a unique function $h^{\infty}(x)$ as $k \to \infty$ with $h^{\infty}(x)$ satisfying (2.3.11). So the graph of $h^{\infty}(x)$ is a μ_h-horizontal slice denoted H^{∞} with $\partial H^{\infty} \subset \partial_v H^1$, since the domain of $h^{\infty}(x)$ is I_x. $\square$

Lemma 2.3.2. Let H be a μ_h-horizontal slab. Let $\bar{H}$ be a μ_h-horizontal slice with $\partial \bar{H} \subset \partial_v H$, and let $\bar{V}$ be a μ_v-vertical slice with $\partial \bar{V} \subset \partial_h H$ such that $0 \le \mu_v \mu_h < 1$. Then $\bar{H}$ and $\bar{V}$ intersect in precisely one point.

PROOF: The lemma will be proved if we show that there is a unique point $x \in I_x$ such that $x = v(h(x))$ where $\tilde{I}_x =$ closure $\{x \in I_x | h(x)$ is in the domain of $v(\cdot)\}$.

Now, since I_x is a closed subset of $\mathbf{R}^n$, it is a complete metric space and, since $\bar{V} \subset H$ $v \circ h$ maps $\tilde{I}_x$ into $\tilde{I}_x$, hence $v \circ h$ maps the complete metric space $\tilde{I}_x$ into itself. If we show that $v \circ h$ is a contraction map, then it follows from the contraction mapping theorem (see Chow and Hale [1982]) that $v \circ h$ has a unique fixed point in I_x and, therefore, the lemma is proved. To show that $v \circ h$ is a contraction map, choose $x_1, x_2 \in \tilde{I}_x$, in which case we have

$$\begin{aligned} |v(h(x_1)) - v(h(x_2))| &\le \mu_v |h(x_1) - h(x_2)| \\ &\le \mu_v \mu_h |x_1 - x_2| . \end{aligned} \qquad (2.3.12)$$

So, since $0 \le \mu_v \mu_h < 1$, $v \circ h$ is a contraction map. $\square$

At this point we would like to comment on the motivation behind our somewhat involved definitions of μ_h-horizontal and μ_v-vertical slabs given in Definitions 2.3.2 and 2.3.5.

The main motivating factor is to define objects which display the properties described in Lemma 2.3.1 and Lemma 2.3.2. So, roughly speaking, horizontal slabs are $n + m$ dimensional objects whose "horizontal sides" are foliated by horizontal slices resulting in the intersection of a countable infinity of nested horizontal slabs being a horizontal slice. Similarly, a vertical slab is an $n + m$ dimensional object whose vertical sides are foliated by vertical slices resulting in the intersection of a countable infinity of nested vertical slabs being a vertical slice. We will see that this property, along with the fact that horizontal and vertical slices intersect in a unique point, is of crucial importance in explicitly constructing the invariant set of the map f.

2.3b. The Main Theorem

We are now at the point where we can give conditions sufficient for the map f to have an invariant set on which it is topologically conjugate to a subshift of finite type.

Let $S = \{1, 2, \ldots, N\}$, $N \geq 2$, and let H_i, $i = 1, \ldots, N$, be a set of disjoint μ_h-horizontal slabs with $D_H \equiv \bigcup_{i=1}^{N} H_i$. We assume that f is one-to-one on D_H and we define

$$f(H_i) \cap H_j \equiv V_{ji}, \qquad \forall\, i, j \in S$$

and

$$H_i \cap f^{-1}(H_j) \equiv f^{-1}(V_{ji}) \equiv H_{ij}, \qquad \forall\, i, j \in S. \tag{2.3.13}$$

Notice the subscripts on the sets V_{ji} and H_{ij}. The first subscript indicates which particular μ_h-horizontal slab the set is in, and the second subscript indicates for the V_{ji} into which μ_h-horizontal slab the set is mapped by f^{-1} and for the H_{ij} into which μ_h-horizontal slab the set is mapped by f.

Let A be an $N \times N$ matrix whose entries are either 0 or 1, i.e., A is a transition matrix (see Section 2.2) which will eventually be used to define symbolic dynamics for f. We have the following two "structural" assumptions for f.

A1. For all $i, j \in S$ such that $(A)_{ij} = 1$, V_{ji} is a μ_v-vertical slab contained in H_j with $\partial_v V_{ji} \subset \partial f(H_i)$ and $0 \leq \mu_v \mu_h < 1$. Moreover, f maps H_{ij} homeomorphically onto V_{ji} with $f^{-1}(\partial_v V_{ji}) \subset \partial_v H_i$.

A2. Let H be a μ_h-horizontal slab which intersects H_j fully. Then $f^{-1}(H) \cap H_i \equiv \tilde{H}_i$ is a μ_h-horizontal slab intersecting H_i fully for all $i \in S$ such that $(A)_{ij} = 1$. Moreover,

$$d(\tilde{H}_i) \leq \nu_h\, d(H) \qquad \text{for some } 0 < \nu_h < 1. \tag{2.3.14}$$

Similarly, let V be a μ_v-vertical slab contained in H_j such that also $V \subset V_{ji}$ for some $i, j \in S$ with $(A)_{ij} = 1$. Then $f(V) \cap H_k \equiv \tilde{V}_k$ is a μ_v-vertical slab contained in H_k for all $k \in S$ such that $(A)_{jk} = 1$. Moreover,

$$d(\tilde{V}_k) \le \nu_v \, d(V) \qquad \text{for some } 0 < \nu_v < 1. \qquad (2.3.15)$$

See Figures 2.3.6 and 2.3.7 for a geometric interpretation of A1 and A2.

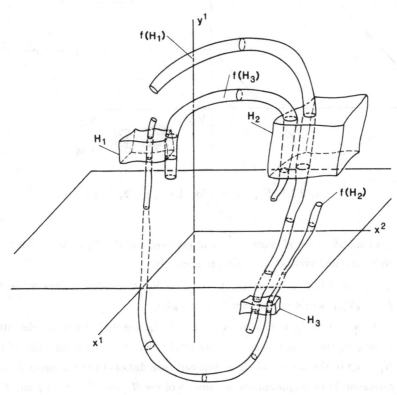

Figure 2.3.6. An Example of Horizontal Slabs and Their Images under f; $A = \begin{pmatrix} 0 & 1 & 1 \\ 1 & 0 & 1 \\ 0 & 1 & 0 \end{pmatrix}$

Let us now make the following remarks concerning A1 and A2.

1. A1 is the global hypothesis dealing with the nonlinear nature of f. It assures that the appropriate boundaries of the images and preimages of the H_i under f are aligned along the appropriate contracting and expanding directions. A2

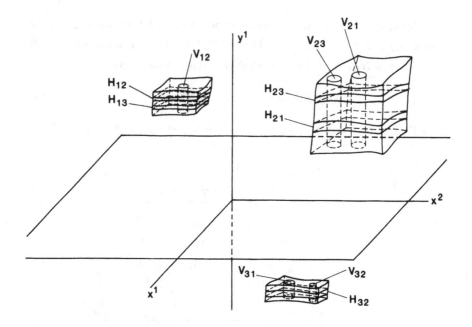

Figure 2.3.7. H_{ij} and V_{ji} for $1 \leq i, j \leq 3$, $(A)_{ij} = 1$.

gives specific rates of contraction and expansion of the H_i under f along the horizontal and vertical directions, respectively.

2. Regarding A1, it is important to have $\partial_v V_{ji} \subset \partial f(H_i)$, since otherwise $f^{-1}(\partial_v V_{ji})$ would not be contained in $\partial_v H_i$.

3. In A2 let $H = H_j$; then it follows that the H_{ij} are μ_h-horizontal slabs intersecting H_i fully. Moreover, the horizontal (resp. vertical) boundaries of the H_{ij} map to the horizontal (resp. vertical) boundaries of the V_{ji} under f. The correspondence of appropriate boundaries of the H_{ij} and V_{ji} under f and f^{-1} is very important.

4. It is important to realize that A1 and A2 are hypotheses which concern only *one* forward and backward iterate of f. We will see that A1 and A2 imply results on *all* iterates of f.

Theorem 2.3.3. *Suppose f satisfies A1 and A2; then f possesses an invariant set of points $\Lambda \subset D_H$ on which it is topologically conjugate to a subshift of finite type with transition matrix A, i.e., there exists a homeomorphism $\phi: \Lambda \rightarrow \Sigma_A^N$ such* *that the following diagram commutes*

$$
\begin{array}{ccc}
\Lambda & \xrightarrow{\;f\;} & \Lambda \\
\phi \downarrow & & \downarrow \phi \\
\Sigma_A^N & \xrightarrow{\;\sigma\;} & \Sigma_A^N
\end{array}
\tag{2.3.16}
$$

Moreover, if A is irreducible then Λ is a Cantor set.

PROOF: The proof is broken down into several steps.

1) Geometrically construct the invariant set, Λ, of f and verify that it is nonempty.

2) Based on the geometrical construction of Λ, define a map $\phi: \Lambda \to \Sigma_A^N$.

3) Prove that ϕ is a homeomorphism.

4) Prove that the diagram (2.3.16) commutes, i.e., $\phi \circ f = \sigma \circ \phi$.

We begin with the first step.

1) Construction of the Invariant set Λ.

The invariant set Λ consists of the points in D_H that remain in D_H under all forward and backward iterations by f. If we denote the set of points that remain in D_H under all backward iterations by $\Lambda_{-\infty}$, and the set of points that remain in D_H under all forward iterations by $\Lambda_{+\infty}$, then the invariant set Λ is the set of points that is common to both $\Lambda_{-\infty}$ and $\Lambda_{+\infty}$ or, in other words, $\Lambda = \Lambda_{-\infty} \cap \Lambda_{+\infty}$. In constructing Λ we will construct and determine the nature of $\Lambda_{-\infty}$ and $\Lambda_{+\infty}$ separately, and then take their intersection in order to obtain Λ.

1a) Construction of $\Lambda_{-\infty}$.

We want to construct and determine the nature of the set of points which remain in D_H under all backwards iterations by f, i.e.,

$$
\left\{ p \in D_H \mid f^{-i}(p) \in H_{s_{-i}}, \quad s_{-i} \in S, \quad i = 0, 1, 2, \ldots, n, \ldots \right\} \equiv \Lambda_{-\infty}.
$$

This will be accomplished via an inductive construction where we construct sequentially the set of points remaining in D_H under $1, 2, \ldots, n, \ldots$ backwards iterations by f, utilize A1 and A2 to determine the nature of the set constructed at each step, and then consider the set obtained in the limit as $n \to \infty$.

We begin by writing down expressions for the set of points which remain in D_H under $1, 2, \ldots, n, \ldots$ backwards iterations by f.

$$\Lambda_{-1} \equiv \bigcup_{\substack{s_i \in S \\ i=0,-1}} \left(f\left(H_{s_{-1}}\right) \cap H_{s_0}\right) \equiv \bigcup_{\substack{s_i \in S \\ i=0,-1}} V_{s_0 s_{-1}}$$

$$= \left\{ p \in D_H \mid p \in H_{s_0}, \quad f^{-1}(p) \in H_{s_{-1}}; \quad s_0, s_{-1} \in S \right\}.$$

$$(2.3.17)$$

$$\Lambda_{-2} \equiv \bigcup_{\substack{s_i \in S \\ i=0,-1,-2}} \left(f\left(V_{s_{-1} s_{-2}}\right) \cap H_{s_0}\right)$$

$$= \bigcup_{\substack{s_i \in S \\ i=0,-1,-2}} \left(f^2\left(H_{s_{-2}}\right) \cap f\left(H_{s_{-1}}\right) \cap H_{s_0}\right) \equiv \bigcup_{\substack{s_i \in S \\ i=0,-1,-2}} V_{s_0 s_{-1} s_{-2}}$$

$$= \left\{ p \in D_H \mid p \in H_{s_0}, f^{-1}(p) \in H_{s_{-1}}, f^{-2}(p) \in H_{s_{-2}}; s_0, s_{-1}, s_{-2} \in S \right\}.$$

$$(2.3.18)$$

$$\vdots \qquad \vdots \qquad \vdots$$

$$\Lambda_{-n} \equiv \bigcup_{\substack{s_i \in S \\ i=0,-1,\ldots,-n}} \left(f\left(V_{s_{-1}\cdots s_{-n}}\right) \cap H_{s_0}\right)$$

$$= \bigcup_{\substack{s_i \in S \\ i=0,-1,\ldots,-n}} \left(f^n\left(H_{s_{-n}}\right) \cap f^{n-1}\left(H_{s_{-n+1}}\right) \cap \cdots \cap H_{s_0}\right) \equiv \bigcup_{\substack{s_i \in S \\ i=0,-1,\ldots,-n}} V_{s_0 \cdots s_{-n}} \quad (2.3.19)$$

$$= \left\{ p \in D_H \mid f^{-i}(p) \in H_{s_{-i}}; \quad s_{-i} \in S, \quad i = 0, 1, \ldots, n \right\}.$$

$$\vdots \qquad \vdots \qquad \vdots$$

Now, by A1, Λ_{-1} consists of a collection of disjoint μ_v-vertical slabs $V_{s_0 s_{-1}}$ contained in H_{s_0} for all $s_0, s_{-1} \in S$ such that $(A)_{s_{-1} s_0} = 1$.

Proceeding to Λ_{-2}, we use the information obtained concerning Λ_{-1} at the previous step and appeal to A2 to conclude:

i) Λ_{-2} consists of a collection of disjoint μ_v-vertical slabs $V_{s_0 s_{-1} s_{-2}}$ contained in H_{s_0} for all $s_0, s_{-1}, s_{-2} \in S$ such that $(A)_{s_{-2} s_{-1}}(A)_{s_{-1} s_0} = 1$.

ii) $d\left(V_{s_0 s_{-1} s_{-2}}\right) \leq \nu_v \, d\left(V_{s_0 s_{-1}}\right).$ $\qquad\qquad (2.3.20)$

iii) By definition of Λ_{-2} and Λ_{-1} it follows that

$$V_{s_0 s_{-1} s_{-2}} \subset V_{s_0 s_{-1}}. \qquad\qquad (2.3.21)$$

To determine the nature of Λ_{-3}, we use the information obtained concerning Λ_{-2} at the previous step and appeal to A2 to conclude:

   i) Λ_{-3} consists of a collection of disjoint μ_v-vertical slabs $V_{s_0 s_{-1} s_{-2} s_{-3}}$ contained in H_{s_0} for all $s_0, s_{-1}, s_{-2}, s_{-3} \in S$ such that

$$(A)_{s_{-3} s_{-2}} (A)_{s_{-2} s_{-1}} (A)_{s_{-1} s_0} = 1.$$

   ii) $d\left(V_{s_0 s_{-1} s_{-2} s_{-3}}\right) \leq \nu_v \, d\left(V_{s_0 s_{-1} s_{-2}}\right) \leq \nu_v^2 \, d\left(V_{s_0 s_{-1}}\right).$ (2.3.22)

   iii) By definition of Λ_{-3}, Λ_{-2} and Λ_{-1} it follows that

$$V_{s_0 s_{-1} s_{-2} s_{-3}} \subset V_{s_0 s_{-1} s_{-2}} \subset V_{s_0 s_{-1}} \, .$$ (2.3.23)

Continuing to argue in this manner, we determine the nature of Λ_{-n} by using the information obtained concerning Λ_{-n+1} at the $(n-1)^{\text{th}}$ step and appeal to A2 to conclude:

   i) Λ_{-n} consists of a collection of disjoint μ_v-vertical slabs $V_{s_0 s_{-1} \cdots s_{-n}}$ contained in H_{s_0} for all $s_0, s_{-1}, \cdots, s_{-n} \in S$ such that $(A)_{s_{-n} s_{-n+1}}$ $\cdots (A)_{s_{-1} s_0} = 1.$

   ii) $d\left(V_{s_0 s_{-1} \cdots s_{-n}}\right) \leq \nu_v \, d\left(V_{s_0 s_{-1} \cdots s_{-n+1}}\right) \leq \cdots \leq \nu_v^{n-1} \, d\left(V_{s_0 s_{-1}}\right).$ (2.3.24)

   iii) By definition of Λ_{-k}, $k = 1, 2, \ldots, n$, we have

$$V_{s_0 s_{-1} \cdots s_{-n}} \subset V_{s_0 s_{-1} \cdots s_{-n+1}} \subset \cdots \subset V_{s_0 s_{-1}} \, .$$ (2.3.25)

Before proceeding to discuss the limit as $n \to \infty$ we make the important remark that, at each stage of the construction process, each μ_v-vertical slab can be labelled uniquely by an admissible string of elements of S determined by the transition matrix A and having a length of one plus the number of the step. Furthermore, all possible admissible strings of the appropriate length are realized at each step due to the assumption A2.

Now in the limit as $n \to \infty$ we obtain the set

$$\Lambda_{-\infty} = \left\{ p \in D_H \mid f^{-i}(p) \in H_{s_{-i}} ; \quad s_{-i} \in S , \quad i = 0, 1, 2, \ldots, n, \ldots \right\} ,$$ (2.3.26)

and we can immediately make the following conclusions concerning the nature of $\Lambda_{-\infty}$.

   i) Each element of $\Lambda_{-\infty}$, $V_{s_0 s_{-1} \cdots s_{-n} \cdots}$, $s_{-i} \in S$, $i = 0, 1, \ldots,$, can be labelled by a unique infinite sequence of elements of S allowed by the transition matrix A. Furthermore, all possible such sequences are realized.

ii) Since $V_{s_0 s_{-1} \cdots s_{-n} \cdots} = \bigcap_{n=1}^{\infty} V_{s_0 \cdots s_{-n}}$ where $V_{s_0 \cdots s_{-n}}$ are μ_v-vertical slabs contained in H_{s_0} for all $s_0, \cdots, s_{-n} \in S$ such that $(A)_{s_{-n} s_{-n+1}} \cdots (A)_{s_{-1} s_0} = 1$ and $V_{s_0 \cdots s_{-n}} \subset V_{s_0 \cdots s_{-n+1}}$ with $d\left(V_{s_0 \cdots s_{-n}}\right) \to 0$ as $n \to \infty$, by Lemma 2.3.1 we can conclude that $\Lambda_{-\infty}$ consists of a set of μ_v-vertical slices, $V_{s_0 \cdots s_{-n} \cdots}$, with $\partial V_{s_0 \cdots s_{-n} \cdots} \subset \partial_h H_{s_0}$. The cardinality of this set is determined by the transition matrix A. In particular, if A is irreducible, $\Lambda_{-\infty}$ consists of an uncountable infinity of μ_v-vertical slices.

1b) Construction of $\Lambda_{+\infty}$.

The construction of $\Lambda_{+\infty}$ is virtually identical to the construction of $\Lambda_{-\infty}$, with the obvious modifications.

We begin by writing down expressions for the set of points that remain in H under $1, 2, \ldots, n, \ldots$ forward iterations by f.

$$\Lambda_1 \equiv \bigcup_{\substack{s_i \in S \\ i=0,1}} \left(f^{-1}\left(H_{s_1}\right) \cap H_{s_0}\right) \equiv \bigcup_{\substack{s_i \in S \\ i=0,1}} H_{s_0 s_1}$$

$$= \{p \in D_H \mid p \in H_{s_0}, \quad f(p) \in H_{s_1}; \quad s_0, s_1 \in S\}.$$

$$(2.3.27)$$

$$\Lambda_2 \equiv \bigcup_{\substack{s_i \in S \\ i=0,1,2}} \left(f^{-1}\left(H_{s_1 s_2}\right) \cap H_{s_0}\right)$$

$$= \bigcup_{\substack{s_i \in S \\ i=0,1,2}} \left(f^{-2}\left(H_{s_2}\right) \cap f^{-1}\left(H_{s_1}\right) \cap H_{s_0}\right) \equiv \bigcup_{\substack{s_i \in S \\ i=0,1,2}} H_{s_0 s_1 s_2}$$

$$(2.3.28)$$

$$= \left\{p \in D_H \mid p \in H_{s_0}, f(p) \in H_{s_1}, f^2(p) \in H_{s_2}; s_0, s_1, s_2 \in S\right\}.$$

$$\vdots \qquad \vdots \qquad \vdots$$

$$\Lambda_n \equiv \bigcup_{\substack{s_i \in S \\ i=0,1,\ldots,n}} \left(f^{-1}\left(H_{s_1 \cdots s_n}\right) \cap H_{s_0}\right)$$

$$= \bigcup_{\substack{s_i \in S \\ i=0,1,\ldots,n}} \left(f^{-n}\left(H_{s_n}\right) \cap f^{-n+1}\left(H_{s_{n-1}}\right) \cap \cdots \cap H_{s_0}\right) \equiv \bigcup_{\substack{s_i \in S \\ i=0,1,\ldots,n}} H_{s_0 \cdots s_n}$$

$$(2.3.29)$$

$$= \left\{p \in D_H \mid f^i(p) \in H_{s_i}; \quad s_i \in S, \quad i=0,1,\ldots,n\right\}.$$

$$\vdots \qquad \vdots \qquad \vdots$$

Now, by A2, Λ_1 consists of a collection of disjoint μ_h-horizontal slabs $H_{s_0 s_1}$ intersecting H_{s_0} fully for all $s_0, s_1 \in S$ such that $(A)_{s_0 s_1} = 1$.

For Λ_2, we use the information obtained concerning Λ_1 at the previous step and appeal to A2 to conclude:

i) Λ_2 consists of a collection of disjoint μ_h-horizontal slabs $H_{s_0 s_1 s_2}$ intersecting H_{s_0} fully for all $s_0, s_1, s_2 \in S$ such that $(A)_{s_0 s_1}(A)_{s_1 s_2} = 1$.

ii) $d(H_{s_0 s_1 s_2}) \leq \nu_h\, d(H_{s_0 s_1})$.   (2.3.30)

iii) By definition of Λ_2 and Λ_1 we have

$$H_{s_0 s_1 s_2} \subset H_{s_0 s_1} . \qquad (2.3.31)$$

Continuing to argue in this manner, we determine the nature of Λ_n by using the information obtained concerning Λ_{n-1} at the $(n-1)^{\text{th}}$ step and appeal to A2 to conclude:

i) Λ_n consists of a collection of disjoint μ_h-horizontal slabs $H_{s_0 \cdots s_n}$ intersecting H_{s_0} fully for all $s_0, \cdots, s_n \in S$ such that $(A)_{s_0 s_1}(A)_{s_1 s_2} \cdots (A)_{s_{n-1} s_n} = 1$.

ii) $d(H_{s_0 s_1 \cdots s_n}) \leq \nu_h\, d(H_{s_0 s_1 \cdots s_{n-1}}) \leq \cdots \leq \nu_h^{n-1}\, d(H_{s_0 s_1})$.   (2.3.32)

iii) By definition of Λ_k, $k = 1, 2, \ldots, n$, we have

$$H_{s_0 \cdots s_n} \subset H_{s_0 \cdots s_{n-1}} \subset \cdots \subset H_{s_0 s_1} . \qquad (2.3.33)$$

As in the construction of $\Lambda_{-\infty}$, before proceeding to discuss the limit as $n \to \infty$ we make the important remark that at each stage of the construction process each μ_h-horizontal slab can be labelled uniquely by an admissible string of elements of S determined by the transition matrix A and having a length of one plus the number of the step. Furthermore, all possible admissible strings of the appropriate length are realized at each step due to the assumption A2.

Now in the limit as $n \to \infty$ we obtain the set

$$\Lambda_{+\infty} = \left\{ p \in D_H \mid f^i(p) \in H_{s_i} ; \quad s_i \in S , \quad i = 0, 1, 2, \ldots, n, \ldots \right\} , \qquad (2.3.34)$$

and we can immediately make the following conclusions concerning the nature of $\Lambda_{+\infty}$.

i) Each element of $\Lambda_{+\infty}$, $H_{s_0 \cdots s_n \cdots}$, $s_i \in S$, $i = 0, 1, \ldots,$, can be labelled by a unique infinite sequence of elements of S allowed by the transition matrix A. Furthermore, all possible such sequences are realized.

ii) Since $H_{s_0 \cdots s_n \cdots} = \bigcap\limits_{n=1}^{\infty} H_{s_0 \cdots s_n}$ where $H_{s_0 \cdots s_n}$ are μ_h-horizontal slabs intersecting H_{s_0} fully for all $s_0, \cdots, s_n \in S$ such that $(A)_{s_0 s_1} \cdots (A)_{s_{n-1} s_n} = 1$
and $H_{s_0 \cdots s_{n+1}} \subset H_{s_0 \cdots s_n}$ with $d(H_{s_0 \cdots s_n}) \to 0$ as $n \to \infty$, by Lemma
2.3.1 we can conclude that $\Lambda_{+\infty}$ consists of a set of μ_h-horizontal slices,
$H_{s_0 \cdots s_n \cdots}$, with $\partial H_{s_0 \cdots s_n \cdots} \subset \partial_v H_{s_0}$. The cardinality of this set is determined by the transition matrix A. In particular, if A is irreducible, $\Lambda_{+\infty}$
consists of an uncountable infinity of μ_h-horizontal slices.

1c) Construction of the Invariant Set $\Lambda \equiv \Lambda_{-\infty} \cap \Lambda_{+\infty}$.

From 1a) and 1b) we have seen that $\Lambda_{+\infty}$ consists of a collection of μ_v-
vertical slices $V_{s_0 \cdots s_{-n} \cdots}$ with $\partial V_{s_0 \cdots s_n \cdots} \subset \partial_h H_{s_0}$, and $\Lambda_{-\infty}$ consists of a
collection of μ_h-horizontal slices $H_{s_0 \cdots s_n \cdots}$ with $\partial H_{s_0 \cdots s_n \cdots} \subset \partial_v H_{s_0}$ where
the subscripts on the slices are infinite sequences of elements of S admitted by the
transition matrix A. Thus, by Lemma 2.3.2, we see that $\Lambda \equiv \Lambda_{-\infty} \cap \Lambda_{+\infty}$ consists
of a set of points corresponding to the intersection of the $V_{s_0 \cdots s_{-n} \cdots}$ and the
$H_{s_0 \cdots s_n \cdots}$. The cardinality of Λ (as well as other properties) depends on the nature
of the transition matrix A. In particular, if A is irreducible, then Λ consists of an
uncountable infinity of points.

2) Definition of the map $\phi: \Lambda \to \Sigma_A^N$.

For any point $p \in \Lambda$ we have

$$p = V_{s_0 \cdots s_{-n} \cdots} \cap H_{s_0 \cdots s_n \cdots} \tag{2.3.35}$$

where $V_{s_0 \cdots s_{-n} \cdots}$ is a μ_v-vertical slice with $\partial V_{s_0 \cdots s_{-n} \cdots} \subset \partial_h H_{s_0}$ defined by

$$V_{s_0 \cdots s_{-n} \cdots} = \{ p \in D_H \mid f^{-i}(p) \in H_{s_{-i}} \,; \quad s_i \in S, \quad i = 0, 1, 2, \ldots \}, \tag{2.3.36}$$

and $H_{s_0 \cdots s_n \cdots}$ is a μ_h-horizontal slice with $\partial H_{s_0 \cdots s_n \cdots} \subset \partial_v H_{s_0}$ defined by

$$H_{s_0 \cdots s_n \cdots} = \{ p \in D_H \mid f^{i}(p) \in H_{s_i} \,; \quad s_i \in S, \quad i = 0, 1, 2, \ldots \}, \tag{2.3.37}$$

and the infinite sequences subscripting $V_{s_0 \cdots s_{-n} \cdots}$ and $H_{s_0 \cdots s_n \cdots}$ satisfy

$$(A)_{s_i s_{i+1}} = 1, \quad \text{for all } i. \tag{2.3.38}$$

Thus, we define a map from Λ into Σ_A^N as follows:

$$\phi: \Lambda \to \Sigma_A^N$$
$$p \mapsto s = \{ \cdots s_{-n} \cdots s_{-1} . s_0 \cdots s_n \cdots \} \tag{2.3.39}$$

where the infinite sequences $s_0 \cdots s_n \cdots$ and $s_{-1} \cdots s_{-n} \cdots$ are obtained from the respective μ_h-horizontal and μ_v-vertical slices whose intersection is p; thus $s \in \Sigma_A^N$. This map is well defined since the H_i are disjoint.

From the definition of $V_{s_0 \cdots s_{-n} \cdots}$ and $H_{s_0 \cdots s_n \cdots}$ we see that the bi-infinite sequence associated to p contains a considerable amount of information concerning the dynamics of p under iteration by f. In particular, if the ith entry of s is s_i then we know that $f^i(p) \in H_{s_i}$.

3) Prove that ϕ is a homeomorphism.

The proof of this is virtually identical to the proof of the similar assertion given in our discussion of the Smale horseshoe map in Section 2.1.

4) Prove that $\phi \circ f = \sigma \circ \phi$.

This is an immediate consequence of the definition of ϕ. Let $p \in \Lambda$ with $\phi(p) = \{\cdots s_{-n} \cdots s_{-1}.s_0 s_1 \cdots s_n \cdots\}$. By the construction of Λ, p is the unique point in H such that $f^i(p) \in H_{s_i}$, $i = 0, \pm 1, \pm 2, \ldots$ and therefore $\phi \circ f(p) = \{\cdots s_{-n} \cdots s_{-1} s_0.s_1 \cdots s_n \cdots\}$. But, by the definition of the shift map σ, we have $\sigma \circ \phi(p) = \{\cdots s_{-n} \cdots s_{-1} s_0.s_1 \cdots s_n \cdots\}$. Thus, we have established that $\phi \circ f(p) = \sigma \circ \phi(p)$ for any $p \in \Lambda$. $\square$

Let us make the following remarks concerning Theorem 2.3.3.

1) Recall the consequences of $f|_\Lambda$ being topologically conjugate to $\sigma|_{\Sigma_A^N}$ (see Section 2.2). In particular, if A is irreducible, Λ is a Cantor set of points and $f|_\Lambda$ exhibits the same rich dynamics as $\sigma|_{\Sigma_A^N}$ described in Section 2.2.

2) It is important to note that, although Theorem 2.3.3 describes a very rich dynamical structure for f, it by no means tells the complete story of the dynamics of f. Many important global questions remain involving the dynamics in a neighborhood of Λ (see the comments at the end of Section 2.1).

3) An obvious question is what does one look for in the phase space of a map in order to show that A1 and A2 hold for that map? One might presume that a nontrivial amount of knowledge concerning the dynamics of the map is needed. In Chapter 3 we will see that special types of orbits, called homoclinic and heteroclinic, often give rise to conditions for which A1 and A2 hold. Regarding the condition A2, for theoretical purposes (e.g., as in proving Theorem 2.3.3) our form of the statement of the condition is often the easiest to utilize. However, for computational purposes, A2 can be difficult to implement, therefore

we next address the question of devising an alternative condition to A2 which is more computationally oriented.

2.3c. Sector Bundles

The condition A2 gives uniform estimates on the contraction of the widths of μ_h-horizontal slabs under f^{-1} and the contraction of the widths of μ_v-vertical slabs under f. Typically, when one thinks of expansion or contraction properties one thinks of the behavior of the jacobian of the map, its eigenvalues and eigenvectors, and how they vary over the region in question.

Recall the geometrical point of view described in Section 1.3 of the derivative of the map at a point acting on tangent vectors emanating from that point. We will quantify the expansion and contraction properties of f by describing how it acts on tangent vectors which are aligned in certain directions or *sectors*, and we now begin the development of these ideas. First, however, we need an additional hypothesis on f.

Recall that in A1 or A2 no differentiability properties were stated for f, and they are obviously needed if we are to consider the derivative of f.

Hypothesis: Let $\mathcal{H} = \bigcup_{\substack{i,j \in S \\ (A)_{ij}=1}} H_{ij}$ and $\mathcal{V} = \bigcup_{\substack{i,j \in S \\ (A)_{ij}=1}} V_{ji}$, then f is C^1 on $\mathcal{H}$ and f^{-1} is C^1 on $\mathcal{V}$.

Note that $f(\mathcal{H}) = \mathcal{V}$.

Now choose a point $z_0 = (x_0, y_0) \in \mathcal{V} \cup \mathcal{H}$. The stable sector at z_0, denoted $S_{z_0}^s$, is defined as follows:

$$S_{z_0}^s = \{(\xi_{z_0}, \eta_{z_0}) \in \mathbf{R}^n \times \mathbf{R}^m \mid |\eta_{z_0}| \leq \mu_h |\xi_{z_0}|\} . \tag{2.3.40}$$

The unstable sector at z_0, denoted $S_{z_0}^u$ is defined as follows:

$$S_{z_0}^u = \{(\xi_{z_0}, \eta_{z_0}) \in \mathbf{R}^n \times \mathbf{R}^m \mid |\xi_{z_0}| \leq \mu_v |\eta_{z_0}|\} . \tag{2.3.41}$$

Geometrically, we think of (ξ_{z_0}, η_{z_0}) as a vector emanating from the point z_0. Thus, $S_{z_0}^s$ and $S_{z_0}^u$ appear geometrically as cones with vertex at z_0, see Figure 2.3.8.

For $z_0 \in \mathcal{H}$ we have $f(z_0) \in \mathcal{V}$. Now recall from Section 1.3 that $Df(z_0)$ maps tangent vectors at z_0 to tangent vectors at $f(z_0)$. Applying this to all vectors

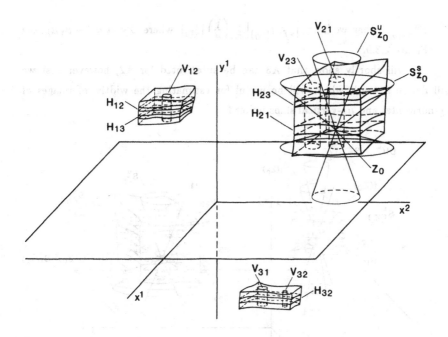

Figure 2.3.8. Stable and Unstable Sectors at z_0.

in $S_{z_0}^u$, we see that $Df(z_0)$ maps the cone $S_{z_0}^u$ to some cone with vertex at $f(z_0)$. Similarly, $Df^{-1}(z_0)$ maps the cone $S_{z_0}^s$ to some cone with vertex at $f^{-1}(z_0)$. We will be interested in the behavior of Df and Df^{-1} on the cones $S_{z_0}^u$ and $S_{z_0}^s$, respectively, as z_0 varies over the regions $\mathcal{H}$ and $\mathcal{V}$, respectively. We define the following sets:

$$S_{\mathcal{H}}^s = \bigcup_{z_0 \in \mathcal{H}} S_{z_0}^s$$

$$S_{\mathcal{V}}^s = \bigcup_{z_0 \in \mathcal{V}} S_{z_0}^s$$

$$S_{\mathcal{H}}^u = \bigcup_{z_0 \in \mathcal{H}} S_{z_0}^u \qquad (2.3.42)$$

$$S_{\mathcal{V}}^u = \bigcup_{z_0 \in \mathcal{V}} S_{z_0}^u .$$

These sets are called *sector bundles* or *cone fields*. We have the following hypothesis:

A3. $Df(S_{\mathcal{H}}^u) \subset S_{\mathcal{V}}^u$ and $Df^{-1}(S_{\mathcal{V}}^s) \subset S_{\mathcal{H}}^s$. Moreover, if $(\xi_{z_0}, \eta_{z_0}) \in S_{z_0}^u$ and $Df(z_0)(\xi_{z_0}, \eta_{z_0}) \equiv (\xi_{f(z_0)}, \eta_{f(z_0)}) \in S_{f(z_0)}^u$, then we have $|\eta_{f(z_0)}| \geq \left(\frac{1}{\mu}\right) |\eta_{z_0}|$. Similarly, if $(\xi_{z_0}, \eta_{z_0}) \in S_{z_0}^s$ and $Df^{-1}(z_0)(\xi_{z_0}, \eta_{z_0}) \equiv (\xi_{f^{-1}(z_0)}, \eta_{f^{-1}(z_0)}) \in$

$S^s_{f^{-1}(z_0)}$, then we have $\left|\xi_{f^{-1}(z_0)}\right| \ge \left(\frac{1}{\mu}\right)\left|\xi_{z_0}\right|$ where $0 < \mu < 1 - \mu_v\mu_h$, see Figure 2.3.9.

We will shortly show that A3 can be substituted for A2, however first we will derive a result which will be useful for estimating the widths of images of μ_h-horizontal and μ_v-vertical slabs under f.

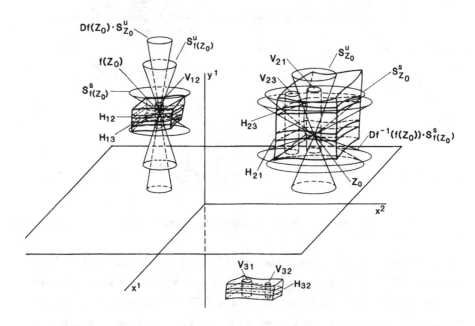

Figure 2.3.9. Images of Sectors under Df and Df^{-1}.

Let H be a μ_h-horizontal slab. Let $\bar{H}_1$ and $\bar{H}_2$ be disjoint μ_h-horizontal slices contained in H with $\partial\bar{H}_1$ and $\partial\bar{H}_2$ contained in $\partial_v H$. We denote the domain of the functions $h_1(x)$ and $h_2(x)$ of which $\bar{H}_1$ and $\bar{H}_2$ are the graphs by I_x. Let $\bar{V}_1$ and $\bar{V}_2$ be disjoint μ_v-vertical slices contained in H with $\partial\bar{V}_1$ and $\partial\bar{V}_2$ contained in $\partial_h H$. We denote the domains of the functions $v_1(y)$ and $v_2(y)$ of which $\bar{V}_1$ and $\bar{V}_2$ are the graphs by I_y^1 and I_y^2, respectively. Let

$$\|h_1 - h_2\| \equiv \sup_{x \in I_x} |h_1(x) - h_2(x)| , \tag{2.3.43}$$

$$\|v_1 - v_2\| \equiv \sup_{y \in I_y^1 \cap I_y^2} |v_1(y) - v_2(y)| . \tag{2.3.44}$$

By Lemma 2.3.2 , $\bar{H}_1$ and $\bar{V}_1$ intersect in a unique point which we call $z_1 \equiv (x_1, y_1)$, and $\bar{H}_2$ and $\bar{V}_2$ intersect in a unique point which we call $z_2 \equiv (x_2, y_2)$. See Figure 2.3.10 for an illustration of the geometry.

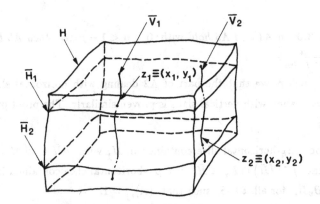

Figure 2.3.10. Intersecting Horizontal and Vertical Slices.

We have the following lemma.

Lemma 2.3.4.

1) $$|x_1 - x_2| \le \frac{1}{1 - \mu_v \mu_h} \left[\|v_1 - v_2\| + \mu_v \|h_1 - h_2\| \right] .$$

2) $$|y_1 - y_2| \le \frac{1}{1 - \mu_v \mu_h} \left[\|h_1 - h_2\| + \mu_h \|v_1 - v_2\| \right] .$$

PROOF: We have

$$|x_1 - x_2| = |v_1(y_1) - v_2(y_2)| \le |v_1(y_1) - v_1(y_2)| + |v_1(y_2) - v_2(y_2)| . \quad (2.3.45)$$

Using (2.3.4) and (2.3.44), (2.3.45) becomes

$$|x_1 - x_2| \le \mu_v |y_1 - y_2| + \|v_1 - v_2\| \quad (2.3.46)$$

We also have

$$|y_1 - y_2| = |h_1(x_1) - h_2(x_2)| \le |h_1(x_1) - h_1(x_2)| + |h_1(x_2) - h_2(x_2)| . \quad (2.3.47)$$

Using (2.3.3) and (2.3.43), (2.3.47) becomes

$$|y_1 - y_2| \le \mu_h |x_1 - x_2| + \|h_1 - h_2\| \tag{2.3.48}$$

Substituting (2.3.48) into (2.3.46) gives 1), and substituting (2.3.46) into (2.3.48) gives 2). $\qquad\square$

Theorem 2.3.5. *If A1 and A3 hold with* $0 < \mu < 1 - \mu_h \mu_v$ *then A2 holds with*
$$\nu_h = \nu_v = \frac{\mu}{1 - \mu_v \mu_h}.$$

PROOF: We will prove that the part of A2 dealing with horizontal slabs holds, since the part dealing with vertical slabs is proven similarly. The proof proceeds in several steps:

1) Let $\bar{H}$ be a μ_h-horizontal slice contained in H_j with $\partial \bar{H} \subset \partial_v H_j$. Then we show that $f^{-1}\left(\bar{H}\right) \cap H_i \equiv \tilde{H}_i$ is a μ_h-horizontal slice contained in H_i with $\partial \tilde{H}_i \subset \partial_v H_i$ for all $i \in S$ such that $(A)_{ij} = 1$.

2) Let H be a μ_h-horizontal slab which intersects H_j fully. Then we use 1) to show that $f^{-1}(H) \cap H_i \equiv \tilde{H}_i$ is a μ_h-horizontal slab intersecting H_i fully for all $i \in S$ such that $(A)_{ij} = 1$.

3) Show that $d\left(\tilde{H}_i\right) \le \frac{\mu}{1 - \mu_v \mu_h} d(H)$.

We begin with step 1). Let $\bar{H}$ be a μ_h-horizontal slice contained in H_j with $\partial \bar{H} \subset \partial_v H_j$. We denote the region in the $y = 0$ plane over which H_j is defined by I_x^j, i.e., I_x^j is the domain of the function $h(x)$ whose graph is $\bar{H}$.

Since $\partial \bar{H} \subset \partial_v H_j$, by Lemma 2.3.2 we know that $\bar{H}$ intersects V_{ji} with $\partial \left(\bar{H} \cap V_{ji}\right) \subset \partial_v V_{ji}$ for all $i \in S$ such that $(A)_{ij} = 1$. Now A1 holds so that $f^{-1}\left(\partial_v V_{ji}\right) \subset \partial_v H_i$; therefore, $f^{-1}\left(\partial \left(\bar{H} \cap V_{ji}\right)\right) \subset \partial_v H_i$. So $f^{-1}\left(\bar{H} \cap V_{ji}\right)$ consists of a collection of n dimensional sets with $\partial \left(f^{-1}\left(\bar{H} \cap V_{ji}\right)\right) \subset \partial_v H_i$, see Figure 2.3.11.

We now show that $f^{-1}\left(\bar{H} \cap V_{ji}\right)$ are μ_h-horizontal slices. By A3, since Df^{-1} maps S_y^s into S_x^s, it follows from the mean value theorem that for any two points $(x_1, y_1), (x_2, y_2) \in f^{-1}\left(\bar{H} \cap V_{ji}\right)$ we have

$$|y_1 - y_2| \le \mu_h |x_1 - x_2| . \tag{2.3.49}$$

This shows that $f^{-1}\left(\bar{H} \cap V_{ji}\right)$ is the graph of some function $\tilde{h}(x)$ defined over I_x^i and satisfying

$$\left|\tilde{h}(x_1) - \tilde{h}(x_2)\right| \le \mu_h |x_1 - x_2| . \tag{2.3.50}$$

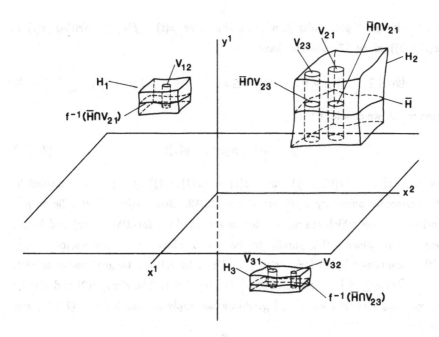

Figure 2.3.11. Image of $\bar{H}$ under f^{-1}.

Step 2). Let H be a μ_h-horizontal slab intersecting H_j fully. Therefore,

$$\partial_v \left(H \cap V_{ji} \right) \subset \partial_v V_{ji}$$

for all $i \in S$ such that $(A)_{ij} = 1$. Now applying the result of step 1) to the horizontal boundary of $H \cap V_{ji}$, we see that the $f^{-1} \left(H \cap V_{ji} \right) \equiv \tilde{H}_i$ are disjoint μ_h-horizontal slabs intersecting H_i fully for each $i \in S$ such that $(A)_{ij} = 1$.

Step 3). We now show that $d \left(\tilde{H}_i \right) \le \frac{\mu}{1 - \mu_v \mu_h} d(H)$. Let H be a μ_h-horizontal slab intersecting H_j fully, and let $f^{-1} \left(H \cap V_{ji} \right) \equiv \tilde{H}_i$. Fix i and let $p_0 = (x_0, y_0)$ and $p_1 = (x_1, y_1)$ be two points on the horizontal boundary of $\tilde{H}_i$ which have the *same x coordinate*, i.e., $x_0 = x_1$, such that

$$d \left(\tilde{H}_i \right) = |p_0 - p_1| = |y_0 - y_1| . \tag{2.3.51}$$

See Figure 2.3.12.

Consider the line

$$p(t) = (1 - t)p_0 + tp_1 , \qquad 0 \le t \le 1 \tag{2.3.52}$$

and the image of $p(t)$ under f, which is the curve $z(t) = f(p(t))$. Writing $z(t) = (x(t), y(t))$, since A3 holds, we have

$$|\dot{y}(t)| \geq \frac{1}{\mu}|\dot{p}(t)| > 0, \qquad 0 \leq t \leq 1, \qquad 0 < \mu < 1 - \mu_v \mu_h, \qquad (2.3.53)$$

from which we conclude that

$$|p_0 - p_1| \leq \mu |y(0) - y(1)| . \qquad (2.3.54)$$

By A1, $z(0) = (x(0), y(0))$ and $z(1) = (x(1), y(1))$ are points contained in the horizontal boundary of H, see Figure 2.3.12. Hence $z(0)$ and $z(1)$ lie on μ_h-horizontal slices which are represented as graphs of the functions $h_0(x)$ and $h_1(x)$, respectively. Since $p(t)$ is parallel to the $x = 0$ plane the tangent vector to $p(t)$, $\dot{p}(t)$, is contained in $S_{\mathcal{X}}^u$ for $0 \leq t \leq 1$. Then, by A3, the tangent vector to $z(t)$, $\dot{z}(t) = Df(p(t))\dot{p}(t)$, is contained in S_y^u for $0 \leq t \leq 1$. Therefore, $z(0)$ and $z(1)$ lie on some μ_v-vertical slice $\bar{V} \subset V_{ji}$. So we can apply Lemma 2.3.4 to (2.3.54) and obtain

$$\begin{aligned} |p_0 - p_1| &\leq \frac{\mu}{1 - \mu_v \mu_h} \|h_0 - h_1\| \\ &\leq \frac{\mu}{1 - \mu_v \mu_h} d(H) . \end{aligned} \qquad (2.3.55)$$

Since $|p_0 - p_1| = d(\tilde{H})$, and this argument holds for all $i \in S$ such that $(A)_{ij} = 1$, this completes the proof. $\qquad\square$

2.3d. Alternate Conditions for Verifying A1 and A2

We now want to give a set of alternate conditions which imply A1 and A2 of Theorem 2.3.3. We will see in Chapter 3 that these conditions are often more easily applied near orbits homoclinic to fixed points of ordinary differential equations than A1, A2, and A3. These conditions are actually higher dimensional generalizations of those given by Afraimovich, Bykov and Silnikov [1984], which they obtained during their study of the Lorenz equations.

First we will need to include a differentiability hypothesis on f.

Hypothesis: f is a C^1 diffeomorphism of D_H onto $f(D_H)$.

As notation, for $(x, y) \in D$ we let

$$f(x, y) = (f_1(x, y), f_2(x, y)) \equiv (\bar{x}, \bar{y}) \qquad (2.3.56)$$

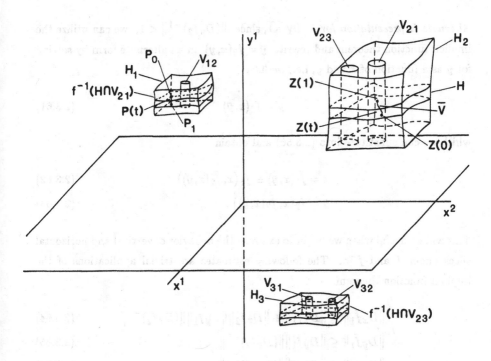

Figure 2.3.12. Width of H under f^{-1}.

and our first condition for f is as follows.

$\overline{\text{A1}}.$

$$\|D_x f_1\| < 1 \tag{2.3.57}$$

$$\|(D_y f_2)^{-1}\| < 1 \tag{2.3.58}$$

$$1 - \|(D_y f_2)^{-1}\|\|D_x f_1\| > 2\sqrt{\|D_y f_1\|\|D_x f_2\|\|(D_y f_2)^{-1}\|^2} \tag{2.3.59}$$

$$1 - (\|D_x f_1\| + \|(D_y f_2)^{-1}\|) + \|D_x f_1\|\|(D_y f_2)^{-1}\| > \|D_x f_2\|\|D_y f_1\|\|(D_y f_2)^{-1}\| \tag{2.3.60}$$

where $\|\cdot\| \equiv \sup\limits_{(x,y)\in D_H} |\cdot|$ and $|\cdot|$ is some matrix norm. $\overline{\text{A1}}$ is the condition dealing with stretching and contraction rates and, in the applications which we address in Chapter 3, they will be readily verifiable.

We next need to introduce some condition governing the behavior of horizontal and vertical slabs and their respective boundaries under f. However, first we will elaborate on some implications of $\overline{\text{A1}}$ which will be needed later on.

Alternate Representation for f. By $\overline{A1}$, since $\left\| (D_y f_2)^{-1} \right\| < 1$, we can utilize the implicit function theorem and rewrite $\bar{y} = f_2(x, y)$ in an alternate form by solving for y as a function of x and $\bar{y}$, i.e., we have

$$y = \tilde{f}_2(x, \bar{y}) \tag{2.3.61}$$

which we can substitute into (2.3.56) and obtain

$$\bar{x} = \tilde{f}_1(x, \bar{y}) \equiv f_1\big(x, \tilde{f}_2(x, \bar{y})\big) \tag{2.3.62}$$

$$\bar{y} = f_2\big(x, \tilde{f}_2(x, \bar{y})\big). \tag{2.3.63}$$

This will be useful when we begin to examine the behavior of vertical and horizontal slices under f and f^{-1}. The following estimates are trivial applications of the implicit function theorem.

$$\left\| D_x \tilde{f}_1 \right\| \leq \left\| D_x f_1 \right\| + \left\| D_x f_2 \right\| \left\| D_y f_1 \right\| \left\| (D_y f_2)^{-1} \right\| \tag{2.3.64}$$

$$\left\| D_{\bar{y}} \tilde{f}_1 \right\| \leq \left\| D_y f_1 \right\| \left\| (D_y f_2)^{-1} \right\| \tag{2.3.65}$$

$$\left\| D_x \tilde{f}_2 \right\| \leq \left\| D_x f_2 \right\| \left\| (D_y f_2)^{-1} \right\| \tag{2.3.66}$$

$$\left\| D_{\bar{y}} \tilde{f}_2 \right\| \leq \left\| (D_y f_2)^{-1} \right\|. \tag{2.3.67}$$

We will now give a series of lemmas whose motivation at present is not obvious but will become more so later on.

Lemma 2.3.6. *Suppose $\overline{A1}$ holds. Then the inequalities*

$$\left\| D_y f_1 \right\| \left\| (D_y f_2)^{-1} \right\| \mu_h^2 - \left(1 - \left\| D_x f_1 \right\| \left\| (D_y f_2)^{-1} \right\|\right) \mu_h + \left\| D_x f_2 \right\| \left\| (D_y f_2)^{-1} \right\| < 0 \tag{2.3.68}$$

$$\left\| D_x f_2 \right\| \left\| (D_y f_2)^{-1} \right\| \mu_v^2 - \left(1 - \left\| D_x f_1 \right\| \left\| (D_y f_2)^{-1} \right\|\right) \mu_v + \left\| D_y f_1 \right\| \left\| (D_y f_2)^{-1} \right\| < 0 \tag{2.3.69}$$

have positive solutions. Moreover, these solutions lie in the intervals $0 < \mu_h^{\min} < \mu_h < \mu_h^{\max}$ and $0 < \mu_v^{\min} < \mu_v < \mu_v^{\max}$, respectively, where, setting $a = \left\| D_x f_1 \right\| \left\| (D_y f_2)^{-1} \right\|$,

$$\mu_h^{\genfrac{}{}{0pt}{}{\max}{\min}} = \frac{1 - a \pm \sqrt{(1-a)^2 - 4 \left\| D_y f_1 \right\| \left\| D_x f_2 \right\| \left\| (D_y f_2)^{-1} \right\|^2}}{2 \left\| D_y f_1 \right\| \left\| (D_y f_2)^{-1} \right\|} \tag{2.3.70}$$

$$\mu_v^{\max}_{\min} = \frac{1 - a \pm \sqrt{(1-a)^2 - 4\|D_y f_1\|\|D_x f_2\|\|(D_y f_2)^{-1}\|^2}}{2\|D_x f_2\|\|(D_y f_2)^{-1}\|}. \qquad (2.3.71)$$

PROOF: This is a trivial calculation noting that by $\overline{A1}$ we have

$$1 - \|D_x f_1\|\|(D_y f_2)^{-1}\| > 0$$

$$(1 - \|D_x f_1\|\|(D_y f_2)^{-1}\|)^2 - 4\|D_y f_1\|\|D_x f_2\|\|(D_y f_2)^{-1}\|^2 > 0.$$

$\square$

Lemma 2.3.7. *Suppose $\overline{A1}$ holds, and let $\mu_h > 0$ satisfy (2.3.68) and $\mu_v > 0$ satisfy (2.3.69). Then it follows that*

$$\mu_h < 1/(\|D_y f_1\|\|(D_y f_2)^{-1}\|) \qquad (2.3.72)$$

$$\mu_v < 1/(\|D_x f_2\|\|(D_y f_2)^{-1}\|). \qquad (2.3.73)$$

PROOF: Equation (2.3.72) follows by examining $\mu_h^{\max}$ and noting that by $\overline{A1}$ the numerator is smaller than 2. A similar argument applied to $\mu_v^{\max}$ gives (2.3.73). $\square$

Lemma 2.3.8. *Suppose $\overline{A1}$ holds. Then there exists μ_h satisfying (2.3.68) and μ_v satisfying (2.3.69) such that*

$$\mu_h < \frac{1 - \|(D_y f_2)^{-1}\|}{\|D_y f_1\|\|(D_y f_2)^{-1}\|} \qquad (2.3.74)$$

$$\mu_v < \frac{1 - \|D_x f_1\|}{\|D_x f_2\|}. \qquad (2.3.75)$$

PROOF: We give the proof for (2.3.74) with the proof for (2.3.75) following a similar line of reasoning.

By Lemma 2.3.6, for (2.3.74) to hold, it is sufficient that (once again setting $a = \|D_x f_1\|\|(D_y f_2)^{-1}\|$)

$$\frac{1 - a - \sqrt{(1-a)^2 - 4\|D_y f_1\|\|D_x f_2\|\|(D_y f_2)^{-1}\|^2}}{2\|D_y f_1\|\|(D_y f_2)^{-1}\|} < \frac{1 - \|(D_y f_2)^{-1}\|}{\|D_y f_1\|\|(D_y f_2)^{-1}\|} \qquad (2.3.76)$$

or

$$\sqrt{(1-a)^2 - 4\|D_y f_1\|\|D_x f_2\|\|(D_y f_2)^{-1}\|^2} > 2\|(D_y f_2)^{-1}\| - a - 1. \qquad (2.3.77)$$

If the right hand side of (2.3.77) is negative, then we are finished. If it is positive, then squaring both sides and subtracting away similar terms leads to (2.3.60).   □

Lemma 2.3.9. *Suppose* $\overline{A1}$ *holds with* μ_h *satisfying (2.3.68). Then there exists* μ_v *satisfying (2.3.69) such that*

$$0 \le \mu_v \mu_h < 1. \tag{2.3.78}$$

PROOF: From (2.3.70) and (2.3.71) it follows that

$$\mu_v^{\max} \mu_h^{\min} = \mu_v^{\min} \mu_h^{\max} = 1 \tag{2.3.79}$$

and from (2.3.79) the lemma follows.   □

We can now state the condition describing how f behaves on horizontal and vertical slabs.

$\overline{A2}$. H_i, $i = 1, \ldots, N$ are μ_h-horizontal slabs with μ_h satisfying (2.3.68) and (2.3.74). For all $i, j \in S$ such that $(A)_{ij} = 1$, V_{ji} is a μ_v-vertical slab with μ_v satisfying (2.3.69), (2.3.75), and (2.3.78). Moreover, we require $\partial_v V_{ji} \subset \partial f(H_i)$ and $f^{-1}(\partial_v V_{ji}) \subset \partial_v H_i$.

Our goal will now be to show that $\overline{A1}$ and $\overline{A2}$ imply A1 and A2 and hence Theorem 2.3.3. However, first we will need two preliminary lemmas.

Lemma 2.3.10. *Suppose* $\overline{A1}$ *and* $\overline{A2}$ *hold, and let* $\bar{H}$ *be a* μ_h-horizontal slice *contained in* H_j *with* $\partial \bar{H} \subset \partial_v H_j$ *with* μ_h *satisfying (2.3.68). Then* $f^{-1}(\bar{H}) \cap H_i$ *is also a* μ_h-horizontal slice *contained in* H_i *with* $\partial(f^{-1}(\bar{H}) \cap H_i) \subset \partial_v H_i$ *for all* $i \in S$ *such that* $(A)_{ij} = 1$ *and with* μ_h *satisfying (2.3.68).*

PROOF: The proof is accomplished in several steps:

1) Describe $f^{-1}(\bar{H}) \cap H_i$.
2) Show that for each $i \in S$ such that $(A)_{ij} = 1$ $f^{-1}(\bar{H}) \cap H_i$ is the graph of a function of x.
3) Show that $f^{-1}(\bar{H}) \cap H_i$ is μ_h-horizontal and that μ_h satisfies (2.3.68).

We begin with Step 1). Note that $f^{-1}(\bar{H}) \cap H_i = f^{-1}(\bar{H} \cap V_{ji})$. Now, since $\partial \bar{H} \subset \partial_v H_j$, we know that $\bar{H}$ intersects V_{ji} with $\partial(\bar{H} \cap V_{ji}) \subset \partial_v V_{ji}$ for all $i \in S$ such that $(A)_{ij} = 1$. Now $\overline{A2}$ holds so that $f^{-1}(\partial_v V_{ji}) \subset \partial_v H_i$; therefore,

$f^{-1}(\partial(\bar{H} \cap V_{ji})) \subset \partial_v H_i$. So $f^{-1}(\bar{H} \cap V_{ji})$ consists of a collection of n dimensional sets with $\partial(f^{-1}(\bar{H} \cap V_{ji})) \subset \partial_v H_i$, see Figure 2.3.11.

Step 2). Fix any $i \in S$ such that $(A)_{ij} = 1$ and consider $\bar{H} \cap V_{ji}$. Now, since $\bar{H}$ is a μ_h-horizontal slice, any two points $(\bar{x}_1, \bar{y}_1)$, $(\bar{x}_2, \bar{y}_2) \in \bar{H} \cap V_{ji}$ satisfy

$$|\bar{y}_1 - \bar{y}_2| \leq \mu_h |\bar{x}_1 - \bar{x}_2| . \tag{2.3.80}$$

Therefore, $\bar{H} \cap V_{ji}$ is the graph of a Lipschitz function $\bar{y} = H(\bar{x})$ with Lipschitz constant μ_h.

Now consider $f^{-1}(\bar{H} \cap V_{ji})$. We want to show that this set is the graph of a function. Now any point $(x, y) \in f^{-1}(\bar{H} \cap V_{ji})$ must satisfy

$$\bar{y} = f_2(x, y) = H(f_1(x, y)) = H(\bar{x}) . \tag{2.3.81}$$

If we denote $X = \{x \in \mathbb{R}^n \mid \exists\, y \in \mathbb{R}^m$ with $(x, y) \in f^{-1}(\bar{H} \cap V_{ji})\}$, then (2.3.81) has at least one solution for each $x \in X$. We now show that this solution is unique. Recall that since $\overline{A1}$ holds, then by the implicit function theorem an alternate expression for the map $f = (f_1, f_2)$ is given by

$$\bar{x} = \tilde{f}_1(x, \bar{y}) \tag{2.3.82}$$
$$y = \tilde{f}_2(x, \bar{y}) . \tag{2.3.83}$$

Now, also by the implicit function theorem, (2.3.82) defines $\bar{y}$ as a function of x under the condition

$$\mu_h \|D_y \tilde{f}_1\| < 1 \tag{2.3.84}$$

or, using (2.3.65), (2.3.84) becomes

$$\mu_h < 1/\|D_y f_1\| \|(D_y f_2)^{-1}\| \tag{2.3.85}$$

which follows from Lemma 2.3.7. Thus, when (2.3.85) holds we have

$$\bar{y} = \bar{y}(x) \tag{2.3.86}$$

and substituting (2.3.86) into (2.3.83) gives

$$y = \tilde{f}_2(x, \bar{y}(x)) \equiv h(x) . \tag{2.3.87}$$

Hence, we have shown that $f^{-1}(\bar{H} \cap V_{ji})$ is the graph of a function $y = h(x)$.

Step 3). We now show that $y = h(x)$ is Lipschitz with Lipschitz constant μ_h.

Let (x_1, y_1), $(x_2, y_2) \in f^{-1}(\bar{H} \cap V_{ji})$ with their respective images under f denoted by $(\bar{x}_1, \bar{y}_1)$, $(\bar{x}_2, \bar{y}_2) \in \bar{H} \cap V_{ji}$. Then the following relations hold.

$$\bar{y}_1 = H(\tilde{f}_1(x_1, \bar{y}_1)) \tag{2.3.88}$$

$$\bar{y}_2 = H(\tilde{f}_1(x_2, \bar{y}_2)) \tag{2.3.89}$$

$$y_1 = \tilde{f}_2(x_1, \bar{y}_1) \tag{2.3.90}$$

$$y_2 = \tilde{f}_2(x_2, \bar{y}_2) . \tag{2.3.91}$$

Using (2.3.88)–(2.3.91) along with (2.3.64)–(2.3.67) we obtain the following estimate

$$|\bar{y}_1 - \bar{y}_2| \leq \frac{\left(\|D_x f_1\| + \|D_x f_2\| \|D_y f_1\| \|(D_y f_2)^{-1}\|\right) \mu_h}{1 - \mu_h \|D_y f_1\| \|(D_y f_2)^{-1}\|} |x_1 - x_2| \tag{2.3.92}$$

which we use to obtain

$$|y_1 - y_2| = |h(x_1) - h(x_2)|$$
$$\leq \frac{\mu_h \|D_x f_1\| \|(D_y f_2)^{-1}\| + \|D_x f_2\| \|(D_y f_2)^{-1}\|}{1 - \mu_h \|D_y f_1\| \|(D_y f_2)^{-1}\|} |x_1 - x_2| . \tag{2.3.93}$$

(Note: positivity of $1 - \mu_h \|D_y f_1\| \|(D_y f_2)^{-1}\|$ follows from Lemma 2.3.7.) Thus, from (2.3.93) we see that $h(x)$ is Lipschitz with Lipschitz constant μ_h provided

$$\frac{\mu_h \|D_x f_1\| \|(D_y f_2)^{-1}\| + \|D_x f_2\| \|(D_y f_2)^{-1}\|}{1 - \mu_h \|D_y f_1\| \|(D_y f_2)^{-1}\|} < \mu_h , \tag{2.3.94}$$

that is, μ_h must satisfy

$$\|D_y f_1\| \|(D_y f_2)^{-1}\| \mu_h^2 - (1 - \|D_x f_1\| \|(D_y f_2)^{-1}\|) \mu_h + \|D_x f_2\| \|(D_y f_2)^{-1}\| < 0$$

which it does by hypothesis.   □

Lemma 2.3.11. *Suppose* $\overline{A1}$ *and* $\overline{A2}$ *hold, and let* $\bar{V}$ *be a* μ_v-*vertical slice contained in* H_j *with* $\partial \bar{V} \subset \partial_h H_j$ *and with* μ_v *satisfying (2.3.69). Then* $f(\bar{V}) \cap H_k$ *is also a* μ_v-*vertical slice contained in* H_k *with* $\partial(f(\bar{V}) \cap H_k) \subset \partial_h H_k$ *for all* $k \in S$ *such that* $(A)_{jk} = 1$ *and with* μ_v *satisfying (2.3.69).*

PROOF: The proof is very similar to the proof of Lemma 2.3.10 and proceeds in three steps.

1) Describe $f(\bar{V}) \cap H_k$.

2) Show that for each $k \in S$ such that $(A)_{jk} = 1$ $f(\bar{V}) \cap H_k$ is the graph of a function of y.

3) Show that $f(\bar{V}) \cap H_k$ is μ_v-vertical and that μ_v satisfies (2.3.69).

We begin with Step 1). Note that $f(\bar{V}) \cap H_k = f(\bar{V} \cap H_{jk})$. Now, since $\partial \bar{V} \subset \partial_h H_j$, we know that $\bar{V}$ intersects H_{jk} with $\partial(\bar{V} \cap H_{jk}) \subset \partial_h H_{jk}$ for all $k \in S$ such that $(A)_{jk} = 1$. Since $\overline{A2}$ holds, we know that $f(\partial_h H_{jk}) \subset \partial_h H_k$; therefore, $f(\partial(\bar{V} \cap H_{jk})) \subset \partial_h H_k$. So $f(\bar{V} \cap H_{jk})$ consists of a collection of m dimensional sets with $\partial(f(\bar{V} \cap H_{jk})) \subset \partial_h H_k$.

Step 2). Fix $k \in S$ such that $(A)_{jk} = 1$, and consider $\bar{V} \cap H_{jk}$. Since $\bar{V}$ is a μ_v-vertical slice any two points $(x_1, y_1), (x_2, y_2) \in \bar{V} \cap H_{jk}$ satisfy

$$|x_1 - x_2| \leq \mu_v |y_1 - y_2| . \tag{2.3.95}$$

Therefore, $\bar{V} \cap H_{jk}$ is the graph of a Lipschitz function $x = v(y)$ with Lipschitz constant μ_v.

We now want to show that $f(\bar{V} \cap H_{jk})$ is the graph of a function. For $(v(y), y) \in \bar{V} \cap H_{jk}$ we have

$$\bar{x} = f_1(v(y), y) \tag{2.3.96}$$

$$\bar{y} = f_2(v(y), y) . \tag{2.3.97}$$

By the implicit function theorem, when

$$\|(D_y f_2)^{-1}\| \|D_x f_2\| \mu_v < 1 \tag{2.3.98}$$

we can solve equation (2.3.97) for y as a function of $\bar{y}$, i.e., $y = y(\bar{y})$ and (2.3.98) holds by Lemma 2.3.7. Thus, substituting $y = y(\bar{y})$ into (2.3.96) gives

$$\bar{x} = f_1\Big(v(y(\bar{y})), y(\bar{y})\Big) \equiv V(\bar{y}) . \tag{2.3.99}$$

Therefore, $f(\bar{V} \cap H_{jk})$ is the graph of $\bar{x} = V(\bar{y})$.

Step 3). We now show that $\bar{x} = V(\bar{y})$ is Lipschitz with Lipschitz constant μ_v.

Let (x_1, y_1), $(x_2, y_2) \in \bar{V} \cap H_{jk}$ and denote their respective images under f by $(\bar{x}_1, \bar{y}_1)$, $(\bar{x}_2, \bar{y}_2) \in f(\bar{V} \cap H_{jk})$. Then the following relations hold.

$$\bar{x}_1 = \tilde{f}_1(v(y_1), \bar{y}_1) \tag{2.3.100}$$

$$\bar{x}_2 = \tilde{f}_1(v(y_2), \bar{y}_2) \tag{2.3.101}$$

$$y_1 = \tilde{f}_2(v(y_1), \bar{y}_1) \tag{2.3.102}$$

$$y_2 = \tilde{f}_2(v(y_2), \bar{y}_2). \tag{2.3.103}$$

Using (2.3.100)–(2.3.103) along with (2.3.64)–(2.3.67) we obtain the following estimate

$$|y_1 - y_2| \leq \frac{\|(D_y f_2)^{-1}\|}{1 - \|D_x f_2\|\|(D_y f_2)^{-1}\|\mu_v} |\bar{y}_1 - \bar{y}_2| \tag{2.3.104}$$

which we use to obtain

$$|\bar{x}_1 - \bar{x}_2| = |V(\bar{y}_1) - V(\bar{y}_2)|$$

$$\leq \frac{\|D_y f_1\|\|(D_y f_2)^{-1}\| + \mu_v\|D_x f_1\|\|(D_y f_2)^{-1}\|}{1 - \|D_x f_2\|\|(D_y f_2)^{-1}\|\mu_v} |\bar{y}_1 - \bar{y}_2|. \tag{2.3.105}$$

So $\bar{x} = V(\bar{y})$ is Lipschitz with Lipschitz constant μ_v provided

$$\frac{\|D_y f_1\|\|(D_y f_2)^{-1}\| + \mu_v\|D_x f_1\|\|(D_y f_2)^{-1}\|}{1 - \|D_x f_2\|\|(D_y f_2)^{-1}\|\mu_v} < \mu_v \tag{2.3.106}$$

or

$$\|D_x f_2\|\|(D_y f_2)^{-1}\|\mu_v^2 - (1 - \|D_x f_1\|\|(D_y f_2)^{-1}\|)\mu_v + \|D_y f_1\|\|(D_y f_2)^{-1}\| < 0 \tag{2.3.107}$$

and (22.3.107) holds by hypothesis. □

We are now ready to prove our main theorem.

Theorem 2.3.12. $\overline{A1}$ and $\overline{A2}$ imply A1 and A2.

PROOF: That $\overline{A2}$ implies A1 is obvious; thus, we only need to show that $\overline{A1}$ and $\overline{A2}$ imply A2.

We begin with the part of A2 dealing with horizontal slabs. The proof consists of two steps.

1) Let H be a μ_h-horizontal slab which intersects H_j fully. Then we show that $f^{-1}(H) \cap H_i \equiv \tilde{H}_i$ is a μ_h-horizontal slab intersecting H_i fully for all $i \in S$ such that $(A)_{ij} = 1$.

2) Show that $d(\tilde{H}_i) \leq \nu_h d(H)$ with $0 < \nu_h < 1$.

We begin with Step 1). This follows immediately by applying Lemma 2.3.10 to the horizontal boundary of H.

Step 2). For fixed i let $p_1 = (x_1, y_1)$, $p_2 = (x_2, y_2)$ be two points on the horizontal boundary of $\tilde{H}_i$ with $\underline{x_1 = x_2}$ such that

$$d(\tilde{H}_i) = |p_1 - p_2| = |y_1 - y_2| . \tag{2.3.108}$$

We denote their respective images under f by $(\bar{x}_1, \bar{y}_1)$ and $(\bar{x}_2, \bar{y}_2)$. Now, by $\overline{A2}$, $(\bar{x}_1, \bar{y}_1)$, $(\bar{x}_2, \bar{y}_2) \in \partial_h H$; hence $\bar{y}_1 = H_1(\bar{x}_1)$ and $\bar{y}_2 = H_2(\bar{x}_2)$ where the graph of the functions $y = H_1(x)$ and $y = H_2(x)$ are μ_h-horizontal slices.

Now we have the relations

$$y_1 = \tilde{f}_2(x_1, \bar{y}_1) \tag{2.3.109}$$

$$y_2 = \tilde{f}_2(x_2, \bar{y}_2) \tag{2.3.110}$$

$$\bar{y}_1 = H_1(\bar{x}_1) = H_1\big(\tilde{f}_1(x_1, \bar{y}_1)\big) \tag{2.3.111}$$

$$\bar{y}_2 = H_2(\bar{x}_2) = H_2\big(\tilde{f}_1(x_2, \bar{y}_2)\big). \tag{2.3.112}$$

Subtracting (2.3.109) from (2.3.110) and using (2.3.67) along with the fact that $x_1 = x_2$ yields the estimate

$$|y_1 - y_2| \leq \big\|(D_y f_2)^{-1}\big\| \, |\bar{y}_1 - \bar{y}_2| . \tag{2.3.113}$$

Subtracting (2.3.111) from (2.3.112) and using (2.3.65) along with the fact that H_1 and H_2 are μ_h-horizontal and $x_1 = x_2$ yields the estimate

$$|\bar{y}_1 - \bar{y}_2| \leq \frac{1}{1 - \mu_h \|D_y f_1\| \|(D_y f_2)^{-1}\|} \, \|H_1 - H_2\| . \tag{2.3.114}$$

Combining (2.3.114), (2.3.113), and (2.3.108) yields

$$d(\tilde{H}_i) \leq \frac{\|(D_y f_2)^{-1}\|}{1 - \mu_h \|D_y f_1\| \|(D_y f_2)^{-1}\|} \|H_1 - H_2\| \leq \frac{\|(D_y f_2)^{-1}\|}{1 - \mu_h \|D_y f_1\| \|(D_y f_2)^{-1}\|} d(H) \tag{2.3.115}$$

and we have

$$\frac{\|(D_y f_2)^{-1}\|}{1 - \mu_h \|D_y f_1\| \|(D_y f_2)^{-1}\|} < 1 , \tag{2.3.116}$$

since μ_h satisfies (2.3.74).

Now the part of the proof dealing with vertical slabs proceeds along the same lines in two steps.

1) Let V be a μ_v-vertical slab contained in H_j such that also $V \subset V_{ji}$ for some $i, j \in S$ with $(A)_{ij} = 1$. Then we show that $f(V) \cap H_k \equiv \tilde{V}_k$ is a μ_v-vertical slab contained in H_k for all $k \in S$ such that $(A)_{jk} = 1$.
2) Show that $d(\tilde{V}_k) \leq \nu_v d(V)$ with $0 < \nu_v < 1$.

We begin with Step 1). This follows immediately by applying Lemma 2.3.11 to the vertical boundary of V.

Step 2). For fixed k let $\bar{p}_1 = (\bar{x}_1, \bar{y}_1)$, $\bar{p}_2 = (\bar{x}_2, \bar{y}_2)$ be two points on the vertical boundary of $\tilde{V}_k$ with $\bar{y}_1 = \bar{y}_2$ such that

$$d(\tilde{V}_k) = |\bar{p}_1 - \bar{p}_2| = |\bar{x}_1 - \bar{x}_2| . \tag{2.3.117}$$

We denote their respective images under f^{-1} by (x_1, y_1) and (x_2, y_2). Now, by $\overline{A2}$, $(x_1, y_1), (x_2, y_2) \in \partial_v V$; hence $x_1 = V_1(y_1)$ and $x_2 = V_2(y_2)$ where the graphs of the functions $x = V_1(y)$ and $x = V_2(y)$ are μ_v-vertical slices. The following relations hold:

$$\bar{x}_1 = \tilde{f}_1(V_1(y_1), \bar{y}_1) \tag{2.3.118}$$
$$\bar{x}_2 = \tilde{f}_1(V_2(y_2), \bar{y}_2) \tag{2.3.119}$$
$$y_1 = \tilde{f}_2(V_1(y_1), \bar{y}_1) \tag{2.3.120}$$
$$y_2 = \tilde{f}_2(V_2(y_2), \bar{y}_2) . \tag{2.3.121}$$

Subtracting (2.3.120) from (2.3.121) and using (2.3.66) along with the fact that $\bar{y}_1 = \bar{y}_2$ yields the estimate

$$|y_1 - y_2| \leq \frac{\|D_x f_2\| \|(D_y f_2)^{-1}\|}{1 - \|D_x f_2\| \|(D_y f_2)^{-1}\| \mu_v} \|V_1 - V_2\| . \tag{2.3.122}$$

Subtracting (2.3.118) from (2.3.119) and using (2.3.64) along with the fact that $\bar{y}_1 = \bar{y}_2$ yields the estimate

$$|\bar{x}_1 - \bar{x}_2| \leq (\|D_x f_1\| + \|D_x f_2\| \|D_y f_1\| \|(D_y f_2)^{-1}\|) (\mu_v |y_1 - y_2| + \|V_1 - V_2\|) . \tag{2.3.123}$$

Combining (2.3.123) and (2.3.122) gives

$$d(\tilde{V}_k) = |\bar{x}_1 - \bar{x}_2| \tag{2.3.124}$$

$$\leq \left(\frac{\|D_y f_1\| \|(D_y f_2)^{-1}\| + \|D_x f_1\| \|(D_y f_2)^{-1}\| \mu_v}{1 - \|D_x f_2\| \|(D_y f_2)^{-1}\| \mu_v} \|D_x f_2\| + \|D_x f_1\| \right) \|V_1 - V_2|$$

$$\leq \left(\mu_v \|D_x f_2\| + \|D_x f_1\| \right) \|V_1 - V_2\| \tag{2.3.125}$$

$$\leq \left(\mu_v \|D_x f_2\| + \|D_x f_1\| \right) d(V) \tag{2.3.126}$$

and $\mu_v \|D_x f_2\| + \|D_x f_1\| < 1$ by (2.3.75).   □

2.3e. Hyperbolic Sets

In Chapter 1 we introduced the idea of a hyperbolic fixed point of a map or flow and, more generally, the idea of a normally hyperbolic invariant manifold. We now want to show that the invariant set Λ constructed in Theorem 2.3.3 shares some properties similar to these invariant sets. We begin by giving the definition of a hyperbolic invariant set of a map.

Definition 2.3.8. Let $f: \mathbf{R}^n \to \mathbf{R}^n$ be a C^r $(r \geq 1)$ diffeomorphism, and let Λ be a closed set which is invariant under f. We say that Λ is a *hyperbolic set* if for each $p \in \Lambda$ there exists a splitting $\mathbf{R}^n = E_p^s \oplus E_p^u$ such that

1)
$$\begin{aligned} Df(p) \cdot E_p^s &= E_{f(p)}^s \\ Df(p) \cdot E_p^u &= E_{f(p)}^u \end{aligned} \tag{2.3.127}$$

2) There exist positive real numbers C and λ with $0 < \lambda < 1$ such that

$$\begin{aligned} \text{if } \xi_p \in E_p^s, \quad &\left| Df^n(p)\xi_p \right| \leq C\lambda^n |\xi_p| \\ \text{if } \eta_p \in E_p^u, \quad &\left| Df^{-n}(p)\eta_p \right| \leq C\lambda^n |\eta_p| \end{aligned} \tag{2.3.128}$$

3) E_p^s and E_p^u vary continuously with p.

Let us make the following remarks concerning Definition 2.3.8.

1) It should be clear from Definition 2.3.8 that the orbits of points in Λ have a well-defined asymptotic behavior. Specifically, orbits whose initial points are slightly displaced from p along directions in E_p^s converge exponentially to the

orbit of p as $n \to \infty$, and orbits whose initial points are slightly displaced from p along directions in E_p^u converge exponentially to the orbit of p as $n \to -\infty$.

2) It is important to note that the constant C does not depend on p; in this case, the set is sometimes said to be *uniformly hyperbolic*. Much of dynamical systems theory has been built around uniform hyperbolicity assumptions; however, recently Pesin [1976], [1977] has developed a theory of *nonuniform hyperbolicity* which is still waiting to be exploited for the purposes of applications.

Finally, we remark that in much of the literature of dynamical systems theory the constant C is taken to be one. This can be done by utilizing a special metric called an *adapted metric*. This trick is due to Mather, and a discussion of it can be found in Hirsch and Pugh [1970]. Although it is of tremendous theoretical use, because the main purpose of this book is to develop techniques which are applicable to specific dynamical systems arising in applications, we will not state the theorems in terms of an adapted metric, since computations with such a metric may be somewhat unwieldy.

3) Continuity of the splitting $\mathbb{R}^n = E_p^s \oplus E_p^u$ can be stated in several equivalent ways. One statement, sufficient for our purposes, is as follows: let p be fixed, and choose a set of basis vectors for E_p^s and E_p^u; then the splitting is said to be continuous if the basis vectors vary continuously with p. More discussion of this point can be found in Nitecki [1971] or Hirsch, Pugh, and Shub [1977].

Note that the idea of a hyperbolic invariant set of a map is developed in terms of the structure of the linearized map. We will see that this structure has implications for the nonlinear map. First we begin with some definitions. For any point $p \in \Lambda$, $\epsilon > 0$, the stable and unstable sets of p of size ϵ are defined as follows:

$$W_\epsilon^s(p) = \left\{ p' \in \Lambda \mid \left| f^n(p) - f^n(p') \right| \le \epsilon \quad \text{for } n \ge 0 \right\}$$
$$W_\epsilon^u(p) = \left\{ p' \in \Lambda \mid \left| f^{-n}(p) - f^{-n}(p') \right| \le \epsilon \quad \text{for } n \ge 0 \right\} . \tag{2.3.129}$$

Now from Theorem 1.3.7, we have seen that if p is a hyperbolic fixed point the following hold.

1) For ϵ sufficiently small, $W_\epsilon^s(p)$ is a C^r manifold tangent to E_p^s at p and having the same dimension as p. $W_\epsilon^s(p)$ is called the local stable manifold of p.

2) The stable manifold of p is defined as follows:

$$W^s(p) = \bigcup_{n=0}^{\infty} f^{-n}\left(W_\epsilon^s(p)\right) . \tag{2.3.130}$$

Similar statements hold for $W_\epsilon^u(p)$.

The invariant manifold theorem for hyperbolic invariant sets (see Hirsch, Pugh, and Shub [1977]) tells us that a similar structure holds for each point in Λ.

Theorem 2.3.13. *Let Λ be a hyperbolic invariant set of a C^r $(r \geq 1)$ diffeomorphism f. Then for $\epsilon > 0$ sufficiently small and for each point $p \in \Lambda$ the following hold.*

1) $W_\epsilon^s(p)$ *and* $W_\epsilon^u(p)$ *are* C^r *manifolds tangent to* E_p^s *and* E_p^u, *respectively, at p and having the same dimension as* E_p^s *and* E_p^u, *respectively.*

2) *There are constants* $C > 0$, $0 < \lambda < 1$, *such that, if* $p' \in W_\epsilon^s(p)$, *then*
   $|f^n(p) - f^n(p')| \leq C\lambda^n |p - p'|$ *for* $n \geq 0$ *and, if* $p' \in W_\epsilon^u(p)$, *then*
   $|f^{-n}(p) - f^{-n}(p')| \leq C\lambda^n |p - p'|$ *for* $n \geq 0$.

3)
$$f\left(W_\epsilon^s(p)\right) \subset W_\epsilon^s(f(p))$$
$$f^{-1}\left(W_\epsilon^u(p)\right) \subset W_\epsilon^u(f^{-1}(p)) . \tag{2.3.131}$$

4) $W_\epsilon^s(p)$ *and* $W_\epsilon^u(p)$ *vary continuously with p.*

PROOF: See Hirsch, Pugh, and Shub [1977].   □

We make the following remarks concerning Theorem 2.3.13.

1) The constants $C > 0$ and $0 < \lambda < 1$ do not necessarily need to be the same as those appearing in Definition 2.3.8.

2) What it means for $W_\epsilon^s(p)$ and $W_\epsilon^u(p)$ to vary continuously with p is best explained within the context of function space topologies, see Hirsch [1976] or Hirsch, Pugh, and Shub [1977].

3) For any point $p \in \Lambda$ the stable and unstable manifolds of p are respectively defined as follows:

$$W^s(p) = \bigcup_{n=0}^{\infty} f^{-n}\left(W_\epsilon^s(f^n(p))\right)$$
$$W^u(p) = \bigcup_{n=0}^{\infty} f^n\left(W_\epsilon^u(f^{-n}(p))\right) \tag{2.3.132}$$

Now, in practice, verifying the conditions of Definition 2.3.8 for an invariant set of a map is quite difficult. Fortunately, there is an equivalent formulation of hyperbolicity due to Newhouse and Palis [1973], which we now describe.

As above, let $f : \mathbb{R}^n \to \mathbb{R}^n$ be a C^r $(r \geq 1)$ diffeomorphism, and let Λ be a closed set which is invariant under f. Let $\mathbb{R}^n = E_p^s \oplus E_p^u$ be a splitting of $\mathbb{R}^n$ for $p \in \Lambda$, and let $\mu(p)$ be a positive real valued function defined on Λ. We define the $\mu(p)$ sector, denoted $S_{\mu(p)}$, as follows:

$$S_{\mu(p)} = \left\{ (\xi_p, \eta_p) \in E_p^s \oplus E_p^u \mid |\xi_p| \leq \mu(p) |\eta_p| \right\} \tag{2.3.133}$$

and we define the complementary sector, $S'_{\mu(p)}$, as follows:

$$S'_{\mu(p)} = \mathbb{R}^n - S_{\mu(p)} . \tag{2.3.134}$$

Then we have the following theorem.

Theorem 2.3.14. *Let* $f : \mathbb{R}^n \to \mathbb{R}^n$ *be a* C^r $(r \geq 1)$ *diffeomorphism, and let* $\Lambda \subset \mathbb{R}^n$ *be a closed set which is invariant under* f. *Then* Λ *is a hyperbolic invariant set if and only if there exists a splitting* $\mathbb{R}^n = E_p^s \oplus E_p^u$ *for each* $p \in \Lambda$, *an integer* $n > 0$, *constants* $C > 0$, $0 < \lambda < 1$ *with* $C\lambda^n < 1$, *and a real valued function* $\mu : \Lambda \to \mathbb{R}^+$ *such that the following conditions are satisfied:*

1) $\displaystyle\sup_{p \in \Lambda} \left\{ \max(\mu(p), \mu(p)^{-1}) \right\} < \infty .$ \hfill (2.3.135)

2) *For each* $p \in \Lambda$, *we have*

    a) $Df^n(p) \cdot S_{\mu(p)} \subset S_{\mu(f^n(p))}$

    and

        b)   *if* $\xi_p \in S'_{\mu(p)}$, $|Df^n(p)\xi_p| \leq C\lambda^n |\xi_p|$

        c)   *if* $\xi_p \in S_{\mu(p)}$, $\left|Df^{-n}(p)\xi_p\right| \leq C\lambda^n |\xi_p| .$ \hfill (2.3.136)

The proof of this theorem can be found in Newhouse and Palis [1973]. Theorem 2.3.14 tells us that in order to establish hyperbolicity for Λ we need only find bundles of sectors $S = \bigcup_{p \in \Lambda} S_{\mu(p)}$, $S' = \bigcup_{p \in \Lambda} S'_{\mu(p)}$, such that Df maps S into S while expanding each vector in S, and Df maps S' into S' while contracting each vector in S'. See Figure 2.3.13 for an illustration of the geometry of Theorem 2.3.14.

Now, regarding our map, the reader should notice that, if A1 and A3 hold, then the invariant set Λ is a hyperbolic invariant set, since the conditions of A3 are weakened versions of the necessary and sufficient conditions for a set to be hyperbolic given in Theorem 2.3.14.

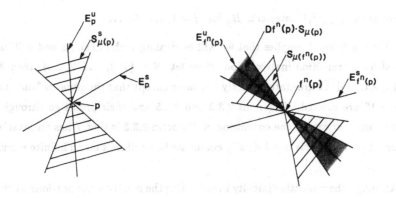

Figure 2.3.13. Geometry of Theorem 2.3.14.

The reason for discussing the concept of hyperbolic invariant sets is that they have played a central role in the development of modern dynamical systems theory. For example, the ideas of Markov partitions, pseudo orbits, shadowing, etc. all utilize crucially the notion of a hyperbolic invariant set. Indeed, the existence of a hyperbolic invariant set is often assumed a priori. This has caused great difficulty to the applied scientist since, in order to utilize many of the techniques or theorems of dynamical systems, theory he must first show that his particular system possesses a hyperbolic invariant set. Theorem 2.3.5 allows one to establish this fact. For more information on the consequences and utilization of hyperbolic invariant sets see Smale [1967], Nitecki [1971], Bowen [1970], [1978], Conley [1978], Shub [1987], and Franks [1982].

2.3f. The Case of an Infinite Number of Horizontal Slabs

In some applications it may arise that our map contains an invariant Cantor set on which it is topologically conjugate to a full shift on an infinite number of symbols. We now discuss the necessary modifications of A1, A2, A3, $\overline{A1}$, and $\overline{A2}$ in order to provide conditions whereby this occurs in a specific map.

Let H_i, $i = 1, 2, \ldots, N, \ldots$, be a collection of μ_h-horizontal slabs contained in the domain, D, of f. Everything will go through if we choose the H_i such that the following two conditions hold.

1) $\lim\limits_{i \to \infty} d(H_i) = 0$.

2) For each i, $f(H_i)$ intersects H_j for $j = 1, \ldots, N, \ldots$.

Note that condition 2) implies that we will be dealing with full shifts and will have no need for a transition matrix. Now, if we let $S = \{1, 2, \ldots, N, \ldots\}$, then A1, A2, A3, $\overline{A1}$, and $\overline{A2}$ are stated exactly the same except that the phrases "such that $(A)_{ij} = 1$" are deleted. Theorems 2.3.3 and 2.3.5 and their proofs go through in the same manner except the conclusion of Theorem 2.3.3 is that f has an invariant Cantor set on which it is topologically conjugate to a full shift on an infinite number of symbols.

Although there is little difficulty in extending the results of the previous sections to the case of an infinite number of symbols, the dynamical consequences require careful interpretation. In practice, dynamical situations in which conditions 1) and 2) hold usually require that the H_i converge to the boundary of D as $i \to \infty$. This is important since, in showing that A1 and A3 hold for the Poincaré map associated with some ordinary differential equation, often the boundary of D corresponds to the stable and unstable manifolds of some invariant set. In this case, orbits starting on ∂D could not return to D under iteration by the Poincaré map. We will see this in some examples in Chapter 3, and there we will be careful to explain the dynamical consequences of the symbol "∞."

2.4. Criteria for Chaos: The Nonhyperbolic Case

In Section 2.3 we gave sufficient conditions for a map to possess a chaotic invariant Cantor set of points. The conditions involved the decomposition of a subset of the domain of the map into contracting (horizontal) directions and expanding (vertical) directions along with more global conditions pertaining to the compatibility of the map with these directions (i.e., horizontal slabs mapping to vertical slabs). In this situation we will derive analogous conditions for the case when not all directions are strongly contracting or expanding, which we will call the "nonhyperbolic case." These conditions can be viewed as a generalization of those given in Wiggins [1986a]. The main difference in our results for the nonhyperbolic case as opposed to the hyperbolic case is that the conditions we give analogous to A1, A2, and A3 in Section 2.3 will lead to our map having a chaotic invariant Cantor set of *surfaces* as opposed to points. An invariant set of surfaces arises due to the fact that, since not all directions are expanding or contracting, the intersection of the set of forward and backward iterates of the map (i.e., the map's invariant set) need not be points. We now begin our development of the criteria, which will parallel as closely as possible the development of the criteria for the hyperbolic case given in Section 2.3.

2.4a. The Geometry of Chaos

We consider a map

$$f : D \times \Omega \to \mathbf{R}^n \times \mathbf{R}^m \times \mathbf{R}^p \qquad (2.4.1)$$

where D is a closed and bounded $n+m$ dimensional set contained in $\mathbf{R}^n \times \mathbf{R}^m$, and Ω is a closed, connected, bounded set contained in $\mathbf{R}^p$. The extra p dimensions will correspond to dimensions experiencing neutral growth behavior. We will discuss continuity and differentiability properties of f as they are needed.

We will now begin the development of the analogs of the horizontal and vertical slabs described in Section 2.3 and then proceed to define widths of the slabs and to discuss various intersection properties of the slabs. Our development of these ideas will closely parallel that of Section 2.3 with the addition of the necessary modifications in order to accommodate the neutral growth directions. We begin with some preliminary definitions. The following sets will be useful for defining the domains of various maps.

$$D_x = \{x \in \mathbf{R}^n \text{ for which there exists a } y \in \mathbf{R}^m \text{ with } (x, y) \in D\}$$
$$D_y = \{y \in \mathbf{R}^m \text{ for which there exists a } x \in \mathbf{R}^n \text{ with } (x, y) \in D\} \qquad (2.4.2)$$

These sets represent the projection of D onto $\mathbf{R}^n$ and $\mathbf{R}^m$, respectively. See Figure 2.4.1 for an illustration of the domain $D \times \Omega$ of f.

We remark that, hereafter, in giving illustrations to describe the analogs of slices, slabs, etc. from Section 2.3 we will not explicitly show $D \times \Omega$ in order to reduce the clutter in the diagram.

Let I_x be a closed simply connected n dimensional set contained in D_x, and let I_y be a closed simply connected m dimensional set contained in D_y. We will need the following definitions.

Definition 2.4.1. A μ_h-*horizontal slice*, $\bar{H}$, is defined to be the graph of a Lipschitz function $h : I_x \times \Omega \to \mathbf{R}^m$ where h satisfies the following two conditions:

1) The set $\bar{H} = \{(x, h(x, z), z) \in \mathbf{R}^n \times \mathbf{R}^m \times \mathbf{R}^p \mid x \in I_x, z \in \Omega\}$ is contained in $D \times \Omega$.

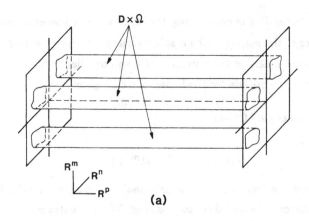

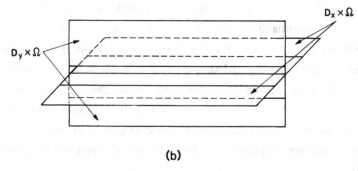

Figure 2.4.1. a) $D \times \Omega$ in $\mathbf{R}^n \times \mathbf{R}^m \times \mathbf{R}^p$, $n = m = p = 1$.

b) $D_x \times \Omega$ and $D_y \times \Omega$ in $\mathbf{R}^n \times \mathbf{R}^m \times \mathbf{R}^p$, $n = m = p = 1$.

2) For every $x_1, x_2 \in I_x$, we have

$$|h(x_1, z) - h(x_2, z)| \leq \mu_h |x_1 - x_2| \tag{2.4.3}$$

for some $0 \leq \mu_h < \infty$ and for all $z \in \Omega$.

Similarly, a μ_v-*vertical slice*, $\bar{V}$, is defined to be the graph of a Lipschitz function $v: I_y \times \Omega \to \mathbf{R}^n$ where v satisfies the following two conditions:

1) The set $\bar{V} = \{(v(y, z), y, z) \in \mathbf{R}^n \times \mathbf{R}^m \times \mathbf{R}^p \mid y \in I_y, z \in \Omega\}$ is contained in D.

2) For every $y_1, y_2 \in I_y$, we have

$$|v(y_1, z) - v(y_2, z)| \leq \mu_v |y_1 - y_2| \tag{2.4.4}$$

for some $0 \leq \mu_v < \infty$ and for all $z \in \Omega$.

So Definition 2.4.1 can be viewed as a parametrized version of Definition 2.3.1 with the conditions (2.4.3) and (2.4.4) holding uniformly in the parameters, see Figure 2.4.2.

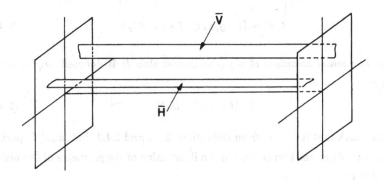

Figure 2.4.2. Horizontal and Vertical Slices in $\mathbb{R}^n \times \mathbb{R}^m \times \mathbb{R}^p$, $n = m = p = 1$.

Next we "fatten up" these horizontal and vertical slices into $n + m + p$ dimensional horizontal and vertical "slabs." We begin with the definition of μ_h-horizontal slabs.

Definition 2.4.2. Fix some μ_h, $0 \leq \mu_h < \infty$. Let $\bar{H}$ be a μ_h-horizontal slice and let $J^m : I_y \times \Omega \to \mathbb{R}^n$ be a continuous map having the property that the graph of J^m intersects $\bar{H}$ in a unique, continuous p dimensional surface. Let $\bar{H}^\alpha$, $\alpha \in I$, be the set of all μ_h-horizontal slices that intersect the boundary of the graph of J^m in continuous p dimensional surfaces and have the same domain as $\bar{H}$, where I is some index set (note: as in Definition 2.3.2, it may be necessary to adjust I_x or equivalently J^m in order for the $\bar{H}^\alpha$ to be contained in $D \times \Omega$).

Consider the following set in $\mathbb{R}^n \times \mathbb{R}^m \times \mathbb{R}^p$.

$S_H = \{ (x, y, z) \in \mathbb{R}^n \times \mathbb{R}^m \times \mathbb{R}^p \mid x \in I_x, \ z \in \Omega$ and y has the property that for each $x \in I_x$, $z \in \Omega$, given any line L through (x, y, z) with L parallel to the $x = 0$, $z = 0$ plane then L intersects the points $(x, h^\alpha(x, z), z)$, $(x, h^\beta(x, z), z)$ for some $\alpha, \beta \in I$ with (x, y, z) between these two points along $L \}$.

Then a μ_h-*horizontal slab* H is defined to be the closure of S_H.

Boundaries of μ_h-horizontal slabs are defined as follows.

Definition 2.4.3. The *vertical boundary* of a μ_h-horizontal slab H is denoted $\partial_v H$ and is defined as

$$\partial_v H \equiv \{(x, y, z) \in H \mid x \in \partial I_x\}. \qquad (2.4.5)$$

The *horizontal boundary* of a μ_h-horizontal slab H is denoted $\partial_h H$ and is defined as

$$\partial_h H \equiv \partial H - \partial_v H. \qquad (2.4.6)$$

We remark that it follows from Definitions 2.4.2 and 2.4.3 that $\partial_v H$ is parallel to the $x = 0$ plane, see Figure 2.4.3 for an illustration of the geometry of Definitions 2.4.2 and 2.4.3.

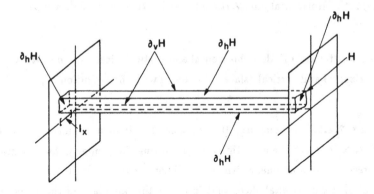

Figure 2.4.3. Horizontal Slab in $\mathbf{R}^n \times \mathbf{R}^m \times \mathbf{R}^p$, $n = m = p = 1$.

For describing the behavior of μ_h-horizontal slabs under maps the following definition will be useful.

Definition 2.4.4. Let H and $\tilde{H}$ be μ_h-horizontal slabs. $\tilde{H}$ is said to intersect H *fully* if $\tilde{H} \subset H$ and $\partial_v \tilde{H} \subset \partial_v H$.

See Figure 2.4.4 for an illustration of Definition 2.4.4.

Now we will define μ_v-vertical slabs.

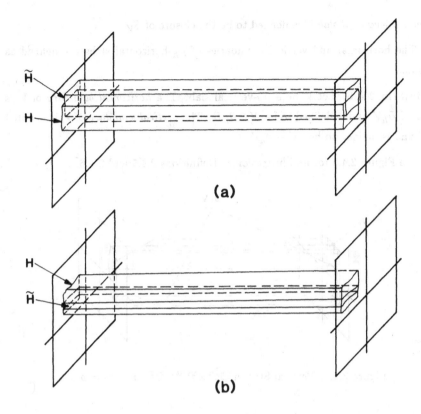

Figure 2.4.4. a) $\tilde{H}$ Does Not Intersect H Fully.

b) $\tilde{H}$ Does Intersect H Fully.

Definition 2.4.5. Fix some μ_v, $0 \leq \mu_v < \infty$. Let H be a μ_h-horizontal slab and let $\bar{V}$ be a μ_v-vertical slice contained in H with $\partial \bar{V} \subset \partial_h H$. Let $J^n: I_x \times \Omega \to \mathbf{R}^m$ be a continuous map having the property that the graph of J^n intersects $\bar{V}$ in a unique, continuous p dimensional surface. Let $\bar{V}^\alpha$, $\alpha \in I$, be the set of all μ_v-vertical slices that intersect the boundary of the graph of J^n with $\partial \bar{V}^\alpha \subset \partial_h H$ where I is some index set. We denote the domain of the function $v_\alpha(y, z)$ whose graph is $\bar{V}^\alpha$ by $I_y^\alpha \times \Omega$.

Consider the following set in $\mathbf{R}^n \times \mathbf{R}^m \times \mathbf{R}^p$.

$S_V = \{ (x, y, z) \in \mathbf{R}^n \times \mathbf{R}^m \times \mathbf{R}^p \mid (x, y, z)$ is contained in the interior of the set bounded by $\bar{V}^\alpha$, $\alpha \in I$, and $\partial_h H \}$.

Then a μ_v-*vertical slab* V is defined to be the closure of S_V.

The horizontal and vertical boundaries of μ_h-horizontal slabs are defined as follows.

Definition 2.4.6. Let V be a μ_v-vertical slab. The *horizontal boundary* of V is denoted $\partial_h V$ and is defined to be $V \cap \partial_h H$. The *vertical boundary* of V is denoted $\partial_v V$ and is defined to be $\partial V - \partial_h V$.

See Figure 2.4.5 for an illustration of Definitions 2.4.5 and 2.4.6.

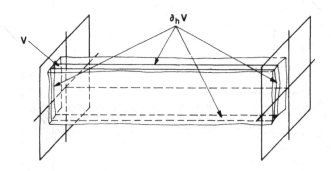

Figure 2.4.5. Vertical Slab in $\mathbb{R}^n \times \mathbb{R}^m \times \mathbb{R}^p$, $n = m = p = 1$.

Let us make a general remark concerning the boundaries of horizontal and vertical slabs. In each case the part of the boundary corresponding to the "ends" of the slabs (i.e., to be more precise, for a horizontal slab $\{ (x,y,z) \in \partial H \mid z \in \partial \Omega \}$ and for a vertical slab $\{ (x,y,z) \in \partial V \mid z \in \partial \Omega \}$) is defined to be in the horizontal boundary. This may appear strange, but it causes no problems, since we will require that this part of the boundary will always behave the same under f and f^{-1}, i.e., by definition it will always map to the part of the boundary corresponding to the ends of the slabs. We remark that, in the important special case in which Ω is a torus, the "ends" of the slabs do not exist, since a torus has no boundary.

Next we define the widths of μ_h-horizontal and μ_v-vertical slabs.

Definition 2.4.7. The width of a μ_h-horizontal slab H is denoted $d(H)$ and is defined as follows:

$$d(H) = \sup_{\substack{(x,z) \in I_x \times \Omega \\ \alpha, \beta \in I}} \left| h_\alpha(x,z) - h_\beta(x,z) \right| . \qquad (2.4.7)$$

Similarly, the width of a μ_v-vertical slab V is denoted by $d(V)$ and is defined as follows:

$$d(V) = \sup_{\substack{(y,z)\in \tilde{I}_y \times \Omega \\ \alpha,\beta \in I}} \left| v_\alpha(y,z) - v_\beta(y,z) \right| \tag{2.4.8}$$

where $\tilde{I}_y = I_y^\alpha \cap I_y^\beta$.

We now give two key lemmas.

Lemma 2.4.1. a) If $H^1 \supset H^2 \supset \cdots \supset H^n \supset \cdots$ is an infinite sequence of μ_h-horizontal slabs with H^{k+1} intersecting H^k fully, $k = 1,2,\ldots$, and $d(H^k) \to 0$ as $k \to \infty$, then $\bigcap_{k=1}^{\infty} H^k \equiv H^\infty$ is a μ_h-horizontal slice with $\partial H^\infty \subset \partial H^1$.

b) Similarly, if $V^1 \supset V^2 \supset \cdots \supset V^n \supset \cdots$ is an infinite sequence of μ_v-vertical slabs contained in a μ_h-horizontal slab H with $d(V^k) \to 0$ as $k \to \infty$, then $\bigcap_{k=1}^{\infty} V^k \equiv V^\infty$ is a μ_v-vertical slice with $\partial V^\infty \subset \partial_h H$.

PROOF: The proof of this lemma is very similar to the proof of Lemma 2.3.2. We leave the details to the reader.   □

Lemma 2.4.2. Let H be a μ_h-horizontal slab. Let $\bar{H}$ be a μ_h-horizontal slice with $\partial \bar{H} \subset \partial H$, and let $\bar{V}$ be a μ_v-vertical slice with $\partial \bar{V} \subset \partial_h H$ such that $0 < \mu_v \mu_h < 1$. Then $\bar{H}$ and $\bar{V}$ intersect in a unique p dimensional Lipschitz surface.

PROOF: The proof of this lemma is similar to the proof of Lemma 2.3.2. We let

$$\begin{aligned}
\bar{H} &= \text{graph } h(x,z) \quad \text{for } x \in I_x, z \in \Omega \\
\bar{V} &= \text{graph } v(y,z) \quad \text{for } y \in I_y, z \in \Omega.
\end{aligned} \tag{2.4.9}$$

Let $\tilde{I}_x = \text{closure } \{ x \in I_x \,|\, h(x,z) \text{ is in the domain of } v(\cdot,\cdot) \text{ for each } z \in \Omega \}$. Now if $\bar{H}$ intersects $\bar{V}$, there must be a point x such that $x = v(y,z)$ with $y = h(x,z)$ for some $z \in \Omega$. So the lemma will be proved if we can show that the equation $x = v(h(x,z),z)$ has a solution for each $z \in \Omega$ and the solution is Lipschitz in z. Now, for each $z \in \Omega$, $v(h(\cdot,z),z)$ maps $\tilde{I}_x$ into $\tilde{I}_x$, since $\bar{V} \subset H$ with $\partial \bar{V} \subset \partial_h H$. Since $\tilde{I}_x$ is a closed subset of the complete metric space $\mathbb{R}^n$ it is likewise a complete metric space. Recall that for each $x_1, x_2 \in \tilde{I}_x$, $z \in \Omega$, we have

$$\begin{aligned}
\left| v(h(x_1,z),z) - v(h(x_2,z),z) \right| &\le \mu_h \left| h(x_1,z) - h(x_2,z) \right| \\
&\le \mu_v \mu_h \left| x_1 - x_2 \right|.
\end{aligned} \tag{2.4.10}$$

Thus, for each $z \in \Omega$, $v(h(\cdot, z), z)$ is a contraction map of the complete metric space $\tilde{I}_x$ into itself since $0 \leq \mu_v \mu_h < 1$. So, by the contraction mapping theorem, the equation $x = v(h(x, z), z)$ has a unique solution for each $z \in \Omega$. Now, since v and h are Lipschitz, it follows from the uniform contraction principle (Chow and Hale [1982]) that this solution depends on z in a Lipschitz manner.   □

We are now at the point where we can give conditions sufficient for our map to possess a chaotic Cantor set. However, first let us recall the role that the analogs of Lemmas 2.4.1 and 2.4.2 in Section 2.3 played in obtaining the result for the hyperbolic case. In the hyperbolic case all directions were either strongly contracting or expanding, and this fact, coupled with Lemmas 2.3.1 and 2.3.2 and the structural assumptions A1 and A2 on f, led to the existence of a chaotic Cantor set of *points*. In the present situation the reader might guess that analogous structural assumptions on f along with Lemmas 2.4.1 and 2.4.2 might instead lead to a chaotic Cantor set of p dimensional *surfaces* for the map. We will shortly show that this is indeed the case.

Let $S = \{1, 2, \ldots, N\}$, $N \geq 2$, and let H_i, $i = 1, \ldots, N$, be a set of disjoint μ_h-horizontal slabs with $D_H = \bigcup_{i=1}^{N} H_i$. We assume that f is one-to-one on D_H, and we define

$$f(H_i) \cap H_j \equiv V_{ji}, \qquad \forall i, j \in S$$

and   (2.4.11)

$$H_i \cap f^{-1}(H_j) \equiv f^{-1}(V_{ji}) \equiv H_{ij} \qquad \forall i, j \in S.$$

Notice the subscripts on the sets V_{ji} and H_{ij}. The first subscript indicates which particular μ_h-horizontal slab the set is in, and the second subscript indicates for the V_{ji} into which μ_h-horizontal slab the set is mapped by f^{-1} and for the H_{ij} into which μ_h-horizontal slab the set is mapped by f.

Let A be an $N \times N$ matrix whose entries are either 0 or 1, i.e., A is a transition matrix (see Section 2.2) which will eventually be used to define symbolic dynamics for f. We have the following two "structural" assumptions for f.

A1. For all $i, j \in S$ such that $(A)_{ij} = 1$ V_{ji} is a μ_v-vertical slab contained in H_j with $\partial_v V_{ji} \subset \partial f(H_i)$ and $0 \leq \mu_v \mu_h < 1$. Moreover, f maps H_{ij} homeomorphically onto V_{ji} with $f^{-1}(\partial_v V_{ji}) \subset \partial_v H_i$.

A2. Let H be a μ_h-horizontal slab which intersects H_j fully. Then $f^{-1}(H) \cap H_i \equiv \tilde{H}_i$ is a μ_h-horizontal slab intersecting H_i fully for all $i \in S$ such that $(A)_{ij} = 1$. Moreover,

$$d(\tilde{H}_i) \leq \nu_h d(H) \qquad \text{for some } 0 < \nu_h < 1. \tag{2.4.12}$$

Similarly, let V be a μ_v-vertical slab contained in H_j such that also $V \subset V_{ji}$ for some $i, j \in S$ with $(A)_{ij} = 1$. Then $f(V) \cap H_k \equiv \tilde{V}_k$ is a μ_v-vertical slab contained in H_k for all $k \in S$ such that $(A)_{jk} = 1$. Moreover,

$$d(\tilde{V}_k) \leq \nu_v d(V) \qquad \text{for some } 0 < \nu_v < 1. \tag{2.4.13}$$

See Figures 2.4.6 and 2.4.7 for an illustration of the geometry of A1 and A2.

The reader may be struck by the fact that A1 and A2 read just as in the hyperbolic case discussed in Section 2.3; however, note that the definition of horizontal and vertical slabs along with the domain of f have been modified to account for neutral growth directions.

2.4b. The Main Theorem

We now state our main theorem which gives sufficient conditions in order for our map to possess a chaotic invariant set.

Theorem 2.4.3. *Suppose f satisfies A1 and A2; then f possesses an invariant set of p dimensional Lipschitz surfaces. Moreover, denoting this set of surfaces by Λ, there exists a homeomorphism $\phi \colon \Lambda \to \Sigma_A^N$ such that the following diagram commutes.*

$$
\begin{array}{ccc}
\Lambda & \xrightarrow{\;f\;} & \Lambda \\
\phi \downarrow & & \downarrow \phi \\
\Sigma_A^N & \xrightarrow{\;\sigma\;} & \Sigma_A^N
\end{array}
\qquad . \tag{2.4.14}
$$

Let us make the following remarks regarding Theorem 2.4.3.

1) If A is irreducible then Λ is a Cantor set of surfaces, see Section 2.2.

2) Let us discuss in more detail the expression $\Lambda \xrightarrow{f} \Lambda$. The phrase "$\Lambda$ is an invariant set of p dimensional Lipschitz surfaces" means that given any point $\tau \in \Lambda$ $f(\tau)$ is also an element of Λ, and hence a p dimensional Lipschitz surface. Thus, the expression $\Lambda \xrightarrow{f} \Lambda$ implies that points in the restricted

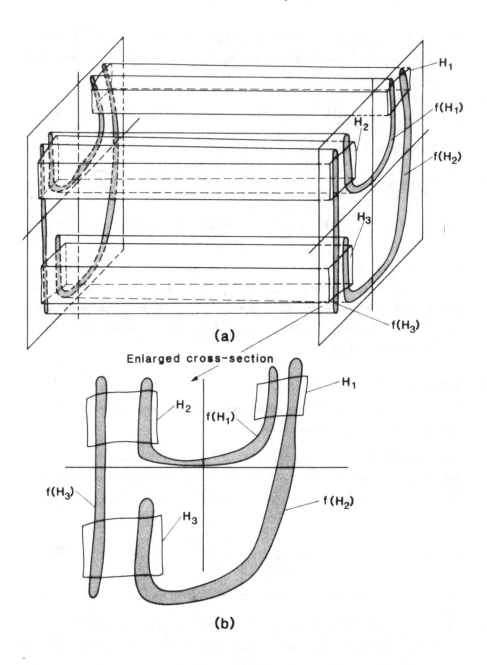

(a)

Enlarged cross-section

(b)

Figure 2.4.6. a) Horizontal Slabs and Their Images under f, $A = \begin{pmatrix} 1 & 1 & 0 \\ 1 & 0 & 1 \\ 0 & 1 & 1 \end{pmatrix}$.

b) Enlarged Cross-sectional View for z Fixed.

domain of f are to be taken as points in Λ, i.e., as p dimensional Lipschitz surfaces (note: in the sense that the domain of a map is part of the definition of the map, it might make more sense to rename f when it is viewed as being restricted to Λ; however, we do not take this approach).

3) In order for ϕ to be a homeomorphism it is necessary to equip Λ with a topology. There are two ways of doing this.

   The first way uses the fact that elements of Λ can be written as the graphs of Lipschitz functions. Let the graphs of the Lipschitz functions $u_1(z)$ and $u_2(z)$, $z \in \Omega$, represent two elements of Λ. Then the distance between the graphs of u_1 and u_2 is defined to be

$$d(u_1, u_2) = \sup_{z \in \Omega} |u_1(z) - u_2(z)| \ .$$

   This metric suffices to define a topology on Λ.

   A second way of equipping Λ with a topology would be simply to "mod out" the z direction and use the quotient topology.

4) An important special case is when Ω is a torus. In this case Λ becomes a set of tori.

5) Suppose A is irreducible; then we can make the following conclusions:
   a) There exists a countable infinity of periodic surfaces in Λ.
   b) There exists an uncountable infinity of nonperiodic surfaces in Λ.
   c) There exists a surface in Λ that at some point along its orbit is arbitrarily close to every other surface in Λ.

Thus, one might think of the dynamics in directions normal to the surfaces as being chaotic in the same sense as in the hyperbolic case described in Section 2.3 while the dynamics in directions tangent to the surfaces is unknown.

PROOF: (of Theorem 2.4.3). The proof of this theorem proceeds precisely the same as the proof of Theorem 2.3.3. See Wiggins [1986a] for details of the case when Ω is a torus.   □

2.4c. Sector Bundles

We now want to give a more computable criterion for verifying the stretching and contraction estimates which appear in A2. This criterion will be analogous to the condition A3 given in Section 2.3 for the hyperbolic case. The condition will

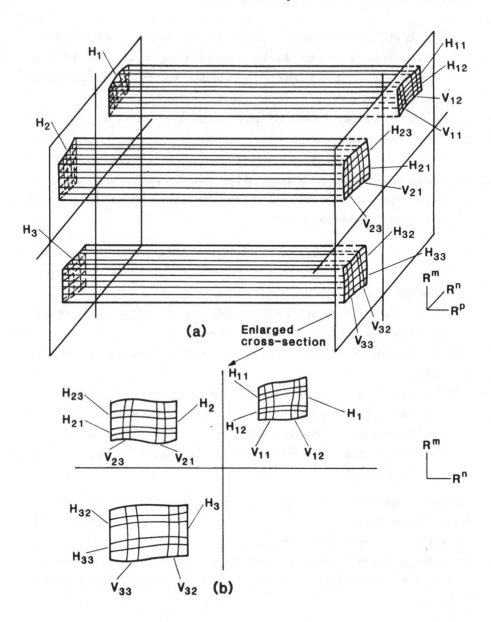

(a)

Enlarged
cross-section

(b)

Figure 2.4.7. a) H_{ij} and V_{ji} for $1 \leq i,\ j \leq 3$, $(A)_{ij} = 1$.
b) Enlarged Cross-sectional View for z Fixed.

likewise be phrased in terms of the action of the derivative of f acting on tangent vectors. The idea will be to give essentially the same conditions on the stretching and contracting directions but make them uniform in z. We will need the following additional requirements on f:

R1. Let $\mathcal{H} = \bigcup_{\substack{i,j,\in S \\ (A)_{ij}=1}} H_{ij}$ and $\mathcal{V} = \bigcup_{\substack{i,j,\in S \\ (A)_{ij}=1}} V_{ji}$; then f is C^1 on $\mathcal{H}$ and f^{-1} is C^1 on $\mathcal{V}$.

R2. Consider the definition of μ_h-horizontal slices and μ_v-vertical slices given in Definition 2.4.1. We strengthen the Lipschitz requirements (2.4.3) and (2.4.4) as follows:

  a) For every $x_1, x_2 \in I_x$, $z_1, z_2 \in \Omega$ we have

$$|h(x_1, z_1) - h(x_2, z_2)| \leq \mu_h |x_1 - x_2| + \bar{\mu}_h |z_1 - z_2| \qquad (2.4.15)$$

  for some $0 \leq \mu_h < \infty$, $0 \leq \bar{\mu}_h \leq \mu_h$.

  b) For every $y_1, y_2 \in I_y$, $z_1, z_2 \in \Omega$ we have

$$|v(y_1, z_1) - v(y_2, z_2)| \leq \mu_v |y_1 - y_2| + \bar{\mu}_v |z_1 - z_2| \qquad (2.4.16)$$

  for some $0 \leq \mu_v < \infty$, $0 \leq \bar{\mu}_v \leq \mu_v$.

We next define stable and unstable sectors at a point.

Choose a point $p_0 \equiv (x_0, y_0, z_0) \in \mathcal{V} \cup \mathcal{H}$. The *stable sector* at p_0, denoted $S_{p_0}^s$, is defined as follows

$$S_{p_0}^s = \left\{ (\xi_{p_0}, \eta_{p_0}, \chi_{p_0}) \in \mathbb{R}^n \times \mathbb{R}^m \times \mathbb{R}^p \mid |\eta_{p_0}| \leq \mu_h |\xi_{p_0}|, \quad |\eta_{p_0}| \leq \bar{\mu}_h |\chi_{p_0}| \right\} \qquad (2.4.17)$$

and the *unstable sector* at p_0, denoted $S_{p_0}^u$, is defined as follows

$$S_{p_0}^u = \left\{ (\xi_{p_0}, \eta_{p_0}, \chi_{p_0}) \in \mathbb{R}^n \times \mathbb{R}^m \times \mathbb{R}^p \mid |\xi_{p_0}| \leq \mu_v |\eta_{p_0}|, \quad |\xi_{p_0}| \leq \bar{\mu}_v |\chi_{p_0}| \right\}. \qquad (2.4.18)$$

See Figure 2.4.8 for an illustration of the geometry.

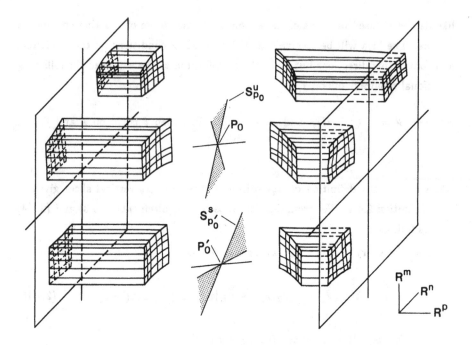

Figure 2.4.8. Stable and Unstable Sectors (Note: We Represent
the Horizontal Slabs as Cut Open for Clarity).

We define *sector bundles* or *cone fields* as follows:

$$S^s_{\mathcal{X}} = \bigcup_{p_0 \in \mathcal{X}} S^s_{p_0}$$

$$S^s_{\mathcal{Y}} = \bigcup_{p_0 \in \mathcal{Y}} S^s_{p_0}$$

$$S^u_{\mathcal{X}} = \bigcup_{p_0 \in \mathcal{X}} S^u_{p_0} \qquad (2.4.19)$$

$$S^u_{\mathcal{Y}} = \bigcup_{p_0 \in \mathcal{Y}} S^u_{p_0}.$$

We have the following hypothesis:

A3. $Df(S^u_{\mathcal{X}}) \subset S^u_{\mathcal{Y}}$ and $Df^{-1}(S^s_{\mathcal{Y}}) \subset S^s_{\mathcal{X}}$. Moreover, if $(\xi_{p_0}, \eta_{p_0}, \chi_{p_0}) \in S^u_{p_0}$ and

$$Df(p_0)\left(\xi_{p_0}, \eta_{p_0}, \chi_{p_0}\right) = \left(\xi_{f(p_0)}, \eta_{f(p_0)}, \chi_{f(p_0)}\right) \in S^u_{f(p_0)}, \text{ then we have}$$

1. $\left|\eta_{f(p_0)}\right| \geq \left(\dfrac{1}{\mu_u}\right)\left|\eta_{p_0}\right|, \qquad 0 < \mu_u \leq 1 - \mu_v\mu_h - \mu_h\bar{\mu}_v - \bar{\mu}_h$

2. $1 < \dfrac{\left|\eta_{f(p_0)}\right|}{\left|\chi_{f(p_0)}\right|}$

for all $p_0 \in \mathcal{H}$, $\left(\xi_{p_0}, \eta_{p_0}, \chi_{p_0}\right) \in S^u_{p_0}$.

Similarly, if $\left(\xi_{p_0}, \eta_{p_0}, \chi_{p_0}\right) \in S^s_{p_0}$ and

$$Df^{-1}(p_0)\left(\xi_{p_0}, \eta_{p_0}, \chi_{p_0}\right) = \left(\xi_{f^{-1}(p_0)}, \eta_{f^{-1}(p_0)}, \chi_{f^{-1}(p_0)}\right) \in S^s_{f^{-1}(p_0)},$$

then we have

1. $\left|\xi_{f^{-1}(p_0)}\right| \geq \left(\dfrac{1}{\mu_s}\right)\left|\xi_{p_0}\right|, \qquad 0 < \mu_s \leq 1 - \mu_v\mu_h - \mu_v\bar{\mu}_h - \bar{\mu}_v$

2. $1 < \dfrac{\left|\xi_{f^{-1}(p_0)}\right|}{\left|\chi_{f^{-1}(p_0)}\right|}$

for all $p_0 \in \mathcal{V}$, $\left(\xi_{p_0}, \eta_{p_0}, \chi_{p_0}\right) \in S^s_{p_0}$.

We make the following remarks concerning A3.

1) The conditions $Df(S^u_{\mathcal{H}}) \subset S^u_{\mathcal{V}}$ and $Df^{-1}(S^s_{\mathcal{V}}) \subset S^s_{\mathcal{H}}$ imply the preservation of the horizontal and vertical directions under f and f^{-1}, respectively, as well as the fact that the images of vertical slices under f and horizontal slices under f^{-1} may not "roll-up" in the z direction.

2) The condition $\left|\eta_{f(p_0)}\right| \geq \left(\dfrac{1}{\mu_u}\right)\left|\eta_{p_0}\right|$ for any $p_0 \in \mathcal{H}$, $\left(\xi_{p_0}, \eta_{p_0}, \chi_{p_0}\right) \in S^u_{p_0}$ implies that the vertical directions are uniformly expanded under f. Similarly, the condition $\left|\xi_{f^{-1}(p_0)}\right| \geq \left(\dfrac{1}{\mu_s}\right)\left|\xi_{p_0}\right|$ for any $p_0 \in \mathcal{V}$, $\left(\xi_{p_0}, \eta_{p_0}, \chi_{p_0}\right) \in S^s_{p_0}$ implies that the horizontal directions are uniformly expanded under f^{-1}.

3) The condition $1 < \left|\eta_{f(p_0)}\right| / \left|\chi_{f(p_0)}\right|$ implies that the "shear" in the z direction experienced by a μ_v-vertical slice under mapping by f is bounded. Similarly, the condition $1 < \left|\xi_{f^{-1}(p_0)}\right| / \left|\chi_{f^{-1}(p_0)}\right|$ implies that the "shear" in the z direction experienced by a μ_h-horizontal slice under mapping by f^{-1} is bounded. We will see that these conditions are important for estimating the widths of images of slabs.

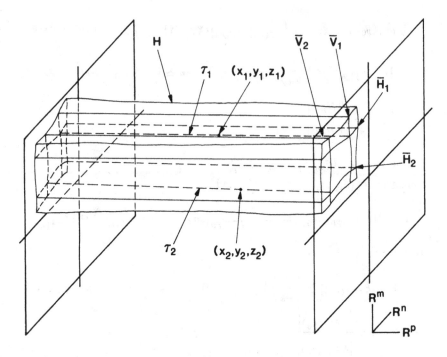

Figure 2.4.9. The Geometry of $\bar{H}_1$, $\bar{H}_2$, $\bar{V}_1$, and $\bar{V}_2$ in H.

Now the idea will be to show that A3 can be substituted for A2; however, first we will derive a preliminary estimate which will be useful for estimating the widths of images of slabs under f.

Let H be a μ_h-horizontal slab. Let $\bar{H}_1$ and $\bar{H}_2$ be disjoint μ_h-horizontal slices contained in H with $\partial \bar{H}_1$ and $\partial \bar{H}_2$ contained in $\partial_v H$. We denote the domain of the functions $h_1(x, z)$ and $h_2(x, z)$ of which $\bar{H}_1$ and $\bar{H}_2$ are the graphs by $I_x \times \Omega$. Let $\bar{V}_1$ and $\bar{V}_2$ be disjoint μ_v-vertical slices contained in H with $\partial \bar{V}_1$ and $\partial \bar{V}_2$ contained in $\partial_h H$. We denote the domains of the functions $v_1(y, z)$ and $v_2(y, z)$ of which $\bar{V}_1$ and $\bar{V}_2$ are the graphs by $I_y^1 \times \Omega$ and $I_y^2 \times \Omega$, respectively. Let

$$\|h_1 - h_2\| = \sup_{(x,z) \in I_x \times \Omega} |h_1(x, z) - h_2(x, z)| \,,$$

$$\|v_1 - v_2\| = \sup_{(y,z) \in (I_y^1 \cap I_y^2) \times \Omega} |v_1(y, z) - v_2(y, z)| \,. \tag{2.4.20}$$

By Lemma 2.4.2, $\bar{H}_1$ and $\bar{V}_1$ intersect in a unique p dimensional continuous Lipschitz surface which we call τ_1, and $\bar{H}_2$ and $\bar{V}_2$ intersect in a unique p dimensional continuous Lipschitz surface which we call τ_2. Let (x_1, y_1, z_1) and (x_2, y_2, z_2) be arbitrary points on τ_1 and τ_2, respectively. See Figure 2.4.9 for an illustration of the geometry.

We have the following lemma.

Lemma 2.4.4.

$$|x_1 - x_2| \leq \frac{1}{1 - \mu_v \mu_h} \left[(\bar{\mu}_v + \mu_v \bar{\mu}_h) |z_1 - z_2| + \mu_v \|h_1 - h_2\| + \|v_1 - v_2\| \right],$$

$$|y_1 - y_2| \leq \frac{1}{1 - \mu_v \mu_h} \left[(\bar{\mu}_h + \bar{\mu}_v \mu_h) |z_1 - z_2| + \mu_h \|v_1 - v_2\| + \|h_1 - h_2\| \right].$$

PROOF: We have

$$|x_1 - x_2| = |v_1(y_1, z_1) - v_2(y_2, z_2)| \leq |v_1(y_1, z_1) - v_1(y_2, z_2)| + |v_1(y_2, z_2) - v_2(y_2, z_2)|$$

$$\leq \mu_v |y_1 - y_2| + \bar{\mu}_v |z_1 - z_2| + \|v_1 - v_2\| \quad (2.4.21)$$

and

$$|y_1 - y_2| = |h_1(x_1, z_1) - h_2(x_2, z_2)| \leq |h_1(x_1, z_1) - h_1(x_2, z_2)| + |h_1(x_2, z_2) - h_2(x_2, z_2)|$$
$$(2.4.22)$$
$$\leq \mu_h |x_1 - x_2| + \bar{\mu}_h |z_1 - z_2| + \|h_1 - h_2\|.$$

Substituting (2.4.22) into (2.4.21) gives the first inequality, and substituting (2.4.21) into (2.4.22) gives the second inequality. $\qquad \square$

Theorem 2.4.5. *If A1 and A3 hold then A2 holds with* $\nu_h = \dfrac{\mu_u}{1 - \mu_v \mu_h - \mu_h \bar{\mu}_v - \bar{\mu}_h}$ *and* $\nu_v = \dfrac{\mu_s}{1 - \mu_v \mu_h - \mu_v \bar{\mu}_h - \bar{\mu}_v}$.

PROOF: The proof proceeds in much the same way as the analogous Theorem 2.3.5 for the hyperbolic case. However, we will include the details, since the geometry associated with the nonstretching directions is somewhat different. We will prove that the part of A2 dealing with vertical slabs holds, since the part dealing with horizontal slabs is proven similarly. The proof proceeds in several steps:

1) Let $\bar{V}$ be a μ_v-vertical slice contained in H_j such that also $\bar{V} \subset V_{ji}$ with $\partial \bar{V} \subset \partial_h H_j$ for some $i, j \in S$ with $(A)_{ij} = 1$. Then we show that $f(\bar{V}) \cap H_k$ is a μ_v-vertical slice with $\partial(f(\bar{V})) \subset \partial_h H_k$ for all $k \in S$ such that $(A)_{jk} = 1$.

2) Let V be a μ_v-vertical slab contained in H_j such that also $V \subset V_{ji}$ for some $i, j \in S$ with $(A)_{ij} = 1$. Then we use 1) to show that $f(V) \cap H_k \equiv \tilde{V}_k$ is a μ_v-vertical slab contained in H_k for all $k \in S$ such that $(A)_{jk} = 1$.

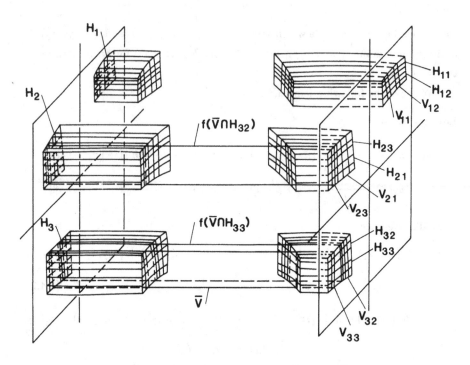

Figure 2.4.10. The Geometry of $\bar{V}$ and $f(\bar{V})$.

3) Show that $d(\tilde{V}_k) \leq \dfrac{\mu_s}{1 - \mu_v\mu_h - \mu_v\bar{\mu}_h - \bar{\mu}_v} d(V)$.

We begin with Step 1). Let $\bar{V}$ be a μ_v-vertical slice contained in H_j such that also $\bar{V} \subset V_{ji}$ with $\partial\bar{V} \subset \partial_h H_j$ for some $i,j \in S$ with $(A)_{ij} = 1$. Then, by Lemma 2.4.2, $\bar{V}$ intersects H_{jk} with $\partial\left(\bar{V} \cap H_{jk}\right) \subset \partial_h H_{jk}$ for all $k \in S$ such that $(A)_{jk} = 1$. Now A1 holds so that $f(\partial_h H_{jk}) \subset \partial_h V_{kj}$; therefore, $f\left(\partial\left(\bar{V} \cap H_{jk}\right)\right) \subset \partial_h V_{kj}$ for each $k \in S$ such that $(A)_{jk} = 1$. So $f\left(\bar{V} \cap H_{jk}\right)$ consists of a collection of $m + p$ dimensional sets with $\partial\left(f\left(\bar{V} \cap H_{jk}\right)\right) \subset \partial_h V_{kj}$, see Figure 2.4.10.

We now argue that $f\left(\bar{V} \cap H_{jk}\right)$ are μ_v-vertical slices. By A3, Df maps $S_{\mathcal{X}}^u$ into $S_{\mathcal{Y}}^u$ for all $p_0 \in \mathcal{X}$. Thus, for any (x_1, y_1, z_1), $(x_2, y_2, z_2) \in f\left(\bar{V} \cap H_{jk}\right)$ we have

$$|x_1 - x_2| \leq \mu_v |y_1 - y_2|,$$
$$|x_1 - x_2| \leq \bar{\mu}_v |z_1 - z_2|. \tag{2.4.23}$$

So (2.4.23) allows us to conclude that, for each $k \in S$ such that $(A)_{jk} = 1$,

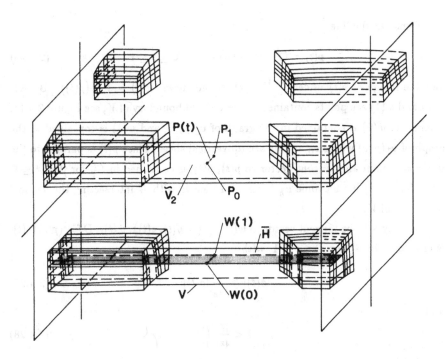

Figure 2.4.11. V and $\tilde{V}_2$ (Note: $\tilde{V}_3$ Has Been Left Out of the Figure for Clarity).

$f\left(\bar{V} \cap H_{jk}\right)$ can be expressed as the graph over the (y, z) variables of a Lipschitz function $\tilde{V}(y, z)$ such that

$$\left|\tilde{V}(y_1, z_1) - \tilde{V}(y_2, z_2)\right| \leq \mu_v |y_1 - y_2| + \bar{\mu}_v |z_1 - z_2| . \qquad (2.4.24)$$

Step 2)   Let V be a μ_v-vertical slab contained in H_j such that also $V \subset V_{ji}$ for some $i, j \in S$ with $(A)_{ij} = 1$. Then $\partial_h \left(V \cap H_{jk}\right) \subset \partial_h H_{jk}$ for all $k \in S$ such that $(A)_{jk} = 1$. Applying the result of Step 1) to the vertical boundaries of each $V \cap H_{jk}$, we see that $f\left(V \cap H_{jk}\right) \equiv \tilde{V}_k$ is a μ_v-vertical slab contained in H_k for each $k \in S$ such that $(A)_{jk} = 1$. Moreover, the $\tilde{V}_k$ are disjoint.

Step 3)   We now show that $d\left(\tilde{V}_k\right) \leq \dfrac{\mu_s}{1 - \mu_v\mu_h - \mu_v\bar{\mu}_h - \bar{\mu}_v} d(V)$. Fix k and let $p_0 = (x_0, y_0, z_0)$ and $p_1 = (x_1, y_1, z_1)$ be two points on the vertical boundary of $\tilde{V}_k$ having the same y and z coordinates, i.e., $y_0 = y_1$ and $z_0 = z_1$, such that

$$d\left(\tilde{V}_k\right) = |p_0 - p_1| = |x_0 - x_1| , \qquad (2.4.25)$$

see Figure 2.4.11.

Consider the line

$$p(t) = (1-t)p_0 + tp_1 , \qquad 0 \le t \le 1 \tag{2.4.26}$$

and the image of $p(t)$ under f^{-1} which is the curve $w(t) = f^{-1}(p(t))$. By A1, $w(0)$ and $w(1)$ are points contained in the vertical boundary of V, see Figure 2.4.11. Therefore, $w(0)$ is contained in the graph of $v_0(y,z)$, and $w(1)$ is contained in the graph of $v_1(y,z)$ where v_0 and v_1 are μ_v-vertical slices. Since $p(t)$ is parallel to the $y = z = 0$ plane the tangent vector to $p(t)$, $\dot{p}(t)$, is contained in S_y^s for $0 \le t \le 1$. Therefore, $w(t)$ lies on some μ_h-horizontal slice $\bar{H}$ with $\bar{H}$ intersecting the vertical boundary of V.

Also by A3, the tangent vector to $w(t) = (x(t), y(t), z(t))$, $\dot{w}(t) = Df^{-1}(p(t))\dot{p}(t)$, is contained in $S_{\mathcal{H}}^s$ for $0 \le t \le 1$ with

$$|\dot{x}(t)| \ge \frac{1}{\mu_s}|\dot{p}(t)| \tag{2.4.27}$$

and

$$1 < \frac{|\dot{x}(t)|}{|\dot{z}(t)|} \tag{2.4.28}$$

for $0 \le t \le 1$.

From (2.4.27) and (2.4.28) we conclude that

$$|p_0 - p_1| \le \mu_s |x(0) - x(1)| \tag{2.4.29}$$

and

$$|z(0) - z(1)| < |x(0) - x(1)| . \tag{2.4.30}$$

Using Lemma 2.4.4 we obtain

$$|x(0) - x(1)| \le \frac{1}{1 - \mu_v \mu_h}\left[(\bar{\mu}_v + \mu_v \bar{\mu}_h)|z(0) - z(1)| + \|v_0 - v_1\|\right] . \tag{2.4.31}$$

Substituting (2.4.30) into (2.4.31) gives

$$|x(0) - x(1)| \le \frac{1}{1 - \mu_v \mu_h - \mu_v \bar{\mu}_h - \bar{\mu}_v}\|v_0 - v_1\| . \tag{2.4.32}$$

So from (2.4.25), (2.4.29), and (2.4.32) we obtain

$$d(\tilde{V}_k) \le \frac{\mu_s}{1 - \mu_v \mu_h - \mu_v \bar{\mu}_h - \bar{\mu}_v}d(V) \tag{2.4.33}$$

which is true for each $k \in S$ such that $(A)_{jk} = 1$. $\qquad\qquad\square$

CHAPTER 3
Homoclinic and Heteroclinic Motions

In this chapter we will study some of the consequences of homoclinic and hetero-clinic orbits in dynamical systems. Part of the motivation for the study of these special orbits comes from the fact that, in recent years, it has become apparent that homoclinic and heteroclinic orbits are often the mechanism for the chaos and tran-sient chaos numerically observed in physical systems. We will comment on specific examples as we go along.

3.1. Examples and Definitions

The purpose of this first section is to introduce the idea of homoclinic and hetero-clinic motions. We will do this by first giving some examples of specific physical systems which exhibit homoclinic and heteroclinic motions so that the reader may develop some intuition. After this we will give specific mathematical definitions for homoclinic and heteroclinic orbits.

EXAMPLE 3.1.1. The Simple Pendulum

We consider a mass m, suspended via a weightless rigid bar of length L from a support and moving under the influence of gravity as shown in Figure 3.1.1 (we also neglect dissipative effects such as wind resistance).

The equation of motion of the pendulum in rescaled, dimensionless variables can be written as

$$\ddot{\theta} + \sin\theta = 0 \tag{3.1.1}$$

or, as a system

$$\begin{aligned} \dot{\theta} &= v \\ \dot{v} &= -\sin\theta \end{aligned} \qquad (\theta, v) \in T^1 \times \mathbb{R}. \tag{3.1.2}$$

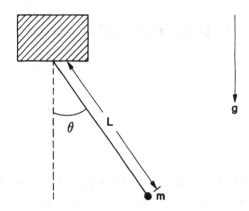

Figure 3.1.1. The Simple Pendulum.

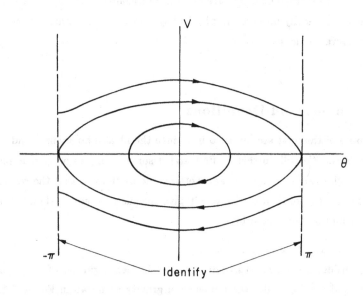

Figure 3.1.2. Phase Space of the Simple Pendulum.

The phase space of the simple pendulum is the cylinder, $T^1 \times \mathbf{R}$, and has the structure as shown in Figure 3.1.2 with $\theta = \pi$ and $\theta = -\pi$ identified.

From Figure 3.1.2 we see that the pendulum has two equilibrium positions, one stable at $(\theta, v) = (0,0)$ corresponding to the mass hanging straight down, and one unstable at $(\theta, v) = (\pi, 0) \equiv (-\pi, 0)$ corresponding to the mass standing upright vertically. Also, we see that there are two orbits which connect the unstable equilibrium to itself. These correspond to trajectories which approach the unstable equilibrium position asymptotically in time (Note: no trajectory may reach the unstable equilibrium position in finite time, since the equilibrium position itself is a solution to (3.1.2), and we have uniqueness of solutions). There are two such trajectories since the pendulum may rotate either clockwise or counterclockwise. These two special orbits are said to be *homoclinic* to the unstable fixed point at $(\theta, v) = (0, 0)$.

It should be apparent that the homoclinic orbits consist of the (nontransverse) intersection of the stable and unstable manifolds of $(\theta, v) = (0, 0)$. We will see that this characterization of homoclinic orbits is quite useful. In this example, the homoclinic orbits do not signal any complicated motions but merely separate two qualitatively distinct motions, namely, the librational motions inside the homoclinic orbits and the rotational motions outside the homoclinic orbits. Recall that in the context of planar ordinary differential equations the name *separatrix* is often given to what we have called the homoclinic orbits. This is because the one dimensional orbits separate the two dimensional phase plane into two disjoint parts.

EXAMPLE 3.1.2. The Buckled Beam

The system consisting of a long slender cantilevered beam buckled in the field of two permanent magnets has been extensively studied both experimentally and theoretically by Moon and Holmes (see Moon [1980], Holmes [1979]). The experimental apparatus is shown in Figure 3.1.3.

It has been shown that, in certain parameter ranges, the first mode of oscillation of the beam is adequately described by the following normalized version of Duffing's equation

$$\ddot{x} - x + x^3 = 0 \tag{3.1.3}$$

or, as a system

$$\begin{aligned} \dot{x} &= y \\ \dot{y} &= x - x^3 \end{aligned} \qquad (x, y) \in \mathbf{R}^1 \times \mathbf{R}^1 . \tag{3.1.4}$$

The phase space of this system appears as in Figure 3.1.4.

The system has an unstable equilibrium point at $(x, y) = (0, 0)$ corresponding to the beam being at a position midway between the two magnets. This unstable equilibrium point is connected to itself by two homoclinic orbits corresponding to motions which approach the unstable equilibrium point asymptotically in both time

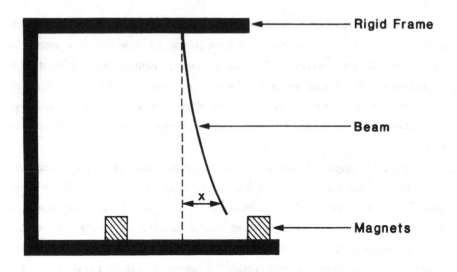

Figure 3.1.3. Elastic Beam in Magnetic Field.

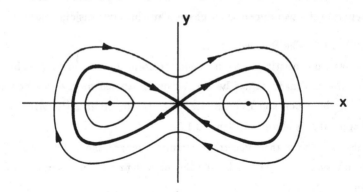

Figure 3.1.4. The Phase Space of the Beam.

directions. Thus, the homoclinic orbits are characterized by the (nontransverse) intersection of the stable and unstable manifolds of $(x, y) = (0, 0)$ (note: there are two homoclinic orbits in this system as a result of the reflectional invariance of the system). As in the case of the simple pendulum, the homoclinic orbits are not indicative of any complicated motions but merely form a boundary (separatrix) between two qualitatively distinct motions.

Suppose, however, that we force this system horizontally with a small amplitude periodic force given by $\gamma \cos \omega t$, γ small. In this case the equation of motion is given by

$$\begin{aligned} \dot{x} &= y \\ \dot{y} &= x - x^3 + \gamma \cos \omega t \end{aligned} \qquad (x,y) \in \mathbb{R}^1 \times \mathbb{R}^1 \qquad (3.1.5)$$

or, as the suspended system

$$\begin{aligned} \dot{x} &= y \\ \dot{y} &= x - x^3 + \gamma \cos \theta \qquad (x,y,\theta) \in \mathbb{R}^1 \times \mathbb{R}^1 \times T^1 . \\ \dot{\theta} &= \omega \end{aligned} \qquad (3.1.6)$$

It can be shown (see Guckenheimer and Holmes [1983] or Chapter 4) that the unstable equilibrium point in the unforced system now becomes an unstable periodic orbit of period $2\pi/\omega$ for γ sufficiently small. Furthermore, for certain values of the parameters γ and ω, the two dimensional stable and unstable manifolds of this unstable periodic orbit may intersect transversely in the phase space $\mathbb{R}^1 \times \mathbb{R}^1 \times T^1$ to yield a picture like that shown in Figure 3.1.5.

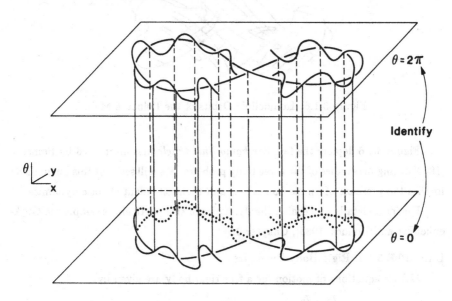

Figure 3.1.5. The Phase Space of the Forced Beam.

In this case the points in the intersections of the stable and unstable manifolds of the unstable periodic orbit lie on orbits which approach the unstable periodic orbit asymptotically in both directions of time and are said to be homoclinic to the unstable periodic orbit.

The resulting complicated geometrical phenomena associated with these homoclinic orbits is made a bit clearer by instead considering an associated two dimensional Poincaré map (see Section 1.6). We construct a two dimensional cross-section Σ to the three dimensional phase space of (3.1.6) as follows

$$\Sigma = \{(x,y,\theta)|\theta = 0 \in (0,2\pi]\}. \tag{3.1.7}$$

Then the Poincaré map of Σ into itself is defined by

$$P: \Sigma \longrightarrow \Sigma$$
$$(x(0),y(0)) \longmapsto (x\left(\frac{2\pi}{\omega}\right), y\left(\frac{2\pi}{\omega}\right)). \tag{3.1.8}$$

In terms of the Poincaré map, the unstable periodic orbit is manifested as an unstable fixed point whose stable and unstable manifolds intersect as in Figure 3.1.6.

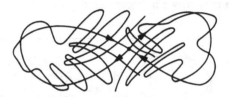

Figure 3.1.6. Homoclinic Orbits of the Poincaré Map.

Figure 3.1.6 depicts the familiar *homoclinic tangle* first discovered by Poincaré [1899] during his studies of the three body problem. We will see that this phenomena implies the presence of Smale horseshoes and their attendant chaotic dynamics.

For more details concerning the dynamics of this particular example see Guckenheimer and Holmes [1983], Chapter 2.

EXAMPLE 3.1.3. Rigid Body Dynamics

Euler's equations of motion for a free rigid body are given by

$$\dot{m}_1 = \frac{I_2 - I_3}{I_2 I_3} m_2 m_3$$
$$\dot{m}_2 = \frac{I_3 - I_1}{I_1 I_3} m_1 m_3 \qquad (m_1, m_2, m_3) \in \mathbb{R}^1 \times \mathbb{R}^1 \times \mathbb{R}^1 \tag{3.1.9}$$
$$\dot{m}_3 = \frac{I_1 - I_2}{I_1 I_2} m_1 m_2$$

where $I_1 \geq I_2 \geq I_3$ are the moments of inertia about the principal, body fixed axes and $m_i = I_i \omega_i$, $i = 1, 2, 3$, where ω_i is the angular velocity about the ith principal axis (see Goldstein [1980]).

These equations have two constants of motion given by

$$H = \frac{1}{2} \left(\frac{m_1^2}{I_1} + \frac{m_2^2}{I_2} + \frac{m_3^2}{I_3} \right) \tag{3.1.10}$$
$$l^2 = m_1^2 + m_2^2 + m_3^2.$$

Thus, the orbits of (3.1.9) are given by the intersection of the ellipsoids $H =$ constant with the spheres $l^2 =$ constant. The flow on the sphere has saddle points at $(0, \pm l, 0)$ and centers at $(0, 0, \pm l)$. The saddles are connected by four orbits, as shown in Figure 3.1.7. These four orbits have the property that trajectories through points on the orbits approach one of the saddles as $t \to +\infty$ and the other saddle as $t \to -\infty$. These orbits are said to be *heteroclinic* to the fixed points $(0, \pm l, 0)$.

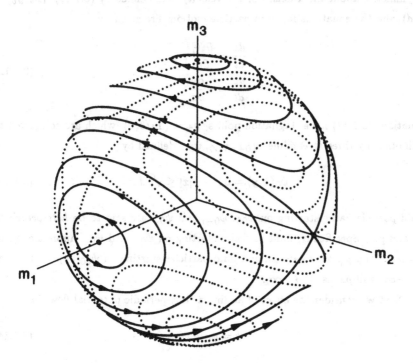

Figure 3.1.7. The Phase Space of (3.1.9), l Fixed.

When the free rigid body is perturbed by adding attachments, chaotic motions may be created similar to those in Example 3.1.2. We refer the reader to Holmes and Marsden [1983], Koiller [1984], and Krishnaprasad and Marsden [1987] for details and examples.

EXAMPLE 3.1.4. Point Vortices in a Time Varying Strain Field

The fluid flow induced by a pair of translating point vortices separated by a distance $2d$ and with circulation $\pm\Gamma$ is sketched in Figure 3.1.8. The motion is viewed in a frame moving with the velocity of the vortices, $v = \Gamma/4\pi d\, \hat{e}_x$.

The stream function for this flow can be found in Lamb [1945] and is given by

$$\psi_0 = -\frac{\Gamma}{4\pi}\log\left[\frac{(x-x_v)^2+(y-y_v)^2}{(x-x_v)^2+(y+y_v)^2}\right] - \frac{\Gamma y}{4\pi d} \qquad (3.1.11)$$

where (x_v, y_v) is the position of the vortex in the upper half plane; also note that ψ_0 is symmetric about the x-axis. For the velocity field defined by (3.1.11) $(x_v, y_v) = (0, d)$, and the equations for fluid particle motions are given by

$$\begin{aligned}\frac{dx}{dt} &= \frac{\partial\psi_0}{\partial y}, \\ \frac{dy}{dt} &= \frac{-\partial\psi_0}{\partial x}.\end{aligned} \qquad (3.1.12)$$

Equations (3.1.12) have stagnation points $p_\pm = (\pm\sqrt{3}, 0)$ which are connected to each other by three streamlines ψ_{0+}, ψ_{00}, ψ_{0-} defined by

$$\psi_0(x, y) = 0, \qquad |x| \le \sqrt{3}\, d. \qquad (3.1.13)$$

Fluid particle paths starting on ψ_{00}, ψ_{0-}, and ψ_{0+} are said to be *heteroclinic* to p_+ and p_-. Specifically, fluid particles starting on ψ_{0+} and ψ_{0-} approach p_- as $t \to +\infty$ and p_+ as $t \to -\infty$, and fluid particles starting on ψ_{00} approach p_- as $t \to -\infty$ and p_+ as $t \to +\infty$.

Next we consider the effect of adding a time-periodic potential flow, i.e.,

$$\psi = \psi_0 + \psi_\epsilon \qquad (3.1.14)$$

with

$$\psi_\epsilon = \epsilon xy\,\omega\,\sin\omega t + v_\epsilon y \qquad (3.1.15)$$

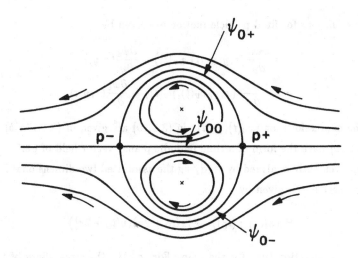

Figure 3.1.8. Flow Induced by a Pair of Translating Vortices.

and where the constant translation speed v_ϵ is included in anticipation of a co-ordinate change and is determined by requiring that the vortices have zero drift velocity. Such a flow satisfies the Euler equations and is produced, for example, by the motion of a vortex pair in a wavy-walled channel. The resulting motion of the vortices is relatively simple. Introducing the dimensionless parameter,

$$a = \frac{\Gamma}{2\pi\omega d^2} \tag{3.1.16}$$

and the dimensionless variables $(x/d, y/d) \to (x, y)$, $\omega t \to t$ and $v_\epsilon/d\omega \to v_\epsilon$, we compute the motion of the vortices with $(x_v(0), y_v(0)) = (0, 1)$ to obtain

$$x_v(t) = \exp(-\epsilon \cos t) \int_0^t \exp(\epsilon \cos t') \left\{ \frac{a}{2} \left[\exp(-\epsilon(\cos t' - 1)) - 1 \right] + v_\epsilon \right\} dt' \tag{3.1.17a}$$

$$y_v(t) = \exp(\epsilon(\cos t - 1)) \tag{3.1.17b}$$

where

$$v_\epsilon = \frac{a}{2} \left[1 - \frac{\exp(\epsilon)}{I_0(\epsilon)} \right] \tag{3.1.18}$$

and where $I_0(\epsilon)$ is the modified Bessel function of order zero.

The equations for fluid particle motion are given by

$$
\begin{aligned}
\dot{x} &= \frac{\partial \psi_0}{\partial y}\big(x, y; x_v(t), y_v(t)\big) + \frac{\partial \psi_\epsilon}{\partial y}(x, y; t) \\
\dot{y} &= \frac{-\partial \psi_0}{\partial x}\big(x, y; x_v(t), y_v(t)\big) - \frac{\partial \psi_\epsilon}{\partial x}(x, y, t)
\end{aligned}
\tag{3.1.19}
$$

where the expressions for $\big(x_v(t), y_v(t)\big)$ in (3.1.19) are given in (3.1.17a, b). Equations (3.1.19) have the form of a time periodic planar vector field of period 2π and is most conveniently analyzed by studying the associated two dimensional Poincaré map (cf. Section 1.6) given by

$$
P: \big(x(t_0), y(t_0)\big) \rightarrow \big(x(t_0 + 2\pi), y(t_0 + 2\pi)\big)
\tag{3.1.20}
$$

where t_0 is the section time for the map. For $\epsilon = 0$ the streamlines of the flow shown in Figure 3.1.9 are the invariant curves of the map. In particular, this map has two hyperbolic saddle points at

$$
p_\pm = (\pm\sqrt{3}, 0)
\tag{3.1.21}
$$

and the unstable manifold of p_+ coincides with the stable manifold of p_-. These manifolds are also heteroclinic orbits and are defined by the streamlines $\psi_{0\pm}$ defined above in dimensional form.

Now for $\epsilon \neq 0$ and small, the fixed points $p_\pm$ persist, denoted $p_{\pm,\epsilon}$. The streamline ψ_{00} persists unbroken; however, the remaining branches of the stable and unstable manifolds of $p_{\pm,\epsilon}$ intersect in a discrete set of points leading to a complicated geometric structure as shown in Figure 3.1.9.

The heteroclinic points of the Poincaré map (i.e., the points asymptotic to $p_{+,\epsilon}$ in positive time and $p_{-,\epsilon}$ in negative time) are responsible for chaotic particle trajectories as well as mixing and transport properties of this flow. For more information on this problem see Rom-Kedar, Leonard, and Wiggins [1988].

EXAMPLE 3.1.5. Traveling Wave Solutions of Partial Differential Equations

Consider a partial differential equation in one space and one time variable denoted x and t, respectively. Then transforming to the variable $z = x + ct$ gives rise to an ordinary differential equation whose solutions represent traveling waves with propagation speed c. Homoclinic or heteroclinic orbits in the ordinary differential equation represent solitary waves in the partial differential equation. For more

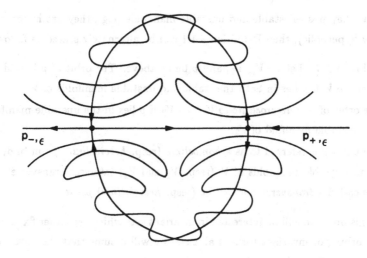

Figure 3.1.9. The Poincaré Map for (3.1.19), $\epsilon \neq 0$, Small.

information see Conley [1975], Feroe [1982], Glendinning [1987], Hastings [1982], Kopell [1977], and Smoller [1983].

EXAMPLE 3.1.6. Phase Transitions

In continuum mechanics homoclinic and heteroclinic orbits often arise as structures separating two distinct phases of the continua. More specifically, they may arise in the phase space of the Euler-Lagrange Equation associated with minimizing some type of energy functional of a system. For more information see Carr [1983], Coullet and Elphick [1987], Slemrod [1983], and Slemrod and Marsden [1985].

We now want to give a general definition for homoclinic and heteroclinic orbits.

Definition 3.1.1. Let V be an invariant set of a dynamical system (map or flow). Let p be a point in the phase space of the dynamical system, and suppose that the orbit of p approaches V asymptotically as $t \to -\infty$ and V asymptotically as $t \to +\infty$; then the orbit of p is said to be *homoclinic to* V.

Let V_1 and V_2 be disjoint invariant sets of a dynamical system, and suppose that the orbit of p approaches V_1 asymptotically as $t \to -\infty$ and V_2 asymptotically as $t \to +\infty$; then the orbit of p is said to be *heteroclinic to* V_1 *and* V_2.

Definition 3.1.1 is too general for us to get an analytical handle on the orbit structure near homoclinic and heteroclinic orbits. However, if V, V_1, and V_2 are such that they possess stable and unstable manifolds (e.g., they are hyperbolic or normally hyperbolic), then Definition 3.1.1 can be alternately stated as follows.

Definition 3.1.2. Let V, V_1, V_2, and p be as above. The orbit of p is said to be *homoclinic to* V if p lies in both the stable and unstable manifolds of V.

The orbit of p is *heteroclinic to* V_1 *and* V_2 if p lies in the unstable manifold of V_1 and the stable manifold of V_2.

The point p is referred to as a *homoclinic* (resp. *heteroclinic*) *point* and, if the stable and unstable manifolds of V (resp. V_1 and V_2) intersect transversely at p, then p is called a *transverse homoclinic* (resp. *heteroclinic*) *point*.

In this book we will be interested in invariant sets which are either fixed points, periodic orbits, or invariant tori. In all cases we will assume that the invariant set has some type of hyperbolic structure.

3.2. Orbits Homoclinic to Hyperbolic Fixed Points of Ordinary Differential Equations

We will now begin our study of the orbit structure near orbits homoclinic to hyperbolic fixed points of ordinary differential equations. We will attempt to answer the following three questions:

1) Does there exist chaotic behavior near the homoclinic orbit?
2) Does the behavior persist for nearby systems, e.g., if system parameters are varied?
3) What are the effects of symmetries in the system?

Only in the simplest cases will we be able to give complete answers to all three questions.

Before beginning our analysis of specific systems, we will describe the general technique of analysis in Section 3.2a. In 3.2b we will derive a classical bifurcation result for planar systems, which can be found in Andronov et al. [1971]. In 3.2c we study third order systems and show how Smale horseshoes may arise near homoclinic orbits. In 3.2d we study two examples of homoclinic orbits in fourth order systems. In 3.2e we study the orbit structure near orbits homoclinic to hyperbolic fixed points in fourth order Hamiltonian systems, and in 3.2f we discuss some known results in dimensions > 4.

3.2a. The Technique of Analysis

Before proceeding to specific systems we want to describe the basic idea behind our method of analysis as well as some general results which will simplify our later work.

We will be considering ordinary differential equations of the form

$$\dot{z} = F(z), \qquad z \in \mathbb{R}^{s+u} \tag{3.2.1}$$

where $F: U \to \mathbb{R}^{s+u}$ is C^r ($r \geq 2$ is adequate) on some open set $U \subset \mathbb{R}^{s+u}$. We have the following assumptions on (3.2.1).

A1. Equation (3.2.1) has a hyperbolic fixed point at $z = z_0$. In particular, we assume that the matrix $DF(z_0)$ has s eigenvalues having negative real parts and u eigenvalues having positive real parts.

A2. Equation (3.2.1) has a homoclinic orbit connecting z_0 to itself, i.e., there exists a solution $\phi(t)$ of (3.2.1) such that $\lim\limits_{t \to +\infty} \phi(t) = \lim\limits_{t \to -\infty} \phi(t) = z_0$.

Our goal will be to study the orbit structure near the homoclinic orbit. This will be accomplished by constructing a Poincaré map near the homoclinic orbit. The Poincaré map will consist of the composition of two maps; one given by the (essentially) linear flow near the fixed point and the other given by an (essentially) rigid motion along the homoclinic orbit outside a neighborhood of the fixed point. Assumptions A1 and A2 alone are not always sufficient to allow for the construction of such a Poincaré map. The homoclinic orbit must be a nonwandering set (cf. Section 1.1k). This is always the case in dimensions two and three but not necessarily in dimensions four and higher. In our present construction we will introduce this as an assumption, and it will be necessary for us to verify this fact for specific systems.

We now want to describe the construction of the Poincaré map in a neighborhood of the homoclinic orbit. This will be accomplished in a series of steps.

Steps 1–3. In this series of steps we transform the fixed point to the origin and show how the local stable and unstable manifolds of the origin can be used as local coordinates.

Step 4. We study the geometry near the origin and set up cross-sections to the vector field.

Step 5. We construct the Poincaré map near the homoclinic orbit.

Step 6. We construct an approximate Poincaré map which we can more readily compute.

Step 7. We give results which show how the dynamics of the approximate Poincaré map are related to the dynamics of the exact Poincaré map.

We now begin the construction.

Step 1. *Transform the Fixed Point to the Origin.*

This is a trivial step which we include for completeness. Under the affine (i.e., linear plus translation) transformation, $w = z - z_0$, (3.2.1) becomes

$$\dot{w} = F(w + z_0) \equiv G(w) , \qquad (3.2.2)$$

and it is clear that (3.2.2) has a fixed point at $w = 0$.

Step 2. *Utilize the Linear Stable and Unstable Eigenspaces as Coordinates.*

By assumption A2, the $(s+u) \times (s+u)$ matrix $DG(0)$ has s eigenvalues having negative real parts and u eigenvalues having positive real parts. Thus, from linear algebra we can find a linear transformation such that $DG(0)$ has the following form

$$DG(0) = \begin{pmatrix} A & O_{su} \\ O_{us} & B \end{pmatrix} \qquad (3.2.3)$$

where A is an $s \times s$ Jordan block such that all the diagonal entries have negative real parts, B is a $u \times u$ Jordan block such that all the diagonal entries have positive real parts, and O_{su} (resp. O_{us}) represents an $s \times u$ (resp. $u \times s$) matrix whose entries are all zero. Utilizing this same linear transformation the nonlinear system (3.2.2) can be put in the form

$$\begin{aligned} \dot{\xi} &= A\xi + F_1(\xi, \eta) \\ \dot{\eta} &= B\eta + F_2(\xi, \eta) \end{aligned} , \qquad (\xi, \eta) \in \mathbf{R}^s \times \mathbf{R}^u \qquad (3.2.4)$$

where F_1 and F_2 are C^{r-1} and satisfy

$$F_1(0,0) = F_2(0,0) = DF_1(0,0) = DF_2(0,0) = 0 . \qquad (3.2.5)$$

(Note: we want to make a remark concerning (3.2.5). F_1 is an s vector, F_2 is a u vector, DF_1 is an $s \times (s + u)$ matrix, and DF_2 is a $u \times (s + u)$ matrix, so strictly speaking (3.2.5) is incorrect, since the equality sign has no meaning. However, (3.2.5) has a symbolic meaning in the sense that F_1, F_2, DF_1, and DF_2 are all equal to the zero element in the appropriate space. Another way of writing (3.2.5) would be $F_1, F_2 = \mathcal{O}\left(|\xi|^2 + |\eta|^2\right) \equiv \mathcal{O}(2).$)

Step 3. *Utilize Stable and Unstable Manifolds as Coordinates.*

Consider the linearized system

$$\dot{\xi} = A\xi$$
$$\dot{\eta} = B\eta .$$

(3.2.6)

From Section 1.3 we know that there exists an s dimensional linear subspace E^s given by $\eta = 0$ and a u dimensional linear subspace E^u given by $\xi = 0$ such that solutions of (3.2.6) starting in E^s decay exponentially to the origin as $t \to +\infty$, and solutions starting in E^u decay exponentially to the origin as $t \to -\infty$. For the nonlinear problem (3.2.4) the stable and unstable manifold theorem (see Theorem 1.3.7) tells us that there exists C^r manifolds W^s and W^u intersecting at the origin and tangent to E^s and E^u, respectively, at the origin which have the properties that solutions of (3.2.4) starting in W^s decay exponentially to the origin as $t \to +\infty$ and solutions of (3.2.4) starting in W^u decay exponentially to the origin as $t \to -\infty$. Since W^s and W^u are tangent to E^s and E^u, respectively, locally they can be represented as graphs, i.e.,

$$W^s_{\text{loc}} = \text{graph } \phi_s(\xi)$$
$$W^u_{\text{loc}} = \text{graph } \phi_u(\eta)$$

(3.2.7)

where $\phi_s(\xi)$ and $\phi_u(\eta)$ are C^r maps of $\mathcal{N}^s \subset \mathbb{R}^s \to \mathbb{R}^u$ and $\mathcal{N}^u \subset \mathbb{R}^u \to \mathbb{R}^s$, respectively, which are defined in sufficiently small neighborhoods, $\mathcal{N}^s$ and $\mathcal{N}^u$, of the origin. Eventually we will be interested in comparing the linear flow generated by (3.2.6) with the nonlinear flow generated by (3.2.4) near the origin. For this purpose it is useful to use W^s_{loc} and W^u_{loc} as coordinates rather than E^s and E^u. This is accomplished by the following transformation

$$(x, y) = \left(\xi - \phi_u(\eta), \eta - \phi_s(\xi)\right) .$$

(3.2.8)

Under (3.2.8), the nonlinear equation (3.2.4) becomes

$$\dot{x} = Ax + f_1(x,y)$$
$$\dot{y} = By + f_2(x,y) \qquad (x,y) \in \mathbf{R}^s \times \mathbf{R}^u \qquad (3.2.9)$$

where f_1 and f_2 are $O(2)$ and also

$$f_1(0,y) = f_2(x,0) = 0. \qquad (3.2.10)$$

Equation (3.2.10) reflects the fact that $y = 0$ is the local stable manifold of the origin and $x = 0$ is the local unstable manifold of the origin. We emphasize the fact that (3.2.10) is only valid locally in some neighborhood $\mathcal{N}^s \times \mathcal{N}^u \subset \mathbf{R}^s \times \mathbf{R}^u$, since the transformation (3.2.8) is only a local transformation defined on $\mathcal{N}^s \times \mathcal{N}^u$. In Sections 3.2b–3.2f we will assume that the equations under consideration have been transformed to the form of (3.2.9). See Figure 3.2.1 for an illustration of the geometry of the transformation (3.2.8).

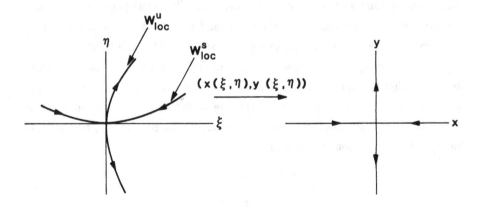

Figure 3.2.1. Geometry of the Transformation (3.2.8).

Step 4. *The Geometry of the Vector Field Near the Origin.*

We will denote the homoclinic trajectory obtained after $\phi(t)$ has undergone the transformations in Steps 1, 2, and 3 above by $\psi(t)$; so we have $\lim\limits_{t \to +\infty} \psi(t) = \lim\limits_{t \to -\infty} \psi(t) = 0$. We remark that the vector field in the local coordinate system (3.2.8) may be joined smoothly to the vector field outside of $\mathcal{N}^s \times \mathcal{N}^u$ by an appropriate choice of "bump functions," see Spivak [1979]. Consider the following $s + u - 1$ dimensional sets

$$C_\epsilon^s = \{ (x,y) \in \mathbf{R}^s \times \mathbf{R}^u \mid |x| = \epsilon, \quad |y| < \epsilon \}$$
$$C_\epsilon^u = \{ (x,y) \in \mathbf{R}^s \times \mathbf{R}^u \mid |x| < \epsilon, \quad |y| = \epsilon \}$$

(3.2.11)

and their associated closures

$$\bar{C}_\epsilon^s = \{ (x,y) \in \mathbf{R}^s \times \mathbf{R}^u \mid |x| = \epsilon, \quad |y| \leq \epsilon \}$$
$$\bar{C}_\epsilon^u = \{ (x,y) \in \mathbf{R}^s \times \mathbf{R}^u \mid |x| \leq \epsilon, \quad |y| = \epsilon \}.$$

(3.2.12)

We assume that ϵ is chosen sufficiently small such that $\bar{C}_\epsilon^s$ and $\bar{C}_\epsilon^u$ are contained in $\mathcal{N}^s \times \mathcal{N}^u$. We define the following neighborhood of the origin

$$\mathcal{N} = \{ (x,y) \in \mathbf{R}^s \times \mathbf{R}^u \mid |x| \leq \epsilon, \quad |y| \leq \epsilon \};$$

(3.2.13)

then it is easy to see that $\bar{C}_\epsilon^s$ and $\bar{C}_\epsilon^u$ form the boundary of $\mathcal{N}$.

Let $n^s(x,y)$ denote the unit vector normal to C_ϵ^s at the point $(x,y) \in C_\epsilon^s$. Similarly let $n^u(x,y)$ denote the unit vector normal to C_ϵ^u at the point $(x,y) \in C_\epsilon^u$. Recall from Section 1.6 that C_ϵ^s and C_ϵ^u are cross-sections of the vector field (3.2.9) provided

$$n^s(x,y) \cdot (Ax + f_1(x,y), By + f_2(x,y)) \neq 0, \quad \forall\, (x,y) \in C_\epsilon^s$$
$$n^u(x,y) \cdot (Ax + f_1(x,y), By + f_2(x,y)) \neq 0, \quad \forall\, (x,y) \in C_\epsilon^u.$$

(3.2.14)

We have the following proposition.

Proposition 3.2.1. *For ϵ sufficiently small, C_ϵ^s and C_ϵ^u are cross-sections to the vector field (3.2.9). Moreover, (3.2.9) points strictly into the interior of $\mathcal{N}$ on C_ϵ^s and (3.2.9) points strictly to the exterior of $\mathcal{N}$ on C_ϵ^u.*

PROOF: This is an easy computation using the fact that A has eigenvalues with negative real parts, and B has eigenvalues with positive real parts. $\square$

In computing the Poincaré map it will be useful to have expressions for the intersection of the stable manifold with C_ϵ^s and the intersection of the unstable manifold with C_ϵ^u. These are given as follows:

$$S_\epsilon^s = \{ (x,y) \in \mathbf{R}^s \times \mathbf{R}^u \mid |x| = \epsilon, |y| = 0 \}$$
$$S_\epsilon^u = \{ (x,y) \in \mathbf{R}^s \times \mathbf{R}^u \mid |x| = 0, |y| = \epsilon \}.$$

(3.2.15)

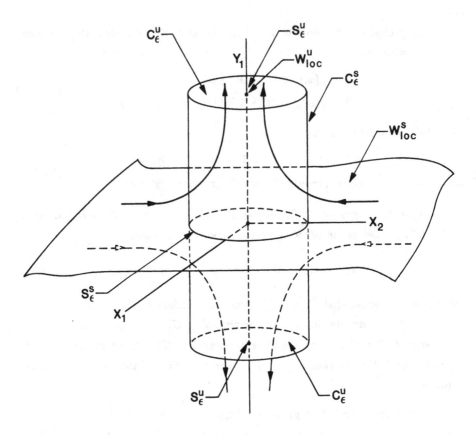

Figure 3.2.2. Geometry of the Vector Field Near the Origin, $s = 2$, $u = 1$.

See Figure 3.2.2 for an illustration of the geometry of the vector field near the origin.

Step 5. *Construction of the Poincaré Map.*

We now will describe the construction of the Poincaré map defined in a neighborhood of the homoclinic orbit. As mentioned earlier, the Poincaré map will consist of the composition of two maps; one defined in a neighborhood of the origin and the other defined outside of a neighborhood of the origin along $\psi(t)$. We will discuss each map separately.

a) The Map Near the Origin.

Consider the sets $C^s_\epsilon - S^s_\epsilon$ and $C^u_\epsilon - S^u_\epsilon$. By Proposition 3.2.1, for ϵ sufficiently small all points in $C^s_\epsilon - S^s_\epsilon$ reach $C^u_\epsilon - S^u_\epsilon$ under the flow generated by (3.2.9).

Let us denote the flow generated by (3.2.9) by

$$\phi(t, x_0, y_0) = \big(x(t, x_0, y_0), y(t, x_0, y_0)\big). \tag{3.2.16}$$

Suppose $(x_0, y_0) \in C_\epsilon^s - S_\epsilon^s$; then (x_0, y_0) reaches $C_\epsilon^u - S_\epsilon^u$ in a time $T = T(x_0, y_0)$ which is a solution of the equation

$$|y(T, x_0, y_0)| = \epsilon. \tag{3.2.17}$$

(Note: $T(x_0, y_0) \to +\infty$ logarithmically as $y_0 \to 0$.)

We define the map

$$\begin{aligned} P_0 : C_\epsilon^s - S_\epsilon^s &\to C_\epsilon^u - S_\epsilon^u \\ (x_0, y_0) &\mapsto \big(x(T(x_0, y_0), x_0, y_0), y(T(x_0, y_0), x_0, y_0)\big) \end{aligned} \tag{3.2.18}$$

where $T(x_0, y_0)$ is the solution of (3.2.17) with (x_0, y_0) regarded as fixed.

b) The Map Along $\psi(t)$ Away from the Origin.

Let α and β denote points of intersection of the homoclinic orbit with C_ϵ^u and C_ϵ^s, respectively. Let U_α be a neighborhood of α in C_ϵ^u, and let U_β be a neighborhood of β in C_ϵ^s. Now, since α and β lie on the homoclinic orbit $\psi(t)$, there exists a finite time τ such that $\phi(\tau, \alpha) = \beta$. Since the flow is C^r ($r \geq 2$) then we can choose U_α sufficiently small such that $\phi(\tau(u), u) \subset U_\beta$, $u \in U_\alpha$ where $\tau(u)$ is the time necessary for a point $u \in U_\alpha$ to reach U_β. Thus, we define the map

$$\begin{aligned} P_1 : U_\alpha &\to U_\beta \\ u &\mapsto \phi(\tau(u), u). \end{aligned} \tag{3.2.19}$$

c) The Poincaré Map.

Now $U_\beta \subset C_\epsilon^s$ and $U_\alpha \in C_\epsilon^u$. Suppose it is possible to choose an open set $V_\beta \subset U_\beta$ such that

$$P_0(V_\beta) \subset U_\alpha. \tag{3.2.20}$$

If this can be done we define the Poincaré map to be

$$P \equiv P_1 \circ P_0 : V_\beta \to U_\beta. \tag{3.2.21}$$

See Figure 3.2.3.

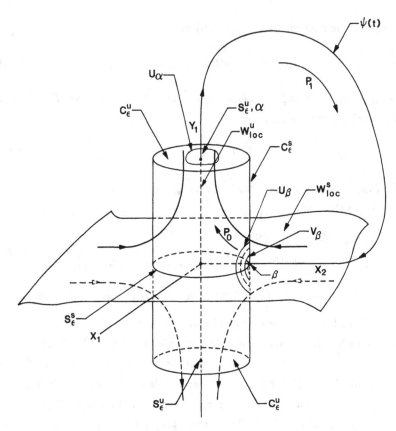

Figure 3.2.3. The Poincaré Map, $P = P_1 \circ P_0$.

The condition (3.2.20) cannot always be satisfied with only the assumptions A1 and A2 given in the beginning of this section (however, it can always be satisfied in dimensions 2 and 3), and in the following sections we will treat its applicability on a case by case basis.

Step 6. *The Approximate Poincaré Map.*

Our method for studying the orbit structure near the homoclinic orbit will consist of constructing a Poincaré map similar to that just described and then studying its dynamics. However, from the definition of the map, it is clear that to construct it we must first solve for the flow generated by (3.2.9). This cannot be done in general. Instead, we will construct an approximate Poincaré map which will reproduce the dynamics of the exact Poincaré map we are interested in studying.

The approximate Poincaré map consists of the composition of two maps.

a) The Approximate Map Near the Origin

The flow generated by the vector field (3.2.9) linearized about the origin is given by

$$\phi(t, x_0, y_0) = (e^{At}x_0, e^{Bt}y_0) ; \qquad (3.2.22)$$

and we define the map

$$P_0^L : C_\epsilon^s - S_\epsilon^s \to C_\epsilon^u - S_\epsilon^u$$
$$(x_0, y_0) \mapsto (e^{AT}x_0, e^{BT}y_0) \qquad (3.2.23)$$

where T solves

$$\left| e^{BT}y_0 \right| = \epsilon . \qquad (3.2.24)$$

Intuitively, it should be reasonable that P_0^L is "close" to P_0 for ϵ sufficiently small, since nearer the origin the vector field looks more and more linear. We will make this precise in Step 7.

b) The Approximate Map Along $\psi(t)$ Away from the Origin

Consider the map P_1 defined in (3.2.19). Taylor expanding P_1 about α gives

$$P_1(\alpha + u') = P_1(\alpha) + DP_1(\alpha)u' + \mathcal{O}(|u'|^2)$$
$$= \beta + DP_1(\alpha)u' + \mathcal{O}(|u'|^2) . \qquad (3.2.25)$$

We define the map

$$P_1^L : U_\alpha \to U_\beta$$
$$\alpha + u' \mapsto \beta + DP_1(\alpha)u' . \qquad (3.2.26)$$

c) The Approximate Poincaré Map

As in Step 5, we suppose that it is possible to choose $V_\beta \subset U_\beta$ such that $P_0(V_\beta) \subset U_\alpha$. Then we define the approximate Poincaré map, P^L, as

$$P^L = P_1^L \circ P_0^L : V_\beta \to U_\beta . \qquad (3.2.27)$$

Next we demonstrate how well P^L approximates P.

Step 7. *The Relation Between the Exact and Approximate Poincaré Maps.*

First we want to show that the map constructed near the origin is approximated to within an error of $\mathcal{O}(\epsilon^2)$ if the flow generated by the linearized vector field is used for its construction.

We begin by rescaling the coordinates as follows

$$x = \epsilon \bar{x}$$
$$y = \epsilon \bar{y} \, . \tag{3.2.28}$$

$0 < \epsilon \ll 1$.

In this case (3.2.9) becomes

$$\dot{\bar{x}} = A\bar{x} + \frac{1}{\epsilon} f_1(\epsilon \bar{x}, \epsilon \bar{y}) \equiv A\bar{x} + \bar{f}_1(\bar{x}, \bar{y}; \epsilon)$$
$$\dot{\bar{y}} = B\bar{y} + \frac{1}{\epsilon} f_2(\epsilon \bar{x}, \epsilon \bar{y}) \equiv B\bar{y} + \bar{f}_2(\bar{x}, \bar{y}; \epsilon) \tag{3.2.29}$$

where

$$\lim_{\epsilon \to 0} \bar{f}_1(\bar{x}, \bar{y}; \epsilon) = 0$$
$$\lim_{\epsilon \to 0} \bar{f}_2(\bar{x}, \bar{y}; \epsilon) = 0 \, . \tag{3.2.30}$$

We denote the flow generated by (3.2.29) by

$$\bar{\phi}(t, \bar{x}_0, \bar{y}_0, \epsilon) = \big(\bar{x}(t, \bar{x}_0, \bar{y}_0, \epsilon), \bar{y}(t, \bar{x}_0, \bar{y}_0, \epsilon) \big) \, , \tag{3.2.31}$$

and it should be clear that

$$\bar{\phi}(t, \bar{x}_0, \bar{y}_0, 0) = \big(e^{At} \bar{x}_0, e^{Bt} \bar{y}_0 \big) \, , \tag{3.2.32}$$

i.e., at $\epsilon = 0$ (3.2.9) reduces to the linearized equations. Thus, the rescaling by ϵ has the effect of "magnifying" a neighborhood of the origin. In the rescaled coordinates the cross-sections to the vector field take the form

$$C_1^s = \{ (\bar{x}, \bar{y}) \in \mathbf{R}^s \times \mathbf{R}^u \mid |\bar{x}| = 1 \, , \ |\bar{y}| < 1 \}$$
$$C_1^u = \{ (\bar{x}, \bar{y}) \in \mathbf{R}^s \times \mathbf{R}^u \mid |\bar{x}| < 1 \, , \ |\bar{y}| = 1 \} \tag{3.2.33}$$

and the intersection of the stable manifold with C_1^s and the unstable manifold with C_1^u are given by

$$S_1^s = \{ (\bar{x}, \bar{y}) \in \mathbf{R}^s \times \mathbf{R}^u \mid |\bar{x}| = 1, \ |y| = 0 \}$$
$$S_1^u = \{ (\bar{x}, \bar{y}) \in \mathbf{R}^s \times \mathbf{R}^u \mid |\bar{x}| = 0, \ |\bar{y}| = 1 \} \, . \tag{3.2.34}$$

Then, in the scaled coordinates, the map near the origin becomes

$$\bar{P}_0 : C_1^s - S_1^s \to C_1^u - S_1^u$$
$$(\bar{x}_0, \bar{y}_0) \mapsto \bar{\phi} \big(T(\bar{x}_0, \bar{y}_0, \epsilon), \bar{x}_0, \bar{y}_0, \epsilon \big) \tag{3.2.35}$$

where $T(\bar{x}_0, \bar{y}_0, \epsilon)$ is the solution of

$$|\bar{y}(T, \bar{x}_0, \bar{y}_0, \epsilon)| = 1 \,. \tag{3.2.36}$$

For $\epsilon = 0$ we denote the map by

$$\bar{P}_0^L : C_1^s - S_1^s \to C_1^u - S_1^u \tag{3.2.37}$$
$$(\bar{x}_0, \bar{y}_0) \mapsto (e^{ATL}\bar{x}_0, e^{BTL}\bar{y}_0)$$

where T_L is the solution of

$$\left| e^{BTL}\bar{y}_0 \right| = 1 \,. \tag{3.2.38}$$

We now want to show that $\left| \bar{P}_0 - \bar{P}_0^L \right| = O(\epsilon)$. However, we will first need some preliminary lemmas.

Lemma 3.2.2. *The solution of (3.2.36), $T(\bar{x}_0, \bar{y}_0, \epsilon)$, is a C^r function of $(\bar{x}_0, \bar{y}_0, \epsilon)$ for $(\bar{x}_0, \bar{y}_0) \in C_1^s - S_1^s$ and for ϵ sufficiently close to zero.*

PROOF: The equation for the time of flight of a point $(\bar{x}_0, \bar{y}_0) \in C_1^s - S_1^s$ to $C_1^u - S_1^u$ is given by

$$h(T, \bar{x}_0, \bar{y}_0, \epsilon) = |\bar{y}(T, \bar{x}_0, \bar{y}_0, \epsilon)| - 1$$
$$= \sqrt{\left(\bar{y}_1(T, \bar{x}_0, \bar{y}_0, \epsilon)\right)^2 + \cdots + \left(\bar{y}_u(T, \bar{x}_0, \bar{y}_0, \epsilon)\right)^2} - 1 = 0 \tag{3.2.39}$$

where $(T, \bar{x}_0, \bar{y}_0, \epsilon) \in \mathbb{R}^1 \times C_1^s - S_1^s \times \mathbb{R}^1$. We will use the implicit function theorem to show that T is a C^r function of $(\bar{x}_0, \bar{y}_0, \epsilon)$.

Now equation (3.2.39) has a solution at $\epsilon = 0$ for each $(\bar{x}_0, \bar{y}_0) \in C_1^s - S_1^s$, namely,

$$h(T_L, \bar{x}_0, \bar{y}_0, 0) = \left| e^{BTL}\bar{y}_0 \right| - 1 = 0 \,. \tag{3.2.40}$$

A simple calculation using (3.2.39) gives that

$$D_t h(T_L, \bar{x}_0, \bar{y}_0, 0) = \frac{\bar{y}(T_L, \bar{x}_0, \bar{y}_0, 0)}{|\bar{y}(T_L, \bar{x}_0, \bar{y}_0, 0)|} \cdot B\bar{y}(T_L, \bar{x}_0, \bar{y}_0, 0) \,. \tag{3.2.41}$$

So, since B has eigenvalues with nonzero real parts, (3.2.41) is nonzero for each $(\bar{x}_0, \bar{y}_0) \in C_1^s - S_1^s$, or, more geometrically, (3.2.41) is non-zero since C_1^u is a cross-section of the vector field. Hence, by the implicit function theorem, for ϵ sufficiently close to zero and for each $(\bar{x}_0, \bar{y}_0) \in C_1^s - S_1^s$ $T = T(x_0, y_0, \epsilon)$ is C^r in (x_0, y_0, ϵ).
$\square$

Lemma 3.2.3. $D_\epsilon T(\bar{x}_0, \bar{y}_0, 0)$ and $D_\epsilon^2 T(\bar{x}_0, \bar{y}_0, 0)$ are bounded in $C_1^s - S_1^s$.

PROOF: We have shown that T is $\bar{C}^r$ in $C_1^s - S_1^s$. However, a problem may arise as S_1^s is approached (i.e., as $|\bar{y}_0| \to 0$), since in this case the time of flight approaches ∞ logarithmically. Thus, for proving the lemma it suffices to show that $\varlimsup_{|\bar{y}_0| \to 0} D_\epsilon T(\bar{x}_0, \bar{y}_0, 0)$ and $\varlimsup_{|\bar{y}_0| \to 0} D_\epsilon^2 T(\bar{x}_0, \bar{y}_0, 0)$ are bounded.

We can compute these derivatives directly using (3.2.39) and the implicit function theorem. We begin by computing $D_\epsilon T(\bar{x}_0, \bar{y}_0, 0)$.

Using Lemma 3.2.2 and (3.2.39), we obtain

$$D_\epsilon T(\bar{x}_0, \bar{y}_0, 0) = -\big[D_t h\big(T(\bar{x}_0, \bar{y}_0, 0), \bar{x}_0, \bar{y}_0, 0\big)\big]^{-1} D_\epsilon h\big(T(\bar{x}_0, \bar{y}_0, 0), \bar{x}_0, \bar{y}_0, 0\big)$$

(3.2.42)

and, from (3.2.41), we obtain

$$\begin{aligned} D_t h\big(T(\bar{x}_0, \bar{y}_0, 0), \bar{x}_0, \bar{y}_0, 0\big) &= \frac{\bar{y}(T_L, \bar{x}_0, \bar{y}_0, 0)}{|\bar{y}(T_L, \bar{x}_0, \bar{y}_0, 0)|} \cdot B\bar{y}(T_L, \bar{x}_0, \bar{y}_0, 0) \\ &= \frac{e^{BT_L}\bar{y}_0}{|e^{BT_L}\bar{y}_0|} \cdot Be^{BT_L}\bar{y}_0 \end{aligned}$$

(3.2.43)

and

$$D_\epsilon h\big(T(\bar{x}_0, \bar{y}_0, 0), \bar{x}_0, \bar{y}_0, 0\big) = \frac{e^{BT_L}\bar{y}_0}{|e^{BT_L}\bar{y}_0|} \cdot D_\epsilon \bar{y}(T_L, \bar{x}_0, \bar{y}_0, 0) \,.$$

(3.2.44)

Thus, using (3.2.43) and (3.2.44) we have

$$D_\epsilon T(\bar{x}_0, \bar{y}_0, 0) = -[e^{BT_L}\bar{y}_0 \cdot Be^{BT_L}\bar{y}_0]^{-1}[e^{BT_L}\bar{y}_0 \cdot D_\epsilon \bar{y}(T_L, \bar{x}_0, \bar{y}_0, 0)].$$

(3.2.45)

Now, in order to show that (3.2.45) is bounded as $|\bar{y}_0| \to 0$, it suffices to show two things:

$$\varlimsup_{|\bar{y}_0| \to 0} D_t h(T_L, \bar{x}_0, \bar{y}_0, 0) \quad \text{is bounded;} \tag{3.2.46a}$$

$$\varlimsup_{|\bar{y}_0| \to 0} D_\epsilon \bar{y}(T_L, \bar{x}_0, \bar{y}_0, 0) \quad \text{is bounded.} \tag{3.2.46b}$$

(3.2.46a) follows from the geometrical fact that, for each $(\bar{x}_0, \bar{y}_0) \in C_1^s - S_1^s$, we have

$$\left| e^{BT_L}\bar{y}_0 \right| = 1 \,.$$

We remark that it is necessary to consider the limit superior rather than the limit for (3.2.46a), since the limit may change as $|\bar{y}_0| \to 0$ along different eigendirections of B.

The fact that (3.2.46b) holds relies heavily on the following:

$$1) \quad \bar{f}_2(\bar{x}, 0, \epsilon) = 0,$$

$$2) \quad \left| e^{At}\bar{x}_0 \right| \le Ke^{-\alpha t} |\bar{x}_0| \qquad t \ge 0,$$

$$\left| e^{Bt}\bar{y}_0 \right| \le Ke^{\alpha t} |\bar{y}_0| \qquad t \le 0.$$

For some constants $K, \alpha > 0$.

Now 2) follows from the fact that the origin is a hyperbolic fixed point (see Hale [1980]), and 1) follows from the choice of the local stable and unstable manifolds as coordinates. Along with the fact that f_2 is C^{r-1}, $r \ge 1$, the latter also implies the existence of a constant $\bar{K} > 0$ such that

$$|f_2(x, y)| \le \bar{K}(|x|\,|y| + |y|^2).$$

For $(x, y) \in \mathcal{N}^s \times \mathcal{N}^u$.

Using (3.2.29) we have

$$\left| \bar{f}_2(\bar{x}, \bar{y}, \epsilon) \right| \le \bar{K}\epsilon[|\bar{x}|\,|\bar{y}| + |\bar{y}|^2].$$

From (3.2.30) we obtain

$$\frac{\left| \bar{f}_2(\bar{x}, \bar{y}, \epsilon) - \bar{f}_2(\bar{x}, \bar{y}, 0) \right|}{\epsilon} \le \bar{K}\left[|\bar{x}|\,|\bar{y}| + |\bar{y}|^2 \right]$$

and hence

$$\left| D_\epsilon \bar{f}_2(\bar{x}, \bar{y}, 0) \right| \le \bar{K}\left[|\bar{x}|\,|\bar{y}| + |\bar{y}|^2 \right] \quad . \tag{3.2.47}$$

Now, using (3.2.47), we can solve for $D_\epsilon \bar{y}(T_L, \bar{x}_0, \bar{y}_0, 0)$ directly using the variation of constants formula (see Arnold [1973] or Hale [1980]) to obtain

$$D_\epsilon y(T_L, \bar{x}_0, \bar{y}_0, 0) = e^{BT_L} \int_0^{T_L} e^{-Bs} D_\epsilon \bar{f}_2(e^{As}\bar{x}_0, e^{Bs}\bar{y}_0, 0)ds . \tag{3.2.48}$$

Using 2), (3.2.47), and (3.2.48) it can easily be shown that (3.2.46b) is bounded as $|y_0| \to 0$.

We next compute $D_\epsilon^2 T(\bar{x}_0, \bar{y}_0, 0)$. Using Lemma 3.2.2 and (3.2.39) we obtain

$$D_\epsilon^2 T(\bar{x}_0, \bar{y}_0, 0) = -[D_t h]^{-1}[D_t^2 h(D_\epsilon T)^2 + 2(D_\epsilon D_t h)D_\epsilon T + D_\epsilon^2 h] \tag{3.2.49}$$

where all derivatives in (3.2.49) are evaluated at $\big(T(\bar{x}_0,\bar{y}_0,0),\bar{x}_0,\bar{y}_0,0\big)$. Now we have already shown that

$$\overline{\lim_{|\bar{y}_0|\,\to\,0}}\ D_t h\big(T(\bar{x}_0,\bar{y}_0,0),\bar{x}_0,\bar{y}_0,0\big)\quad\text{is bounded;}\tag{3.2.50}$$

and

$$\overline{\lim_{|\bar{y}_0|\,\to\,0}}\ D_\epsilon T(\bar{x}_0,\bar{y}_0,0)\quad\text{is bounded.}\tag{3.2.51}$$

Simple calculations using (3.2.39) give

$$D_t^2 h|_{\epsilon=0}=\left(\frac{Be^{BT_L}\bar{y}_0}{\left|e^{BT_L}\bar{y}_0\right|}-\frac{e^{BT_L}\bar{y}_0\left(\frac{e^{BT_L}\bar{y}_0}{\left|e^{BT_L}\bar{y}_0\right|}\cdot Be^{BT_L}\bar{y}_0\right)}{\left|e^{BT_L}\bar{y}_0\right|^2}\right)\cdot Be^{BT_L}\bar{y}_0\tag{3.2.52}$$

$$D_\epsilon D_t h=D_\epsilon\left(\frac{\bar{y}\big(T(\bar{x}_0,\bar{y}_0,\epsilon),\bar{x}_0,\bar{y}_0,\epsilon\big)}{\left|\bar{y}\big(T(\bar{x}_0,\bar{y}_0,\epsilon),\bar{x}_0,\bar{y}_0,\epsilon\big)\right|}\cdot B\bar{y}\big(T(\bar{x}_0,\bar{y}_0,\epsilon),\bar{x}_0,\bar{y}_0,\epsilon\big)\right)\tag{3.2.53}$$

$$D_\epsilon^2 h=D_\epsilon\left(\frac{\bar{y}\big(T(\bar{x}_0,\bar{y}_0,\epsilon),\bar{x}_0,\bar{y}_0,\epsilon\big)}{\left|\bar{y}\big(T(\bar{x}_0,\bar{y}_0,\epsilon),\bar{x}_0,\bar{y}_0,\epsilon\big)\right|}\cdot D_\epsilon\bar{y}\big(T(\bar{x}_0,\bar{y}_0,\epsilon),\bar{x}_0,\bar{y}_0,\epsilon\big)\right).\tag{3.2.54}$$

Now, using arguments similar to those given above (i.e., $\bar{f}_2(\bar{x},0,\epsilon)=0$, hyperbolicity of the fixed point, and the variation of constants formula), we can conclude that the limit superiors as $|\bar{y}_0|\to 0$ (3.2.52), (3.2.53), and (3.2.54) are bounded; hence,

$$\overline{\lim_{|\bar{y}_0|\,\to\,0}}\ D_\epsilon^2 T(\bar{x}_0,\bar{y}_0,\epsilon)\quad\text{is bounded.}\tag{3.2.55}$$

We leave the details to the reader. $\square$

We now use Lemmas 3.2.2 and 3.2.3 to prove the following proposition.

Proposition 3.2.4. $\left|\bar{P}_0-\bar{P}_0^L\right|=\mathcal{O}(\epsilon)$.

PROOF: By Lemma 3.2.2, for each $(\bar{x}_0,\bar{y}_0)\in C_1^s-S_1^s$ we can Taylor expand T as follows

$$T(\bar{x}_0,\bar{y}_0,\epsilon)=T(\bar{x}_0,\bar{y}_0,0)+\epsilon T_1(\bar{x}_0,\bar{y}_0,0)+\mathcal{O}(\epsilon^2)\tag{3.2.56}$$

where $T_1(\bar{x}_0,\bar{y}_0,0)\equiv D_\epsilon T(\bar{x}_0,\bar{y}_0,0)$ and $T(\bar{x}_0,\bar{y}_0,0)=T_L$.

Now, using the expression (3.2.56) in $\bar{P}_0(\bar{x}_0,\bar{y}_0)$ and Taylor expanding about $\epsilon=0$, we obtain

$$\bar{P}_0(\bar{x}_0,\bar{y}_0,\epsilon)=\big(\bar{x}(T_L,\bar{x}_0,\bar{y}_0,0),\bar{y}(T_L,\bar{x}_0,\bar{y}_0,0)\big)$$
$$+\epsilon\big(D_\epsilon\bar{x}(T_L,\bar{x}_0,\bar{y}_0,0)+T_1 D_t\bar{x}(T_L,\bar{x}_0,\bar{y}_0,0),D_\epsilon\bar{y}(T_L,\bar{x}_0,\bar{y}_0,0)+T_1 D_t\bar{y}(T_L,\bar{x}_0,\bar{y}_0,0)\big)$$
$$+\mathcal{O}(\epsilon^2).$$
$$\tag{3.2.57}$$

From (3.2.32) we have $\left(\bar{x}(T_L, \bar{x}_0, \bar{y}_0, 0), \bar{y}(T_L, \bar{x}_0, \bar{y}_0, 0)\right) = \left(e^{AT_L}\bar{x}_0, e^{BT_L}\bar{y}_0\right)$, so that (3.2.57) can be written as

$$\bar{P}_0(\bar{x}_0, \bar{y}_0, \epsilon) = \bar{P}_0^L(\bar{x}_0, \bar{y}_0)$$
$$+ \epsilon\left(D_\epsilon \bar{x}(T_L, \bar{x}_0, \bar{y}_0, 0) + T_1 D_t \bar{x}(T_L, \bar{x}_0, \bar{y}_0, 0), D_\epsilon \bar{y}(T_L, \bar{x}_0, \bar{y}_0, 0) + T_1 D_t \bar{y}(T_L, \bar{x}_0, \bar{y}_0, 0)\right)$$
$$+ \mathcal{O}(\epsilon^2).$$

$$(3.2.58)$$

Now Lemma 3.2.3 and the mean value theorem applied to (3.2.58) proves the proposition. $\qquad\square$

Transforming back to the unscaled, original coordinates we obtain the following result.

Proposition 3.2.5. $\left|P_0 - P_0^L\right| = \mathcal{O}(\epsilon^2).$

PROOF: This is an obvious consequence of Proposition 3.2.4 and the rescaling. $\square$

We also have the following important result.

Proposition 3.2.6. $\left|D\bar{P}_0 - D\bar{P}_0^L\right| = \mathcal{O}(\epsilon), \left|DP_0 - DP_0^L\right| = \mathcal{O}(\epsilon^2).$

PROOF: $\left|D\bar{P}_0 - D\bar{P}_0^L\right| = \mathcal{O}(\epsilon)$ follows a proof similar to that given in Proposition 3.2.4. The main step is to show that DT and D^2T are bounded as $|y_0| \to 0$ analogous to Lemma 3.2.3. We leave the details to the reader. The relation $\left|DP_0 - DP_0^L\right| = \mathcal{O}(\epsilon^2)$ follows by transforming back to the original unscaled coordinates. $\qquad\square$

Now the relationship between the approximate and exact maps along $\psi(t)$ is relatively trivial. Recall that in Step 5 we defined a map P_1 along $\psi(t)$ from a neighborhood $U_\alpha \subset C_\epsilon^u - S_\epsilon^u$ into $U_\beta \subset C_\epsilon^s - S_\epsilon^s$. Taylor expanding P_1 about the point $u = \alpha$ gave

$$P_1(\alpha + u') = P_1(\alpha) + DP_1(\alpha)u' + \mathcal{O}(u'^2)$$
$$= \beta + DP_1(\alpha)u' + \mathcal{O}(u'^2).$$

$$(3.2.59)$$

We defined an approximation to P_1 by

$$P_1^L : U_\alpha \to U_\beta$$
$$\alpha + u' \to \beta + DP_1(\alpha)u',$$

$$(3.2.60)$$

and we have the following result.

Proposition 3.2.7. $\left| P_1 - P_1^L \right| = O(\epsilon^2).$

PROOF: This is a trivial consequence of the fact that the diameter of the sets C_ϵ^s and C_ϵ^u is $O(\epsilon)$. $\qquad\square$

We are now in a position to show the relationship between the exact and approximate Poincaré maps. We assume that it is possible to choose $V_\beta \subset U_\beta$ such that

$$P_0(V_\beta) \subset U_\alpha \quad \text{and} \quad P_0^L(V_\beta) \subset U_\alpha, \qquad (3.2.61)$$

and we have defined the Poincaré maps

$$P \equiv P_1 \circ P_0 : V_\beta \to U_\beta \qquad (3.2.62)$$

$$P^L \equiv P_1^L \circ P_0^L : V_\beta \to U_\beta. \qquad (3.2.63)$$

We have the following result.

Proposition 3.2.8. $\left| P - P^L \right| = O(\epsilon^2), \ \left| DP - DP^L \right| = O(\epsilon^2).$

PROOF: This is a simple consequence of Propositions 3.2.5, 3.2.6, and 3.2.7. $\qquad\square$

The results that we need, which will relate the dynamics of P^L to P, are as follows.

Proposition 3.2.9. *Suppose P^L has a hyperbolic fixed point of (x_0, y_0). Then, for ϵ sufficiently small, P has a hyperbolic fixed point of the same stability type at $(x_0, y_0) + O(\epsilon^2)$.*

PROOF: This follows from an application of the implicit function theorem to the map in the scaled coordinates $(\bar{x}, \bar{y})$. $\qquad\square$

Proposition 3.2.10. *Suppose P^L satisfies A1 and A2, A1 and A3, or $\overline{A1}$ and $\overline{A2}$ of Section 2.3. Then for ϵ sufficiently small, P also satisfies A1 and A2, A1 and A3, or $\overline{A1}$ and $\overline{A2}$.*

PROOF: This follows from the definition of A1, A2, A3, $\overline{A1}$, and $\overline{A2}$ and the fact that the maps as well as their first derivatives are close as described in Proposition 3.2.8. $\qquad\square$

Finally, we remark that this entire analysis goes through in the case where (3.2.1) depends in a C^r $(r \geq 2)$ manner on the parameters.

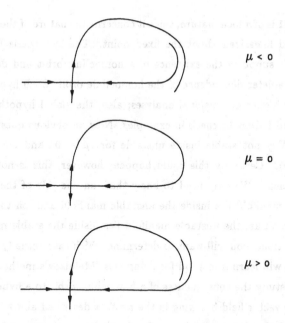

$\mu < 0$

$\mu = 0$

$\mu > 0$

Figure 3.2.4. Behavior of the Homoclinic Orbit as μ is Varied.

3.2b. Planar Systems

Consider the ordinary differential equation

$$\begin{aligned} \dot{x} &= \alpha x + f_1(x, y; \mu) \\ \dot{y} &= \beta y + f_2(x, y; \mu) \end{aligned} \qquad (x, y, \mu) \in \mathbb{R}^1 \times \mathbb{R}^1 \times \mathbb{R}^1 \qquad (3.2.64)$$

with $f_1, f_2 = \mathcal{O}(|x|^2 + |y|^2)$ and $C^r, r \geq 2$ and where μ is regarded as a parameter. We make the following hypotheses on (3.2.64).

H1. $\alpha < 0$, $\beta > 0$, and $\alpha + \beta \neq 0$.

H2. At $\mu = 0$ (3.2.64) possesses a homoclinic orbit connecting the hyperbolic fixed point $(x, y) = (0, 0)$ to itself, and on both sides of $\mu = 0$ the homoclinic orbit is broken. Furthermore, the homoclinic orbit breaks in a transverse manner in the sense that the stable and unstable manifolds have different orientations on different sides of $\mu = 0$. For definiteness, we will assume that, for $\mu < 0$, the stable manifold lies outside the unstable manifold, for $\mu > 0$, the stable manifold lies inside the unstable manifold and, for $\mu = 0$, they coincide see Figure 3.2.4.

The hypothesis H1 is of a local nature, since it concerns the nature of the eigenvalues of the vector field linearized about the fixed point. The hypothesis H2 is global in nature, since it supposes the existence of a homoclinic orbit and describes the nature of the parameter dependence of the homoclinic orbit. Such hypotheses will be typical of our higher dimensional analyses; also, the global hypothesis will be more intricate and harder to check in examples. Now an obvious question is, why this scenario? Why not stable inside unstable for $\mu < 0$ and unstable inside stable for $\mu < 0$? Certainly this could happen; however, this is not important to us at the moment. We only need to know that, on one side of the bifurcation value, the stable manifold lies inside the unstable manifold and, on the other side of the bifurcation value, the unstable manifold lies inside the stable manifold. Of course, in applications, you will want to determine which case actually occurs and, in Chapter 4, we will learn a method for doing this (Melnikov's method); however, now we will just study the consequences of a homoclinic orbit to a hyperbolic fixed point of a planar vector field breaking in the manner described above.

Let us remark that it is certainly possible for the eigenvalues α and β to depend on the parameter μ. However, this will be of no consequence provided H1 is satisfied for each parameter value, and this is true for μ sufficiently close to zero.

The question we ask is the following: *What is the nature of the orbit structure near the homoclinic orbit for μ near $\mu = 0$?* We will answer this question by computing a Poincaré map near the homoclinic orbit as described in Section 3.2a and studying the orbit structure of the Poincaré map.

From 3.2a the analysis will proceed in several steps.

Step 1. Set up the domains for the Poincaré map.

Step 2. Compute P_0^L.

Step 3. Compute P_1^L.

Step 4. Examine the dynamics of $P^L = P_1^L \circ P_0^L$.

We begin with Step 1. *Set up the domains for the Poincaré map.*

For the domain of P_0^L we choose

$$\Pi_0 = \{ (x,y) \in C_\epsilon^s \mid x = \epsilon > 0 , \, y > 0 \} , \qquad (3.2.65)$$

and for the domain of P_1^L we choose

$$\Pi_1 = \{ (x,y) \in C_\epsilon^u \mid x > 0 , \, y = \epsilon > 0 \} \qquad (3.2.66)$$

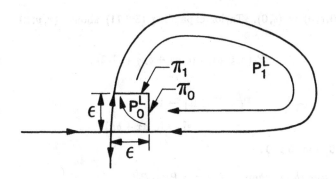

Figure 3.2.5. Π_0 and Π_1.

where C_ϵ^s and C_ϵ^u are defined in (3.2.11), see Figure 3.2.5.

Step 2. *Compute P_0^L.*

The flow defined by the linearization of (3.2.64) about the origin is given by

$$x(t) = x_0 e^{\alpha t}$$
$$y(t) = y_0 e^{\beta t} .$$
(3.2.67)

The time of flight, T, needed for a point $(\epsilon, y_0) \in \Pi_0$ to reach Π_1 under the action of (3.2.67) is given by solving

$$\epsilon = y_0 e^{\beta T}$$
(3.2.68)

to get

$$T = \frac{1}{\beta} \log \frac{\epsilon}{y_0} .$$
(3.2.69)

Thus, P_0^L is given by

$$P_0^L : \Pi_0 \to \Pi_1$$
$$(\epsilon, y_0) \mapsto \left(\epsilon \left(\frac{\epsilon}{y_0}\right)^{\alpha/\beta}, \epsilon\right).$$
(3.2.70)

Step 3. *Compute P_1^L.*

From Step 5, part b of 3.2a, by smoothness of the flow with respect to initial conditions and the fact that it only takes a finite time to flow from Π_1 to Π_0 along the homoclinic orbit, we can find a neighborhood $U \subset \Pi_1$ which is mapped onto Π_0 under the flow generated by (3.2.64). We denote this map by

$$P_1(x, y; \mu) = \left(P_{11}(x, y; \mu), P_{12}(x, y; \mu)\right) : U \subset \Pi_1 \to \Pi_0$$
(3.2.71)

where $P_1(0, \epsilon; 0) = (\epsilon, 0)$. Taylor expanding (3.2.71) about $(x, y; \mu) = (0, \epsilon; 0)$ gives

$$P_1(x, y; \mu) = (\epsilon, ax + b\mu) + \mathcal{O}(2) . \qquad (3.2.72)$$

So we have

$$P_1^L : U \subset \Pi_1 \to \Pi_0 \qquad\qquad (3.2.73)$$
$$(x, \epsilon) \mapsto (\epsilon, ax + b\mu)$$

where $a > 0$ and $b > 0$.

Step 4. *Examine the dynamics of* $P^L = P_1^L \circ P_0^L$.

We have

$$P^L = P_1^L \circ P_0^L : V \subset \Pi_0 \to \Pi_0$$
$$(\epsilon, y_0) \mapsto \left(\epsilon, a\epsilon \left(\frac{\epsilon}{y_0}\right)^{\alpha/\beta} + b\mu\right) \qquad (3.2.74)$$

where $V = (P_0^L)^{-1}(U)$, or

$$P^L(y; \mu) : y \to A y^{|\alpha/\beta|} + b\mu$$

where $A \equiv a\epsilon^{1+(\alpha/\beta)} > 0$ (we have left the subscript "0" off the y_0 for the sake of a less cumbersome notation).

Let $\delta = |\alpha/\beta|$; then $\alpha + \beta \neq 0$ implies $\delta \neq 1$. We will seek fixed points of the Poincaré map, i.e., $y \in \Pi_0$ such that

$$P^L(y; \mu) = Ay^\delta + b\mu = y . \qquad (3.2.75)$$

The fixed points can be displayed graphically as the intersection of the graph of $P^L(y; \mu)$ with the line $y = P^L(y; \mu)$ for fixed μ.

There are two distinct cases.

Case 1. $|\alpha| > |\beta|$ *or* $\delta > 1$.

For this case $D_y P^L(0; 0) = 0$, and the graph of P^L appears as in Figure 3.2.6 for $\mu > 0$, $\mu = 0$, and $\mu < 0$.

So for $\mu > 0$ and small (3.2.75) has a fixed point. The fixed point is stable and hyperbolic, since $0 < D_y P^L < 1$ for μ sufficiently small. Appealing to Proposition 3.2.9 we can conclude that this fixed point corresponds to an attracting periodic orbit of (3.2.64), see Figure 3.2.7. We remark that if the homoclinic orbit were to

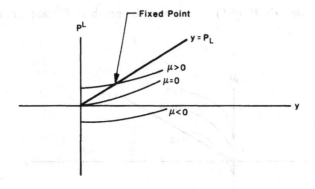

Figure 3.2.6. Graph of P^L for $\mu > 0$, $\mu = 0$, and $\mu < 0$ with $\delta > 1$.

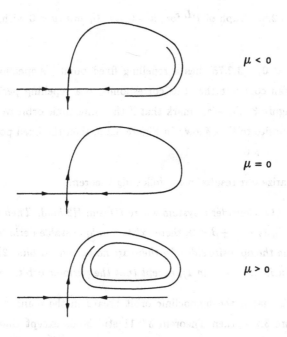

Figure 3.2.7. Phase Plane of (3.2.64) for $\delta > 1$.

break in the manner opposite to that shown in Figure 3.2.7, then the fixed point of (3.2.75) would occur for $\mu < 0$.

Case 2. $|\alpha| < |\beta|$ *or* $\delta < 1$.

For this case, $D_y P^L(0;0) = \infty$, and the graph of P^L appears as in Figure 3.2.8.

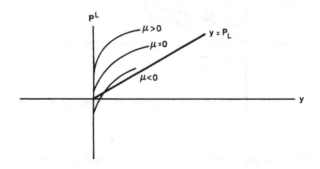

Figure 3.2.8. Graph of P^L for $\mu > 0, \mu = 0$, and $\mu < 0$ with $\delta < 1$.

So for $\mu < 0$, (3.2.75) has a repelling fixed point. Appealing to Proposition 3.2.9 we can conclude that this corresponds to a repelling periodic orbit for (3.2.64), see Figure 3.2.9. We remark that if the homoclinic orbit were to break in the manner opposite to that shown in Figure 3.2.9, then the fixed point of (3.2.75) would occur for $\mu > 0$.

We summarize our results in the following theorem.

Theorem 3.2.11. *Consider a system where H1 and H2 hold. Then we have, for μ sufficiently small, 1) If $\alpha + \beta < 0$, there exists a unique stable periodic orbit on one side of $\mu = 0$; on the opposite side of μ there are no periodic orbits. 2) If $\alpha + \beta > 0$, the same conclusion holds as in 1), except that the periodic orbit is unstable.*

We remark that if the homoclinic orbit breaks in the manner opposite that shown in Figure 3.2.4, then Theorem 3.2.11 still holds except that the periodic orbits occur for μ values having the opposite sign as those given in Theorem 3.2.11. Theorem 3.2.11 is a classical result which can be found in Andronov et al. [1971]. Additional proofs can be found in Guckenheimer and Holmes [1983] and Chow and Hale [1982].

An interesting situation arises if (3.2.64) is invariant under the coordinate change $(x,y) \to (-x,-y)$. In this case (3.2.64) is symmetric with respect to $180°$

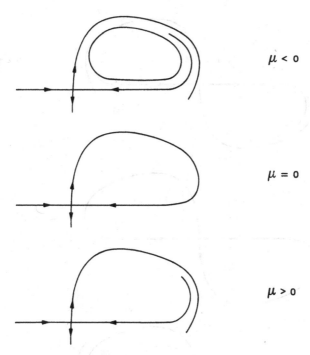

$\mu < 0$

$\mu = 0$

$\mu > 0$

Figure 3.2.9. Phase Plane of (3.2.64) for $\delta < 1$.

rotations about the origin and therefore must possess an additional homoclinic orbit. Then H2 would be modified as shown in Figure 3.2.10.

Similar conclusions as those in Theorem 3.2.11 hold with the provision that two periodic orbits may be formed, one for each homoclinic orbit. There is an important additional effect due to the symmetry. The symmetry enables us to compute a Poincaré map outside of the homoclinic orbit which consists of the composition of four maps, two maps through a neighborhood of the saddle point and one map around each homoclinic orbit. In this case a periodic orbit may bifurcate from the homoclinic orbit which completely surrounds the stable and unstable manifolds of the fixed point. We leave the details to the interested reader, but in Figure 3.2.11 we show the scenario supposing that H1 and H2 hold with $|\alpha| > |\beta|$.

We end our study of planar systems with the following remarks.

1. *The Case* $\alpha + \beta = 0$. In this case, it should be clear that our methods fail. Andronov et al. [1971] state that, in this case, multiple limit cycles will

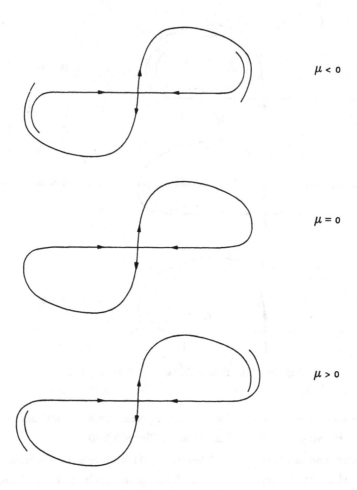

Figure 3.2.10. Behavior of the Symmetric Homoclinic Orbits as μ is Varied.

bifurcate from the homoclinic orbit and present some results for special cases. Dangelmayr and Guckenheimer [1987] have developed techniques which can be used in this situation.

2. *Multiple Homoclinic Orbits without Symmetry*. See Dangelmayr and Gucken-heimer [1987].

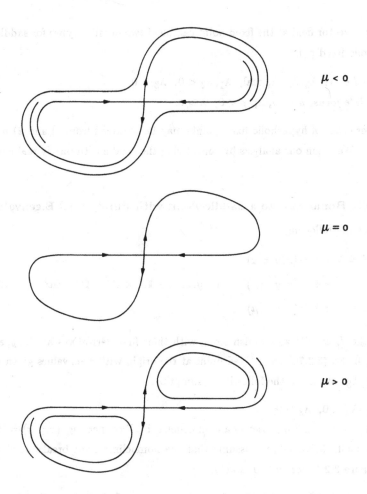

$\mu < 0$

$\mu = 0$

$\mu > 0$

Figure 3.2.11. Bifurcations to Periodic Orbits in the Symmetric Case.

3.2c. Third Order Systems

Now we will consider 3-dimensional vector fields possessing an orbit homoclinic to a fixed point and study the orbit structure in the neighborhood of the homoclinic orbit. We will see that the nature of the orbit structure depends considerably on two important properties:

1) The nature of the eigenvalues of the linearized vector field at the fixed point.
2) The existence of symmetries.

Regarding condition 1) above, it should be clear that the three eigenvalues of the

linearized vector field at the fixed point can be of two possible types for saddle type
hyperbolic fixed points:

1) *Saddle* $\lambda_1, \lambda_2, \lambda_3, \lambda_i$ real, $\lambda_1, \lambda_2 < 0$, $\lambda_3 > 0$.
2) *Saddle-focus* $\rho \pm i\omega, \lambda$; $\rho < 0$, $\lambda > 0$.

All other cases of hyperbolic fixed points may be obtained from 1) and 2) by time
reversal. We begin our analysis by considering the saddle with purely real eigenvalues.

i) Orbits Homoclinic to a Saddle-Point with Purely Real Eigenvalues

Consider the following:

$$\dot{x} = \lambda_1 x + f_1(x, y, z; \mu)$$
$$\dot{y} = \lambda_2 y + f_2(x, y, z; \mu) \qquad (x, y, z, \mu) \in \mathbf{R}^1 \times \mathbf{R}^1 \times \mathbf{R}^1 \times \mathbf{R}^1 \qquad (3.2.76)$$
$$\dot{z} = \lambda_3 z + f_3(x, y, z; \mu)$$

where the f_i are C^2 and vanish along with their first derivatives at $(x, y, z, \mu) = (0, 0, 0, 0)$. So (3.2.76) has a fixed point at the origin with eigenvalues given by λ_1, λ_2, and λ_3. We make the following assumptions.

H1. $\lambda_1, \lambda_2 < 0$, $\lambda_3 > 0$.

H2. At $\mu = 0$ (3.2.76) possesses a homoclinic orbit Γ connecting $(x, y, z) = (0, 0, 0)$
    to itself. Moreover, we assume that the homoclinic orbit breaks as shown in
    Figure 3.2.12 for $\mu > 0$ and $\mu < 0$.

We remark that Figure 3.2.12 is drawn for the case of $\lambda_2 > \lambda_1$ so that the homoclinic orbit enters a neighborhood of the origin on a curve which is tangent to the
y axis at the origin. We assume that (3.2.76) has no symmetries, i.e., the system is
generic.

We will analyze the orbit structure in a neighborhood of Γ in the standard way
by computing a Poincaré map on an appropriately chosen cross-section. We choose
two rectangles transverse to the flow which are defined as follows:

$$\Pi_0 = \{ (x, y, z) \in \mathbb{R}^3 \mid |x| \leq \epsilon, \quad y = \epsilon, \quad 0 < z \leq \epsilon \}$$
$$\Pi_1 = \{ (x, y, z) \in \mathbb{R}^3 \mid |x| \leq \epsilon, \quad |y| \leq \epsilon, \quad z = \epsilon \} \qquad (3.2.77)$$

for some $\epsilon > 0$, see Figure 3.2.13.

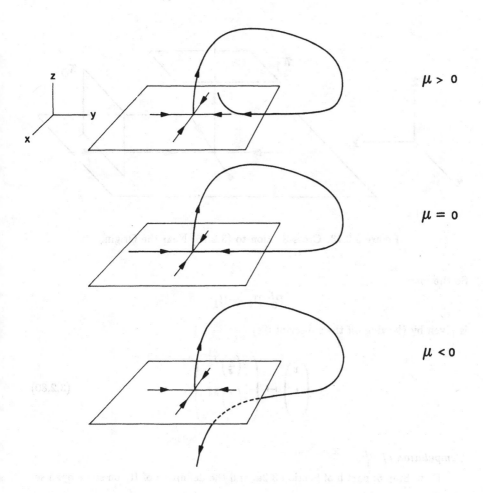

$\mu > 0$

$\mu = 0$

$\mu < 0$

Figure 3.2.12. Behavior of the Homoclinic Orbit Near $\mu = 0$.

Computation of P_0^L.

The flow linearized in a neighborhood of the origin is given by

$$
\begin{aligned}
x(t) &= x_0 e^{\lambda_1 t} \\
y(t) &= y_0 e^{\lambda_2 t} \\
z(t) &= z_0 e^{\lambda_3 t}
\end{aligned}
\tag{3.2.78}
$$

and the time of flight from Π_0 to Π_1 is given by

$$
t = \frac{1}{\lambda_3} \log \frac{\epsilon}{z_0}.
\tag{3.2.79}
$$

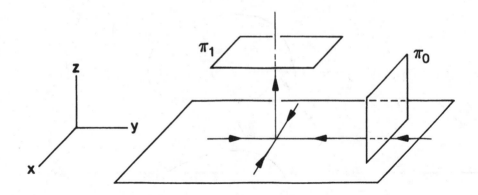

Figure 3.2.13. Cross-Section to (3.2.76) Near the Origin.

So the map

$$P_0^L : \Pi_0 \to \Pi_1$$

is given by (leaving off the subscript 0's)

$$\begin{pmatrix} x \\ \epsilon \\ z \end{pmatrix} \mapsto \begin{pmatrix} x\left(\frac{\epsilon}{z}\right)^{\frac{\lambda_1}{\lambda_3}} \\ \epsilon\left(\frac{\epsilon}{z}\right)^{\frac{\lambda_2}{\lambda_3}} \\ \epsilon \end{pmatrix}. \tag{3.2.80}$$

Computation of P_1^L.

From Step 5, part b of Section 3.2a, and the definition of Π_1 on some open set $U \subset \Pi_1$, we have

$$P_1^L : U \subset \Pi_1 \to \Pi_0$$

$$\begin{pmatrix} x \\ y \\ \epsilon \end{pmatrix} \mapsto \begin{pmatrix} 0 \\ \epsilon \\ 0 \end{pmatrix} + \begin{pmatrix} a & b & 0 \\ 0 & 0 & 0 \\ c & d & 0 \end{pmatrix} \begin{pmatrix} x \\ y \\ 0 \end{pmatrix} + \begin{pmatrix} e\mu \\ 0 \\ f\mu \end{pmatrix} \tag{3.2.81}$$

where a, b, c, d, e, and f are constants. Note from Figure 3.2.12 that we have $f > 0$, so we may rescale the parameter μ so that $f = 1$. Henceforth, we will assume that this has been done.

The Poincaré Map $P^L \equiv P_1^L \circ P_0^L$.

Forming the composition of P_0^L and P_1^L, we obtain the Poincaré map defined in a neighborhood of the homoclinic orbit having the following form.

$$P^L \equiv P_1^L \circ P_0^L : V \subset \Pi_0 \to \Pi_0$$

$$\begin{pmatrix} x \\ z \end{pmatrix} \mapsto \begin{pmatrix} ax\left(\frac{\epsilon}{z}\right)^{\frac{\lambda_1}{\lambda_3}} + b\epsilon\left(\frac{\epsilon}{z}\right)^{\frac{\lambda_2}{\lambda_3}} + e\mu \\ cx\left(\frac{\epsilon}{z}\right)^{\frac{\lambda_1}{\lambda_3}} + d\epsilon\left(\frac{\epsilon}{z}\right)^{\frac{\lambda_2}{\lambda_3}} + \mu \end{pmatrix} \tag{3.2.82}$$

where $V = (P_0^L)^{-1}(U)$.

Calculation of Fixed Points of P^L.

Now we look for fixed points of the Poincaré map (which will correspond to periodic orbits of (3.2.76)). First some notation; let

$$A = a\epsilon^{\frac{\lambda_1}{\lambda_3}}, \qquad B = b\epsilon^{1+\frac{\lambda_2}{\lambda_3}}, \qquad C = c\epsilon^{\frac{\lambda_1}{\lambda_3}}, \qquad D = d\epsilon^{1+\frac{\lambda_2}{\lambda_3}}.$$

Then the condition for fixed points of (3.2.82) is

$$x = Axz^{\frac{|\lambda_1|}{\lambda_3}} + Bz^{\frac{|\lambda_2|}{\lambda_3}} + e\mu \tag{3.2.83a}$$

$$z = Cxz^{\frac{|\lambda_1|}{\lambda_3}} + Dz^{\frac{|\lambda_2|}{\lambda_3}} + \mu. \tag{3.2.83b}$$

Solving (3.2.83a) for x as a function of z gives

$$x = \frac{Bz^{\frac{|\lambda_2|}{\lambda_3}} + e\mu}{1 - Az^{\frac{|\lambda_1|}{\lambda_3}}}. \tag{3.2.84}$$

We will restrict ourselves to a sufficiently small neighborhood of the homoclinic orbit so that z can be taken sufficiently small in order that the denominator of (3.2.84) can be taken to be 1. Substituting this expression for x into (3.2.83b) gives the following condition for fixed points of (3.2.82) in terms of z and μ only.

$$z - \mu = CBz^{\frac{|\lambda_1+\lambda_2|}{\lambda_3}} + Ce\mu z^{\frac{|\lambda_1|}{\lambda_3}} + Dz^{\frac{|\lambda_2|}{\lambda_3}}. \tag{3.2.85}$$

We will graphically display the solutions of (3.2.85) for μ sufficiently small and near zero by graphing the left hand side of (3.2.85) and the right hand side of (3.2.85) and seeking intersections of the curves.

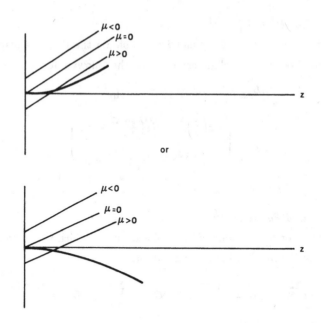

Figure 3.2.14. Graphs of the Right Versus the Left Side of (3.2.85)

for $\mu < 0, \mu = 0$, and $\mu > 0$. The Zero Slope Situations.

First, we want to examine the slope of the right hand side of (3.2.85) at $z = 0$. This is given by the following expression:

$$\frac{d}{dz}\left(CBz^{\frac{|\lambda_1+\lambda_2|}{\lambda_3}} + Ce\mu z^{\frac{|\lambda_1|}{\lambda_3}} + Dz^{\frac{|\lambda_2|}{\lambda_3}}\right)$$

$$= \frac{|\lambda_1+\lambda_2|}{\lambda_3}CBz^{\frac{|\lambda_1+\lambda_2|}{\lambda_3}-1} + \frac{|\lambda_1|}{\lambda_3}Ce\mu z^{\frac{|\lambda_1|}{\lambda_3}-1} + \frac{|\lambda_2|}{\lambda_3}Dz^{\frac{|\lambda_2|}{\lambda_3}-1}. \tag{3.2.86}$$

Now recall that P_1^L is invertible so that $ad-bc \neq 0$. This implies that $AD-BC \neq 0$ so that C and D cannot both be zero. Therefore, at $z = 0$, (3.2.86) takes the values

$$\infty \quad \text{if} \quad |\lambda_1| < \lambda_3 \text{ or } |\lambda_2| < \lambda_3$$

$$0 \quad \text{if} \quad |\lambda_1| > \lambda_3 \text{ and } |\lambda_2| > \lambda_3.$$

There are four possible cases, two each for both the ∞-slope and 0-slope situations. The differences in these situations depend mainly on global effects, i.e., the relative signs of A, B, C, D, e, and μ. We will consider this more carefully shortly. Figure 3.2.14 illustrates the graphical solution of (3.2.85) in the zero slope case.

The two zero slope cases illustrated in Figure 3.2.14 give the same result, namely, that for $\mu > 0$ a periodic orbit bifurcates from the homoclinic orbit.

In the infinite slope case the two possible situations are illustrated in Figure 3.2.15.

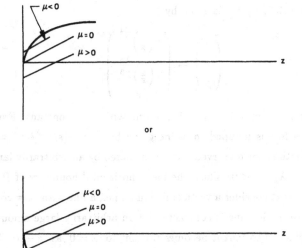

Figure 3.2.15. Graphs of the Right Versus the Left Side of (3.2.85) for $\mu < 0, \mu = 0$, and $\mu > 0$. The Infinite Slope Situations.

Interestingly, in the infinite slope case we get two different results; namely, in one case we get a periodic orbit for $\mu < 0$, and in the other case a periodic orbit for $\mu > 0$. So what's going on? As we will shortly see, there is a global effect in this case which our local analysis does not detect. Now we want to explain this global effect.

Let τ be a tube beginning and ending on Π_0 and Π_1, respectively, which contains Γ. Then $\tau \cap W^s(0)$ is a two dimensional strip which we denote as $\mathcal{R}$. Suppose, *without twisting* $\mathcal{R}$, that we join together the two ends of $\mathcal{R}$. Then there are two possibilities: 1) $W^s(0)$ experiences an even number of half-twists inside τ, in which case, when the ends of $\mathcal{R}$ are joined together it is homeomorphic to a cylinder or 2) $W^s(0)$ experiences an odd number of half-twists inside τ, in which case, when the ends of $\mathcal{R}$ are joined together it is homeomorphic to a Mobius strip, see Figure 3.2.16.

We now want to discuss the dynamical consequences of these two situations. First consider the rectangle $\mathcal{D} \subset \Pi_0$ shown in Figure 3.2.17a which has its lower horizontal boundary in $W^s(0)$. We want to consider the shape of the image of D under P_0^L. From (3.2.80) P_0^L is given by

$$\begin{pmatrix} x \\ \epsilon \\ z \end{pmatrix} \mapsto \begin{pmatrix} x\left(\frac{\epsilon}{z}\right)^{\frac{\lambda_1}{\lambda_3}} \\ \epsilon\left(\frac{\epsilon}{z}\right)^{\frac{\lambda_2}{\lambda_3}} \\ \epsilon \end{pmatrix}. \tag{3.2.87}$$

Now consider a horizontal line in $\mathcal{D}$, i.e., a line with $z = \text{constant}$. From (3.2.87) we see that this line is mapped to a line given by $y = \epsilon(\epsilon/z)^{\lambda_2/\lambda_3} = \text{constant}$. However, its length is not preserved but is contracted by an arbitrarily large amount as $z \to 0$ since $\lambda_2/\lambda_3 < 0$. Thus, the lower horizontal boundary of $\mathcal{D}$ is mapped into the origin. Next consider a vertical line in $\mathcal{D}$, i.e., a line with $x = \text{constant}$. By (3.2.87), as $z \to 0$ this line is contracted by an arbitrarily large amount in the y direction and pinched so that it becomes tangent to $x = 0$ as $z \to 0$. The upshot of this is that $\mathcal{D}$ gets mapped into a "half bowtie" shape. This process is illustrated geometrically in Figure 3.2.17b.

Now under the map P_1^L the half bowtie $P_0^L(\mathcal{D})$ is mapped back around Γ with the sharp tip of $P_0^L(\mathcal{D})$ coming back near $\Gamma \cap \Pi_0$. In the case where $\mathcal{R}$ is homeomorphic to a cylinder, $P_0^L(\mathcal{D})$ twists around an even number of times in its journey around Γ and comes back to Π_0 lying above $W^s(0)$. In the case where $\mathcal{R}$ is homeomorphic to a mobius strip, $P_0^L(\mathcal{D})$ twists around an odd number of times in its journey around Γ and returns to Π_0 lying below $W^s(0)$, see Figure 3.2.18.

Now we will go back to the four different cases which arose in locating the bifurcated periodic orbits and see which particular global effect occurs.

Recall that the z components of the fixed points were obtained by solving

$$z = CBz^{\frac{|\lambda_1+\lambda_2|-\lambda_3}{\lambda_3}} + C e \mu z^{\frac{|\lambda_1|-\lambda_3}{\lambda_3}} + Dz^{\frac{|\lambda_2|-\lambda_3}{\lambda_3}} + \mu. \tag{3.2.88}$$

The right hand side of this equation thus represents the z-component of the first return of a point to Π_0. Then, at $\mu = 0$, the first return will be positive if we have a cylinder (C) and negative if we have a mobius band (M). Using this remark, we can go back to the four cases and label them as in Figure 3.2.19.

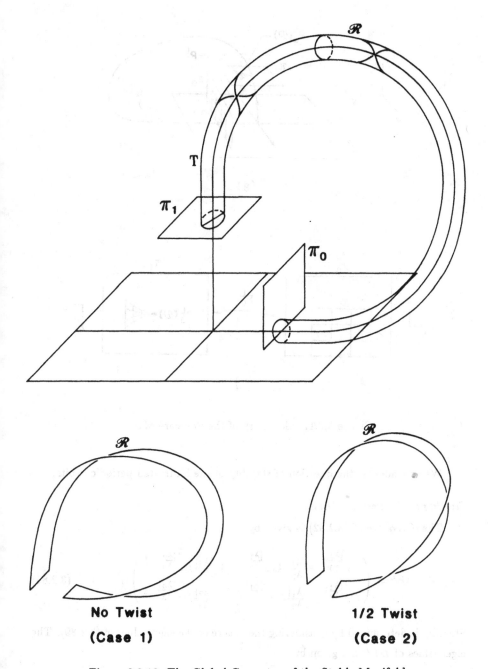

Figure 3.2.16. The Global Geometry of the Stable Manifold.

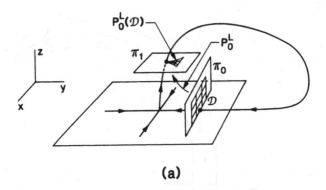

(a)

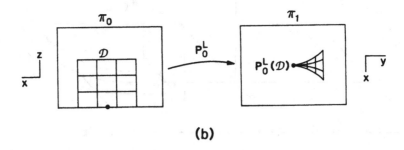

(b)

Figure 3.2.17. Geometry of the Poincaré Map.

We now address the question of stability of the bifurcated periodic orbits.

Stability of the Periodic Orbits.

The derivative of (3.2.82) is given by

$$DP^L = \begin{pmatrix} Az^{\frac{|\lambda_1|}{\lambda_3}} & \frac{|\lambda_1|}{\lambda_3}Axz^{\frac{|\lambda_1|}{\lambda_3}-1} + \frac{|\lambda_2|}{\lambda_3}Bz^{\frac{|\lambda_2|}{\lambda_3}-1} \\ Cz^{\frac{|\lambda_1|}{\lambda_3}} & \frac{|\lambda_1|}{\lambda_3}Cxz^{\frac{|\lambda_1|}{\lambda_3}-1} + \frac{|\lambda_2|}{\lambda_3}Dz^{\frac{|\lambda_2|}{\lambda_3}-1} \end{pmatrix}.$$ (3.2.89)

Stability is determined by considering the nature of the eigenvalues of (3.2.89). The eigenvalues of DP^L are given by

$$\gamma_{1,2} = \frac{\text{trace } DP^L}{2} \pm \frac{1}{2}\sqrt{(\text{trace } DP^L)^2 - 4\det(DP^L)}$$ (3.2.90)

Mobius Band

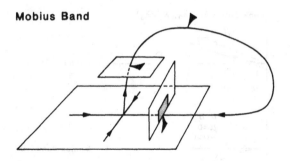

Cylinder

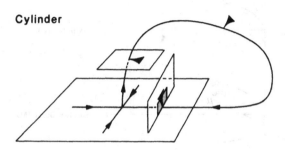

Figure 3.2.18 The Global Effect Due to the Twisting of the Stable Manifold.

where

$$\det DP^L = \frac{|\lambda_2|}{\lambda_3}(AD - BC)z^{\frac{|\lambda_1+\lambda_2|-\lambda_3}{\lambda_3}}$$

$$\text{trace } DP^L = Az^{\frac{|\lambda_1|}{\lambda_3}} + \frac{|\lambda_1|}{\lambda_3}Cxz^{\frac{|\lambda_1|}{\lambda_3}-1} + \frac{|\lambda_2|}{\lambda_3}Dz^{\frac{|\lambda_2|}{\lambda_3}-1}.$$

(3.2.91)

Substituting equation (3.2.84) for x at a fixed point into the expression for traceDP^L gives

$$\text{trace } DP^L = Az^{\frac{|\lambda_1|}{\lambda_3}} + \frac{|\lambda_1|}{\lambda_3}CBz^{\frac{|\lambda_1+\lambda_2|}{\lambda_3}-1} + \frac{|\lambda_2|}{\lambda_3}Dz^{\frac{|\lambda_2|}{\lambda_3}-1} + \frac{|\lambda_1|}{\lambda_3}Ce\mu z^{\frac{|\lambda_1|}{\lambda_3}-1}.$$

(3.2.92)

Let us note the following important facts.

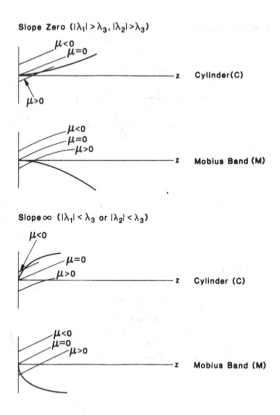

Figure 3.2.19. The z-components of the Fixed Points
and the Associated Global Effect.

For z sufficiently small

$$\det DP^L \text{ is } \begin{cases} \text{a) arbitrarily large} & \text{for } |\lambda_1 + \lambda_2| < \lambda_3; \\ \text{b) arbitrarily small} & \text{for } |\lambda_1 + \lambda_2| > \lambda_3. \end{cases}$$

$$\text{trace } DP^L \text{ is } \begin{cases} \text{a) arbitrarily large} & \text{for } |\lambda_1| < \lambda_3 \text{ or } |\lambda_2| < \lambda_3; \\ \text{b) arbitrarily small} & \text{for } |\lambda_1| > \lambda_3 \text{ and } |\lambda_2| > \lambda_3. \end{cases}$$

Using these facts along with (3.2.90) we can conclude:

1) For $|\lambda_1| > \lambda_3$ and $|\lambda_2| > \lambda_3$ both eigenvalues of DP^L can be made arbitrarily small by taking z sufficiently small.

2) For $|\lambda_1 + \lambda_2| > \lambda_3$ and $|\lambda_1| < \lambda_3$ and/or $|\lambda_2| < \lambda_3$ one eigenvalue can be made arbitrarily small and the other eigenvalue can be made arbitrarily large by taking z sufficiently small.

3) For $|\lambda_1 + \lambda_2| < \lambda_3$ both eigenvalues can be made arbitrarily large by taking z sufficiently small.

We summarize our results in the following theorem.

Theorem 3.2.12. *For* $\mu \neq 0$ *and sufficiently small, a periodic orbit bifurcates from* Γ *in (3.2.76). The periodic orbit is a*

1) *Sink for* $|\lambda_1| > \lambda_3$ *and* $|\lambda_2| > \lambda_3$;

2) *Saddle for* $|\lambda_1 + \lambda_2| > \lambda_3$, $|\lambda_1| < \lambda_3$ *and/or* $|\lambda_2| < \lambda_3$;

3) *Source for* $|\lambda_1 + \lambda_2| < \lambda_3$.

We remark that the construction of the Poincaré map used in the proof of Theorem 3.2.12 was for the case $\lambda_2 > \lambda_1$ (see Figure 3.2.12); however, the same result holds for $\lambda_2 < \lambda_1$ and $\lambda_1 = \lambda_2$. We leave the details to the reader.

Next we consider the case of two homoclinic orbits connecting the saddle type fixed point to itself and show how under certain conditions chaotic dynamics may arise.

Two Orbits Homoclinic to a Fixed Point having Real Eigenvalues.

We consider the same system as before; however, we now replace H2 with H2′ given below.

H2′ (3.2.76) has a pair of orbits, Γ_r, Γ_l, homoclinic to $(0,0,0)$ at $\mu = 0$, and Γ_r and Γ_l lie in separate branches of the unstable manifold of $(0,0,0)$. There are thus two possible pictures illustrated in Figure 3.2.20.

Note that the coordinate axes in Figure 3.2.20 have been rotated with respect to those in Figure 3.2.12. This is merely for artistic convenience. We will only consider the configuration of case a) in Figure 3.2.19. However, the same analysis (and most of the resulting dynamics) will go through the same for case b). Our goal will be to establish that the Poincaré map constructed near the homoclinic orbits contains the chaotic dynamics of the Smale horseshoe or, more specifically, that it contains an invariant Cantor set on which it is homeomorphic to the full shift on two symbols (see Section 2.2).

We begin by constructing the local cross-sections to the vector field near the

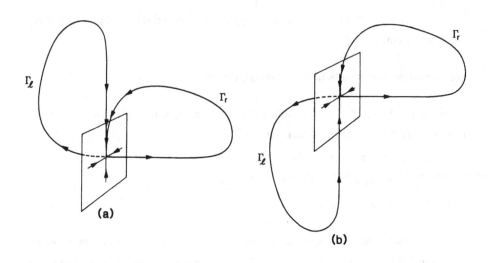

Figure 3.2.20 Possible Scenarios for Two Orbits Homoclinic to the Origin.

origin. We define

$$
\begin{aligned}
\Pi_0^r &= \{\, (x,y,z) \in \mathbf{R}^3 \mid y = \epsilon, \quad |x| \le \epsilon, \quad 0 < z \le \epsilon \,\} \\
\Pi_0^l &= \{\, (x,y,z) \in \mathbf{R}^3 \mid y = \epsilon, \quad |x| \le \epsilon, \quad -\epsilon \le z < 0 \,\} \\
\Pi_1^r &= \{\, (x,y,z) \in \mathbf{R}^3 \mid z = \epsilon, \quad |x| < \epsilon, \quad 0 < y \le \epsilon \,\} \\
\Pi_1^l &= \{\, (x,y,z) \in \mathbf{R}^3 \mid z = -\epsilon, \quad |x| < \epsilon, \quad 0 < y \le \epsilon \,\}
\end{aligned}
\tag{3.2.93}
$$

for $\epsilon > 0$ and small. See Figure 3.2.21 for an illustration of the geometry near the origin.

Now recall the global twisting of the stable manifold of the origin. We want to consider the effect of this in our construction of the Poincaré map. Let τ_r (resp. τ_l) be a tube beginning and ending on Π_1^r (resp. Π_1^l) and Π_0^r (resp. Π_0^l) which contains Γ_r (resp. Γ_l) (see Figure 3.2.15). Then $\tau_r \cap W^s(0)$ (resp. $\tau_l \cap W^s(0)$)

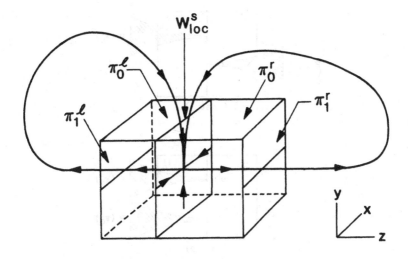

Figure 3.2.21. Local Cross-Sections to the Vector Field Near Origin.

is a two dimensional strip which we denote as $\mathcal{R}_r$ (resp. $\mathcal{R}_l$). If we join together the two ends of $\mathcal{R}_r$ (resp. $\mathcal{R}_l$) *without twisting* $\mathcal{R}_r$ *(resp.* $\mathcal{R}_l$), then $\mathcal{R}_r$ (resp. $\mathcal{R}_l$) is homeomorphic to either a cylinder or a mobius strip (see Figure 3.2.15). Thus this global effect gives rise to three distinct possibilities.

1) $\mathcal{R}_r$ and $\mathcal{R}_l$ are homeomorphic to cylinders.

2) $\mathcal{R}_r$ is homeomorphic to a cylinder and $\mathcal{R}_l$ is homeomorphic to a Mobius strip.

3) $\mathcal{R}_r$ and $\mathcal{R}_l$ are homeomorphic to Mobius strips.

These three cases manifest themselves in the Poincaré map as shown in Figure 3.2.22.

We now want to motivate how we might expect a horseshoe to arise in these situations. Consider case 1). Suppose we vary the parameter μ so that the homoclinic orbits break resulting in the images of Π_0^r and Π_0^l moving in the manner shown in Figure 3.2.23. The question of whether or not we would expect such behavior in a one parameter family of three dimensional vector fields will be addressed shortly.

From Figure 3.2.23 one can begin to see how we might get horseshoe-like dynamics in this system. We can choose μ_h-horizontal slabs in Π_0^r and Π_0^l which are mapped over themselves in μ_v-vertical slabs as μ is varied as shown in Figure 3.2.24.

Note that no horseshoe behavior is possible at $\mu = 0$. Of course many things

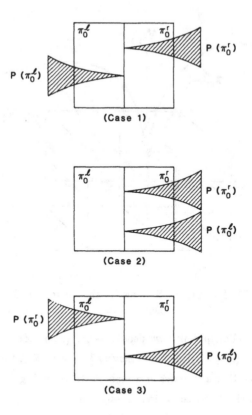

Figure 3.2.22. Geometry of the Poincaré Map, the Three Cases.

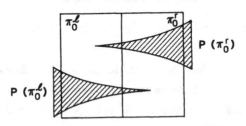

Figure 3.2.23. Geometry of the Poincaré Map for $\mu \neq 0$.

need to be justified in Figure 3.2.24, namely, the stretching and contraction rates and also that the little triangles behave correctly as the homoclinic orbits are broken.

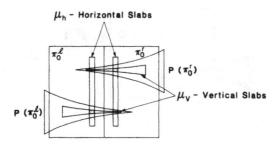

Figure 3.2.24. Horizontal and Vertical Slabs.

However, rather than go through the three cases individually, we will settle for studying a specific example and refer the reader to Afraimovich, Bykov, and Silnikov [1984] for detailed discussions of the general case. However, first we want to discuss the role of parameters.

In a three dimensional vector field one would expect that varying a parameter would result in the destruction of a particular homoclinic orbit. In the case of two homoclinic orbits we cannot expect that the behavior of both homoclinic orbits can be controlled by a single parameter resulting in the behavior shown in Figure 3.2.23. We would need two parameters where each parameter can be thought of as "controlling" a particular homoclinic orbit. In the language of bifurcation theory this is a global codimension two bifurcation problem. However, if the vector field contains a symmetry, e.g., (3.2.76) is invariant under the change of coordinates $(x, y, z) \rightarrow (-x, y, -z)$ which represents a 180° rotation about the y axis, then the existence of one homoclinic orbit necessitates the existence of another so that one parameter controls both. For simplicity we will treat the symmetric case and refer the reader to Afraimovich, Bykov, and Silnikov [1984] for a discussion of the non-symmetric cases. The symmetric case is of historical interest, since this is precisely the situation that arises in the much studied Lorenz equations, see Sparrow [1982].

The case we will consider is characterized by the following properties.

H1'. $0 < -\lambda_2 < \lambda_3 < -\lambda_1$, $d \neq 0$.

H2'. (3.2.76) is invariant under the coordinate transformation $(x, y, z) \rightarrow (-x, y, -z)$ and the homoclinic orbits break for μ near zero in the manner shown in Figure 3.2.25.

The property H1' insures that the Poincaré map has a strongly contracting

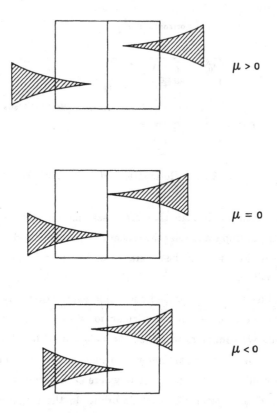

Figure 3.2.25. Dependence of the Homoclinic Orbits
on the Scalar Parameter μ.

direction and a strongly expanding direction (recall from (3.2.81) that d is an entry
in the matrix defining P_1^L).

Now the Poincaré map P^L of $\Pi_0^r \cup \Pi_0^l$ into $\Pi_0^r \cup \Pi_0^l$ consists of two parts

$$P_r^L : \Pi_0^r \to \Pi_0^r \cup \Pi_0^l \qquad\qquad (3.2.94)$$

with P_r^L given by (3.2.82) and

$$P_l^L : \Pi_0^l \to \Pi_0^r \cup \Pi_0^l \qquad\qquad (3.2.95)$$

where by the symmetry we have

$$P_l^L(x, z) = -P_r^L(-x, -z) . \qquad\qquad (3.2.96)$$

Our goal is to show that, for $\mu < 0$, P^L contains an invariant Cantor set on which it is topologically conjugate to the full shift on two symbols. This is done in the following theorem.

Theorem 3.2.13. *There exists $\mu_0 < 0$ such that, for $\mu_0 < \mu < 0$, P^L possesses an invariant Cantor set on which it is topologically conjugate to the full shift on two symbols.*

PROOF: It suffices to show that $\overline{A1}$ and $\overline{A2}$ hold from Section 2.3d. Then the result will follow from Theorem 2.3.12 and Theorem 2.3.3.

$\overline{A1}$. From (3.2.89) with $P^L(x, z) = \left(P_1^L(x, z), P_2^L(x, z)\right)$ we have

$$
\begin{aligned}
D_x P_1^L &= A z^{\frac{|\lambda_1|}{\lambda_3}} \\
D_z P_2^L &= \frac{|\lambda_1|}{\lambda_3} C x z^{\frac{|\lambda_1|}{\lambda_3}-1} + \frac{|\lambda_2|}{\lambda_3} D z^{\frac{|\lambda_2|}{\lambda_3}-1} \\
D_z P_1^L &= \frac{|\lambda_1|}{\lambda_3} A x z^{\frac{|\lambda_1|}{\lambda_3}-1} + \frac{|\lambda_2|}{\lambda_3} B z^{\frac{|\lambda_2|}{\lambda_3}-1} \\
D_x P_2^L &= C z^{\frac{|\lambda_1|}{\lambda_3}} \, .
\end{aligned}
\tag{3.2.97}
$$

Now, by H1', $|\lambda_1|/\lambda_3 > 1$ and $|\lambda_2|/\lambda_3 < 1$, so we have

$$
\begin{aligned}
&\lim_{z \to 0} \left\| D_x P_1^L \right\| = 0 \\
&\lim_{z \to 0} \left\| D_x P_2^L \right\| = 0 \\
&\lim_{z \to 0} \left\| (D_z P_2^L)^{-1} \right\| = 0 \qquad \text{since } d \neq 0 \\
&\lim_{z \to 0} \left\| D_z P_1^L \right\| \left\| (D_z P_2^L)^{-1} \right\| = \left| \frac{B}{D} \right| < \infty, \qquad \text{since } d \neq 0 .
\end{aligned}
\tag{3.2.98}
$$

So, for z sufficiently small, $\overline{A1}$ is satisfied.

$\overline{A2}$. Fix $\mu < 0$. We choose μ_h-horizontal slabs $H_r \subset \Pi_0^r$ and $H_l \subset \Pi_0^l$ with "horizontal" sides parallel to the x axis and "vertical sides" parallel to the z axis such that $P^L(H_r)$ and $P^L(H_l)$ intersect both horizontal boundaries of H_r and H_l. This is always possible for z sufficiently small since $\lim_{z \to 0} \left\| (D_z P_2^L)^{-1} \right\| = 0$, see Figure 3.2.26.

By our previous discussion of the image of Π_0 under P^L it should be evident that the horizontal and vertical boundaries of H_r and H_l satisfy $\overline{A2}$. In particular,

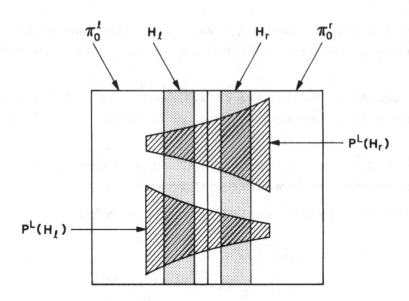

Figure 3.2.26. Image of H_l and H_r under P^L.

H_r and H_l are chosen such that $\mu_h = 0$ and the vertical boundaries of H_r and H_l are μ_v-vertical slices with $\mu_v = 0$. Therefore, μ_h satisfies (2.3.68) and (2.3.74) and, by Lemma 2.3.11 and Lemma 2.3.8, the Lipschitz constant of the vertical boundaries of $P^L(H_r)$ and $P^L(H_l)$ satisfy (2.3.69), (2.3.75), and (2.3.78). So $\overline{A2}$ holds. $\square$

The dynamical consequences of Theorem 3.2.13 are stunning. For $\mu \geq 0$ there is nothing spectacular associated with the dynamics near the (broken) homoclinic orbits. However, for $\mu < 0$ the horseshoes and their attendant chaotic dynamics appear seemingly out of nowhere. This particular type of global bifurcation has been called a *homoclinic explosion*.

Observations and Additional References.

We have barely scratched the surface of the possible dynamics associated with orbits homoclinic to a fixed point having real eigenvalues in a third order ordinary differential equation. There are several issues which deserve a more thorough investigation.

Two Homoclinic Orbits without Symmetry. See Afraimovich, Bykov, and Silnikov [1984] and the references therein.

The Existence of Strange Attractors. Horseshoes are chaotic invariant sets, yet all the orbits in the horseshoe are unstable of saddle type. Nevertheless, it should be clear that horseshoes may exhibit a striking effect on the dynamics of any system. In particular, they are often the chaotic heart of numerically observed strange attractors. For work on the "strange attractor problem" associated with orbits homoclinic to fixed points having real eigenvalues in a third order ordinary differential equation see Afraimovich, Bykov, and Silnikov [1984]. Most of the work done on such systems has been in the context of the Lorenz equations. References for Lorenz attractors include Sparrow [1982], Guckenheimer and Williams [1980], and Williams [1980].

Bifurcations Creating the Horseshoe. In the homoclinic explosion an infinite number of periodic orbits of all possible periods are created. The question arises concerning precisely how these periodic orbits were created and how they are related to each other. This question also has relevance to the strange attractor problem.

In recent years Birman, Williams, and Holmes have been using the *knot type* of a periodic orbit as a bifurcation invariant in order to understand the appearance, disappearance, and interrelation of periodic orbits in third order ordinary differential equations. Roughly speaking, a periodic orbit in three dimensions can be thought of as a knotted closed loop. As system parameters are varied, the periodic orbit may never intersect itself due to uniqueness of solutions. Hence, the knot type of a periodic orbit cannot change as parameters are varied. The knot type is therefore a bifurcation invariant as well as a key tool for developing a classification scheme for periodic orbits. For references see Birman and Williams [1985a,b], Holmes [1986], [1987], and Holmes and Williams [1985].

ii) Orbits Homoclinic to a Saddle-Focus

We now consider the dynamics near an orbit homoclinic to a fixed point of saddle-focus type of a third order ordinary differential equation. This has become known as the *Silnikov phenomena* since it was first studied by Silnikov [1965].

We consider an equation of the following form

$$\dot{x} = \rho x - \omega y + P(x, y, z)$$
$$\dot{y} = \omega x + \rho y + Q(x, y, z) \qquad (3.2.99)$$
$$\dot{z} = \lambda z + R(x, y, z)$$

where P, Q, R are C^2 and $O(2)$ at the origin. It should be clear that $(0,0,0)$ is a fixed point and that the eigenvalues of (3.2.99) linearized about $(0,0,0)$ are given by $\rho \pm i\omega$, λ (note that there are no parameters in this problem at the moment; we will consider bifurcations of (3.2.99) later). We make the following hypotheses on the system (3.2.99).

H1. (3.2.99) possesses a homoclinic orbit Γ connecting $(0,0,0)$ to itself.

H2. $\lambda > -\rho > 0$.

Thus, $(0,0,0)$ possesses a 2-dimensional stable manifold and a 1-dimensional unstable manifold which intersect nontransversely. See Figure 3.2.27.

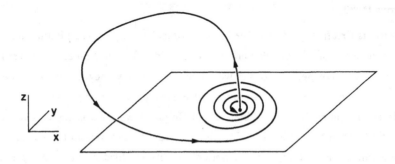

Figure 3.2.27. The Homoclinic Orbit in (3.2.99).

In order to determine the nature of the orbit structure near Γ we construct a Poincaré map defined near Γ in the usual manner, see Section 3.2a.

Computation of P_0^L.

Let Π_0 be a rectangle lying in the x-z plane, and let Π_1 be a rectangle parallel to the x-y plane at $z = \epsilon$, see Figure 3.2.27. As opposed to the case of purely real eigenvalues Π_0 will require a more detailed description. However, in order to do this we need to better understand the dynamics of the flow near the origin.

The flow of (3.2.99) linearized about the origin is given by

$$
\begin{aligned}
x(t) &= e^{\rho t}(x_0 \cos \omega t - y_0 \sin \omega t) \\
y(t) &= e^{\rho t}(x_0 \sin \omega t + y_0 \cos \omega t) \\
z(t) &= z_0 e^{\lambda t}.
\end{aligned}
\tag{3.2.100}
$$

The time of flight for points starting on Π_0 to reach Π_1 is found by solving

$$
\epsilon = z_0 e^{\lambda T}
\tag{3.2.101}
$$

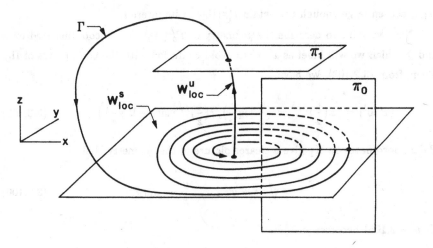

Figure 3.2.28. Cross Sections to (3.2.99) Near the Origin.

or

$$T = \frac{1}{\lambda} \log \frac{\epsilon}{z_0} . \qquad (3.2.102)$$

Thus, P_0^L is given by (omitting the subscript 0's)

$$P_0^L \; : \; \Pi_0 \to \Pi_1$$
$$\begin{pmatrix} x \\ 0 \\ z \end{pmatrix} \mapsto \begin{pmatrix} x\left(\frac{\epsilon}{z}\right)^{\rho/\lambda} \cos\left(\frac{\omega}{\lambda} \log \frac{\epsilon}{z}\right) \\ x\left(\frac{\epsilon}{z}\right)^{\rho/\lambda} \sin\left(\frac{\omega}{\lambda} \log \frac{\epsilon}{z}\right) \\ \epsilon \end{pmatrix} . \qquad (3.2.103)$$

We now consider Π_0 more carefully. For Π_0 arbitrarily chosen it is possible for points on Π_0 to intersect Π_0 many times before reaching Π_1. In this case, P_0^L would not map Π_0 homeomorphically onto $P_0^L(\Pi_0)$. We want to avoid this situation, since the conditions for a map to possess the dynamics of the shift map described in Chapter 2 are given for homeomorphisms. According to (3.2.100) it takes time $t = 2\pi/\omega$ for a point starting in the x-z plane with $x > 0$ to return to the x-z plane with $x > 0$. Now let $x = \epsilon$, $0 < z \le \epsilon$ be the right hand boundary of Π_0. Then if we choose $x = \epsilon e^{2\pi\rho/\omega}$, $0 < z \le \epsilon$ to be the left hand boundary of Π_0, no point starting in the interior of Π_0 returns to Π_0 before reaching Π_1. We take this as the definition of Π_0.

$$\Pi_0 = \left\{ (x, y, z) \in \mathbb{R}^3 \mid y = 0, \;\; \epsilon e^{2\pi\rho/\omega} \le x \le \epsilon, \;\; 0 < z \le \epsilon \right\} \qquad (3.2.104)$$

Π_1 is chosen large enough to contain $P_0^L(\Pi_0)$ in its interior.

Now we want to describe the geometry of $P_0^L(\Pi_0)$. Π_1 is coordinatized by x and y, which we will label as x', y' to avoid confusion with the coordinates of Π_0. Then, from (3.2.103), we have

$$(x',y') = \left(x\left(\frac{\epsilon}{z}\right)^{\rho/\lambda}\cos\left(\frac{\omega}{\lambda}\log\frac{\epsilon}{z}\right), x\left(\frac{\epsilon}{z}\right)^{\rho/\lambda}\sin\left(\frac{\omega}{\lambda}\log\frac{\epsilon}{z}\right)\right). \tag{3.2.105}$$

Polar coordinates on Π_1 give a clearer picture of the geometry. Let

$$r = \sqrt{x'^2 + y'^2}, \qquad \frac{y'}{x'} = \tan\theta \tag{3.2.106}$$

then (3.2.105) becomes

$$(r,\theta) = \left(x\left(\frac{\epsilon}{z}\right)^{\rho/\lambda}, \frac{\omega}{\lambda}\log\frac{\epsilon}{z}\right). \tag{3.2.107}$$

Now consider a vertical line in Π_0, i.e., a line with $x = $ constant. By (3.2.107) it gets mapped into a logarithmic spiral. A horizontal line in Π_0, i.e., a line with $z = $ constant, gets mapped onto a radial line emanating from $(0,0,\epsilon)$. Consider the rectangles

$$R_k = \{\,(x,y,z) \in \mathbb{R}^3 \mid y = 0, \quad \epsilon e^{\frac{2\pi\rho}{\omega}} \le x \le \epsilon, \quad \epsilon e^{\frac{-2\pi(k+1)\lambda}{\omega}} \le z \le \epsilon e^{\frac{-2\pi k\lambda}{\omega}}\,\}. \tag{3.2.108}$$

Then we have

$$\Pi_0 = \bigcup_{k=0}^{\infty} R_k. \tag{3.2.109}$$

We consider the image of the rectangles R_k by determining the behavior of its horizontal and vertical boundaries under P_0^L. We denote these four line segments as

$$h^u = \{\,(x,y,z) \in \mathbb{R}^3 \mid y = 0,\ z = \epsilon e^{\frac{-2\pi k\lambda}{\omega}},\ \epsilon e^{\frac{2\pi\rho}{\omega}} \le x \le \epsilon\,\}$$

$$h^l = \{\,(x,y,z) \in \mathbb{R}^3 \mid y = 0,\ z = \epsilon e^{\frac{-2\pi(k+1)\lambda}{\omega}},\ \epsilon e^{\frac{2\pi\rho}{\omega}} \le x \le \epsilon\,\}$$

$$v^r = \{\,(x,y,z) \in \mathbb{R}^3 \mid y = 0,\ x = \epsilon,\ \epsilon e^{\frac{-2\pi(k+1)\lambda}{\omega}} \le z \le e^{\frac{-2\pi k\lambda}{\omega}}\,\}$$

$$v^l = \{\,(x,y,z) \in \mathbb{R}^3 \mid y = 0,\ x = \epsilon e^{\frac{2\pi\rho}{\omega}},\ \epsilon e^{\frac{-2\pi(k+1)\lambda}{\omega}} \le x \le e^{\frac{-2\pi k\lambda}{\omega}}\,\}. \tag{3.2.110}$$

See Figure 3.2.29. The images of these line segments under P_0^L are given by

$$P_0^L(h^u) = \{ (r,\theta,z) \in \mathbb{R}^3 \mid z = \epsilon,\ \theta = 2\pi k,\ \epsilon e^{\frac{2\pi(k+1)\rho}{\omega}} \le r \le \epsilon e^{\frac{2\pi k\rho}{\omega}} \}$$

$$P_0^L(h^l) = \{ (r,\theta,z) \in \mathbb{R}^3 \mid z = \epsilon,\ \theta = 2\pi(k+1),\ \epsilon e^{\frac{2\pi(k+2)\rho}{\omega}} \le r \le \epsilon e^{\frac{2\pi(k+1)\rho}{\omega}} \}$$

$$P_0^L(v^r) = \{ (r,\theta,z) \in \mathbb{R}^3 \mid z = \epsilon,\ 2\pi k \le \theta \le 2\pi(k+1),\ r(\theta) = \epsilon e^{\frac{\rho\theta}{\omega}} \}$$

$$P_0^L(v^l) = \{ (r,\theta,z) \in \mathbb{R}^3 \mid z = \epsilon,\ 2\pi k \le \theta \le 2\pi(k+1),\ r(\theta) = \epsilon e^{\frac{\rho(2\pi+\theta)}{\omega}} \}$$
$$(3.2.111)$$

so that $P_0^L(R_k)$ appears as in Figure 3.2.29.

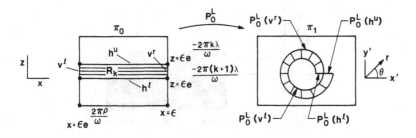

Figure 3.2.29. R_k and the Geometry of its Image under P_0^L.

The geometry of Figure 3.2.29 should give a strong indication that horseshoes may arise in this system.

Computation of P_1^L.

From Step 5, part b, of Section 3.2a, on some open set $U \subset \Pi_1$ we have

$$P_1^L: U \subset \Pi_1 \to \Pi_0$$

$$\begin{pmatrix} x \\ y \\ \epsilon \end{pmatrix} \mapsto \begin{pmatrix} a & b & 0 \\ c & d & 0 \\ 0 & 0 & 0 \end{pmatrix} \begin{pmatrix} x \\ y \\ z \end{pmatrix} + \begin{pmatrix} \bar{x} \\ 0 \\ 0 \end{pmatrix} \tag{3.2.112}$$

where $(\bar{x},0,0,) \equiv \Gamma \cap \Pi_0$ with $\bar{x} = \epsilon \frac{1+e^{\frac{2\pi\rho}{\omega}}}{2}$.

The Poincaré Map $P^L = P_1^L \circ P_0^L$.

From (3.2.103) and (3.2.112) we have

$$P^L: P_1^L \circ P_0^L: V \subset \Pi_0 \to \Pi_0$$

$$\begin{pmatrix} x \\ z \end{pmatrix} \mapsto \begin{pmatrix} x\left(\frac{\epsilon}{z}\right)^{\frac{\rho}{\lambda}} \left[a\cos\left(\frac{\omega}{\lambda}\log\frac{\epsilon}{z}\right) + b\sin\left(\frac{\omega}{\lambda}\log\frac{\epsilon}{z}\right) \right] + \bar{x} \\ x\left(\frac{\epsilon}{z}\right)^{\frac{\rho}{\lambda}} \left[c\cos\left(\frac{\omega}{\lambda}\log\frac{\epsilon}{z}\right) + d\sin\left(\frac{\omega}{\lambda}\log\frac{\epsilon}{z}\right) \right] \end{pmatrix} \tag{3.2.113}$$

where $V = (P_0^L)^{-1}(U)$.

So, if we choose Π_0 sufficiently small, then $P^L(\Pi_0)$ appears as in Figure 3.2.30.

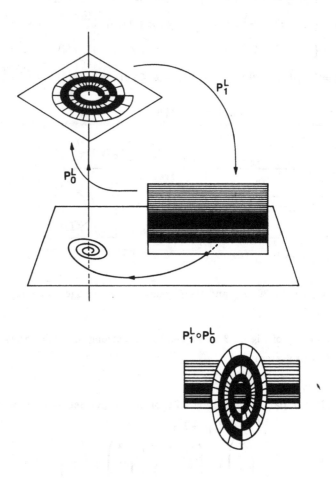

Figure 3.2.30. The Poincaré Map.

We now want to show that P^L contains an invariant Cantor set on which it is topologically conjugate to the shift map. The possibility of horseshoe-like behavior should be apparent from Figure 3.2.30; however, this needs justification. In particular, we need to verify $\overline{A1}$ and $\overline{A2}$ of Section 2.3d. First we will need a preliminary result.

Consider the rectangle R_k in Figure 3.2.31. In order to verify the proper

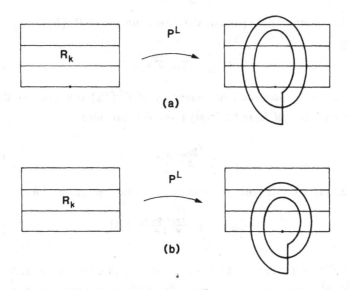

Figure 3.2.31. Two Possibilities for the Image of R_k under P^L.

behavior of horizontal and vertical slabs in R_k, it will be necessary to verify that the inner and outer boundaries of $P^L(R_k)$ both intersect the upper boundary of R_k as shown in Figure 3.2.31a. Or, in other words, the upper horizontal boundary of R_k intersects (at least) two points of the inner boundary of $P^L(R_k)$. Additionally, it will be useful to know how many rectangles above R_k that $P^L(R_k)$ also intersects in this manner. We have the following lemma.

Lemma 3.2.14. *Consider R_k for fixed k sufficiently large. Then the inner boundary of $P^L(R_k)$ intersects the upper horizontal boundary of R_i in (at least) two points for $i \geq k/\alpha$ where $1 \leq \alpha < -\lambda/\rho$. Moreover, the preimage of the vertical boundaries of $P^L(R_k) \cap R_i$ is contained in the vertical boundary of R_k.*

PROOF: The z coordinate of the upper horizontal boundary of R_i is given by

$$\bar{z} = \epsilon e^{\frac{-2\pi i \lambda}{\omega}} . \tag{3.2.114}$$

The point on the inner boundary of $P_0^L(R_k)$ closest to $(0, 0, \epsilon)$ is given by

$$r_{\min} = \epsilon e^{\frac{4\pi \rho}{\omega}} e^{\frac{2\pi k \rho}{\omega}} . \tag{3.2.115}$$

Since P_1^L is a linear map the point on the inner boundary of $P^L(R_k) = P_1^L \circ P_0^L(R_k)$ is given by

$$\bar{r}_{\min} = K\epsilon e^{\frac{4\pi\rho}{\omega}} e^{\frac{2\pi k\rho}{\omega}} \qquad (3.2.116)$$

for some $K > 0$. Now the inner boundary of $P^L(R_k)$ will intersect the upper horizontal boundary of R_i in (at least) two points provided

$$\frac{\bar{r}_{\min}}{\bar{z}} > 1 . \qquad (3.2.117)$$

Using (3.2.114) and (3.2.116), we compute this ratio explicitly and find

$$\frac{\bar{r}_{\min}}{\bar{z}} = K e^{\frac{4\pi\rho}{\omega}} e^{\frac{2\pi}{\omega}(k\rho + i\lambda)}. \qquad (3.2.118)$$

Now $Ke^{4\pi\rho/\omega}$ is a fixed constant, so the size of (3.2.118) is controlled by the $e^{(2\pi/\omega)(k\rho+i\lambda)}$ term. In order to make (3.2.118) larger than one, it is sufficient that $k\rho + i\lambda$ is taken sufficiently large. By H2 we have $\lambda + \rho > 0$, so for $i \geq k/\alpha$, $1 \leq \alpha < -\lambda/\rho$, $k\rho + i\lambda$ is positive, and for k sufficiently large (3.2.118) is larger than one.

We now describe the behavior of the vertical boundaries of R_k. Recall Figure 3.2.29. Under P_0^L the vertical boundaries of R_k map to the inner and outer boundaries of an annulus-like object. Now P_1^L is an invertible affine map. Hence, the inner and outer boundaries of $P_0^L(R_k)$ correspond to the inner and outer boundaries of $P^L(R_k) = P_1^L \circ P_0^L(R_k)$. Therefore, the preimage of the vertical boundary of $P^L(R_k) \cap R_i$ is contained in the vertical boundary of R_k. $\square$

Lemma 3.2.14 points out the necessity of H2 since, if we instead had $-\rho > \lambda > 0$, then the image of R_k would fall below R_k for k sufficiently large, as shown in Figure 3.2.31b.

We now want to show that $\overline{A2}$ holds at all points of Π_0 except for possibly on a countable number of horizontal lines which can be avoided if necessary. If we use the notation $P^L = (P_1^L, P_2^L)$ (note: P_1^L, P_2^L stand for the two of components P^L and should not be confused with the map P_1^L along the homoclinic orbit outside of a neighborhood of the origin), then from (3.2.113) we have

$$DP^L = \begin{pmatrix} D_x P_1^L & D_z P_1^L \\ D_x P_2^L & D_z P_2^L \end{pmatrix} \qquad (3.2.119)$$

where

$$D_x P_1^L \equiv \epsilon^{\frac{x}{\lambda}} z^{\frac{-\rho}{\lambda}} \left[a \cos\left(\frac{\omega}{\lambda} \log \frac{\epsilon}{z}\right) + b \sin\left(\frac{\omega}{\lambda} \log \frac{\epsilon}{z}\right) \right]$$

$$D_z P_1^L \equiv \frac{-x}{\lambda} \epsilon^{\frac{x}{\lambda}} z^{-(1+\frac{\rho}{\lambda})} \left\{ \rho \left[a \cos\left(\frac{\omega}{\lambda} \log \frac{\epsilon}{z}\right) + b \sin\left(\frac{\omega}{\lambda} \log \frac{\epsilon}{z}\right) \right] \right.$$
$$\left. + \omega \left[-a \sin\left(\frac{\omega}{\lambda} \log \frac{\epsilon}{z}\right) + b \cos\left(\frac{\omega}{\lambda} \log \frac{\epsilon}{z}\right) \right] \right\}$$

$$D_x P_2^L \equiv \epsilon^{\frac{x}{\lambda}} z^{\frac{-\rho}{\lambda}} \left[c \cos\left(\frac{\omega}{\lambda} \log \frac{\epsilon}{z}\right) + d \sin\left(\frac{\omega}{\lambda} \log \frac{\epsilon}{z}\right) \right]$$

$$D_z P_2^L \equiv \frac{-x}{\lambda} \epsilon^{\frac{x}{\lambda}} z^{-(1+\frac{\rho}{\lambda})} \left\{ \rho \left[c \cos\left(\frac{\omega}{\lambda} \log \frac{\epsilon}{z}\right) + d \sin\left(\frac{\omega}{\lambda} \log \frac{\epsilon}{z}\right) \right] \right.$$
$$\left. + \omega \left[-c \sin\left(\frac{\omega}{\lambda} \log \frac{\epsilon}{z}\right) + d \cos\left(\frac{\omega}{\lambda} \log \frac{\epsilon}{z}\right) \right] \right\}.$$
$$(3.2.120)$$

We have the following lemma.

Lemma 3.2.15. $\overline{A1}$ *holds everywhere on* Π_0 *with the possible exception of a count-able number of horizontal lines. Moreover, these "bad" horizontal lines can be avoided if necessary.*

PROOF: By H2, $1 + \rho/\lambda > 0$ so we have

$$\lim_{z \to 0} \left\| D_x P_1^L \right\| = 0$$

$$\lim_{z \to 0} \left\| D_x P_2^L \right\| = 0 \qquad (3.2.121)$$

$$\lim_{z \to 0} \left\| (D_z P_2^L)^{-1} \right\| = 0 .$$

We need to worry about the term

$$\left\| D_z P_1^L \right\| \left\| (D_z P_1^L)^{-1} \right\| , \qquad z \text{ small.} \qquad (3.2.122)$$

We need to show that (3.2.122) is bounded, which may not be the case if

$$\rho \left[c \cos\left(\frac{\omega}{\lambda} \log \frac{\epsilon}{z}\right) + d \sin\left(\frac{\omega}{\lambda} \log \frac{\epsilon}{z}\right) \right] + \omega \left[-c \sin\left(\frac{\omega}{\lambda} \log \frac{\epsilon}{z}\right) + d \cos\left(\frac{\omega}{\lambda} \log \frac{\epsilon}{z}\right) \right] = 0$$
$$(3.2.123)$$

or, equivalently, if

$$\frac{c}{d} = \frac{\rho \sin\left(\frac{\omega}{\lambda} \log \frac{\epsilon}{z}\right) + \omega \cos\left(\frac{\omega}{\lambda} \log \frac{\epsilon}{z}\right)}{-\rho \cos\left(\frac{\omega}{\lambda} \log \frac{\epsilon}{z}\right) + \omega \sin\left(\frac{\omega}{\lambda} \log \frac{\epsilon}{z}\right)} . \qquad (3.2.124)$$

However, suppose there exists a z value such that (3.2.124) is satisfied. Then, by periodicity there exists a countable infinity of such z values. Now in practice we are not interested in (3.2.122) on all of Π_0 but rather on a countable set of disjoint horizontal slabs contained in Π_0. So there would be no problem if the "bad" z values fell between our chosen horizontal slabs. We can always insure this by changing the cross-sections Π_0 and/or Π_1 slightly, which results in a change in c/d. Thus, (3.2.121) and boundedness of (3.2.122) on appropriately chosen μ_h-horizontal slabs implies that $\overline{A1}$ holds.   □

We now address the issue of the appropriate choice of μ_h-horizontal slabs and their behavior under P^L, i.e., we must verify $\overline{A2}$. We begin with a preliminary lemma.

Lemma 3.2.16. *Consider R_k for fixed k sufficiently large. Then $P^L(R_k)$ intersects R_i in two disjoint μ_v-vertical slabs with μ_v satisfying (2.3.69), (2.3.75), and (2.3.78) for $i \geq k/\alpha$ where $1 \leq \alpha < -\lambda/\rho$. Moreover, the preimage of the boundaries of these μ_v-vertical slabs lies in the vertical boundary of R_k.*

PROOF: By Lemma 3.2.14 $P^L(R_k)$ intersects R_i, $i \geq k/\alpha$, in two disjoint components with the preimage of the vertical boundaries of these components lying in the vertical boundaries of R_k. Therefore, we need only show that these components are μ_v-vertical slices with μ_v satisfying (2.3.69), (2.3.75), and (2.3.78).

By construction R_k is a μ_h-horizontal slab with $\mu_h = 0$, and the vertical sides of R_k are μ_v-vertical slices with $\mu_v = 0$. So, by Lemma 2.3.11 and Lemma 2.3.8, the vertical boundaries of $P^L(R_k) \cap R_i$ are μ_v-vertical slices with μ_v satisfying (2.3.69), (2.3.75), and (2.3.78).   □

We now use Lemma 3.2.16 to show how we can find two μ_h-horizontal slabs in each R_k, k sufficiently large, such that $\overline{A2}$ is satisfied. Consider $P^L(R_k) \cap \left[\bigcup_{i \geq k/\alpha} R_i \right]$. By Lemma 3.2.16 this consists of two disjoint μ_v-vertical slabs with μ_v satisfying (2.3.69), (2.3.75), and (2.3.78). The preimage of these μ_v-vertical slabs consists of two disjoint components contained in R_k whose vertical boundaries lie in the vertical boundary of R_k. Moreover, since the horizontal boundaries of the μ_v-vertical slabs are μ_h-horizontal with $\mu_h = 0$, it follows from Lemma 2.3.10 and Lemma 2.3.8 that the horizontal boundaries of the two components of the preimage of the μ_v-vertical slabs are μ_h-horizontal with μ_h satisfying (2.3.68) and (2.3.74).

We label these two μ_h-horizontal slabs H_{+k} and H_{-k} and associate to each the symbols $+k$ and $-k$, respectively. Thus $\overline{A2}$ holds on H_{+k} and H_{-k}. $\square$

We now put these results together to show that P^L contains an invariant Cantor set on which it is topologically conjugate to the shift map. There are two distinct possibilities which we will treat separately.

The Full Shift on 2N Symbols.

For k sufficiently large choose N rectangles $R_k, \ldots, R_{k+N}$ where N is chosen such that $k \geq (k+N)/\alpha$ with $1 \leq \alpha < -\lambda/\rho$. Since $k \geq (k+N)/\alpha$ is equivalent to $k(\alpha - 1) \geq N$ with $\alpha \geq 1$, for $\alpha > 1$ and for fixed N it is always possible to choose k large enough so that this condition is satisfied. Then, as discussed above, we choose μ_h-horizontal slabs H_{+i}, H_{-i} in R_i, $i = k, \ldots, k+N$, such that $P^L(H_{+i})$ and $P^L(H_{-i})$ intersects $R_k, \ldots, R_{k+N}$ in μ_v-vertical slabs where $\overline{A2}$ is satisfied. Then, since $\overline{A1}$ and $\overline{A2}$ are satisfied, it follows from Theorem 2.3.12 and Theorem 2.3.3 that P^L possesses an invariant Cantor set on which it is topologically conjugate to the full shift on $2N$ symbols. See Figure 3.2.32 for an illustration of the geometry.

The Subshift of Finite Type on an Infinite Number of Symbols.

Consider the space of symbol sequences

$$\Sigma^{\infty, \alpha} = \left\{ \underline{s} = \{s_i\}_{i=-\infty}^{\infty} \mid s_i \in \pm k, \cdots, k \in \mathbb{Z} - 0 \quad \text{and} \quad |s_{i+1}| \geq \frac{|s_i|}{\alpha} \right\}.$$
$$(3.2.125)$$

For k sufficiently large, in each R_i, $i = k, \ldots$, choose two μ_h-horizontal slabs H_{+i} and H_{-i} to which we associate the symbols $+i$ and $-i$, respectively. By Lemma 3.2.16, $P^L(H_{+i})$ and $P^L(H_{-i})$ intersect R_k for $k \geq i/\alpha$. As above, the H_{+i}, H_{-i}, $i = k, \ldots$, can be chosen such that $\overline{A1}$ and $\overline{A2}$ hold, in which case a simple modification of Theorem 2.3.3 allows us to conclude that P^L contains a Cantor set on which it is topologically conjugate to the shift map acting on $\Sigma^{\infty, \alpha}$. Note that the symbols $\pm\infty$ correspond to orbits on $W^s(0)$. Hence, in this case, some orbits may "leak out" of the Cantor set. See Figure 3.2.33 for an illustration of the geometry.

We summarize our results in the following theorem.

Theorem 3.2.17. a) *For each even positive integer N there exists a map*

$$\phi^N : \Sigma^N \to \Pi_0$$

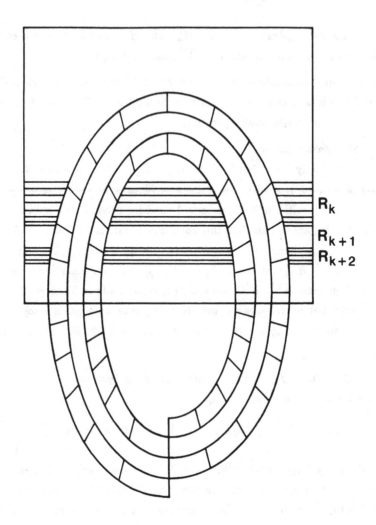

Figure 3.2.32. Image of the R_k under P^L.

which is a homeomorphism of Σ^N onto $O^N \equiv \phi^N(\Sigma^N)$ such that

$$P^L\big|_{O^N} = \phi^N \circ \sigma \circ (\phi^N)^{-1}.$$

b) For each real α with $1 \leq \alpha < -\lambda/\rho$, there exists a map

$$\phi^{\infty,\alpha} \colon \Sigma^{\infty,\alpha} \to \Pi_0$$

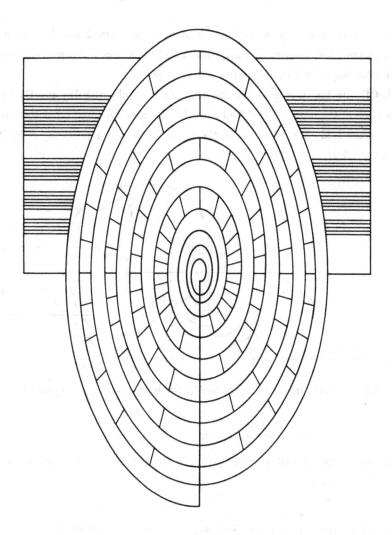

Figure 3.2.33. Image of the R_k under P^L.

which is a homeomorphism of $\Sigma^{\infty,\alpha}$ onto $O^{\infty,\alpha} \equiv \phi^{\infty,\alpha}(\Sigma^{\infty,\alpha})$ such that

$$P^L\big|_{O^{\infty,\alpha}} = \phi^{\infty,\alpha} \circ \sigma \circ (\phi^{\infty,\alpha})^{-1}.$$

Persistence Under Perturbation.

Notice a major difference between the case of purely real eigenvalues at the fixed point and the present case. In the latter case, it was necessary to break the homoclinic orbit in a specific way in order to get horseshoes. In the present case, horseshoes were present in a neighborhood of the homoclinic orbit. It is natural to ask what happens if this situation is perturbed.

Let Ω_0 be the intersection of the unstable manifold with Π_0 and Ω_1 be the intersection of the unstable manifold with Π_1. We consider small C^2 perturbations of the vector field. We denote by Z the z coordinate of the point $\Omega_0 = P_1^L(\Omega_1)$, see Figure 3.2.34.

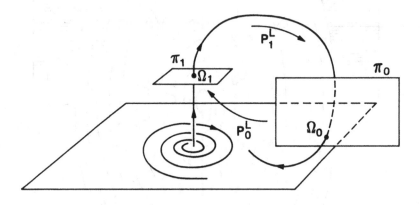

Figure 3.2.34. Intersections of the Unstable Manifold with Π_0 and Π_1.

Then we have the following theorem.

Theorem 3.2.18. *For $|Z|$ small enough, one can find $M > 1$ and, for each N with $1 < N \leq M$, a map*

$$\phi^N : \Sigma^N \to \Pi_0$$

which is a homeomorphism onto its image $O^N = \phi^N(\Sigma^N)$ such that

$$P^L\big|_{O^N} = \phi^N \circ \sigma \circ (\phi^N)^{-1}.$$

PROOF: We leave the details to the reader but see Tresser [1984].   □

Thus Theorem 3.2.18 tells us that for sufficiently small C^2 perturbations a finite number of horseshoes are preserved, see Figure 3.2.35.

The Bifurcation Analysis of Glendinning and Sparrow.

Now that we have seen how complicated the orbit structure is in the neighborhood of an orbit homoclinic to a fixed point of saddle-focus type, we want to get

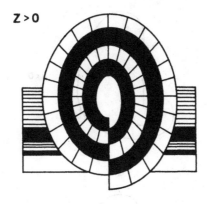

Z > 0

Z < 0

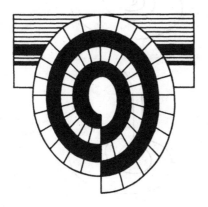

Figure 3.2.35. Perturbed Horseshoes.

an understanding of how this situation occurs as the homoclinic orbit is created. In this regard, the analysis given by Glendinning and Sparrow [1984] is insightful.

Suppose that the homoclinic orbit in (3.2.99) depends on a scalar parameter μ in the manner shown in Figure 3.2.36.

We construct a parameter dependent Poincaré map in the same manner as when we discussed the case of a fixed point with all real eigenvalues. This map is

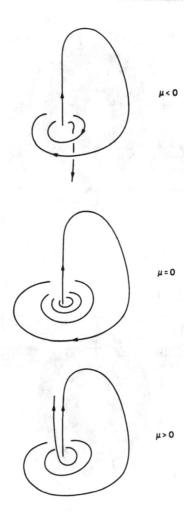

Figure 3.2.36. Behavior of the Homoclinic Orbit with Respect
to the Parameter μ.

given by

$$\begin{pmatrix} x \\ z \end{pmatrix} \mapsto \begin{pmatrix} x\left(\frac{\varepsilon}{z}\right)^{\frac{\rho}{\lambda}}\left[a\cos\frac{\omega}{\lambda}\log\frac{\varepsilon}{z}+b\sin\frac{\omega}{\lambda}\log\frac{\varepsilon}{z}\right]+e\mu+x_0 \\ x\left(\frac{\varepsilon}{z}\right)^{\frac{\rho}{\lambda}}\left[c\cos\frac{\omega}{\lambda}\log\frac{\varepsilon}{z}+d\sin\frac{\omega}{\lambda}\log\frac{\varepsilon}{z}\right]+f\mu \end{pmatrix} \qquad (3.2.126)$$

where from Figure 3.2.36 we have $f > 0$. We have already seen that this map
possesses a countable infinity of horseshoes at $\mu = 0$, and we know that each
horseshoe contains periodic orbits of all periods. To study how the horseshoes are

formed in this situation as the homoclinic orbit is formed is a difficult (and unsolved) problem. We will tackle a more modest problem which will still give us a good idea about some things which are happening; namely, we will study the fixed points of the above map. Recall that the fixed points correspond to periodic orbits which pass through a neighborhood of the origin *once* before closing up. First we put the map in a form which will be easier to work with. The map can be written in the form

$$\begin{pmatrix} x \\ z \end{pmatrix} \mapsto \begin{pmatrix} x\left(\frac{\epsilon}{z}\right)^{\frac{\rho}{\lambda}} p \cos\left(\frac{\omega}{\lambda} \log \frac{\epsilon}{z} + \phi_1\right) + e\mu + x_0 \\ x\left(\frac{\epsilon}{z}\right)^{\frac{\rho}{\lambda}} q \cos\left(\frac{\omega}{\lambda} \log \frac{\epsilon}{z} + \phi_2\right) + \mu \end{pmatrix} \tag{3.2.127}$$

where we have rescaled μ so that $f = 1$ (note that f must be positive).

Now let

$$-\delta = \frac{\rho}{\lambda}, \qquad \alpha = p\epsilon^{-\delta}, \qquad \beta = q\epsilon^{-\delta},$$

$$\xi = -\frac{\omega}{\lambda}, \qquad \Phi_1 = \frac{\omega}{\lambda} \log \epsilon + \phi_1, \qquad \Phi_2 = \frac{\omega}{\lambda} \log \epsilon + \phi_2. \tag{3.2.128}$$

Then the map takes the form

$$\begin{pmatrix} x \\ z \end{pmatrix} \mapsto \begin{pmatrix} \alpha x z^\delta \cos(\xi \log z + \Phi_1) + e\mu + x_0 \\ \beta x z^\delta \cos(\xi \log z + \Phi_2) + \mu \end{pmatrix}. \tag{3.2.129}$$

Now we will study the fixed points of this map and their stability and bifurcations.

Fixed Points.

The fixed points are found by solving

$$x = \alpha x z^\delta \cos(\xi \log z + \Phi_1) + e\mu + x_0, \tag{3.2.130a}$$

$$z = \beta x z^\delta \cos(\xi \log z + \Phi_2) + \mu. \tag{3.2.130b}$$

Solving (3.2.130a) for x as a function of z gives

$$x = \frac{e\mu + x_0}{1 - \alpha z^\delta \cos(\xi \log z + \Phi_1)}. \tag{3.2.131}$$

Substituting (3.2.131) into (3.2.130b) gives

$$(z - \mu)\left(1 - \alpha z^\delta \cos(\xi \log z + \Phi_1)\right) = (e\mu + x_0)\beta z^\delta \cos(\xi \log z + \Phi_2). \tag{3.2.132}$$

Solving (3.2.132) gives us the z-component of the fixed point; substituting this into (3.2.131) gives us the x-component of the fixed point. In order to get an idea about the solutions of (3.2.132) we will assume that z is so small that

$$1 - \alpha z^\delta \cos(\xi \log z + \Phi_1) \sim 1. \tag{3.2.133}$$

Case 1: $\delta < 1$

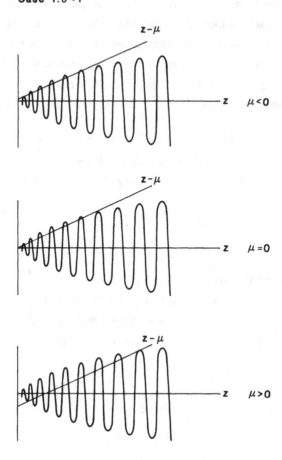

Figure 3.2.37. Case 1: $\delta < 1$.

Then the equation for the z component of the fixed point will be

$$(z - \mu) = (e\mu + x_0)\beta z^{\delta} \cos(\xi \log z + \Phi_2) . \qquad (3.2.134)$$

There are various cases shown in Figure 3.2.37.

So in the case $\delta < 1$ we have

$\mu < 0$: finite number of fixed points.

$\mu = 0$: countable infinity of fixed points.

$\mu > 0$: finite number of fixed points.

Case 2: $\delta > 1$

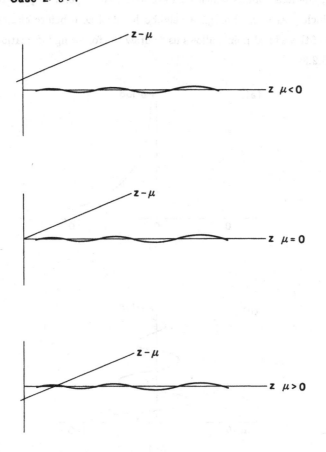

Figure 3.2.38. Case 2: $\delta > 1$.

The next case is $\delta > 1$, i.e., H2 does not hold. We show the results in Figure 3.2.38.

So in the case $\delta > 1$ we have

$\underline{\mu \leq 0}$: There are no fixed points except the one at $z = \mu = 0$ (i.e., the homoclinic orbit).

$\underline{\mu > 0}$: For $z > 0$, there is one fixed point for each μ. This can be seen as follows: the slope of the wiggly curve is of order $z^{\delta - 1}$, which is small for z small since $\delta > 1$. Thus, the $z - \mu$ line only intersects it once.

Again, the fixed points which we have found correspond to periodic orbits of (3.2.99) which pass *once* through a neighborhood of zero before closing up. Our knowledge of these fixed points allows us to draw the following bifurcation diagrams in Figure 3.2.39.

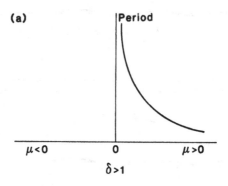

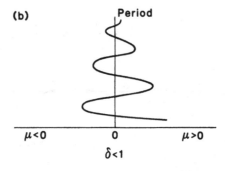

Figure 3.2.39. Dependence of the Period of the Bifurcated Periodic Orbits on μ. a) $\delta > 1$, b) $\delta < 1$.

The $\delta > 1$ diagram should be clear; however, the $\delta < 1$ diagram may be confusing. The wiggly curve in the diagram above represents periodic orbits. It should be clear from Figure 3.2.37 that periodic orbits are born in pairs and the one with the lower z value has the higher period (since it passes closer to the fixed point). We will worry more about the structure of this curve as we proceed.

Stability of the Fixed Points.

The Jacobian of the map is given by

$$\begin{pmatrix} A & C \\ D & B \end{pmatrix}$$

where

$$
\begin{aligned}
A &= \alpha z^\delta \cos(\xi \log z + \Phi_1) \\
B &= \beta x z^{\delta-1} [\delta \cos(\xi \log z + \Phi_2) - \xi \sin(\xi \log z + \Phi_2)] \\
C &= \alpha x z^{\delta-1} [\delta \cos(\xi \log z + \Phi_1) - \xi \sin(\xi \log z + \Phi_1)] \\
D &= \beta z^\delta \cos(\xi \log z + \Phi_2) .
\end{aligned}
\tag{3.2.135}
$$

The eigenvalues of the matrix are given by

$$
\lambda_{1,2} = \frac{1}{2} \left\{ (A+B) \pm \sqrt{(A+B)^2 - 4(AB - CD)} \right\} .
\tag{3.2.136}
$$

$\underline{\delta > 1}$: For $\delta > 1$, it should be clear that the eigenvalues will be small if z is small (since both z^δ and $z^{\delta-1}$ are small). Hence, the one periodic orbit existing for $\mu > 0$ for $\delta > 1$ is stable for μ small, and the homoclinic orbit at $\mu = 0$ is an attractor.

The case $\delta < 1$ is more complicated.

$\underline{\delta < 1}$: First notice that the determinant of the matrix given by $AB - CD$ only contains terms of order $z^{2\delta-1}$, so the map will be

> *area contracting* $1/2 < \delta < 1$,
>
> *area expanding* $0 < \delta < 1/2$,

for z sufficiently small.

So we would expect different results in these two different δ ranges.

Now recall that the wiggly curve whose intersection with $z - \mu$ gave the fixed points was given by

$$(e\mu + x_0)\beta z^\delta \cos(\xi \log z + \Phi_2).$$

Thus, a fixed point corresponding to a maximum of this curve corresponds to $B = 0$, and a fixed point corresponding to a zero crossing of this curve corresponds to $D = 0$. We want to look at the stability of fixed points satisfying these conditions.

$\boxed{D = 0}$ In this case $\lambda_1 = A$, $\lambda_2 = B$. So for z small, λ_1 is small and λ_2 is always large; thus the fixed point is a saddle. Note in particular that, for $\mu = 0$, D is very close to zero; hence all periodic orbits will be saddles as expected.

$\boxed{B = 0}$ The eigenvalues are given by

$$\lambda_{1,2} = A \pm \sqrt{A^2 + 4CD}\,,$$

and both eigenvalues will have large or small modulus depending on whether CD is large or small, since

$A^2 \sim z^{2\delta}$ can be neglected compared to $CD \sim z^{2\delta-1}$.

$A \sim z^{\delta}$ can be neglected compared to $\sqrt{CD} \sim z^{\delta-(1/2)}$.

Whether or not CD is small depends on whether $0 < \delta < 1/2$ or $1/2 < \delta < 1$. So we have

Stable fixed points for $1/2 < \delta < 1$.

Unstable fixed points for $0 < \delta < 1/2$.

Now we want to put everything together for other z values (i.e., for z such that B, $D \neq 0$).

Consider Figure 3.2.40 below which is a blow-up of Figure 3.2.37 for various parameter values and where the intersection of the two curves gives us the z coordinate of the fixed points.

Now we describe what happens at each parameter value shown in Figure 3.2.40.

$\underline{\mu = \mu_6}$: At this point we have a tangency, and we know that a saddle-node pair will be born in a saddle-node bifurcation.

$\underline{\mu = \mu_5}$: At this point we have two fixed points; the one with the lower z value has the larger period. Also, the one at the maximum of the curve has $B = 0$; therefore, it is stable for $\delta > 1/2$, unstable for $\delta < 1/2$. The other fixed point is a saddle.

$\underline{\mu = \mu_4}$: At this point the stable (unstable) fixed point has become a saddle since $D = 0$. Therefore, it must have changed its stability type via a period doubling bifurcation.

$\underline{\mu = \mu_3}$: At this point $B = 0$ again; therefore, the saddle has become either purely stable or unstable again. This must have occurred via a reverse period doubling bifurcation.

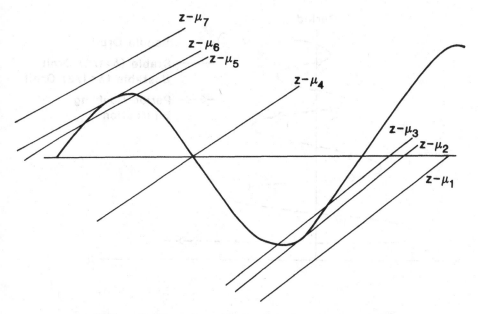

Figure 3.2.40. $\mu_1 > \mu_2 > \mu_3 > \mu_4 > 0 > \mu_5 > \mu_6 > \mu_7$.

$\underline{\mu = \mu_2}$: A saddle-node bifurcation occurs.

So finally we arrive at Figure 3.2.41.

Next we want to get an idea of the size of the "wiggles" in Figure 3.2.41 because, if the wiggles are small, that implies that the 1-loop periodic orbits are only visible for a narrow range of parameters. If the wiggles are large, we might expect there to be a greater likelihood of observing the periodic orbits.

Let us denote the parameter values at which the tangent to the curve in Figure 3.2.41 is vertical by

$$\mu_i, \mu_{i+1}, \ldots, \mu_{i+n}, \ldots \to 0 \qquad (3.2.137)$$

where the μ_i alternate in sign. Now recall that the z component of the fixed point was given by the solutions to the equations

$$z - \mu = (e\mu + x_0)\beta z^\delta \cos(\xi \log z + \Phi_2) . \qquad (3.2.138)$$

So we have

$$z_i - \mu_i = (e\mu_i + x_0)\beta z_i^\delta \cos(\xi \log z_i + \Phi_2) . \qquad (3.2.139a)$$

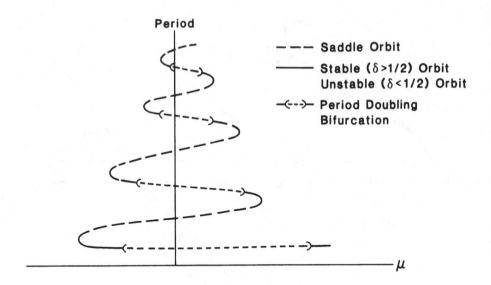

Figure 3.2.41. Stability Diagram for the Bifurcated Periodic Orbits.

$$z_{i+1} - \mu_{i+1} = (e\mu_{i+1} + x_0)\beta z_{i+1}^\delta \cos(\xi \log z_{i+1} + \Phi_2) \,. \qquad (3.2.139b)$$

From $(3.2.139a)$ and $(3.2.139b)$ we obtain

$$\mu_i = \frac{z_i - x_0\beta z_i^\delta \cos(\xi \log z_i + \Phi_2)}{1 + e\beta z_i^\delta \cos(\xi \log z_i + \Phi_2)} \,, \qquad (3.2.140a)$$

$$\mu_{i+1} = \frac{z_{i+1} - x_0\beta z_{i+1}^\delta \cos(\xi \log z_{i+1} + \Phi_2)}{1 + e\beta z_{i+1}^\delta \cos(\xi \log z_{i+1} + \Phi_2)} \,. \qquad (3.2.140b)$$

Now note that we have

$$\xi \log z_{i+1} - \xi \log z_i \approx \pi \qquad \Rightarrow \frac{z_{i+1}}{z_i} \approx \exp\frac{\pi}{\xi} \qquad (3.2.141)$$

and we assume that $z \ll 1$ so that

$$1 + e\beta z_{i(i+1)}^\delta \cos(\xi \log z_{i(i+1)} + \Phi_2) \sim 1 \,. \qquad (3.2.142)$$

So finally we get

$$\frac{\mu_{i+1}}{\mu_i} = \frac{z_{i+1} + [x_0\beta \cos(\xi \log z_i + \Phi_2)]z_{i+1}^\delta}{z_i - [x_0\beta \cos(\xi \log z_i + \Phi_2)]z_i^\delta} \,. \qquad (3.2.143)$$

Now in the limit as $z \to 0$, (3.2.143) becomes

$$\frac{\mu_{i+1}}{\mu_i} \approx -\left(\frac{z_{i+1}}{z_i}\right)^\delta \approx -\exp\left(\frac{\pi\delta}{\xi}\right). \tag{3.2.144}$$

Recall that $\delta = -\rho/\lambda$, $\xi = -\omega/\lambda$, so we get

$$\lim_{i \to \infty} \frac{\mu_{i+1}}{\mu_i} = -\exp\frac{\rho\pi}{\omega}. \tag{3.2.145}$$

This quantity governs the size of the oscillations which we see in Figure 3.2.41.

Subsidiary Homoclinic Orbits.

Now we will show that, as we break our original homoclinic orbit (the *principal* homoclinic orbit), other homoclinic orbits of a different nature arise, and the Silnikov picture is repeated for these new homoclinic orbits. This phenomena was first noted by Hastings [1982], Evans et al. [1982], Gaspard [1983], and Glendinning and Sparrow [1984]. We follow the argument of Gaspard.

When we break the homoclinic orbit, the unstable manifold intersects Π_0 at the point $(e\mu + x_0, \mu)$. Thus, if $\mu > 0$, this point can be used as an initial condition for our map. Now if the z component of the image of this point is zero, we will have found a new homoclinic orbit which passes once through a neighborhood of the origin before falling back into the origin. This condition is given by

$$0 = \beta(e\mu + x_0)\mu^\delta \cos(\xi \log \mu + \Phi_2) + \mu \tag{3.2.146}$$

or

$$-\mu = \beta(e\mu + x_0)\mu^\delta \cos(\xi \log \mu + \Phi_2). \tag{3.2.147}$$

We find the solutions for this graphically for $\delta > 1$ and $\delta < 1$ in the same manner as we investigated the equations for the fixed points; see Figure 3.2.42.

So for $\underline{\delta > 1}$, the only homoclinic orbit is the principal homoclinic orbit which exists at $\mu = 0$.

For $\underline{\delta < 1}$, we get a countable infinity of μ values

$$\mu_i, \mu_{i+1}, \ldots, \mu_{i+n}, \ldots \to 0 \tag{3.2.148}$$

for which these *subsidiary* or *double pulse* homoclinic orbits exist, as shown in Figure 3.2.43.

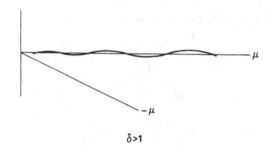

$$\delta > 1$$

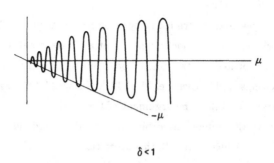

$$\delta < 1$$

Figure 3.2.42. Graphical Solution of (3.2.147).

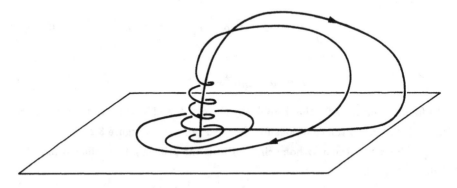

Figure 3.2.43. Double Pulse Homoclinic Orbit.

Note for each of these homoclinic orbits, we can reconstruct our original Silnikov picture of a countable infinity of horseshoes.

For a reference dealing with double pulse homoclinic orbits for the case of real eigenvalues see Yanagida [1986].

The Consequences of Symmetry.

   In the case of a fixed point with real eigenvalues, we saw that the presence of a symmetry resulted in dramatic dynamical consequences. In particular, the symmetry implied the presence of an additional homoclinic orbit which resulted in horseshoes. We now want to examine the dynamical consequences of symmetry in (3.2.99).

   We suppose that H1 and H2 hold and also that (3.2.99) is invariant under the change of coordinates

$$(x, y, z) \to (-x, -y, -z) . \qquad (3.2.149)$$

This is the only symmetry of (3.2.99) which allows homoclinic orbits, see Tresser [1984]. In this case, (3.2.99) has a pair of homoclinic orbits Γ_u, Γ_l as shown in Figure 3.2.44.

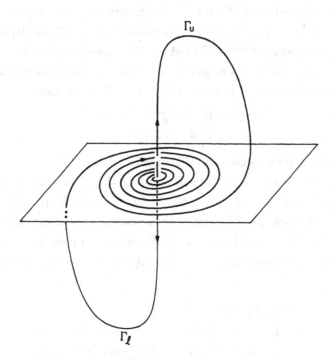

Figure 3.2.44. A Symmetric Pair of Homoclinic Orbits.

Now our previous analysis can be applied to Γ_u and Γ_l separately with Theorem 3.2.17 applying. However, we are interested in the dynamics near the "figure eight" $\Gamma_u \cup \Gamma_l$. The analysis is very similar to that of the nonsymmetric case.

We begin by constructing cross sections to the vector field near the origin. We define

$$\Pi_0^u = \{\, (x,y,z) \in \mathbf{R}^3 \mid \epsilon e^{\frac{2\pi\rho}{\omega}} \le x \le \epsilon,\ y = 0,\ |z| \le \epsilon \,\}$$

$$\Pi_0^l = \{\, (x,y,z) \in \mathbf{R}^3 \mid -\epsilon\left(\frac{e^{-\frac{\rho\pi}{\omega}} + e^{\frac{\rho\pi}{\omega}}}{2}\right) \le x \le -\epsilon\left(\frac{e^{\frac{\rho\pi}{\omega}} + e^{\frac{3\rho\pi}{\omega}}}{2}\right),\ y = 0,\ |z| \le \epsilon \,\}$$

$$\Pi_0^{u,+} = \{\, (x,y,z) \in \Pi_0^u \mid 0 < z \le \epsilon \,\}$$

$$\Pi_0^{u,-} = \{\, (x,y,z) \in \Pi_0^u \mid -\epsilon \le z < 0 \,\}$$

$$\Pi_0^{l,+} = \{\, (x,y,z) \in \Pi_0^l \mid 0 < z \le \epsilon \,\}$$

$$\Pi_0^{l,-} = \{\, (x,y,z) \in \Pi_0^l \mid -\epsilon \le z < 0 \,\} .$$

(3.2.150)

We construct maps $P_0^{L,u}$, $P_0^{L,l}$ on $\Pi_0^{u,+}$ and $\Pi_0^{l,-}$, respectively, in a manner identical to the construction of P_0^L in the nonsymmetric case onto $\Pi_1^u = \{\, (x,y,z) \mid z = \epsilon \,\}$ and $\Pi_1^l = \{\, (x,y,z) \mid z = -\epsilon \,\}$, respectively, where Π_1^u and Π_1^l are chosen large enough to contain $P_0^{L,u}(\Pi_0^{u,+})$ and $P_0^{L,l}(\Pi_0^{l,-})$, respectively. Maps along the homoclinic orbits outside of a neighborhood of the origin are constructed also in a manner identical to that of the nonsymmetric case. Thus, we have

$$P^{L,u} = P_1^{L,u} \circ P_0^{L,u} : \Pi_0^{u,+} \to \Pi_0^u$$
$$P^{L,l} = P_1^{L,l} \circ P_0^{L,l} : \Pi_0^{l,-} \to \Pi_0^l .$$

(3.2.151)

See Figure 3.2.45 for an illustration of the geometry.

We are now in a position to construct the Poincaré map in a neighborhood of $\Gamma_u \cup \Gamma_l$. Let $\phi_t(\cdot)$ denote the linearized flow generated by (3.2.99). Then, by construction, for each $p \in \Pi_0^u$ (resp. Π_0^l) one and only one of the points $\phi_{\pi/\omega}(p)$ and $\phi_{-\pi/\omega}(p)$ belongs to Π_0^l (resp. Π_0^u). We denote this point by $\phi(p)$. Let $z(p)$ denote the z coordinate of any point $p \in \Pi_0^u \cup \Pi_0^l$. Then the Poincaré map is defined as

$$P : \Pi_0^u \cup \Pi_0^l \to \Pi_0^u \cup \Pi_0^l \qquad\qquad (3.2.152)$$

$$p \mapsto \begin{cases} P(p) = P^{L,u}(p) & \text{if } z(p)z\big(P^{L,u}(p)\big) > 0; \\ P(p) = P^{L,l}(p) & \text{if } z(p)z\big(P^{L,l}(p)\big) > 0; \\ P(p) = \phi\big(P^{L,u}(p)\big) & \text{if } z(p)z\big(P^{L,u}(p)\big) < 0; \\ P(p) = \phi\big(P^{L,l}(p)\big) & \text{if } z(p)z\big(P^{L,l}(p)\big) < 0. \end{cases}$$

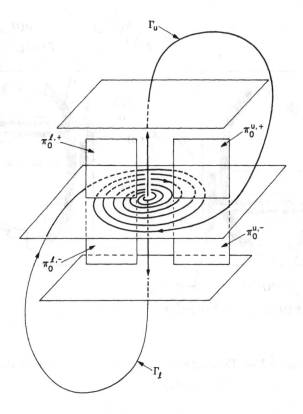

Γ_u

$\pi_0^{l,+}$

$\pi_0^{u,+}$

$\pi_0^{u,-}$

$\pi_0^{l,-}$

Γ_l

Figure 3.2.45. Local Cross-Sections for the Symmetric Case.

We choose a sequence of rectangles $R_k^u \in \Pi_0^{u,+}$, $R_k^l \in \Pi_0^{l,-}$ such that $\Pi_0^{u,+} = \bigcup_{k=0}^{\infty} R_k^u$ and $\Pi_0^{l,-} = \bigcup_{k=0}^{\infty} R_k^u$ in exactly the same manner as in (3.2.108). Using the same arguments as in the nonsymmetric case, we can show that, for k sufficiently large, we can choose two μ_h-horizontal slabs H_{+k}^u, $H_{-k}^u \in R_k^u$ and H_{+k}^l, $H_{-k}^l \in R_k^l$ such that $\overline{A1}$ and $\overline{A2}$ of Section 2.3d hold. See Figure 3.2.46 for an illustration of the geometry.

We now set up the symbolic dynamics.

Let $A^N = \{\pm k, \ldots, \pm(k+N)\}$ for $k, N \in \mathbb{Z}^+$, and let $S^N = A^N \times \{u, l\}$. Then $\Sigma_{u,l}^N$ denotes the set of bi-infinite sequences where each element of the sequence is contained in S^N. For $s = (a, u) \in S^N$ or $s = (a, l) \in S^N$ we define $|s| = |a|$

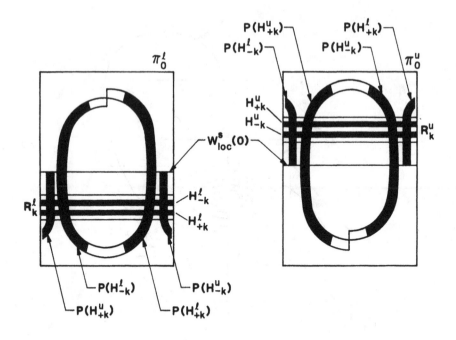

Figure 3.2.46. Horizontal Slabs and Their Images under the
Poincaré Map, P.

and we have

$$\Sigma_{u,l}^{\infty,\alpha} = \{\, \underline{s} = \{s_i\}_{i=-\infty}^{\infty} \mid s_i \in S^{\infty} \text{ and } |s_i| \geq \frac{|s_i|}{\alpha} \,\}.$$

The main theorem can be stated as follows.

Theorem 3.2.19. *a) For each positive integer N there exists a map*

$$\phi^N : \Sigma_{u,l}^N \to \Pi_0^{u,+} \cup \Pi_0^{l,-}$$

which is a homeomorphism of $\Sigma_{u,l}^N$ onto $O_{u,l}^N = \phi^N(\Sigma_{u,l}^N)$ such that

$$P\big|_{O_{u,l}^N} = \phi^N \circ \sigma \circ (\phi^N)^{-1}.$$

b) For each real α with $1 \leq \alpha < -\lambda/\rho$ there exists a map

$$\phi^{\infty,\alpha} : \Sigma_{u,l}^{\infty,\alpha} \to \Pi_0^{u,+} \cup \Pi_0^{l,-}$$

which is a homeomorphism of $\Sigma_{u,l}^{\infty,\alpha}$ onto $O_{u,l}^{\infty,\alpha} = \phi^{\infty,\alpha}(\Sigma_{u,l}^{\infty,\alpha})$ such that

$$P\mid_{O_{u,l}^{\infty,\alpha}}: \phi^{\infty,\alpha} \circ \sigma \circ (\phi^{\infty,\alpha})^{-1}.$$

Now suppose that H2 does not hold, i.e., instead we have

$$-\rho > \lambda > 0. \tag{3.2.153}$$

In the nonsymmetric case, we saw that there were no horseshoes if (3.2.153) holds, and the bifurcation analysis of Glendinning and Sparrow showed that the homoclinic orbit is an attractor and, when it breaks, an attracting periodic orbit is created. However, in the case of a symmetric vector field an interesting effect occurs.

Let $\Sigma^{2,+}$ denote the set of *infinite* sequences of 1's and 2's. An element of $\Sigma^{2,+}$ is written as $\underline{s} = \{s_i\}_{i=0}^{\infty}$, $s_i \in \{1,2\}$. Choose a point p in a neighborhood of the origin. We want to consider the forward orbit of p to which we assign an infinite sequence of 1's and 2's by the following rule. The first entry of the sequence is a 1 (resp. 2) if p moves around Γ_u (resp. Γ_l) and then re-enters a neighborhood of the origin. The second entry in the sequence is a 1 (resp. 2) if p subsequently moves around Γ_u (resp. Γ_l) and re-enters a neighborhood of the origin. We continue constructing the sequence in this manner. An obvious question is whether any such forward orbits actually exist. We have the following theorem.

Theorem 3.2.20. *If (3.2.153) holds then in each neighborhood of* $\Gamma^u \cup \Gamma^l$ *there exist sets of orbits in one to one correspondence with elements of* $\Sigma^{2,+}$. *The dynamics encoded in the sequences is such that a 1 corresponds to a circuit around* Γ^u *and a 2 corresponds to a circuit around* Γ^l.

PROOF: See Holmes [1980]. □

Although there are no horseshoes if (3.2.153) holds, Theorem 3.2.20 tells us that the *approach* of orbits to $\Gamma^u \cup \Gamma^l$ is chaotic.

Observations and Additional References
Comparison Between the Saddle with Real Eigenvalues and the Saddle-Focus. Before leaving three dimensions we want to reemphasize the main differences between the two cases studied.

Real Eigenvalues. In order to have horseshoes it was necessary to start with two homoclinic orbits. Even so, there were no horseshoes near the homoclinic orbit until the homoclinic orbits were broken such as might happen by varying a parameter. It was necessary to know the global twisting of orbits around the homoclinic orbits in order to determine how the horseshoe was formed.

Complex Eigenvalues. One homoclinic orbit is sufficient for a countable infinity of horseshoes whose existence does not require first breaking the homoclinic connection. Knowledge of global twisting around the homoclinic orbit is unnecessary, since the spiralling associated with the imaginary part of the eigenvalues tends to "smear" trajectories uniformly around the homoclinic orbit.

There exists an extensive amount of work concerning Silnikov's phenomenon, yet there are still some open problems.

Strange Attractors. Silnikov type attractors have not attracted the great amount of attention that has been given to Lorenz attractors. The topology of the spiralling associated with the imaginary parts of the eigenvalues makes the Silnikov problem more difficult.

Creation of the Horseshoes and Bifurcation Analysis. We have given part of the bifurcation analysis of Glendinning and Sparrow [1984]. Their paper also contains some interesting numerical work and conjectures. See also Gaspard, Kapral, and Nicolis [1984]. Knot theory has not been applied to this problem.

Nonhyperbolic Fixed Points. There appear to be little or no results concerning orbits homoclinic to nonhyperbolic fixed points in three dimensions.

Applications. The Silnikov phenomenon arises in a variety of applications. See, for example, Arneodo, Coullet, and Tresser [1981a,b]], [1985], Arneodo, Coullet, Spiegel, and Tresser [1985], Arneodo, Coullet, and Spiegel [1982], Gaspard and Nicolis [1983], Hastings [1982], Pilovskii, Rabinovich, and Trakhtengerts [1979], Rabinovich [1978], Rabinovich and Fabrikant [1979], Roux, Rossi, Bachelart, and Vidal [1981], and Vyskind and Rabinovich [1976].

3.2d. Fourth Order Systems

We will now study two examples of an orbit homoclinic to a hyperbolic fixed point of a fourth order ordinary differential equation. The jump from three to four dimensions brings in a large number of new difficulties and, before proceeding to the examples, we want to give a brief overview.

1. *More Cases to Consider*. In three dimensions there were essentially only two
   cases to consider (the others could be obtained via time reversal). In four
   dimensions there are five distinct cases to consider according to the different
   possibilities for the eigenvalues of the linearized vector field at a hyperbolic
   fixed point. They are

   Real Eigenvalues: 1) $\lambda_1, \lambda_2 > 0, \ \lambda_3, \lambda_4 < 0$.

   　　　　　　　　　　2) $\lambda_1 > 0, \ \lambda_2, \lambda_3, \lambda_4 < 0$.

   Complex Eigenvalues: 1) $\rho_1 \pm i\omega_1, \ \rho_2 \pm i\omega_2; \ \rho_1 > 0, \ \rho_2 < 0, \ \omega_1, \ \omega_2 \neq 0$.

   Real and Complex Eigenvalues: 1) $\rho_1 \pm i\omega_1, \ \lambda_1, \lambda_2 > 0; \ \rho_1 < 0, \omega_1 \neq 0$.

   　　　　　　　　　　　　　　　　2) $\rho_1 \pm i\omega_1, \ \lambda_1 < 0, \ \lambda_2 > 0; \ \rho_1 < 0, \omega_1 \neq 0$.

   Other cases may be obtained from these via time reversal. If nonhyperbolic
   fixed points are considered, then even more cases must be considered.

   We will study the example having all complex eigenvalues and the example
   having real and complex eigenvalues having a one dimensional unstable manifold.

2. *More General Horseshoes*. In our three dimensional examples we reduced the
   problem to the study of a two dimensional Poincaré map. The horseshoes contained in a two dimensional map had one expanding direction, one contracting
   direction, and one folding "direction" around an axis normal to the plane.

   For fourth order systems we will be studying a three dimensional Poincaré
   map. Horseshoes contained in three dimensional maps may have one expanding
   direction and two contracting directions or vice versa. Additionally, they may
   either have one or two folding directions. The various possibilities are governed
   by the nature of the eigenvalues at the fixed point. For the most part, the
   different possibilities will not be of much concern to us since our main goal is
   simply to prove the existence of horseshoes. However, these properties would
   be of interest in studying the more "global" aspects of the horseshoes and, in
   particular, in finding conditions under which they formed the chaotic hearts of
   strange attractors. See Figure 3.2.47 for an illustration of some different types
   of three dimensional horseshoes.

3. *Computation of the Time of Flight from Π_0 to Π_1*. In two and three dimensions
   solving for the time of flight was a trivial matter. However (except for the case
   of complex eigenvalues), in four dimensions when the unstable manifold is two

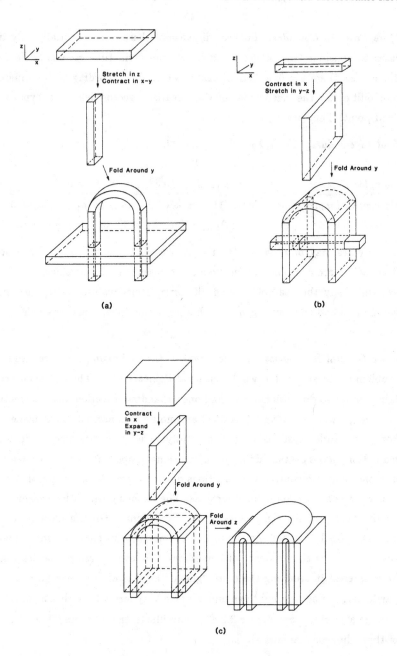

Figure 3.2.47. Examples of Three Dimensional Horseshoes. a) One Expanding, Two
Contracting, and One Folding Direction. b) One Contracting, Two
Expanding, and One Folding Directions. c) One Contracting, Two
Expanding, and Two Folding Directions.

dimensional, the equation for the time of flight is a difficult transcendental equation which requires a more subtle analysis. We will see an example of this when we study four dimensional Hamiltonian systems.

4. *The Presence of Symmetries.* In two and three dimensions it was possible to more or less guess the possible symmetries of the vector field which allowed the presence of homoclinic orbits. Moreover, they were all discrete symmetries. In four dimensions the situation is complicated by the possibility of continuous symmetries. This would allow for the existence of manifolds of homoclinic orbits (see Armbruster, Guckenheimer, and Holmes [1987]).

i) A Complex Conjugate Pair and Two Real Eigenvalues

We consider an equation of the following form

$$\begin{aligned}
\dot{x} &= \rho x - \omega y + P(x, y, z, w)\\
\dot{y} &= \omega x + \rho y + Q(x, y, z, w)\\
\dot{z} &= \lambda z + R(x, y, z, w)\\
\dot{w} &= \nu w + S(x, y, z, w)
\end{aligned} \qquad (x, y, z, w) \in \mathbf{R}^4 \qquad (3.2.154)$$

where $\rho, \lambda < 0$, $\omega, \nu > 0$, and P, Q, R, and S are C^2 and $\mathcal{O}(2)$ at the origin. It should be clear that $(x, y, z, w) = (0, 0, 0, 0)$ is a hyperbolic fixed point of (3.2.154) with the eigenvalues of the vector field linearized about the origin given by $\rho \pm i\omega$, λ, ν. Hence, the origin has a three dimensional stable manifold and a one dimensional unstable manifold. We make the following additional assumptions on (3.2.154).

H1. Equation (3.2.154) has a homoclinic orbit Γ connecting $(0, 0, 0, 0)$ to itself.

H2. $\nu > -\rho > 0$, $-\lambda \neq \nu$.

Our goal is to study the orbit structure of (3.2.154) near Γ. In order to do this we will follow our standard procedure of computing a local Poincaré map defined in a neighborhood of Γ.

Computation of P_0^L.

We will assume that $\rho > \lambda$. In this case the homoclinic orbit is generically tangent to the x-y plane at the origin. This assumption is merely for geometrical convenience in constructing cross-sections to the vector field; it will not affect any of our final results concerning the dynamics of (3.2.154) when H1 and H2 hold.

We define the following cross-sections to (3.2.154)

$$\Pi_0 = \left\{ (x, y, z, w) \in \mathbf{R}^4 \,\middle|\, \epsilon e^{\frac{2\pi\rho}{\omega}} \leq x \leq \epsilon, \ y = 0, \ 0 < w \leq \epsilon, \ -\epsilon \leq z \leq \epsilon \right\}.$$

$$(3.2.155)$$

As in the case of an orbit homoclinic to a saddle-focus in $\mathbf{R}^3$, we choose the x width of Π_0 so that orbits starting on Π_0 do not reintersect Π_0 before leaving a neighborhood of the origin. Additionally, we define

$$\Pi_1 = \left\{ (x, y, z, w) \in \mathbf{R}^4 \,\middle|\, w = \epsilon \right\}.$$

$$(3.2.156)$$

Π_1 will just be chosen large enough to contain the image of Π_0 under the map P_0^L which we now describe. See Figure 3.2.48 for an illustration of the geometry.

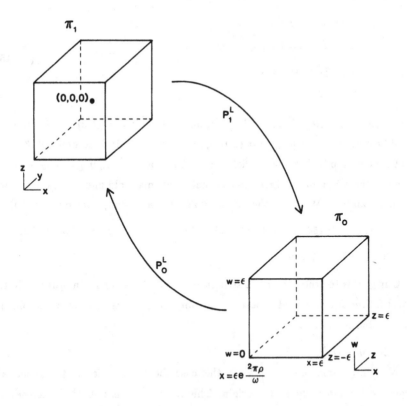

Figure 3.2.48. Cross-Sections Near the Origin.

The linearized flow generated by (3.2.154) is given by

$$x(t) = e^{\rho t}(x_0 \cos \omega t - y_0 \sin \omega t)$$
$$y(t) = e^{\rho t}(x_0 \sin \omega t + y_0 \cos \omega t)$$
$$z(t) = z_0 e^{\lambda t}$$
$$w(t) = w_0 e^{\nu t} .$$

$$(3.2.157)$$

The time of flight T from Π_0 to Π_1 is found by solving

$$\epsilon = w_0 e^{\nu T} \qquad (3.2.158)$$

from which we obtain

$$T = \frac{1}{\nu} \log \frac{\epsilon}{w_0} . \qquad (3.2.159)$$

Using (3.2.157) and (3.2.159), the map P_0^L from Π_0 to Π_1 is found to be (leaving off the subscript 0's)

$$P_0^L : \Pi_0 \to \Pi_1$$

$$\begin{pmatrix} x \\ 0 \\ z \\ w \end{pmatrix} \mapsto \begin{pmatrix} x\left(\frac{\epsilon}{w}\right)^{\rho/\nu} \cos\left(\frac{\omega}{\nu} \log \frac{\epsilon}{w}\right) \\ x\left(\frac{\epsilon}{w}\right)^{\rho/\nu} \sin\left(\frac{\omega}{\nu} \log \frac{\epsilon}{w}\right) \\ z\left(\frac{\epsilon}{w}\right)^{\lambda/\nu} \\ \epsilon \end{pmatrix} . \qquad (3.2.160)$$

See Figure 3.2.48 for an illustration of the geometry.

We now want to get an idea of the geometry of $P_0^L(\Pi_0)$. For this, it will be useful to consider a foliation of Π_0 by slabs, as in the case of the saddle-focus in $\mathbb{R}^3$. We define

$$R_k = \{ (x, y, z, w) \in \mathbb{R}^4 \mid \epsilon e^{\frac{2\pi\rho}{\omega}} \le x \le \epsilon, \, y = 0, \, 0 < z \le \epsilon,$$
$$\epsilon e^{\frac{-2\pi(k+1)\nu}{\omega}} \le w \le \epsilon e^{\frac{-2\pi k\nu}{\omega}} \} . (3.2.161)$$

Then we have

$$\Pi_0 = \bigcup_{k=0}^{\infty} R_k . \qquad (3.2.162)$$

It will be useful to coordinatize the x-y part of Π_1 by polar coordinates. Denoting the x, y, z coordinates on Π_1 by x', y', z' in order to avoid confusion with the coordinates on Π_0, we have

$$r = \sqrt{x'^2 + y'^2} , \qquad \tan \theta = \frac{y'}{x'} \qquad (3.2.163)$$

and, in these coordinates, P_0^L is written

$$P_0^L : \Pi_0 \to \Pi_1$$

$$\begin{pmatrix} x \\ 0 \\ z \\ w \end{pmatrix} \mapsto \begin{pmatrix} x\left(\frac{\epsilon}{w}\right)^{\rho/\nu} \\ \frac{w}{\nu} \log \frac{\epsilon}{w} \\ z\left(\frac{\epsilon}{w}\right)^{\lambda/\nu} \\ \epsilon \end{pmatrix} = \begin{pmatrix} r \\ \theta \\ z' \\ \epsilon \end{pmatrix}. \tag{3.2.164}$$

Now consider an $R_k \subset \Pi_0$ for fixed k. Using (3.2.164) we make the following observations concerning $P_0^L(R_k)$.

1) The two dimensional sheets $w = $ constant contained in R_k are mapped to $\theta = $ constant under P_0^L.
2) The two vertical boundaries of R_k that are parallel to the w-z plane are mapped to two dimensional logarithmic spirals.
3) The two dimensional sheet $z = 0$ contained in R_k is mapped to $z' = 0$ in Π_1.
4) The ratio z'/w goes to zero as $w \to 0$ for $\lambda < -\nu$ and goes to infinity for $\lambda > -\nu$.

Using these four remarks the two possibilities for the image of R_k, k fixed, under P_0^L are shown in Figure 3.2.49.

Computation of P_1^L.

For some open set $U \subset \Pi_1$ we have

$$P_1^L : U \subset \Pi_1 \to \Pi_0$$

$$\begin{pmatrix} x \\ y \\ z \\ \epsilon \end{pmatrix} \mapsto \begin{pmatrix} a & b & c & 0 \\ d & e & f & 0 \\ g & h & i & 0 \\ 0 & 0 & 0 & 0 \end{pmatrix} \begin{pmatrix} x \\ y \\ z \\ 0 \end{pmatrix} + \begin{pmatrix} \bar{x} \\ 0 \\ 0 \\ 0 \end{pmatrix} \tag{3.2.165}$$

where $\left(\bar{x} = \epsilon((1 + e^{2\pi\rho})/2), 0, 0, 0\right) = \Gamma \cap \Pi_0$.

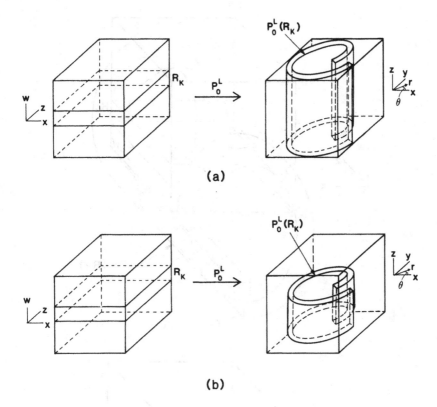

Figure 3.2.49. Geometry of $P_0^L(R_k)$. a) $\lambda > -\nu$. b) $\lambda < -\nu$.

The Poincaré Map $P^L = P_1^L \circ P_0^L$.

Using (3.2.160) and (3.2.165) we have

$$P^L : P_1^L \circ P_0^L : V \subset \Pi_0 \to \Pi_0$$

$$\begin{pmatrix} x \\ z \\ w \end{pmatrix} \mapsto \begin{pmatrix} ax\left(\frac{\epsilon}{w}\right)^{\frac{\ell}{\nu}} \cos\left(\frac{\omega}{\nu} \log \frac{\epsilon}{w}\right) + bx\left(\frac{\epsilon}{w}\right)^{\frac{\ell}{\nu}} \sin\left(\frac{\omega}{\nu} \log \frac{\epsilon}{w}\right) + cz\left(\frac{\epsilon}{w}\right)^{\frac{\lambda}{\nu}} + \bar{x} \\ dx\left(\frac{\epsilon}{w}\right)^{\frac{\ell}{\nu}} \cos\left(\frac{\omega}{\nu} \log \frac{\epsilon}{w}\right) + ex\left(\frac{\epsilon}{w}\right)^{\frac{\ell}{\nu}} \sin\left(\frac{\omega}{\nu} \log \frac{\epsilon}{w}\right) + cz\left(\frac{\epsilon}{w}\right)^{\frac{\lambda}{\nu}} \\ gx\left(\frac{\epsilon}{w}\right)^{\frac{\ell}{\nu}} \cos\left(\frac{\omega}{\nu} \log \frac{\epsilon}{w}\right) + hx\left(\frac{\epsilon}{w}\right)^{\frac{\ell}{\nu}} \sin\left(\frac{\omega}{\nu} \log \frac{\epsilon}{w}\right) + iz\left(\frac{\epsilon}{w}\right)^{\frac{\lambda}{\nu}} + \bar{x} \end{pmatrix}$$
$$(3.2.166)$$

where $V = (P_0^L)^{-1}(U)$.

So, if we choose Π_0 sufficiently small, then $P^L(R_k)$ appears as in Figure 3.2.50. Now we want to show that P^L contains horseshoes. From Figure 3.2.50, it

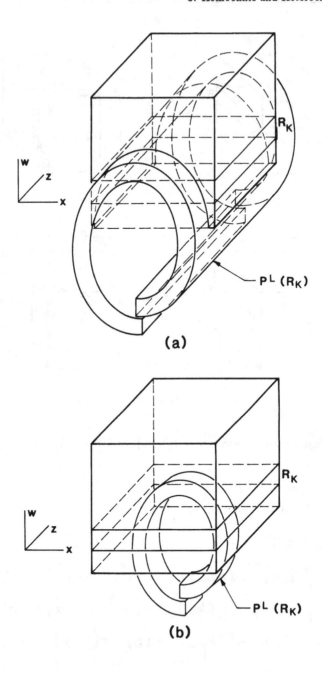

Figure 3.2.50. Geometry of $P^L(R_k)$. a) $\lambda > -\nu$. b) $\lambda < -\nu$.

should be evident that this is a strong possibility; however, certain properties must be satisfied. In particular, we are concerned with two main effects.

1) The ability to find horizontal slabs which map over themselves into vertical slabs with proper behavior of the boundaries.

2) The existence of sufficiently large stretching and contraction rates along appropriate directions.

These effects are formalized in Properties A1 and A2, A1 and A3, or $\overline{A1}$ and $\overline{A2}$ of Chapter 2, which imply Theorem 2.3.3 and the resulting chaotic dynamics associated with the shift map. In our study of an orbit homoclinic to a saddle-focus in $\mathbb{R}^3$ we verified $\overline{A1}$ and $\overline{A2}$ in great detail. However, in this example, we will only indicate the key points leading to the verification of $\overline{A1}$ and $\overline{A2}$, since the complete details are very similar to those given for the saddle-focus in $\mathbb{R}^3$.

Existence of Horizontal Slabs Mapping to Vertical Slabs with Proper Boundary Behavior.

Consider $R_k \subset \Pi_0$, k fixed. It can then be shown that $P^L(R_k)$ intersects R_i in two disjoint μ_v vertical slabs with μ_v satisfying (2.3.69), (2.3.75), and (2.3.78) for $i \geq k/\alpha$ where $1 \leq \alpha < -\nu/\rho$ and k sufficiently large. This relies crucially on the properties $\nu > -\rho > 0$ and $-\lambda \neq \nu$, and the argument is very similar to that given in Lemma 3.2.16.

Stretching and Contraction Rates.

An argument similar to that given in Lemma 3.2.15 can be used to show that $\overline{A1}$ holds everywhere on Π_0 with the possible exception of a countable set of (avoidable) $w = $ constant sheets.

If $-\lambda > \nu$ there are two contracting directions and one expanding direction and if $-\lambda < \nu$, there is one contracting direction and two expanding directions.

Thus, P^L contains a countable infinity of horseshoes and a theorem identical to Theorem 3.2.17 holds. However, despite the rich dynamics which this describes, we have barely scratched the surface of the possible dynamics of P^L and much remains to be discovered.

ii) Silnikov's Example in $\mathbb{R}^4$

The following system was first studied by Silnikov [1967].

We consider an equation of the following form

$$
\begin{aligned}
\dot{x}_1 &= -\rho_1 x_1 - \omega_1 x_2 + P(x_1, x_2, y_1, y_2) \\
\dot{x}_2 &= \omega_1 x_1 - \rho_1 x_2 + Q(x_1, x_2, y_1, y_2) \\
\dot{y}_1 &= \rho_2 y_1 + \omega_2 y_2 + R(x_1, x_2, y_1, y_2) \\
\dot{y}_2 &= -\omega_2 y_1 + \rho_2 y_2 + S(x_1, x_2, y_1, y_2)
\end{aligned}
\qquad (x_1, x_2, y_1, y_2) \in \mathbb{R}^4 \qquad (3.2.167)
$$

where $\rho_1, \rho_2, \omega_1, \omega_2 > 0$ and P, Q, R, and S are C^2 and $\mathcal{O}(2)$ at the origin. Thus, $(x_1, x_2, y_1, y_2) = (0,0,0,0)$ is a fixed point of (3.2.167) and the eigenvalues of (3.2.167) linearized about the origin are given by $-\rho_1 \pm i\omega_1$, $\rho_2 \mp i\omega_2$. Therefore, the origin has a two dimensional stable manifold and a two dimensional unstable manifold. Additionally, we make the following assumptions on (3.2.167).

H1. Equation (3.2.167) has a homoclinic orbit Γ connecting $(0,0,0,0)$ to itself.

H2. $\rho_1 \neq \rho_2$.

So the two dimensional stable and unstable manifolds of the origin intersect non-transversely along Γ. Our goal is to study the orbit structure in a neighborhood of Γ.

Computation of P_0^L.

We compute the map near the origin given by the linearized flow. For this it is more convenient to use polar coordinates. Letting

$$
\begin{aligned}
x_1 &= r_1 \cos \theta_1 \\
x_2 &= r_1 \sin \theta_1 \\
y_1 &= r_2 \cos \theta_2 \\
y_2 &= r_2 \sin \theta_2
\end{aligned}
\qquad (3.2.168)
$$

the linearized vector field is given by

$$
\begin{aligned}
\dot{r}_1 &= -\rho_1 r_1 \\
\dot{\theta}_1 &= \omega_1 \\
\dot{r}_2 &= \rho_2 r_2 \\
\dot{\theta}_2 &= -\omega_2 .
\end{aligned}
\qquad (3.2.169)
$$

The flow generated by (3.2.169) is easily found to be

$$
\begin{aligned}
r_1(t) &= r_{10}e^{-\rho_1 t} \\
\theta_1(t) &= \omega_1 t + \theta_{10} \\
r_2(t) &= r_{20}e^{\rho_2 t} \\
\theta_2(t) &= -\omega_2 t + \theta_{20}\,.
\end{aligned}
\tag{3.2.170}
$$

We define the usual cross-sections to the vector field near the origin

$$
\begin{aligned}
\Pi_0 &= \{\,(r_1,\theta_1,r_2,\theta_2)\,|\,r_1 = \epsilon\,\} \\
\Pi_1 &= \{\,(r_1,\theta_1,r_2,\theta_2)\,|\,r_2 = \epsilon\,\}\,.
\end{aligned}
\tag{3.2.171}
$$

Note that Π_0 and Π_1 have the structure of three dimensional solid tori with the local stable manifold, i.e., $r_2 = 0$, being the center circle of Π_0 and the local unstable manifold, i.e., $r_1 = 0$, being the center circle of Π_1. We let $p_1 = (0,0,\epsilon,\bar{\theta}) = \Gamma \cap W_{\mathrm{loc}}^u$ and $p_0 = (\epsilon,\bar{\theta},0,0) = \Gamma \cap W_{\mathrm{loc}}^s$. See Figure 3.2.51 for an illustration of the geometry.

The time of flight T from Π_0 to Π_1 is found by solving

$$
\epsilon = r_{20}e^{\rho_2 T}
\tag{3.2.172}
$$

to obtain

$$
T = \frac{1}{\rho_2} \log \frac{\epsilon}{r_{20}}\,.
\tag{3.2.173}
$$

Using (3.2.170) and (3.2.173), the map P_0^L is given by (leaving off the subscript 0's)

$$
P_0^L \colon \Pi_0 \to \Pi_1
$$

$$
\begin{pmatrix} \epsilon \\ \theta_1 \\ r_2 \\ \theta_2 \end{pmatrix}
\mapsto
\begin{pmatrix}
\epsilon\left(\dfrac{r_2}{\epsilon}\right)^{\frac{\rho_1}{\rho_2}} \\[4pt]
\theta_1 + \dfrac{\omega_1}{\rho_2}\log\dfrac{\epsilon}{r_2} \\[4pt]
\epsilon \\[4pt]
\theta_2 - \dfrac{\omega_2}{\rho_2}\log\dfrac{\epsilon}{r_2}
\end{pmatrix}.
\tag{3.2.174}
$$

We now want to get an idea of the geometry of the image of Π_0 under P_0^L. Consider an infinite sequence of solid annuli contained in Π_0 defined as follows:

$$
A_k = \Big\{\,(r_1,\theta_1,r_2,\theta_2)\,|\,r_1 = \epsilon,\ \bar{\theta}_1 - \alpha \le \theta_1 \le \bar{\theta}_1 + \alpha,\ \epsilon e^{\frac{-2\pi(k+1)\rho_2}{\omega_1}} \le r_2 \le \epsilon e^{\frac{-2\pi k \rho_2}{\omega_1}},
$$

$$
0 \le \theta_2 \le 2\pi\,\Big\}
\tag{3.2.175}
$$

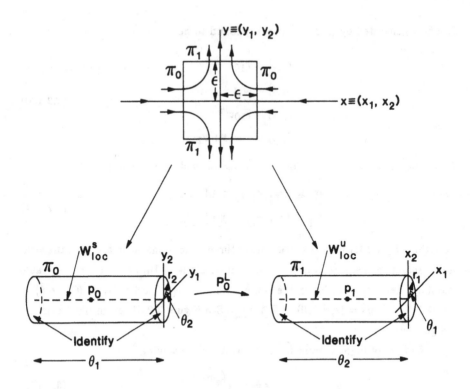

Figure 3.2.51. Geometry of the Flow and the Cross-Sections Π_0 and Π_1
Near the Origin.

for some $\alpha > 0$ and $k = 0, 1, 2, \ldots$; see Figure 3.2.52.

We want to study the geometry of $P_0^L(A_k)$ for fixed $k \geq 0$. In particular, we are interested in the behavior of the boundary of A_k under P_0^L. The boundary of A_k is made up of the union of the two "endcaps," denoted E_k^l and E_k^r, and the inner and outer surfaces, denoted S_k^i and S_k^o. More specifically, we have

$$E_k^l = \{(r_1, \theta_1, r_2, \theta_2) \mid r_1 = \epsilon, \ \theta_1 = \bar{\theta}_1 - \alpha, \ \epsilon e^{(k+1)c} \leq r_2 \leq \epsilon e^{kc}, \ 0 \leq \theta_2 \leq 2\pi \}$$
$$E_k^r = \{(r_1, \theta_1, r_2, \theta_2) \mid r_1 = \epsilon, \ \theta_1 = \bar{\theta}_1 + \alpha, \ \epsilon e^{(k+1)c} \leq r_2 \leq \epsilon e^{kc}, \ 0 \leq \theta_2 \leq 2\pi \}$$
$$S_k^i = \{(r_1, \theta_1, r_2, \theta_2) \mid r_1 = \epsilon, \bar{\theta}_1 - \alpha \leq \theta_1 \leq \bar{\theta}_1 + \alpha, r_2 = \epsilon e^{(k+1)c}, 0 \leq \theta_2 \leq 2\pi \}$$
$$S_k^o = \{(r_1, \theta_1, r_2, \theta_2) \mid r_1 = \epsilon, \ \bar{\theta}_1 - \alpha \leq \theta_1 \leq \bar{\theta}_1 + \alpha, \ r_2 = \epsilon e^{kc}, \ 0 \leq \theta_2 \leq 2\pi \},$$
$$(3.2.176)$$

where $c = -2\pi\rho_2/\omega_1$. See Figure 3.2.53 for an illustration of the geometry.

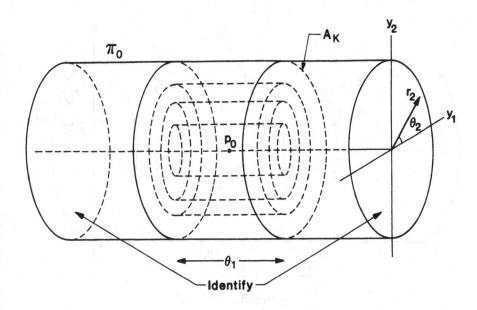

Figure 3.2.52. Geometry of the $A_k \subset \Pi_0$.

Now, using (3.2.174), we can conclude the following

$$P_0^L(E_k^l) = \{ (r_1, \theta_1, r_2, \theta_2) \mid \epsilon e^{\frac{-2\pi(k+1)\rho_1}{\omega_1}} \leq r_1 \leq \epsilon e^{\frac{-2\pi k \rho_1}{\omega_1}},$$
$$\bar{\theta}_1 - \alpha + 2\pi k \leq \theta_1 \leq \bar{\theta}_1 - \alpha + 2\pi(k+1),\ r_2 = \epsilon, 0 \leq \theta_2 \leq 2\pi \}$$

$$P_0^L(E_k^r) = \{ (r_1, \theta_1, r_2, \theta_2) \mid \epsilon e^{\frac{-2\pi(k+1)\rho_1}{\omega_1}} \leq r_1 \leq \epsilon e^{\frac{-2\pi k \rho_1}{\omega_1}},$$
$$\bar{\theta}_1 + \alpha + 2\pi k \leq \theta_1 \leq \bar{\theta}_1 + \alpha + 2\pi(k+1),\ r_2 = \epsilon, 0 \leq \theta_2 \leq 2\pi \}$$

$$P_0^L(S_k^i) = \{ (r_1, \theta_1, r_2, \theta_2) \mid r_1 = \epsilon e^{\frac{-2\pi(k+1)\rho_1}{\omega_1}},$$
$$\bar{\theta}_1 - \alpha + 2\pi(k+1) \leq \theta_1 \leq \bar{\theta}_1 + \alpha + 2\pi(k+1),\ r_2 = \epsilon, 0 \leq \theta_2 \leq 2\pi \}$$

$$P_0^L(S_k^o) = \{ (r_1, \theta_1, r_2, \theta_2) \mid r_1 = \epsilon e^{\frac{-2\pi k \rho_1}{\omega_1}},$$
$$\bar{\theta}_1 - \alpha + 2\pi k \leq \theta_1 \leq \bar{\theta}_1 + \alpha + 2\pi k,\ r_2 = \epsilon, 0 \leq \theta_2 \leq 2\pi \}.$$

$$(3.2.177)$$

Putting these together we see that $P_0^L(A_k)$ appears as in Figure 3.2.54.

The Map P_1^L.

An expression for the affine map P_1^L can be computed in the usual way. How-

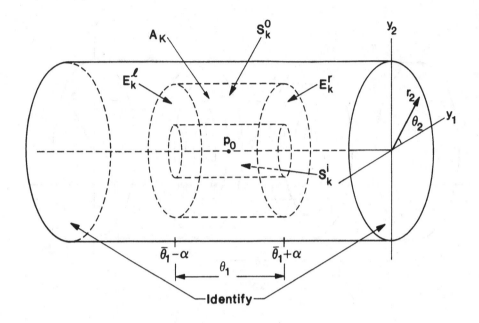

Figure 3.2.53. Geometry of the Boundary of A_k.

ever, we will not do this but rather will describe the relevant features of the geometry of the map. In particular, since p_1 maps to p_0 under the action of the flow generated by (3.2.167), by continuous dependence on initial conditions we can find a neighborhood of p_1 which is mapped onto a neighborhood of p_0. So, for k sufficiently large, a part of $P_0^L(A_k)$ is mapped over A_k, as shown in Figure 3.2.55.

Horseshoes in P^L.

We now point out the relevant features which insure the presence of horseshoes in P^L. The details are similar to those given in our three dimensional examples and are left to the reader. We consider A_k for fixed $k \geq 0$ and $P^L(A_k)$.

Proper Behavior of Boundaries under P^L. For k sufficiently large, $P^L(A_k)$ completely cuts through A_k. Moreover, part of the image of the endcaps of A_k under P^L intersect A_k essentially parallel to their preimage, see Figure 3.2.55. Thus, μ_h horizontal slabs can be found in A_k which map over themselves in μ_v vertical slabs with proper behavior of horizontal and vertical boundaries and with μ_v and μ_h satisfying the necessary requirements (see Section 2.3).

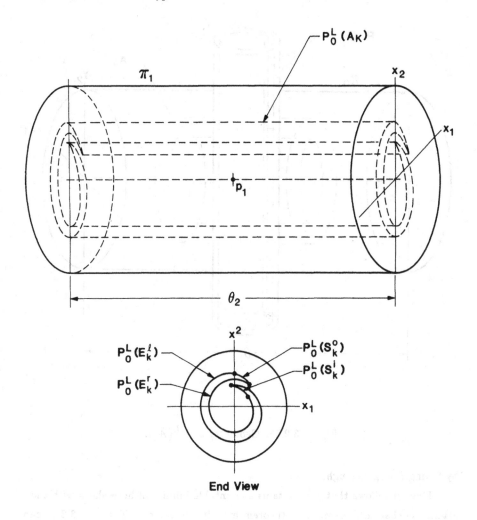

Figure 3.2.54. Geometry of $P_0^L(A_k)$.

Expanding and Contracting Directions. For $\rho_2 > \rho_1$, P^L expands along the r_2 direction, and for $\rho_1 < \rho_2$, P^L contracts along the r_2 direction. Lines parallel to W_{loc}^s are contracted, and a glance at Figure 3.2.54 shows that lines connecting S_k^i and S_k^o are stretched under P^L. Thus, P^L contains two expanding directions and one contracting direction if $\rho_2 > \rho_1$, or one expanding direction and two contracting directions if $\rho_1 < \rho_2$. These growth rates can be made arbitrarily large

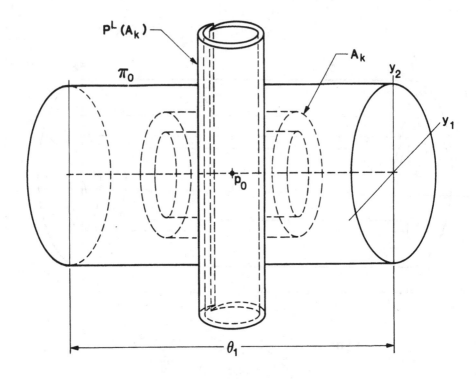

Figure 3.2.55. Geometry of $P^L(A_k)$.

by taking k large enough.

Thus, it follows that P^L contains a countable infinity of horseshoes with their attendant chaotic dynamics. A theorem exactly the same as Theorem 3.2.17 can be stated for P^L.

Observations and References.

Much work remains to be done on homoclinic orbits in four dimensions, particularly concerning detailed analyses of the Poincaré maps and the existence of strange attractors.

Our second example can be found in Glendinning and Tresser [1985]. The existence of subsidiary homoclinic orbits in this example for parametrized systems is considered by Glendinning [1987]; see also Fowler and Sparrow [1984].

Finally, we remark that we have treated no four dimensional examples having multiple homoclinic orbits. One might guess that a saddle point having real eigenvalues with two homoclinic orbits might possess horseshoes in much the same manner as the Lorenz equations in $\mathbb{R}^3$. We work out this example for the Hamiltonian case (see 3.2e.ii)) and remark that the same techniques should work in the general case. However, the precise details remain to be worked out.

3.2e. Orbits Homoclinic to Fixed Points of 4-Dimensional Autonomous Hamiltonian Systems

We now will study orbits homoclinic to fixed points in autonomous Hamiltonian systems. Since all two dimensional (one degree of freedom) Hamiltonian systems are integrable, and therefore do not possess complicated dynamics associated with homoclinic orbits, it is natural to begin our study with four dimensional systems.

Let H be a scalar valued function defined on $\mathbb{R}^4$ which is at least C^3.

$$H(x,y)\colon \mathbb{R}^4 \to \mathbb{R}^1, \qquad (x,y) \equiv (x_1,x_2,y_1,y_2) \in \mathbb{R}^4. \tag{3.2.178}$$

Consider the vector field

$$\begin{pmatrix} \dot{x}_1 \\ \dot{x}_2 \\ \dot{y}_1 \\ \dot{y}_2 \end{pmatrix} = JDH(x_1,x_2,y_1,y_2) \tag{3.2.179}$$

where

$$J = \begin{pmatrix} 0 & 0 & 1 & 0 \\ 0 & 0 & 0 & 1 \\ -1 & 0 & 0 & 0 \\ 0 & -1 & 0 & 0 \end{pmatrix}.$$

This vector field is a *Hamiltonian vector field*. It is an easy calculation to see that $H(x,y) =$ constant is invariant for the flow defined by the vector field.

Suppose that (3.2.179) has a fixed point at $(x,y) = (0,0)$. Then, by Liouville's Theorem (Arnold [1978]), the eigenvalues of the vector field linearized about $(0,0)$ must add up to zero. Therefore, there are two types of hyperbolic fixed points for Hamiltonian systems. They are

1) $\pm \rho \pm i\omega$ – saddle-focus, and
2) $\pm k, \pm l$ – saddle.

We will study both cases; however, first we want to make some general remarks concerning the orbit structure for fourth order Hamiltonian systems.

1) As mentioned above, away from fixed points of (3.2.179), orbits of (3.2.179) lie on three dimensional invariant manifolds defined by $H(x,y) = $ constant.

2) By Liouville's Theorem, the stable and unstable manifolds of fixed points of (3.2.179) have the same dimensions.

3) Suppose (3.2.179) has a hyperbolic fixed point having two dimensional stable and unstable manifolds which intersect along a one dimensional homoclinic orbit. Then, generically, this intersection is transversal (see Robinson [1970]) in the three dimensional surface $H(x,y) = $ constant. This statement is not true for non-Hamiltonian systems.

We are going to assume that in the above two cases we have a homoclinic orbit Γ connecting the fixed point to itself. This homoclinic orbit will lie in the transversal intersection of the stable and unstable manifolds of the origin. Our goal is to study the orbit structure in a neighborhood of Γ. We will follow our usual procedure of constructing a Poincaré map near the homoclinic orbit and studying the orbit structure of the map. However, some modifications must be made due to the fact that orbits lie on invariant three dimensional manifolds. We begin our study with the saddle-focus.

i) Saddle-focus

This problem was first studied by Devaney [1976]. Suppose (3.2.179) has a fixed point at the origin having a homoclinic orbit connecting it to itself and the vector field linearized about the origin is given by

$$
\begin{aligned}
\dot{x}_1 &= \rho x_1 - \omega x_2 \\
\dot{x}_2 &= \omega x_1 + \rho x_2 \\
\dot{y}_1 &= -\rho y_1 + \omega y_2 \\
\dot{y}_2 &= -\omega y_1 - \rho y_2
\end{aligned}
\qquad \rho, \omega > 0
\qquad (3.2.180)
$$

with flow

$$
\begin{aligned}
x_1(t) &= e^{\rho t}(x_{10}\cos\omega t - x_{20}\sin\omega t) \\
x_2(t) &= e^{\rho t}(x_{10}\sin\omega t + x_{20}\cos\omega t) \\
y_1(t) &= e^{-\rho t}(y_{10}\cos\omega t + y_{20}\sin\omega t) \\
y_2(t) &= e^{-\rho t}(-y_{10}\sin\omega t + y_{20}\cos\omega t) \, .
\end{aligned}
\qquad (3.2.181)
$$

Without loss of generality, we can assume that the local stable and unstable manifolds of the origin are given by

$$W_{\text{loc}}^s(0) = \{ (x,y) \mid x = 0 \}$$
$$W_{\text{loc}}^u(0) = \{ (x,y) \mid y = 0 \}$$

(3.2.182)

(note: $(x,y) \equiv (x_1, x_2, y_1, y_2)$).

We study the orbit structure in a neighborhood of (3.2.179) in the same manner as we have done previously. Namely, we compute a Poincaré map on some appropriately chosen cross-section to the flow into itself. Normally, for a four dimensional system a cross section would be three dimensional; however, in the Hamiltonian case, since we are restricted to remain on a 3-d surface, this cross section will be two dimensional (this is a major simplification). Our map will consist of the composition of two maps; one in a neighborhood of the origin given by the linearized vector field and a global one along Γ, which is essentially a rigid motion (just as we have done previously). We will now describe the geometry in a neighborhood of the origin.

For small enough ϵ the following surfaces are cross-sections to the vector field

$$\Pi_0 = \{ (x,y) \mid |x| \le \epsilon, \ |y| = \epsilon \},$$
$$\Pi_1 = \{ (x,y) \mid |x| = \epsilon, \ |y| \le \epsilon \}.$$

(3.2.183)

These surfaces are solid tori ($S^1 \times \mathbb{R}^2$). We also consider the intersection of these surfaces with the three dimensional energy surface

$$\Sigma_0^s = \Pi_0 \cap H^{-1}(0),$$
$$\Sigma_0^u = \Pi_1 \cap H^{-1}(0).$$

(3.2.184)

Let σ^s (resp. σ^u) be the intersection of the local stable (resp. unstable) manifold with Σ_0^s (resp. Σ_0^u). So σ^s and σ^u are the center circles of the solid tori Π_0, Π_1. Finally, given a transverse homoclinic orbit Γ, we denote $q^s = \Gamma \cap \sigma^s$ and $q^u = \Gamma \cap \sigma^u$. We will attempt to illustrate the geometry in a neighborhood of the origin in Figure 3.2.56. In this figure we identify the two ends of the cylinders in order to get the tori. Σ_0^s and Σ_0^u are represented as two dimensional surfaces inside Π_0 and Π_1, respectively. D^s and D^u represent two dimensional neighborhoods of q^s and q^u in Σ_0^s and Σ_0^u, respectively.

Computation of P_0^L.

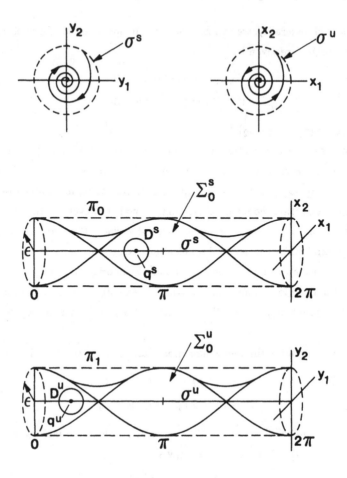

Figure 3.2.56. Geometry of the Flow and Cross-Sections
Near the Origin.

We now want to construct a mapping, P_0^L, of $D^s - \sigma^s$ into $D^u - \sigma^u$ (note: P_0^L cannot be defined on σ^s since these points are on $W^s(0)$). A priori, there is no reason to expect that points in $D^s - \sigma^s$ should map into $D^u - \sigma^u$; however, we shall see that this does happen as a result of the eigenvalues having nonzero imaginary part.

First we transform the linearized flow into polar coordinates. Letting

$$x_1 = r_u \cos \theta_u$$
$$x_2 = r_u \sin \theta_u$$
$$y_1 = r_s \cos \theta_s$$
$$y_2 = r_s \sin \theta_s$$

(3.2.185)

the linearized vector field becomes

$$\dot{r}_u = \rho r_u$$
$$\dot{\theta}_u = \omega$$
$$\dot{r}_s = -\rho r_s$$
$$\dot{\theta}_s = -\omega$$

(3.2.186)

and the flow generated by (3.2.186) is given by

$$r_u(t) = r_u^o e^{\rho t}$$
$$\theta_u(t) = \omega t + \theta_u^o$$
$$r_s(t) = r_s^o e^{-\rho t}$$
$$\theta_s(t) = -\omega t + \theta_s^o .$$

(3.2.187)

Now we want to show that P_0^L maps curves transverse to σ^s in D^s into curves which spiral infinitely often around Σ_0^u and are $C^1 - \epsilon$ close (for some given $\epsilon > 0$) to σ^u. This is illustrated in Figure 3.2.57.

Now how do you see this? Recall the expression for the flow. The length of time necessary for points on Σ_0^s to reach Σ_0^u approaches ∞ as σ^s is approached (we will compute an exact expression for this time shortly). So, as t increases, r_s shrinks, r_u grows, and θ_u and θ_s increase monotonically (mod 2π). So, if we view θ_u as being a coordinate for Σ_0^u, we can see that the image of a curve S is wrapped infinitely often around σ^u.

The time of flight from Σ_0^s to Σ_0^u is found by solving

$$\epsilon = r_u^o e^{\rho T}$$

(3.2.188)

for

$$T = \frac{1}{\rho} \log \frac{\epsilon}{r_u^o} .$$

(3.2.189)

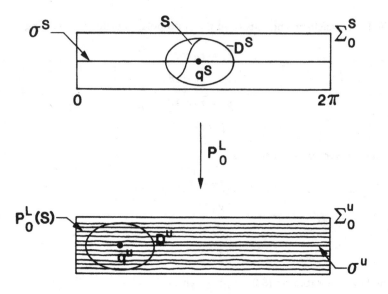

Figure 3.2.57. The Image of a Curve S under P_0^L.

A point on Σ_0^s can be labelled (leaving off the subscript 0's) by θ_s, x_1, x_2, and a point on Σ_0^u can be labelled by θ_u, y_1, y_2. (Note: the "extra" coordinate used to label points on a 2-d surface results from the fact that we do not have an intrinsic coordinate system for $\Sigma_0^{s,u}$; this will not matter.) So the map in Cartesian coordinates is given by

$$P_0^L : D^s - \sigma^s \to \Sigma_0^u \qquad (3.2.190)$$

$$\begin{pmatrix} x_1 \\ x_2 \\ y_1 \\ y_2 \end{pmatrix} \mapsto \begin{pmatrix} \frac{\epsilon}{r_u}\left(x_1 \cos\left(\frac{\omega}{\rho}\log\frac{\epsilon}{r_u}\right) - x_2 \sin\left(\frac{\omega}{\rho}\log\frac{\epsilon}{r_u}\right)\right) \\ \frac{\epsilon}{r_u}\left(x_1 \sin\left(\frac{\omega}{\rho}\log\frac{\epsilon}{r_u}\right) + x_2 \cos\left(\frac{\omega}{\rho}\log\frac{\epsilon}{r_u}\right)\right) \\ \frac{r_u}{\epsilon}\left(y_1 \cos\left(\frac{\omega}{\rho}\log\frac{\epsilon}{r_u}\right) + y_2 \sin\left(\frac{\omega}{\rho}\log\frac{\epsilon}{r_u}\right)\right) \\ \frac{r_u}{\epsilon}\left(-y_1 \sin\left(\frac{\omega}{\rho}\log\frac{\epsilon}{r_u}\right) + y_2 \cos\left(\frac{\omega}{\rho}\log\frac{\epsilon}{r_u}\right)\right) \end{pmatrix}$$

where $r_u = \sqrt{x_1^2 + x_2^2}$ and in polar coordinates by

$$\begin{pmatrix} r_u \\ \theta_u \\ \epsilon \\ \theta_s \end{pmatrix} \mapsto \begin{pmatrix} \epsilon \\ \frac{\omega}{\rho}\log\frac{\epsilon}{r_u} + \theta_u \\ r_u \\ \frac{\omega}{\rho}\log\frac{r_u}{\epsilon} + \theta_s \end{pmatrix}. \qquad (3.2.191)$$

Now let $\beta : I \mapsto D^s$ be a parametrized curve in D^s which intersects σ^s at q, where $I = (-\tau, \tau)$ and $\beta(0) = q$ for some $\tau > 0$; see Figure 3.2.58 (note: in Figure 3.2.58

the right hand side of the Figure represents Σ_0^s and Σ_0^u removed from π_0 and π_1, respectively, and "flattened out").

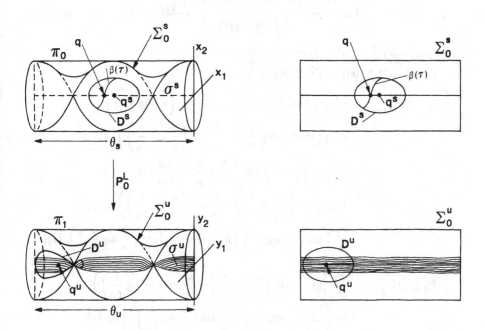

Figure 3.2.58. The Image of the Curve $\beta(\tau)$ under P_0^L.

Note from Figure 3.2.58 that $\beta(\tau)$ has components $\beta_{\theta_s}(\tau)$, $\beta_{x_1}(\tau)$, and $\beta_{x_2}(\tau)$ with $\beta_{r_s}(\tau) = 0$ with respect to the coordinates $(x_1, x_2, \theta_s, r_s)$, and that $P_0^L(\beta(\tau))$ has components $P_0^L(\beta(\tau))_{\theta_u}$, $P_0^L(\beta(\tau))_{y_1}$, and $P_0^L(\beta(\tau))_{y_2}$ with $P_0^L(\beta(\tau))_{r_u} = 0$ with respect to the coordinates $(r_u, \theta_u, y_1, y_2)$. This coordinatization will be particularly useful. We have the following lemma.

Lemma 3.2.21. For τ sufficiently small $P_0^L(\beta(\tau))$ is C^1 ϵ close to σ^u in Σ_0^u. Furthermore, tangent vectors normal to σ^s are stretched by an arbitrarily large amount and vectors tangent to σ^s are shrunk by an arbitrarily large amount as $\tau \to 0$.

PROOF: We already know that β is mapped around Σ_0^u, accumulating on σ^u as

$\tau \to 0$. We need to show that the tangent vector of $P_0^L(\beta(\tau))$ is ϵ-close to the tangent vector of σ^u, for some given $\epsilon > 0$, and for D^s small enough. This will be true if

$$\lim_{\tau \to 0} \frac{\left|\left(\frac{d}{d\tau}P_0^L(\beta(\tau))_{y_1}, \frac{d}{d\tau}P_0^L(\beta(\tau))_{y_2}\right)\right|}{\left|\frac{d}{d\tau}P_0^L(\beta(\tau))_{\theta_u}\right|} = 0. \qquad (3.2.192)$$

Note that $\frac{d}{d\tau}P_0^L(\beta(\tau)) = DP_0^L(\beta(\tau))\frac{d\beta}{d\tau}$.

So using (3.2.190) and (3.2.191) gives

$$\begin{aligned}
\frac{d}{d\tau}P_0^L(\beta(\tau))_{y_1} =\ & \frac{x_1}{\epsilon r_u}\left[y_1\left(\cos\left(\frac{\omega}{\rho}\log\frac{\epsilon}{r_u}\right) + \frac{\omega}{\rho}\sin\left(\frac{\omega}{\rho}\log\frac{\epsilon}{r_u}\right)\right)\right. \\
& \left. + y_2\left(\sin\left(\frac{\omega}{\rho}\log\frac{\epsilon}{r_u}\right) - \frac{\omega}{\rho}\cos\left(\frac{\omega}{\rho}\log\frac{\epsilon}{r_u}\right)\right)\right]\dot{\beta}_{x_1}(\tau) \\
& + \frac{x_2}{\epsilon r_u}\left[y_1\left(\cos\left(\frac{\omega}{\rho}\log\frac{\epsilon}{r_u}\right) + \frac{\omega}{\rho}\sin\left(\frac{\omega}{\rho}\log\frac{\epsilon}{r_u}\right)\right)\right. \\
& \left. + y_2\left(\sin\left(\frac{\omega}{\rho}\log\frac{\epsilon}{r_u}\right) - \frac{\omega}{\rho}\cos\left(\frac{\omega}{\rho}\log\frac{\epsilon}{r_u}\right)\right)\right]\dot{\beta}_{x_2}(\tau) \\
& + \frac{r_u}{\epsilon}\left(\cos\left(\frac{\omega}{\rho}\log\frac{\epsilon}{r_u}\right)\right)\dot{\beta}_{y_1}(\tau) + \frac{r_u}{\epsilon}\left(\sin\left(\frac{\omega}{\rho}\log\frac{\epsilon}{r_u}\right)\right)\dot{\beta}_{y_2}(\tau)
\end{aligned}$$
$$(3.2.193)$$

$$\begin{aligned}
\frac{d}{d\tau}P_0^L(\beta(\tau))_{y_2} =\ & \frac{x_1}{\epsilon r_u}\left[y_1\left(-\sin\left(\frac{\omega}{\rho}\log\frac{\epsilon}{r_u}\right) + \frac{\omega}{\rho}\cos\left(\frac{\omega}{\rho}\log\frac{\epsilon}{r_u}\right)\right)\right. \\
& \left. + y_2\left(\cos\left(\frac{\omega}{\rho}\log\frac{\epsilon}{r_u}\right) + \frac{\omega}{\rho}\sin\left(\frac{\omega}{\rho}\log\frac{\epsilon}{r_u}\right)\right)\right]\dot{\beta}_{x_1}(\tau) \\
& + \frac{x_2}{\epsilon r_u}\left[y_1\left(-\sin\left(\frac{\omega}{\rho}\log\frac{\epsilon}{r_u}\right) + \frac{\omega}{\rho}\cos\left(\frac{\omega}{\rho}\log\frac{\epsilon}{r_u}\right)\right)\right. \\
& \left. + y_2\left(\cos\left(\frac{\omega}{\rho}\log\frac{\epsilon}{r_u}\right) + \frac{\omega}{\rho}\sin\left(\frac{\omega}{\rho}\log\frac{\epsilon}{r_u}\right)\right)\right]\dot{\beta}_{x_2}(\tau) \\
& + \frac{r_u}{\epsilon}\left(-\sin\left(\frac{\omega}{\rho}\log\frac{\epsilon}{r_u}\right)\right)\dot{\beta}_{y_1}(\tau) + \left(\frac{r_u}{\epsilon}\cos\left(\frac{\omega}{\rho}\log\frac{\epsilon}{r_u}\right)\right)\dot{\beta}_{y_2}(\tau)
\end{aligned}$$
$$(3.2.194)$$

and

$$\frac{d}{d\tau}P_0(\beta(\tau))_{\theta_u} \equiv \frac{-\omega}{\rho r_u}\dot{\beta}(\tau)_{r_u} + \dot{\beta}(\tau)_{\theta_u} \qquad (3.2.195)$$

where "$\cdot$" $= \frac{d}{d\tau}$. Now, as $\tau \to 0$, $r_u, x_1, x_2 \to 0$, and we can assume $\left|\frac{d}{d\tau}\beta(\tau)\right|$ is bounded so that $\frac{d}{d\tau}P_0^L(\beta(\tau))_{y_1}$ and $\frac{d}{d\tau}P_0^L(\beta(\tau))_{y_2}$ are bounded as $\tau \to 0$, and $\frac{d}{d\tau}P_0^L(\beta(\tau))_{\theta_u} \to \infty$ as $\tau \to 0$. Therefore,

$$\lim_{\tau \to 0} \frac{\left|\left(\frac{d}{d\tau}P_0^L(\beta(\tau))_{y_1}, \frac{d}{d\tau}P_0^L(\beta(\tau))_{y_2}\right)\right|}{\left|\frac{d}{d\tau}P_0^L(\beta(\tau))_{\theta_u}\right|} = 0 \qquad (3.2.196)$$

which means that the tangent vector to the image of $\beta(\tau)$ approaches the tangent vector of σ^u.

Next, we check the stretching and contraction rates for different directions. Let

$$C = \cos\left(\frac{\omega}{\rho}\log\frac{\epsilon}{r_u}\right)$$

$$S = \sin\left(\frac{\omega}{\rho}\log\frac{\epsilon}{r_u}\right).$$

$$T_1 = x_1 C - x_2 S$$
$$T_2 = x_1 S + x_2 C$$
$$T_3 = y_1 C + y_2 S$$
$$T_4 = -y_1 S + y_2 C.$$

Then, in Cartesian coordinates, DP_0^L is given by

$$
\begin{pmatrix}
\frac{-\epsilon x_1}{r_u^3}T_1 + \frac{\epsilon\omega x_1}{\rho r_u^3}T_2 + \frac{\epsilon}{r_u}C & \frac{-\epsilon x_2}{r_u^3}T_1 + \frac{\epsilon\omega x_2}{\rho r_u^3}T_2 - \frac{\epsilon}{r_u}S & 0 & 0 \\
\frac{-\epsilon x_1}{r_u^3}T_2 - \frac{\epsilon\omega x_1}{\rho r_u^3}T_1 + \frac{\epsilon}{r_u}S & \frac{-\epsilon x_2}{r_u^3}T_2 - \frac{\epsilon\omega x_2}{\rho r_u^3}T_1 + \frac{\epsilon}{r_u}C & 0 & 0 \\
\frac{x_1}{\epsilon r_u}T_3 - \frac{\omega x_1}{\epsilon\rho r_u}T_4 & \frac{x_2}{\epsilon r_u}T_3 - \frac{\omega x_2}{\epsilon\rho r_u}T_4 & \frac{r_u}{\epsilon}C & \frac{r_u}{\epsilon}S \\
\frac{x_1}{\epsilon r_u}T_4 + \frac{\omega x_1}{\epsilon\rho r_u}T_3 & \frac{x_2}{\epsilon r_u}T_4 + \frac{\omega x_2}{\epsilon\rho r_u}T_4 & \frac{-r_u}{\epsilon}S & \frac{r_u}{\epsilon}C
\end{pmatrix}
$$
$$(3.2.197)$$

Now look at

$$DP_0^L\begin{pmatrix}\dot{\beta}_{x_1}\\\dot{\beta}_{x_2}\\\dot{\beta}_{y_1}\\\dot{\beta}_{y_2}\end{pmatrix} \equiv \begin{pmatrix}\ddot{\beta}_{x_1}\\\ddot{\beta}_{x_2}\\\ddot{\beta}_{y_1}\\\ddot{\beta}_{y_2}\end{pmatrix}. \qquad (3.2.198)$$

Using 3.2.197 it is (relatively) easy to see that the vector $(0,0,\dot{\beta}_{y_1},\dot{\beta}_{y_2})$ corresponds to a tangent vector in the θ^s direction and that the length of the image of this vector goes to zero as $\tau \to 0$ (or equivalently, $r_u \to 0$). Similarly, $(\dot{\beta}_{x_1},\dot{\beta}_{x_2},0,0)$ corresponds to a tangent vector in the direction perpendicular to σ^s, and we have that the length of

$$DP_0^L\begin{pmatrix}\dot{\beta}_{x_1}\\\dot{\beta}_{x_2}\\0\\0\end{pmatrix} \qquad (3.2.199)$$

goes to ∞ as $\tau \to 0$ $(r_u \to 0)$. $\qquad\qquad\qquad\qquad\qquad\qquad\qquad\square$

The Map P_1^L.

We now describe the map P_1^L which takes q^u into q^s and, consequently, some neighborhood of q^u into q^s. Recall that $W^s(0)$ and $W^s(0)$ intersect transversely in $H^{-1}(0)$ along Γ. Therefore, $W^s(0)$ intersects D^u transversely at q^u, and $W^u(0)$ intersects D^s transversely at q^s. This is the main feature we need to describe the full Poincaré map; see Figure 3.2.59.

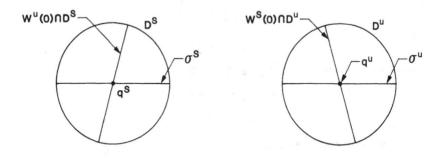

Figure 3.2.59. The Geometry of $W^u(0) \cap D^s$ and
$W^s(0) \cap D^u$.

The Poincaré Map $P^L = P_1^L \circ P_0^L$.

We want to show that P^L contains a horseshoe, so we need to find μ_h horizontal slabs which behave properly under P^L and, to verify the stretching and contraction conditions (see Section 2.3).

We choose a horizontal slab H in D^s with horizontal sides "parallel" to σ^s and vertical boundaries "parallel" to $W^u(0) \cap D^s$, as shown in Figure 3.2.60.

By Lemma 3.2.21, $P_0^L(H)$ is stretched in the direction of $W^u(0)$, contracted in the direction of $W^s(0)$, and wrapped around Σ_0^u many times, as shown in Figure 3.2.61, with the vertical boundaries of $P_0^L(H)$ C^1 ϵ close to σ^u.

So, for D^s and H appropriately chosen, P_1^L maps $P_0^L(H)$ over H as shown in Figure 3.2.62, with the vertical boundaries of H C^1 ϵ close to $W^u(0) \cap D^s$.

Thus, we can choose a sequence of horizontal slabs and conclude that P^L contains an invariant Cantor set on which it is topologically conjugate to a full shift on a countable set of symbols. In other words, we have proven the following theorem of Devaney.

Theorem 3.2.22 (Devaney [1976]). *Consider a two degree of freedom Hamil-*

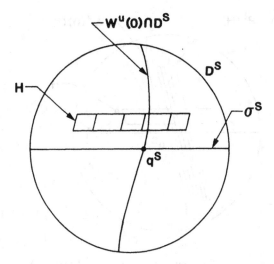

Figure 3.2.60. Horizontal Slab in D^s.

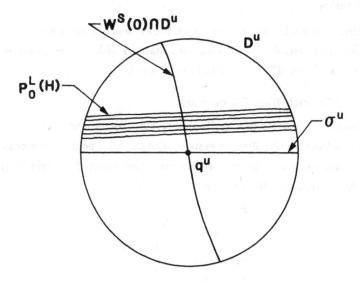

Figure 3.2.61. The Image of H under P_0^L.

tonian system having a transverse homoclinic orbit to a fixed point of saddle-focus type (i.e., the eigenvalues are of the form $\pm\rho\pm i\omega$). Then an associated Poincaré map defined on an appropriately chosen cross-section to the homoclinic orbit contains a

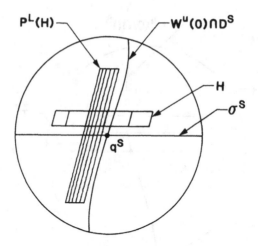

Figure 3.2.62. The Geometry of $P^L(H) \cap H$.

Smale horseshoe.

Finally, we remark that we have shown the existence of horseshoes on the level set $H^{-1}(0)$. However, due to the structural stability of horseshoes, they will also exist on the level sets $H^{-1}(\epsilon)$ for ϵ sufficiently small.

ii) The Saddle with Real Eigenvalues

This problem was first studied by Holmes [1980]. We assume that we are given a 2-degree of freedom Hamiltonian system having a fixed point at the origin with purely real eigenvalues. The vector field linearized about the origin O (after a possible linear transformation) is given by

$$\dot{x}_1 = lx_1$$
$$\dot{x}_2 = kx_2$$
$$\dot{y}_1 = -ly_1 \qquad , l, k > 0 \qquad\qquad (3.2.200)$$
$$\dot{y}_2 = -ky_2$$

with flow

$$x_1(t) = x_{10}e^{lt}$$
$$x_2(t) = x_{20}e^{kt}$$
$$y_1(t) = x_{10}e^{-lt} \qquad\qquad (3.2.201)$$
$$y_2(t) = x_{20}e^{-kt} .$$

So (just as in the saddle-focus case) we have

$$W^s_{loc}(0) = \{(x,y) \mid x = 0\}$$
$$W^u_{loc}(0) = \{(x,y) \mid y = 0\}$$
$$(x,y) \equiv (x_1, x_2, y_1, y_2). \qquad (3.2.202)$$

We now make the following assumption.

Assumption 1: There are two homoclinic orbits, Γ_a, Γ_b connecting O to itself. Γ_a leaves O in the first quadrant of $W^u_{loc}(0)$ and re-enters in the second quadrant of $W^s_{loc}(0)$; Γ_b leaves O in the third quadrant of $W^u_{loc}(0)$ and re-enters in the fourth quadrant of $W^s_{loc}(0)$, see Figure 3.2.63.

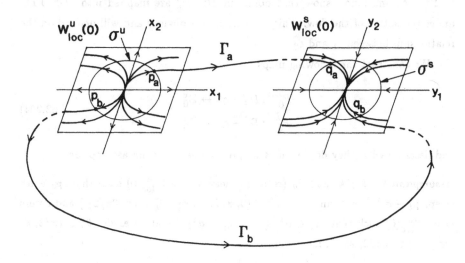

Figure 3.2.63. The Geometry of the Homoclinic Orbits
Near the Origin.

Now the method of analysis will be the usual one; namely, we construct a map in the neighborhood of the origin given by the linear flow and compose it with a "global" map along a homoclinic orbit in order to get a Poincaré map on some appropriately chosen cross-section. However, in this situation the Poincaré map will consist of the union of two maps, one along each homoclinic orbit.

Construction of P^L_0.

First we construct maps in a neighborhood of the origin. It will be necessary to modify our definitions of Π_0 and Π_1 slightly. We define

$$
\begin{aligned}
\Pi_0 &= \{\, (x,y) \mid |x| \leq \epsilon_1 \,, \quad |y| = \epsilon_2 \,\}, \\
\Pi_1 &= \{\, (x,y) \mid |x| = \epsilon_2 \,, \quad |y| \leq \epsilon_1 \,\}.
\end{aligned}
\tag{3.2.203}
$$

Unlike our previous examples, it will be necessary to choose ϵ_1 and ϵ_2 carefully in order to obtain the desired behavior.

Σ_0^s, Σ_0^u, σ^s, and σ^u will be as previously defined in the saddle-focus case. Let Γ_a, Γ_b intersect σ^u, σ^s, at the points p_a, p_b, q_a, q_b. Let D_a^u, D_b^u be neighborhoods of p_a, p_b, respectively, in Σ_0^u, and let D_a^s, D_b^s be neighborhoods of q_a, q_b, respectively, in Σ_0^s. We must now show that points in D_a^s, D_b^s are mapped into $D_a^u \cup D_b^u$ under the action of the flow. This is by no means obvious and will depend on the relationship between ϵ_1 and ϵ_2.

We construct the following maps

$$
\begin{aligned}
P_0^{L,a} &: D_a^s - \sigma^s \rightarrow \Sigma_0^u \\
P_0^{L,b} &: D_b^s - \sigma^s \rightarrow \Sigma_0^u
\end{aligned}
\tag{3.2.204}
$$

and describe what they do, but first we give our second main assumption.

Assumption 2: $l > k$ and Γ_a (resp. Γ_b) leaves O in $W_{\text{loc}}^u(0)$ such that $p_a \in \sigma^u$ (resp. $p_b \in \sigma^u$) lies at an angle $\theta_a^u \in (0, \pi/2)$ (resp. $\theta_b^u \in (\pi, 3\pi/2)$) and enters O in $W_{\text{loc}}^s(0)$ such that $q_a \in \sigma^s$ (resp. $q_b \in \sigma^s$) lies at an angle $\theta_a^s \in (\pi/2, \pi)$ (resp. $\theta_b^s \in (3\pi/2, 2\pi)$); moreover

$$
|(\tan \theta_u)(\tan \theta_s)| < -\delta + l/k[\exp(\tfrac{l}{k} - 1) \log \tfrac{\epsilon_1}{\epsilon_2}]
\tag{3.2.205}
$$

for $\delta > 0$ and small, where θ_s represents the θ_s coordinate of points in S_a (resp. S_b), and θ_u the θ_u coordinates of the image of these same points under P_0^L. S_a and S_b are rectangles in D_a^s and D_b^s, respectively, see Figures 3.2.64 and 3.2.65. (Note: the right side of Figure 3.2.64 represents the two dimensional sheets, Σ_0^s and Σ_0^u, removed from π_0 and π_1, respectively, and "flattened out".)

Let S_a^+ be (resp. S_b^+) the part of S_a (resp. S_b) above σ_s and S_a^- (resp. S_b^-) be the part of S_a (resp. S_b) below σ_s, see Figure 3.2.65.

The following lemma describes the orbit structure of the flow near the origin.

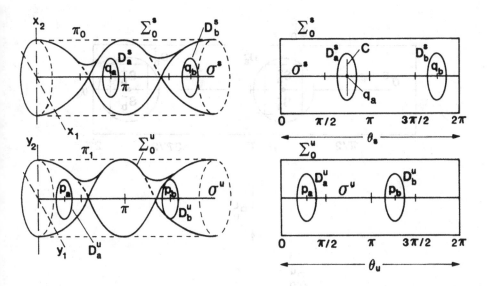

Figure 3.2.64. Geometry of the Flow and Cross-Sections
Near the Origin.

Lemma 3.2.23. $P_0^{L,a}(S_a)$ *consists of two components,* $P_0^{L,a}(S_a^+)$ *and* $P_0^{L,a}(S_a^-)$ *lying across* D_b^u *and* D_b^u, *respectively, with the horizontal boundaries of* $P_0^{L,a}(S_a^+)$ *and* $P_0^{L,a}(S_a^-)$ C^1 ϵ *close to* σ^u *for* S_a *sufficiently small. Similarly,* $P_0^{L,b}(S_b)$ *consists of two components,* $P_0^{L,b}(S_b^+)$ *and* $P_0^{L,b}(S_b^-)$ *lying across* D_a^u *and* D_b^u, *respectively, with the horizontal boundaries of* $P_0^{L,b}(S_b^+)$ *and* $P_0^{L,b}(S_b^-)$ C^1 ϵ *close to* σ^u *for* S_b *sufficiently small. See Figure 3.2.65.*

PROOF: *Step 1.* We first find conditions on ϵ_1 and ϵ_2 so that the images of $S_a^{+,-}$, $S_b^{+,-}$ lie in Σ_0^u. To do this we choose a curve $C \subset S_a$ transverse to σ^s. Now

$$P_0^{L,a}(C) \subset \Sigma_0^u \quad \text{implies} \quad y_1^2(t) + y_2^2(t) < \epsilon_1^2 \qquad (3.2.206)$$

where $y_1(t)$, $y_2(t)$ are the y_1, y_2 components of the image of a point on C.

But, using (3.2.201), we get

$$y_1^2(t) + y_2^2(t) = y_{10}^2 e^{-2lt} + y_{20}^2 e^{-2kt} < (y_{10}^2 + y_{20}^2)e^{-2kt} . \qquad (3.2.207)$$

Since $C \subset \Sigma_0^s$, we have

$$y_{10}^2 + y_{20}^2 = \epsilon_2^2 . \qquad (3.2.208)$$

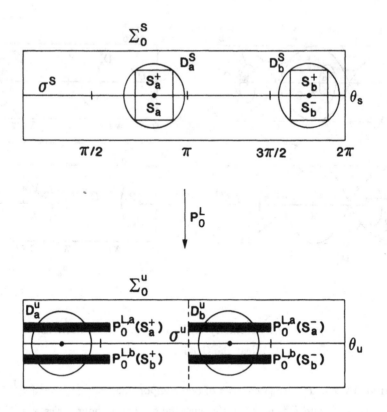

Figure 3.2.65. The Images of Rectangles.

Combining (3.2.206), (3.2.207), and (3.2.208), we see that (3.2.206) will be satisfied provided

$$\epsilon_2^2 e^{-2kt} < \epsilon_1^2 , \qquad (3.2.209)$$

or

$$\frac{\epsilon_2}{\epsilon_1} < e^{kt} . \qquad (3.2.210)$$

Equation (3.2.210) tells us that points on Σ_0^s actually do map to Σ_0^u under the action of the linearized flow provided ϵ_1 and ϵ_2 (i.e., Π_0 and Π_1) are chosen appropriately.

Step 2. From (3.2.210) we estimate t, or more particularly, the minimum time required for points of C to be mapped to Σ_0^u. This occurs for points at the extrema of C since points close to σ^s take arbitrarily long to get to Σ_0^u. At the extrema of

C we have

$$x_{10}^2 + x_{20}^2 = \epsilon_1^2. \tag{3.2.211}$$

So, using (3.2.201), we get the equation

$$x_1^2(t) + x_2^2(t) = x_{10}^2 e^{2lt} + x_{20}^2 e^{2kt} = \epsilon_2^2. \tag{3.2.212}$$

Using the linearized Hamiltonian, $H = lx_1y_1 + kx_2y_2$, on the $H = 0$ surface we have

$$lx_{10}y_{10} = -kx_{20}y_{20}, \tag{3.2.213}$$

so we can use (3.2.211) and (3.2.213) to express x_{10}, x_{20} in terms of ϵ_1 to get

$$\epsilon_2^2 = \epsilon_1^2 \left(\frac{e^{2lt}}{1+a^2} + \frac{e^{2kt}}{1+a^{-2}} \right) \quad \text{where } a = \frac{ly_{10}}{ky_{20}}, \tag{3.2.214}$$

so

$$\frac{\epsilon_2^2}{\epsilon_1^2} = \frac{e^{2lt}}{1+a^2} + \frac{e^{2kt}}{1+a^{-2}}. \tag{3.2.215}$$

Now a^2 is a positive number between 0 and ∞ and $l > k$, so we have

$$\frac{e^{2kt}}{1+a^2} + \frac{e^{2kt}}{1+a^{-2}} < \frac{e^{2lt}}{1+a^2} + \frac{e^{2kt}}{1+a^{-2}} < \frac{e^{2lt}}{1+a^2} + \frac{e^{2lt}}{1+a^{-2}}. \tag{3.2.216}$$

$$\left(\text{Note that } \frac{1}{1+a^2} + \frac{1}{1+a^{-2}} = 1. \right)$$

Combining (3.2.215) and (3.2.216) we obtain

$$e^{kt} < \frac{\epsilon_2}{\epsilon_1} < e^{lt} \tag{3.2.217}$$

or

$$\frac{1}{l} \log \frac{\epsilon_2}{\epsilon_1} < t_{\min} < \frac{1}{k} \log \frac{\epsilon_2}{\epsilon_1}. \tag{3.2.218}$$

Step 3. We now find out how far the image of C extends around σ_u. The angle, θ_u, of the image of a point is given by $\tan^{-1}(x_2(t)/x_1(t))$.

Now points on C arbitrarily close to σ^s take arbitrarily long to reach Σ_0^u. Thus we would have

$$\frac{x_2(t)}{x_1(t)} = \frac{x_{20}e^{kt}}{x_{10}e^{lt}} \sim 0 \quad \text{for } t \text{ large since } l > k. \tag{3.2.219}$$

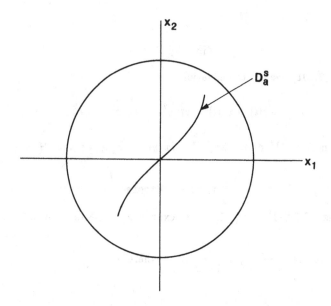

Figure 3.2.66. An End-Wise View of D_a^s along the Torus Π_0.

So these points would arrive on Σ_0^u with points near $\theta_u = 0$ and $\theta_u = \pi$. Now the question is, which points go where? We will look at this situation more closely.

We want to know what happens to C under $P_0^{L,a}$ and, more specifically, which points go to $\theta_u \sim 0$ and which go to $\theta_u \sim \pi$.

We consider points on C arbitrarily close to σ^s. Now let us examine D_a^s by "looking down" the "cut open" torus and seeing how it projects on the $x_1 - x_2$ plane, see Figure 3.2.66.

D_a^s must appear as in Figure 3.2.66 for the following reason: On the $H = 0$ energy surface, using the Hamiltonian, we have seen that the following condition holds

$$\frac{x_{10}}{x_{20}} = \frac{-k}{l}\frac{y_{20}}{y_{10}}, \qquad (l, k > 0). \tag{3.2.220}$$

Now, all points in D_a^s are such that their θ_s values satisfy $\theta_s \in (\pi/2, \pi)$, but $\theta_s = \tan^{-1}(y_{20}/y_{10})$ implies that y_{20} and y_{10} have opposite signs. Therefore, x_{10} and x_{20} must have the same sign, either $+, +$ or $-, -$; thus, we have justified Figure 3.2.66.

Now we can answer our original question, namely, where do the points on C

near σ^s go? So now we know there are two possibilities,

$$\theta_u = \tan^{-1}\left[\frac{x_{20}}{x_{10}}e^{(k-l)t}\right].\tag{3.2.221}$$

The points with x_{10}, x_{20} positive are mapped near $\theta_u = 0$, and the points with x_{10}, x_{20} negative are mapped near $\theta_u = \pi$. Furthermore, since we can assume that all points on C have essentially the same (y_1, y_2) values (draw a picture) and the sign of (y_1, y_2) cannot change during their time evolution (i.e., $y_1(t) = y_{20}e^{-kt}$), then all points of C at least have the same sign as (y_1, y_2) (the y_1, y_2 values are essentially equal for all points on C if it is chosen correctly). It is also important to note that, for points further out on C from σ^s (i.e., the extrema of C), t decreases, which results in θ_u increasing (this is seen by examining $\theta_u = \tan^{-1}\left[\frac{x_{20}}{x_{10}}e^{(k-l)t}\right]$). So, finally, we have established that Figure 3.2.67 holds.

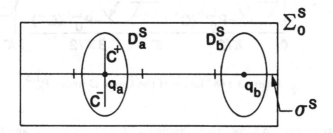

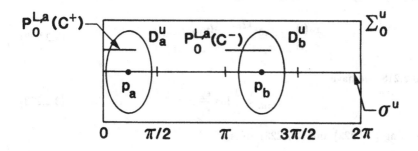

Figure 3.2.67. The Image of C under $P_0^{L,a}$.

Using similar arguments, you can show that a curve in D_b^s, transverse to σ^s, is mapped by $P_0^{L,b}$ as in Figure 3.2.68.

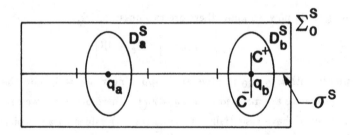

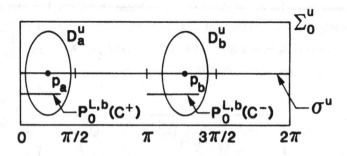

Figure 3.2.68. Image of the Curve C under $P_0^{L,b}$.

Next we need to know how far around σ^u the curves extend, and then we need to fatten them up into strips.

From (3.2.201) and (3.2.220) we obtain

$$\frac{x_2(t)}{x_1(t)} = \frac{x_{20}e^{kt}}{x_{10}e^{lt}} = \frac{-ly_{10}}{ky_{20}}e^{(k-l)t}. \tag{3.2.222}$$

From (3.2.218) we have

$$t_{\min} < \frac{1}{k}\log\frac{\epsilon_2}{\epsilon_1}. \tag{3.2.223}$$

So combining (3.2.222) and (3.2.223) gives

$$\left|\frac{x_2 y_{20}}{x_1 y_{10}}\right| > \frac{l}{k}\exp\left[\left(1-\frac{l}{k}\right)\log\frac{\epsilon_2}{\epsilon_1}\right] \tag{3.2.224}$$

or

$$|\tan\theta_u \tan\theta_s| > \frac{l}{k}\exp\left[\left(1-\frac{l}{k}\log\frac{\epsilon_2}{\epsilon_1}\right)\right]. \tag{3.2.225}$$

So, since the images of C close to σ^s are mapped to $\theta_u \sim 0$ or π, by continuity each component of the image of C will extend from 0 (resp. π) to an angle θ_u (resp. $\theta_u + \pi$) given by the above inequality. Now, if q_a, p_a, q_b, p_b lie on σ^s, σ^u satisfying

$$|\tan \theta_u \tan \theta_s| < \frac{l}{k} \exp \left[\left(1 - \frac{l}{k}\right) \log \frac{\epsilon_2}{\epsilon_1} \right] , \qquad (3.2.226)$$

then we can be sure that the image of C under $P_0^{L,a}$ "pushes past" p_a and p_b as shown in Figure 3.2.67.

Furthermore, if for some fixed, small $\delta > 0$, the following is satisfied (everything is still defined the same)

$$|\tan \theta_u \tan \theta_s| < -\delta + \frac{l}{k} \exp \left[\left(1 - \frac{l}{k}\right) \log \frac{\epsilon_2}{\epsilon_1} \right] . \qquad (3.2.227)$$

(This allows us to vary θ_s slightly.) Then we can fatten up C into a strip, and Figure 3.2.65 will hold.

The analogous situation holds for a strip in D_b^s. This proves the lemma. $\qquad \square$

The map P_1^L.

We now discuss the maps along the homoclinic orbits outside of a neighborhood of the origin. In the usual manner, for D_a^u, D_b^u sufficiently small, we define

$$\begin{aligned} P_1^{L,a} &: D_a^u \to D_a^s, \quad \text{with} \quad P_1^{L,a}(p_a) = q_a , \\ P_1^{L,b} &: D_b^u \to D_b^s, \quad \text{with} \quad P_1^{L,b}(p_b) = q_b . \end{aligned} \qquad (3.2.228)$$

Recall that $W^s(0)$ intersects $W^u(0)$ transversely along Γ_a and Γ_b. So the images of $W^u(0) \cap D_a^u$ and $W^u(0) \cap D_b^u$ under $P_1^{L,a}$ and $P_1^{L,b}$, respectively, intersect $W^s(0)$ transversely at q_a and q_b, respectively. See Figure 3.2.69.

The Poincaré Map P^L.

We define

$$P^L : (D_a^s \cup D_b^s) - \sigma_s \to \Sigma_0^u \qquad (3.2.229)$$

as $P^{L,a} \cup P^{L,b}$ where $P^{L,a} = P_1^{L,a} \circ P_0^{L,a}$ and $P^{L,b} = P_1^{L,b} \circ P_0^{L,b}$. We argue that P^L contains an invariant set on which it is topologically conjugate to a subshift of finite type.

Consider the four horizontal slabs S_a^+, S_a^-, S_b^+, and S_b^-. By $P_0^{L,a}$ and $P_0^{L,b}$

their vertical boundaries are contracted and their horizontal boundaries are expanded by an arbitrary amount depending on the size of D_a^s and D_b^s. Under $P_1^{L,a}$

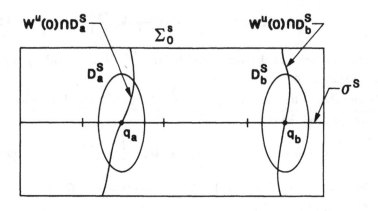

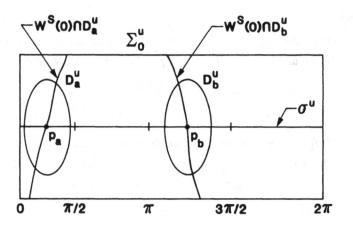

Figure 3.2.69. The Geometry of $W^u(0) \cap D_a^s$, $W^u(0) \cap D_b^s$,
$W^s(0) \cap D_a^u$, and $W^s(0) \cap D_b^u$.

and $P_1^{L,b}$ these four sets are mapped back over D_a^s and D_b^s with their vertical boundaries C^1 ϵ close to $P_1^{L,a}(W^u(0) \cap D_a^u)$ and $P_1^{L,b}(W^u(0) \cap D_b^u)$. See Figure 3.2.70.

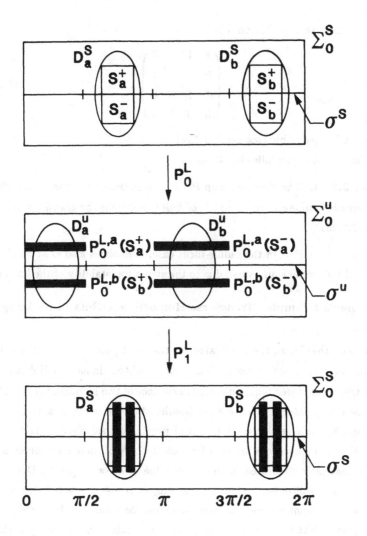

Figure 3.2.70. The Geometry of Horizontal Slabs under P^L.

More specifically, we have

$$P^L(S_a^+) \text{ intersects } S_a^+ \text{ and } S_a^-$$
$$P^L(S_a^-) \text{ intersects } S_b^+ \text{ and } S_b^-$$
$$P^L(S_b^+) \text{ intersects } S_a^+ \text{ and } S_a^-$$
$$P^L(S_b^-) \text{ intersects } S_b^+ \text{ and } S_b^- .$$

(3.2.230)

Thus, the transition matrix would be

$$A = \begin{pmatrix} 1 & 1 & 0 & 0 \\ 0 & 0 & 1 & 1 \\ 1 & 1 & 0 & 0 \\ 0 & 0 & 1 & 1 \end{pmatrix}. \tag{3.2.231}$$

Note that A is irreducible (see Section 2.2c).

We have proven the following theorem.

Theorem 3.2.24. *The Poincaré map P^L has an invariant Cantor set on which it is topologically conjugate to a subshift of finite type with the transition matrix A given in (3.2.231).*

We remark that, as in the saddle-focus case, horseshoes also exist on the level sets $H^{-1}(\epsilon)$ for ϵ sufficiently small due to the structural stability of the horseshoes.

iii) Devaney's Example: Transverse Homoclinic Orbits in an Integrable System

We have seen that when the eigenvalues at the fixed point are real, then the existence of horseshoes near a homoclinic orbit is subtle. Indeed in Holmes' [1980] example there were three requirements; 1) existence of two homoclinic orbits, 2) the homoclinic orbits enter and leave a neighborhood of the origin in specific angular ranges, and 3) the eigenvalues at the fixed point must satisfy a certain relation. The fact that the horseshoes may not be present if any of these requirements is not met is dramatically illustrated in an example due to Devaney [1978]. Devaney has constructed a Hamiltonian system on real projective n-space which has a unique hyperbolic fixed point having real eigenvalues and $2n$ transverse homoclinic orbits. He then proves that the system is completely integrable and hence, by a theorem of Moser [1973], no horseshoes may exist.

3.2f. Higher Dimensional Results

We now describe two results of Silnikov [1968], [1970] which are valid for arbitrary (but finite) dimensions. The first result deals with the saddle-focus case.

We consider the system

$$\begin{aligned} \dot{x} &= Ax + f(x, y) \\ \dot{y} &= By + g(x, y) \end{aligned} \qquad (x, y) \in \mathbb{R}^m \times \mathbb{R}^n \tag{3.2.232}$$

where A is an $m \times m$ matrix having eigenvalues $\lambda_1, \ldots, \lambda_m$ with negative real parts, B is an $n \times n$ matrix having eigenvalues $\gamma_1, \ldots, \gamma_n$ with positive real parts, and f and g are analytic and $\mathcal{O}(2)$ at the origin. Thus, (3.2.232) has a hyperbolic fixed point at the origin having an m dimensional stable manifold, $W^s(0)$, and an n dimensional unstable manifold, $W^u(0)$. We make the following additional assumptions.

A1. (3.2.232) possesses a homoclinic orbit Γ connecting the origin to itself. Moreover, we assume that $W^s(0)$ and $W^u(0)$ intersect simply along Γ in the sense $\dim(T_p W^s(0) \cap T_p W^u(0)) = 1 \ \forall \ p \in \Gamma$, where $T_p W^{s,u}(0)$ denotes the tangent spaces at the point p.

A2. γ_1 and γ_2 are complex conjugate and $Re(\gamma_1) < -Re(\lambda_i)$, $i = 1, \ldots, m$.

A3. $Re(\gamma_1) < Re(\gamma_j)$, $j = 3, \ldots, n$.

A4. A certain matrix is nonsingular.

Silnikov has proven the following theorem.

Theorem 3.2.25. *If A1 through A4 are satisfied, then an appropriately defined Poincaré map near Γ contains an invariant set on which it is topologically conjugate to the subshift acting on $\Sigma^{\infty,\delta}$ where $\delta = -Re(\lambda_1)/Re(\gamma_1)$.*

We remark on the somewhat mysterious A4. This assumption insures that the closure of the connected part of $W^u(0)$ in a sufficiently small neighborhood of Γ is locally disconnected. Computation of the matrix is involved, and we refer the reader to Silnikov [1970] for the details.

Next we give a result of Silnikov [1968] describing the bifurcation of periodic orbits from homoclinic orbits.

We consider the system

$$\dot{z} = Z(z,\mu), \qquad z \in \mathbb{R}^{m+n}, \quad \mu \in \mathbb{R}^1 \qquad (3.2.233)$$

where Z is analytic. We assume that (3.2.233) has a hyperbolic fixed point at $z = 0$ for $\mu \in [-\mu_0, \mu_0]$ and that the eigenvalues of (3.2.233) linearized about $z = 0$ are given by $\lambda_1(\mu), \ldots, \lambda_m(\mu)$ having negative real parts and $\gamma_1(\mu), \ldots, \gamma_n(\mu)$ having positive real parts. We further assume

A1. At $\mu = 0$ (3.2.233) has a homoclinic orbit Γ connecting $z = 0$ to itself. Moreover, $\dim(T_p W^s(0) \cap T_p W^u(0)) = 1 \ \forall \ p \in \Gamma$.

A2. $\gamma_1(0)$ is real and $\gamma_1(0) < -Re(\lambda_i)$, $i = 1, \ldots, m$.

A3. $\gamma_1(0) < Re(\gamma_j(0))$, $j = 2, \ldots, n$.

A4. A certain matrix is nonsingular.

Silnikov has proven the following result.

Theorem 3.2.26. *If A1–A4 hold, then only one periodic motion may bifurcate from Γ as μ is varied. The periodic orbit is stable if $n = 1$ and a saddle if $n > 1$.*

We remark that, similar to the first result, A4 deals with the geometry of the closure of the unstable manifold with a sufficiently small neighborhood of the origin. The computation of the matrix is involved, and we refer the reader to Silnikov [1968] for details. For both results it should be possible to reduce the differentiability of the vector field from analytic to C^r, r finite. This would be useful for center manifold type applications. Much work remains to be done in higher dimensions.

3.3. Orbits Heteroclinic to Hyperbolic Fixed Points of Ordinary Differential Equations

We will now examine two examples that show how orbits heteroclinic to fixed points may be a mechanism for the creation of horseshoes. Recall that, in our examples of homoclinic orbits, horseshoes were created as a result of the violent stretching and contraction that a region of phase space experienced as it passed near the saddle point, with the homoclinic orbit providing the mechanism for the region of phase space to eventually return near to where it started. Now a heteroclinic orbit does not provide a mechanism for a part of phase space to eventually return near to where it started. However, two or more heteroclinic orbits may provide this mechanism.

Definition 3.3.1. Let $p_1, p_2, \ldots, p_N$ be hyperbolic fixed points of an ordinary differential equation with $\Gamma_{1,2}, \Gamma_{2,3}, \ldots, \Gamma_{N-1,N}$ and $\Gamma_{N,1}$ heteroclinic orbits connecting p_1 and p_2, p_2 and p_3, $\ldots$, p_{N-1} and p_N, and p_N and p_1, respectively. Then $\{p_1\} \cup \Gamma_{1,2} \cup \{p_2\} \cup \Gamma_{2,3} \cup \{p_3\} \cup \cdots \cup \{p_{N-1}\} \cup \Gamma_{N-1,N} \cup \{p_N\} \cup \Gamma_{N,1}$ is called a *heteroclinic cycle* if it is a nonwandering set. See Figure 3.3.1.

Our analysis of the orbit structure near heteroclinic cycles will be the same as our analysis of the orbit structure near homoclinic orbits, with the main difference being that the Poincaré map is the composition of at least four maps, one for a

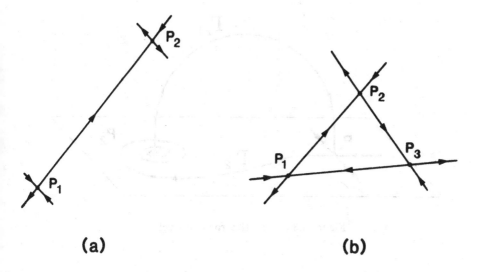

(a)   **(b)**

Figure 3.3.1. a) Heteroclinic Orbit. b) Heteroclinic Cycle.

neighborhood of each fixed point and one for each heteroclinic orbit outside of a neighborhood of the fixed points. We begin with our first example.

i) A Heteroclinic Cycle in $\mathbb{R}^3$

This example was first studied by Tresser [1984]. Suppose we have a third order C^2 ordinary differential equation which possesses two fixed points, p_1, p_2, having eigenvalues of the following type:

p_1 : $\lambda_1 > 0$, $\lambda_3 < \lambda_2 < 0$. *saddle*,

p_2 : $\lambda > 0$, $\rho \pm i\omega$, $\rho < 0$. *saddle-focus*.

So each fixed point is hyperbolic and possesses a two dimensional stable manifold and a one dimensional unstable manifold. Next, we want to hypothesize that these coincide in such a way as to form a heteroclinic cycle; however, this can occur in a variety of ways, not all of which result in nonwandering sets. We will assume the set-up in Figure 3.3.2.

Now, if p_1, p_2, Γ_{12} are such that they lie in the same plane, then it is easy to see that $\Gamma \equiv \Gamma_{12} \cup \Gamma_{21} \cup \{p_1\} \cup \{p_2\}$ is a nonwandering set in this geometrical configuration; we will assume that this is the case.

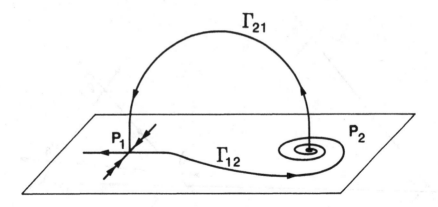

Figure 3.3.2. The Heteroclinic Cycle.

We now proceed in the usual way; namely, we construct a Poincaré map, P^L, defined in a neighborhood of Γ and study its properties. Construction of maps associated with the fixed point having real eigenvalues is similar to those in Section 3.2c,i), and construction of maps associated with the fixed point having complex eigenvalues is similar to those in Section 3.2c,ii). We will refer back to these examples for certain details.

Construction of P_{01}^L and P_{02}^L.

We construct cross-sections Π_{01} and Π_{11} in a neighborhood of p_1 as in (3.2.77) and cross-sections Π_{02} and Π_{12} in a neighborhood of p_2 as in (3.2.104), see Figure 3.3.3.

We assume that in a neighborhood of p_1 the linearized flow in an appropriate local coordinate system is given by

$$
\begin{aligned}
x_1(t) &= x_{10}e^{\lambda_1 t} \\
y_1(t) &= y_{10}e^{\lambda_2 t} \\
z_1(t) &= z_{10}e^{\lambda_3 t} .
\end{aligned}
\tag{3.3.1}
$$

The time of flight from Π_{01} to Π_{11} is given by $t = \frac{1}{\lambda_1} \log \frac{\epsilon}{x_1}$. So the map, P_{01}^L, is

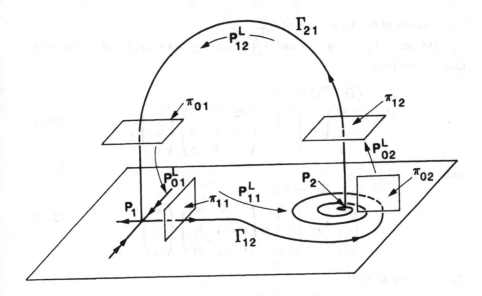

Figure 3.3.3. Geometry of Cross-Sections Near the Origin.

given by

$$P_{01}^L : \Pi_{01} \to \Pi_{11}$$

$$\begin{pmatrix} x_1 \\ \epsilon \\ z_1 \end{pmatrix} \mapsto \begin{pmatrix} \epsilon \\ \epsilon \left(\frac{\epsilon}{x_1} \right)^{\frac{\lambda_2}{\lambda_1}} \\ z_1 \left(\frac{\epsilon}{x_1} \right)^{\frac{\lambda_3}{\lambda_1}} \end{pmatrix}.$$

(3.3.2)

Likewise, we assume that in an appropriate local coordinate system the linearized flow in a neighborhood of p_2 is given by

$$x_2(t) = e^{\rho t}(x_{20} \cos \omega t - y_{20} \sin \omega t)$$
$$y_2(t) = e^{\rho t}(x_{20} \sin \omega t + y_{20} \cos \omega t)$$
$$z_2(t) = z_{20} e^{\lambda t} .$$

(3.3.3)

The time of flight from Π_{02} to Π_{12} is given by $t = \frac{1}{\lambda} \log \frac{\epsilon}{z_2}$, so P_{02}^L is given by

$$P_{02}^L : \Pi_{02} \to \Pi_{12}$$

$$\begin{pmatrix} x_2 \\ 0 \\ z_2 \end{pmatrix} \mapsto \begin{pmatrix} x_2 \left(\frac{\epsilon}{z_2} \right)^{\frac{\rho}{\lambda}} \cos \frac{\omega}{\lambda} \log \frac{\epsilon}{z_2} \\ x_2 \left(\frac{\epsilon}{z_2} \right)^{\frac{\rho}{\lambda}} \sin \frac{\omega}{\lambda} \log \frac{\epsilon}{z_2} \\ \epsilon \end{pmatrix}.$$

(3.3.4)

Construction of the Maps P_{11}^L and P_{12}^L.

The maps P_{11}^L along Γ_{12} and P_{12}^L along Γ_{21} are constructed in the usual manner. We have

$$P_{12}^L : \Pi_{12} \to \Pi_{01}$$

$$\begin{pmatrix} x_2 \\ y_2 \\ \epsilon \end{pmatrix} \mapsto \begin{pmatrix} 0 \\ 0 \\ \epsilon \end{pmatrix} + \begin{pmatrix} a_2 & b_2 & 0 \\ c_2 & d_2 & 0 \\ 0 & 0 & 0 \end{pmatrix} \begin{pmatrix} x_2 \\ y_2 \\ 0 \end{pmatrix}, \tag{3.3.5}$$

and

$$P_{11}^L : \Pi_{11} \to \Pi_{02}$$

$$\begin{pmatrix} \epsilon \\ y_1 \\ z_1 \end{pmatrix} \mapsto \begin{pmatrix} \epsilon \\ 0 \\ 0 \end{pmatrix} + \begin{pmatrix} 0 & 0 & 0 \\ 0 & a_1 & b_1 \\ 0 & c_1 & d_1 \end{pmatrix} \begin{pmatrix} 0 \\ y_1 \\ z_1 \end{pmatrix}. \tag{3.3.6}$$

The Poincaré Map P^L.

We can now compute the Poincaré map

$$P^L \equiv P_{11}^L \circ P_{01}^L \circ P_{12}^L \circ P_{02}^L : \Pi_{02} \to \Pi_{02}, \tag{3.3.7}$$

but first we want to simplify the notation slightly. Let

$$a_2 x_2 \left(\frac{\epsilon}{z_2}\right)^{\frac{\ell}{\lambda}} \cos\frac{\omega}{\lambda}\log\frac{\epsilon}{z_2} + b_2 x_2 \left(\frac{\epsilon}{z_2}\right)^{\frac{\ell}{\lambda}} \sin\frac{\omega}{\lambda}\log\frac{\epsilon}{z_2} = k_1 x_2 \left(\frac{\epsilon}{z_2}\right)^{\frac{\ell}{\lambda}} \cos(\theta + \phi_1)$$

$$c_2 x_2 \left(\frac{\epsilon}{z_2}\right)^{\frac{\ell}{\lambda}} \cos\frac{\omega}{\lambda}\log\frac{\epsilon}{z_2} + d_2 x_2 \left(\frac{\epsilon}{z_2}\right)^{\frac{\ell}{\lambda}} \sin\frac{\omega}{\lambda}\log\frac{\epsilon}{z_2} = k_2 x_2 \left(\frac{\epsilon}{z_2}\right)^{\frac{\ell}{\lambda}} \sin(\theta + \phi_2)$$

where $\theta = \frac{\omega}{\lambda}\log\frac{\epsilon}{z_2}$, $\phi = -\tan^{-1}\frac{b_2}{a_2}$, $k_1 = \sqrt{a_2^2 + b_2^2}$, and $k_2 = \sqrt{c_2^2 + d_2^2}$.

Then, we get

$$P_{12}^L \circ P_{02}^L(x_2, z_2) = \begin{pmatrix} k_1 x_2 \left(\frac{\epsilon}{z_2}\right)^{\frac{\ell}{\lambda}} \cos(\theta + \phi_1) \\ k_2 x_2 \left(\frac{\epsilon}{z_2}\right)^{\frac{\ell}{\lambda}} \sin(\theta + \phi_2) \end{pmatrix} \tag{3.3.8}$$

and

$$P_{01}^L \circ P_{12}^L \circ P_{02}^L(x_2, z_2) =$$

$$\begin{pmatrix} \epsilon^{1+\frac{\lambda_2}{\lambda_1}} \left[k_1 x_2 \left(\frac{\epsilon}{z_2}\right)^{\frac{\ell}{\lambda}} \cos(\theta + \phi_1) \right]^{\left|\frac{\lambda_2}{\lambda_1}\right|} \\ \epsilon^{\frac{\lambda_3}{\lambda_1}} \left[k_2 x_2 \left(\frac{\epsilon}{z_2}\right)^{\frac{\ell}{\lambda}} \sin(\theta + \phi_2) \right] \left[k_1 x_2 \left(\frac{\epsilon}{z_2}\right)^{\frac{\ell}{\lambda}} \cos(\theta + \phi_1) \right]^{\left|\frac{\lambda_3}{\lambda_1}\right|} \end{pmatrix} \tag{3.3.9}$$

and, finally,

$$P^L = P_{11}^L \circ P_{01}^L \circ P_{12}^L \circ P_{02}^L(x_2, z_2) =$$

$$\left(\begin{matrix} a_1 \epsilon^{1+\frac{\lambda_2}{\lambda_1}} \left[\hat{k}_1 \cos(\theta + \phi_1) \right]^{\left|\frac{\lambda_2}{\lambda_1}\right|} + b_1 \epsilon^{\frac{\lambda_3}{\lambda_1}} \left[\hat{k}_2 \sin(\theta + \phi_2) \right] \left[\hat{k}_1 \cos(\theta + \phi_1) \right]^{\left|\frac{\lambda_3}{\lambda_1}\right|} \\ c_1 \epsilon^{1+\frac{\lambda_2}{\lambda_1}} \left[\hat{k}_1 \cos(\theta + \phi_1) \right]^{\left|\frac{\lambda_2}{\lambda_1}\right|} + d_1 \epsilon^{\frac{\lambda_3}{\lambda_1}} \left[\hat{k}_2 \sin(\theta + \phi_2) \right] \left[\hat{k}_1 \cos(\theta + \phi_1) \right]^{\left|\frac{\lambda_3}{\lambda_1}\right|} \end{matrix} \right)$$

$$(3.3.10)$$

where

$$\hat{k}_1 = k_1 x_2 \left(\frac{\epsilon}{z_2} \right)^{\frac{\rho}{\lambda}}$$

$$\hat{k}_2 = k_2 x_2 \left(\frac{\epsilon}{z_2} \right)^{\frac{\rho}{\lambda}} .$$

The analysis of P^L is very similar to the analysis given in the case of an orbit homoclinic to a saddle-focus in $\mathbb{R}^3$. We will leave the details to the reader and only show the relevant features which give rise to horseshoes.

For z_2 small, since $|\lambda_3| > |\lambda_2|$, this map is essentially

$$\begin{pmatrix} x_2 \\ z_2 \end{pmatrix} \mapsto \begin{pmatrix} \bar{k}_1 x_2^{\left|\frac{\lambda_2}{\lambda_1}\right|} z_2^{\frac{\rho\lambda_2}{\lambda\lambda_1}} \cos^{\left|\frac{\lambda_2}{\lambda_1}\right|}(\theta + \phi_1) \\ \bar{k}_2 x_2^{\left|\frac{\lambda_2}{\lambda_1}\right|} z_2^{\frac{\rho\lambda_2}{\lambda\lambda_1}} \cos^{\left|\frac{\lambda_2}{\lambda_1}\right|}(\theta + \phi_1) \end{pmatrix} \qquad (3.3.11)$$

since $|\lambda_3| > |\lambda_2|$ where $\bar{k}_1 = a_1 \epsilon^{1+(\lambda_2/\lambda_1)+(\rho\lambda_2/\lambda\lambda_1)} k_1^{|\lambda_2/\lambda_1|}$ and $\bar{k}_2 = c_1 \epsilon^{1+(\lambda_2/\lambda_1)+(\rho\lambda_2/\lambda\lambda_1)} k_1^{|\lambda_2/\lambda_1|}$. This looks somewhat similar to the map which we derived in the original Silnikov situation for homoclinic orbits in $\mathbb{R}^3$. Figure 3.3.4 should make clear the similarities.

From Figure 3.3.4 it can be seen that a rectangle is rolled into a logarithmic spiral by the saddle-focus and then pinched into two pieces by the saddle; however, we still can find horizontal strips which are mapped into vertical strips, so the situation is essentially the same as the Silnikov situation for homoclinic orbits in $\mathbb{R}^3$.

This map can be analyzed in exactly the same way as the map obtained for a homoclinic orbit to a saddle-focus in $\mathbb{R}^3$, and essentially all of the same conclusions will hold. (We leave out the details.) Notice that the quantity $\rho\lambda_2/\lambda\lambda_1$ plays the same role as the quantity $-\rho/\lambda$ in the case of an orbit homoclinic to a saddle-focus in $\mathbb{R}^3$.

We have the following results.

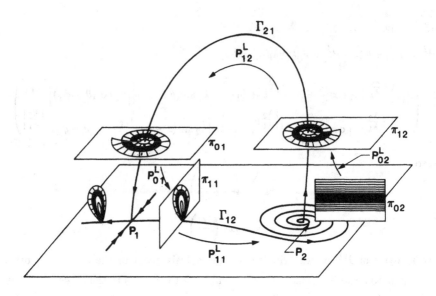

Figure 3.3.4. Geometry of the Poincaré Map.

Theorem 3.3.1. *The Poincaré map P^L possesses a countable infinity of horseshoes provided*

$$\frac{\rho\lambda_2}{\lambda\lambda_1} < 1.$$

PROOF: Left to reader.   □

Tresser [1984] discusses in more detail the coding of orbits in the horseshoes via symbolic dynamics. If $(\rho\lambda_2/\lambda\lambda_1) > 1$ then P^L no longer contains horseshoes; however, a theorem similar to Theorem 3.2.20 can be proven indicating that the heteroclinic cycle is an attractor with chaotic transients nearby. If the heteroclinic cycle is broken via perturbations (such as may occur by varying a parameter in parametrized systems), then a finite number of horseshoes are preserved for sufficiently small perturbations.

ii) A Heteroclinic Cycle in $\mathbb{R}^4$

This example can be found in Glendinning and Tresser [1985]. We consider an autonomous ordinary differential equation in $\mathbb{R}^4$ which possesses two fixed points, p_1, p_2, having eigenvalues of the following type:

$$\begin{aligned}
p_1 &: (\lambda_1, -\rho_1 \pm i\omega_1, -\nu_1) \\
p_2 &: (\lambda_2, -\rho_2 \pm i\omega_2, -\nu_2)
\end{aligned} \tag{3.2.12}$$

with $\nu_j > \lambda_j > \rho_j > 0,\ j = 1,2$.

We suppose that in appropriate local coordinate systems near p_1 and p_2 the vector field takes the form

$$
\begin{aligned}
\dot{x}_i &= -\rho_i x_i + \omega_i y_i \\
\dot{y}_i &= -\omega_i x_i - \rho_i y_i \\
\dot{z}_i &= \lambda_i z_i \\
\dot{w}_i &= -\nu_i w_i
\end{aligned}
\qquad i = 1, 2 .
\qquad (3.3.13)
$$

So p_1 and p_2 possess 3-dimensional stable manifolds (with spiralling in the x_i, y_i directions) and 1-dimensional unstable manifolds. Furthermore, we suppose that there exist two heteroclinic orbits, Γ_{12} connecting p_1 to p_2 and Γ_{21} connecting p_2 to p_1. Γ_{12} leaves p_1 along the z_1 axis and Γ_{21} leaves p_2 along the z_2 axis. We give a rough illustration of the geometry by suppressing the y_1, y_2 directions in Figure 3.3.5.

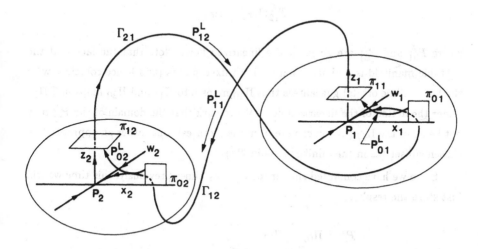

Figure 3.3.5. Geometry of the Heteroclinic Orbits and Cross-Sections
Near the Origin.

We construct the following 3-dimensional cross-sections to the flow,

$$\Pi_{01} = \{ (x_1, y_1, z_1, w_1) \mid y_1 = 0 \}$$
$$\Pi_{11} = \{ (x_1, y_1, z_1, w_1) \mid z_1 = \epsilon \}$$
$$\Pi_{02} = \{ (x_2, y_2, z_2, w_2) \mid y_2 = 0 \}$$
$$\Pi_{12} = \{ (x_2, y_2, z_2, w_2) \mid z_2 = \epsilon \}$$

$$(3.3.14)$$

where these cross-sections are defined as in the example of an orbit homoclinic to a fixed point with eigenvalues $\rho \pm i\omega$, λ, ν $(\rho, \lambda < 0)$ in Section 3.2d,i).

Now we will construct a Poincaré map in a neighborhood of the heteroclinic cycle $\Gamma \equiv \Gamma_{12} \cup \Gamma_{21} \cup \{p_1\} \cup \{p_2\}$ (note: it should be checked that Γ is a nonwandering set) and study its properties in the usual way. The Poincaré map will be the composition of four maps

$$P_{01}^{L} : \Pi_{01} \rightarrow \Pi_{11}$$
$$P_{11}^{L} : \Pi_{11} \rightarrow \Pi_{02}$$
$$P_{02}^{L} : \Pi_{02} \rightarrow \Pi_{12}$$
$$P_{12}^{L} : \Pi_{12} \rightarrow \Pi_{01}$$

$$(3.3.15)$$

where P_{01}^{L} and P_{01}^{L} are given by the linearized flow. Note that the fact that the unstable manifold is 1-dimensional for both fixed points (and hence coincides with the heteroclinic orbits) guarantees that Π_{10} maps into Π_{11} and Π_{20} maps into Π_{21} under the action of the linearized flow. We remark that the domain of the P_{ij}^{L} may not be all of the Π_{ij} (as you can see from the pictures), but rather an appropriately chosen subset (as in the Silnikov case in $\mathbb{R}^3$).

Since we have computed similar maps many times previously, this time we will just state the results.

$$P_{01}^{L} : \Pi_{01} \rightarrow \Pi_{11}$$
$$\begin{pmatrix} x_1 \\ 0 \\ z_1 \\ w_1 \end{pmatrix} \mapsto \begin{pmatrix} x_1 \left(\frac{z_1}{\epsilon} \right)^{\frac{\rho_1}{\lambda_1}} \cos \left(\frac{\omega_1}{\lambda_1} \log \frac{\epsilon}{z_1} \right) \\ x_1 \left(\frac{z_1}{\epsilon} \right)^{\frac{\rho_1}{\lambda_1}} \sin \left(\frac{\omega_1}{\lambda_1} \log \frac{\epsilon}{z_1} \right) \\ \epsilon \\ w_1 \left(\frac{z_1}{\epsilon} \right)^{\frac{\nu_1}{\lambda_1}} \end{pmatrix}$$

$$(3.3.16)$$

$$P_{02}^L : \Pi_{02} \to \Pi_{12}$$

$$\begin{pmatrix} x_2 \\ 0 \\ z_2 \\ w_2 \end{pmatrix} \mapsto \begin{pmatrix} x_2 \left(\frac{z_2}{\epsilon}\right)^{\frac{\rho_2}{\lambda_2}} \cos\left(\frac{\omega_2}{\lambda_2} \log \frac{\epsilon}{z_2}\right) \\ x_2 \left(\frac{z_2}{\epsilon}\right)^{\frac{\rho_2}{\lambda_2}} \sin\left(\frac{\omega_2}{\lambda_2} \log \frac{\epsilon}{z_2}\right) \\ \epsilon \\ w_2 \left(\frac{z_2}{\epsilon}\right)^{\frac{\nu_2}{\lambda_2}} \end{pmatrix} \tag{3.3.17}$$

$$P_{11}^L : \Pi_{11} \to \Pi_{02}$$

$$\begin{pmatrix} x_1 \\ y_1 \\ \epsilon \\ w_1 \end{pmatrix} \mapsto \begin{pmatrix} \epsilon \\ 0 \\ 0 \\ 0 \end{pmatrix} + \begin{pmatrix} 0 & 0 & 0 & a_1 \\ 0 & 0 & 0 & 0 \\ A_1 & B_1 & 0 & 0 \\ C_1 & D_1 & 0 & 0 \end{pmatrix} \begin{pmatrix} x_1 \\ y_1 \\ 0 \\ w_1 \end{pmatrix} \tag{3.3.18}$$

$$P_{12}^L : \Pi_{12} \to \Pi_{01}$$

$$\begin{pmatrix} x_2 \\ y_2 \\ \epsilon \\ w_2 \end{pmatrix} \mapsto \begin{pmatrix} \epsilon \\ 0 \\ 0 \\ 0 \end{pmatrix} + \begin{pmatrix} A_2 & B_2 & 0 & 0 \\ 0 & 0 & 0 & 0 \\ 0 & 0 & 0 & a_2 \\ C_2 & D_2 & 0 & 0 \end{pmatrix} \begin{pmatrix} x_2 \\ y_2 \\ 0 \\ w_2 \end{pmatrix} \tag{3.3.19}$$

where we have chosen Π_{02} and Π_{01} in such a way that Γ_{12} and Γ_{21} intersect these hypersurfaces at $(\epsilon, 0, 0, 0)$, respectively. We assume that our coordinate system is such that the matrices assume the above given block diagonal form. This form expresses the following two geometrical assumptions:

P_{11}^L maps the x_1-y_1 plane $\cap \Pi_{11}$ into the z_2-w_2 plane $\cap \Pi_{02}$;
P_{12}^L maps the x_2-y_2 plane $\cap \Pi_{12}$ into the x_1-w_1 plane $\cap \Pi_{01}$.

Now we want to give a step by step geometrical picture of what the maps do.

Step 1.

$$P_{01}^L : \Pi_{01} \to \Pi_{11}$$

$$\begin{pmatrix} x_1 \\ 0 \\ z_1 \\ w_1 \end{pmatrix} \mapsto \begin{pmatrix} x_1 \left(\frac{z_1}{\epsilon}\right)^{\frac{\rho_1}{\lambda_1}} \cos\left(\frac{\omega_1}{\lambda_1} \log \frac{\epsilon}{z_1}\right) \\ x_1 \left(\frac{z_1}{\epsilon}\right)^{\frac{\rho_1}{\lambda_1}} \sin\left(\frac{\omega_1}{\lambda_1} \log \frac{\epsilon}{z_1}\right) \\ \epsilon \\ w_1 \left(\frac{z_1}{\epsilon}\right)^{\frac{\nu_1}{\lambda_1}} \end{pmatrix}. \tag{3.3.20}$$

Assume P_{01}^L is defined on a piece of Π_{10} as shown in Figure 3.3.6.

Note that P_{01}^L is not defined on $z_1 = 0$, since these points are on $W^s(p_1)$. By examining the expression for P_{01}^L given above, the x_1 and z_1 directions are twisted

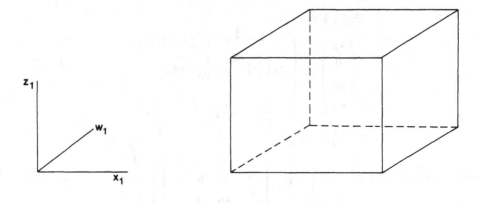

Figure 3.3.6.

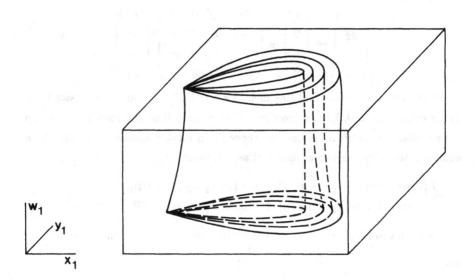

Figure 3.3.7.

into logarithmic spirals with the w_1 coordinate shrinking to zero as $z_1 \to 0$, see Figure 3.3.7.

Step 2. $P_{11}^L : \Pi_{11} \to \Pi_{01}^L$ maps $P_{01}^L(\Pi_{01})$ as shown in Figure 3.3.8.

Step 3. $P_{02} : \Pi_{02} \to \Pi_{12}$ twists $P_{11}^L \circ P_{01}^L(\Pi_{01})$ around in much the same manner

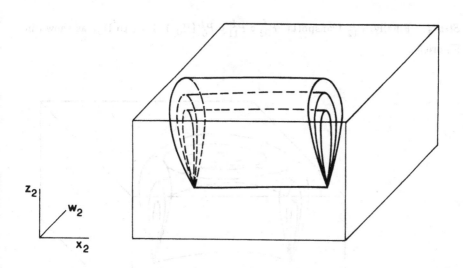

Figure 3.3.8.

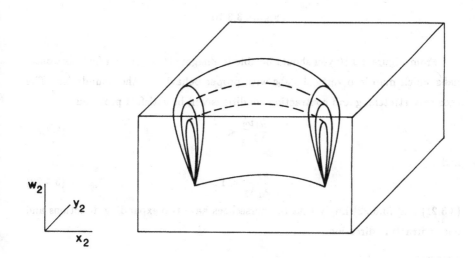

Figure 3.3.9.

as P_{01}^L deforms Π_{01}, see Figure 3.3.9.

Step 4. Finally, P_{12}^L transports $P_{02}^L \circ P_{11}^L \circ P_{01}^L (\Pi_{01})$ back to Π_{01} as shown in Figure 3.3.10.

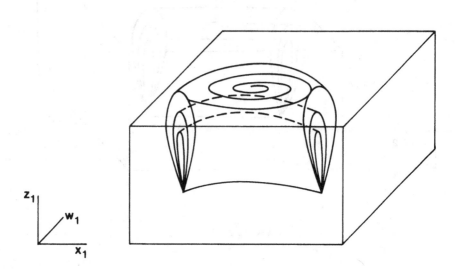

Figure 3.3.10.

From Figure 3.3.10 you should be able to imagine that you can find horizontal slabs which map into vertical slabs with proper behavior of the boundaries. The necessary stretching and contraction conditions will be satisfied provided

$$\frac{\rho_1 \nu_2}{\lambda_1 \lambda_2} < 1 \qquad (3.3.21)$$

and

$$\frac{\rho_2 \nu_1}{\lambda_1 \lambda_2} < 1. \qquad (3.3.22)$$

(3.3.21) and (3.3.22) imply that the horseshoes have two expanding directions and one contracting direction.

Remarks.

1) *Parametrized Systems.* In both of our examples horseshoes were found in a neighborhood of the heteroclinic cycle. However, if there is no spiralling (i.e., only real eigenvalues) it may be necessary to break one or more of the heteroclinic orbits in order to obtain horseshoes (see 3.2c,i)). This is possible

in parametrized systems. However, if there are no symmetries present, typically it is necessary to have the number of parameters equal to the number of heteroclinic orbits in order to insure the proper behavior.

2) *Other Results.* Tresser [1984] has generalized our first example to the situation of multiple hyperbolic fixed points in $\mathbb{R}^3$ with one-dimensional unstable manifolds (or one-dimensional stable manifolds under time reversal). Devaney [1976] discusses some heteroclinic cycles in Hamiltonian systems. Heteroclinic cycles frequently arise in applications. For example, they appear to be the mechanism giving rise to "bursting" in a model for the interaction of eddies in the boundary layer of fluid flow near a wall, see Aubry, Holmes, Lumley, and Stone [1987] and Guckenheimer and Holmes [1987].

3.4. Orbits Homoclinic to Periodic Orbits and Invariant Tori

In Section 3.2 we studied a variety of examples of orbits homoclinic to hyperbolic fixed points of ordinary differential equations. We will now study the dynamics associated with an orbit homoclinic to a hyperbolic periodic orbit or a normally hyperbolic invariant torus of an ordinary differential equation. Unlike the case of orbits homoclinic to hyperbolic fixed points of ordinary differential equations where the results depended on a variety of factors such as dimension of the system, the nature of the eigenvalues at the fixed point, the existence of symmetries, etc., we will derive a general result implying the existence of "horseshoe-like" dynamics which is independent of these considerations (though these factors may give rise to important dynamical effects which are not captured by our theorem).

The spirit of our analysis will be the same as for orbits homoclinic to hyperbolic fixed points of ordinary differential equations; however, there will be some important technical differences. The main difference is that we will not deal at all with an ordinary differential equation but rather with a map. This causes no difficulty in applying our results to ordinary differential equations, for recall from Section 1.6 that the study of the orbit structure near a periodic orbit of an ordinary differential equation could be reduced to the study of the orbit structure near a fixed point of the associated Poincaré map. Similarly, the study of the orbit structure near an $l+1$ dimensional invariant torus of an ordinary differential equation could be reduced to the study of the orbit structure of an l-dimensional invariant torus of the associated Poincaré map. Thus, results for maps describing the dynamics of orbits homoclinic to l-tori have an immediate interpretation for the dynamics of orbits homoclinic to $l + 1$-tori in ordinary differential equations (note: a 0-torus is fixed point and a 1-torus is a periodic orbit for an ordinary differential equation). Technicalities aside, the spirit of our analysis will be the same in the sense that we will look for a region near the homoclinic orbit which is mapped back over itself by some iterate of the map in a "horseshoe-like manner." In particular, we look for horizontal slabs which are mapped to vertical slabs with proper behavior of the boundaries and sufficient stretching and contraction. As in the case for hyperbolic fixed points of ordinary differential equations, the homoclinic orbit provides the mechanism for the global folding of the phase space and the invariant set to which the orbit is homoclinic (i.e., the l-torus) provides the mechanism for the stretching and contraction. Before giving specific hypotheses we want to give an intuitive description of the ideas.

Orbits Homoclinic to a Hyperbolic Fixed Point.

Suppose we have a diffeomorphism of $\mathbb{R}^2$, f, possessing a hyperbolic fixed point p_0 whose stable and unstable manifolds intersect transversely at some point p, as shown in Figure 3.4.1. We remark that unlike the case of orbits homoclinic to hyperbolic fixed points of ordinary differential equations it is possible for the stable and unstable manifolds of a hyperbolic fixed point of a map to intersect in a discrete set of points without violating uniqueness of solutions. This is because orbits of maps are infinite sequences of discrete points whereas orbits of ordinary differential equations are smooth curves.

Also, recall the definition of transversal intersection of manifolds in Section 1.4. The importance of transversality will become important when we rigorously justify the following heuristic arguments.

Now p lies simultaneously in the invariant manifolds $W^s(p_0)$ and $W^u(p_0)$; hence, the orbit of p must lie in both $W^s(p_0)$ and $W^u(p_0)$. Thus, iterating Figure 3.4.1 gives us the *homoclinic tangle* part of which is shown in Figure 3.4.2. p is called a *transverse homoclinic point*. So one transverse homoclinic point implies the existence of a countable infinity of transverse homoclinic points due to the invariance of $W^s(p_0)$ and $W^u(p_0)$. Note from Figure 3.4.2 that $W^s(p_0)$ and $W^u(p_0)$ appear to accumulate on themselves. We will justify this analytically shortly. For a more detailed and careful discussion of Figure 3.4.2, we refer the reader to Abraham

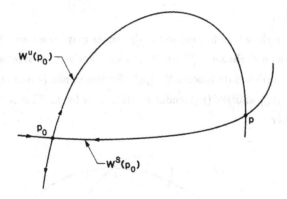

Figure 3.4.1. Intersection of the Stable and Unstable Manifolds of p_0.

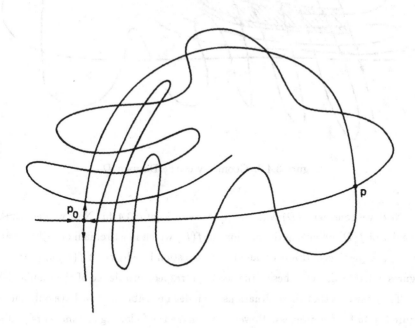

Figure 3.4.2. The Homoclinic Tangle.

and Shaw [1984].

Our goal is to show how horseshoe-like dynamics may arise from this situation. Consider the domain D shown in Figure 3.4.3 whose left vertical side lies in $W^u(p_0)$ and whose right vertical side touches $W^s(p_0)$. By invariance D must maintain this contact with $W^s(p_0)$ and $W^u(p_0)$ under all iterations by f. This is an important point to remember.

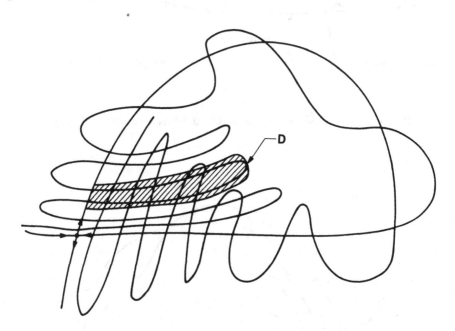

Figure 3.4.3. Geometry of the Domain D.

Next we consider $f(D)$ which appears as in Figure 3.4.4a. Now we deduce the behavior of $f(D)$ by noting the portions of $f(D)$ which must remain on $W^u(p_0)$ and $W^s(p_0)$, respectively. However, an obvious question is, why can't $f(D)$ appear as in Figures 3.4.4b, c, d, since these situations still respect invariance of the manifolds?

The answer is that these situations are indeed possible, and we have only chosen Figure 3.4.4a for definiteness. However, we make the following comments regarding the remaining figures.

Figures 3.4.4b, d. These situations cannot occur if f preserves orientation. Recall from Section 1.6 that Poincaré maps arising from ordinary differential equations

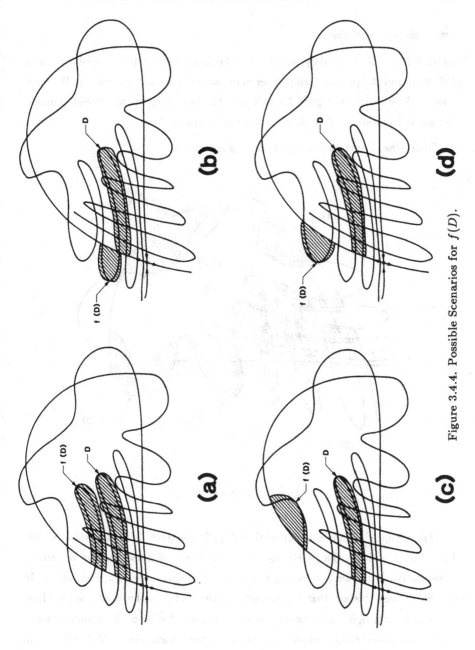

Figure 3.4.4. Possible Scenarios for $f(D)$.

must preserve orientation.

Figure 3.4.4c. It is certainly possible for the image of a "lobe" formed by pieces of $W^s(p_0)$ and $W^u(p_0)$ to "jump" over many other lobes under iteration. We have chosen the situation in Figure 3.4.4a where the lobe goes to the nearest possible lobe under iteration by f while preserving orientation.

Thus other iterates of D appear as in Figure 3.4.5.

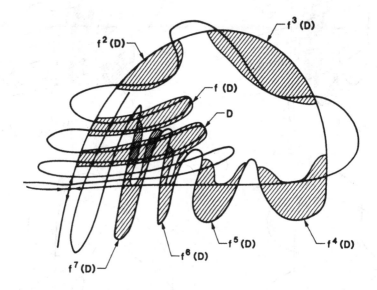

Figure 3.4.5. Iterates of D.

From Figure 3.4.5 it is apparent that $f^7(D)$ intersects D in a horseshoe shape. This should indicate the possibility of the existence of horseshoe-like dynamics; however, much remains to be rigorously justified which we will do shortly. In particular, we will show that the geometry allows us to find horizontal slabs whose image intersects them in vertical slabs with proper behavior of the boundaries and the necessary stretching and contraction rates. Thus, some iterate of f will contain an invariant Cantor set on which it is topologically conjugate to a full shift on a countable set of symbols.

Orbits Homoclinic to Normally Hyperbolic Invariant Tori.

Now consider the situation of a diffeomorphism of $\mathbb{R}^3$, f, having a normally hyperbolic invariant 1-torus, τ_0, (i.e., a circle) whose stable and unstable manifolds intersect transversely in a 1-torus, τ, see Figure 3.4.6.

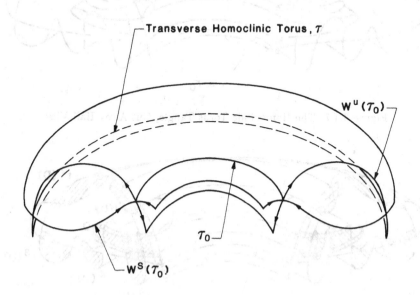

Transverse Homoclinic Torus, τ

$W^u(\tau_0)$

τ_0

$W^s(\tau_0)$

Figure 3.4.6. A Transverse Homoclinic Torus, Cut Away Half View.

We call τ a *transverse homoclinic torus*. By invariance of $W^s(\tau_0)$ and $W^u(\tau_0)$ the orbit of τ must always lie in both $W^s(\tau_0)$ and $W^u(\tau_0)$. Hence, one transverse homoclinic torus implies the existence of a countable infinity of transverse homoclinic tori. Thus, iterating Figure 3.4.6 gives Figure 3.4.7.

Using arguments similar to those given in the previous case for a hyperbolic fixed point, we can find a region D which is mapped over itself by some iterate of f in a horseshoe-like shape as shown in Figure 3.4.8.

However, in this case we will get a circle's worth of horseshoes. The normal hyperbolicity insures that the dynamics normal to the invariant torus dominate the dynamics on the torus so that the region D does not "kink up" in the direction of the invariant torus as it is being mapped back onto itself by some iterate. Thus, we should be able to find horizontal slabs in D which are mapped over themselves in vertical slabs with proper behavior of the boundaries and the necessary stretching

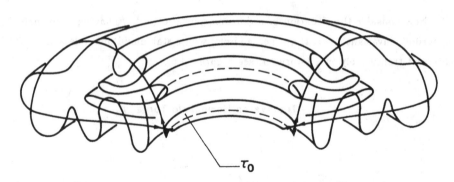

Figure 3.4.7. The Homoclinic Torus Tangle, Cut Away Half View.

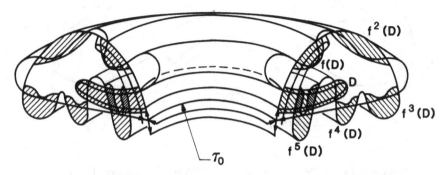

Figure 3.4.8. The Region D and Its Iterates, Cut Away Half View.

and contraction rates. Then Theorem 2.4.3 would imply the existence of a chaotic invariant set in D having the structure of a Cartesian product of a Cantor set with a torus. We remark that the dynamics along the direction of the torus is unknown.

We now turn to the rigorous justification of these examples for arbitrary (but finite) dimension.

We now state our assumptions precisely.

Let $f: M \to M$ be a C^r diffeomorphism ($r \geq 2$) of the C^∞ manifold M, where $\dim M = N$, leaving the compact boundaryless submanifold V invariant (i.e., $f(V) = V$). We make the following "structural" assumptions:

1) V is diffeomorphic to the l-dimensional torus, T^l,

2) V is normally hyperbolic,

3) V has an $l + n$ dimensional stable manifold, $W^s(V)$, and an $l + m$ dimensional

unstable manifold, $W^u(V)$, with $l + m + n = N$, and

4) $W^s(V)$ and $W^u(V)$ intersect transversely in an l dimensional torus τ, i.e.,

$$\dim\big(T_p W^s(V) \cap T_p W^u(V)\big) = l \quad \forall\, p \in \tau,$$

where $T_p W^{s,u}(V)$ denotes the tangent space of $W^{s,u}(V)$ at $p \in \tau$. We remark that τ is an l-dimensional torus which is both forwards and backwards asymptotic to V under the action of f; we refer to τ as a *transverse homoclinic torus*.

Before stating our main theorem, we need to make precise the notion of V being normally hyperbolic. Roughly, it means that the directions normal to V are expanded or contracted more sharply than directions tangent to V under the action of f. We denote the derivative of f at $p \in M$ by $D_p f$, and we assume that M is equipped with a Riemannian metric, $|\cdot|$. Let $T_V M$ be the tangent bundle of M over V with the Df invariant splitting (with respect to $|\cdot|$)

$$T_V M = \mathcal{N}^s \oplus \mathcal{N}^u \oplus TV$$

with $\mathcal{N}^s \oplus TV$ tangent to $W^s(V)$ at V and $\mathcal{N}^u \oplus TV$ tangent to $W^u(V)$ at V. With respect to this splitting, the derivative of f at $p \in M$ can be written as

$$D_p f = \mathcal{N}_p^s f \oplus \mathcal{N}_p^u f \oplus V_p f$$

where

$$\mathcal{N}_p^s f \equiv Df|_{\mathcal{N}_p^s},$$
$$\mathcal{N}_p^u f \equiv Df|_{\mathcal{N}_p^u},$$
$$V_p f \equiv Df|_{T_p V}.$$

We assume the V is normally hyperbolic in the following sense:

$\exists\, 0 < \lambda < 1$ such that for $v \in \mathcal{N}_p^s$, $u \in \mathcal{N}_p^u$

$$\left| (\mathcal{N}_p^s f)^n v \right| \le \lambda^n |v|$$

$$\left| (\mathcal{N}_p^u f)^{-n} u \right| \le \lambda^n |u| \tag{3.4.1}$$

$$\left\| \mathcal{N}_p^s f \right\| < \left\| V_p f \right\| < \left\| \mathcal{N}_p^u f \right\| \qquad \forall\, p \in V, \quad n \ge 0.$$

We now state our main result.

Theorem 3.4.1. *Let f satisfy assumptions 1) through 4) given above. Then, in a neighborhood of τ, f^n has an invariant Cantor set of tori, Λ, for some $n \geq 1$. Moreover, there exists a homeomorphism, ϕ, taking tori in Λ to bi-infinite sequences of N symbols such that the following diagram commutes*

$$
\begin{array}{ccc}
\Lambda & \xrightarrow{f^n} & \Lambda \\
\phi \downarrow & & \downarrow \phi \\
\Sigma & \xrightarrow{\sigma} & \Sigma
\end{array}
$$

where Σ denotes the space of bi-infinite sequences of N symbols and σ is the shift mapping on this space.

Before beginning the proof of Theorem 3.4.1 we need to get some preliminary considerations out of the way.

We assume that in a tubular neighborhood B of V, there exists a local coordinate system in which f takes the following form:

$$
f: \quad
\begin{aligned}
x &\mapsto A(\theta)x + g_1(x, y, \theta) \\
y &\mapsto B(\theta)y + g_2(x, y, \theta) \qquad (x, y, \theta) \in \mathbb{R}^n \times \mathbb{R}^m \times T^l , \\
\theta &\mapsto g_3(\theta)
\end{aligned}
\qquad (3.4.2)
$$

where $g_1 = g_{1x} = g_{1y} = g_{1\theta} = g_2 = g_{2x} = g_{2y} = g_{2\theta} = 0$ at $x = y = 0$, see Samoilenko [1972] or Sell [1979]. (Note: for notational compactness we will denote partial derivatives such as $D_x g_1$ by g_{1x}, etc.)

Furthermore, the normal hyperbolicity hypotheses given in (3.4.1) are sufficient for the existence of the splitting

$$
T_p M = \mathcal{N}_p^s \oplus \mathcal{N}_p^u \oplus T_p V , \qquad p \in V
$$

for all points $p \in V$. Thus, $W^s(V)$ and $W^u(V)$ can be represented in B as graphs of the functions $y = G^s(x, \theta)$ and $x = G^u(y, \theta)$, respectively, with $G^s(0, \theta) = G_x^s(0, \theta) = G_\theta^s(0, \theta) = G^u(0, \theta) = G_y^u(0, \theta) = G_\theta^u(0, \theta) = 0$ in B. (Note: the normal hyperbolicity assumptions 3.4.1 also guarantee that $G^{u,s}$ are at least C^2; this follows from the C^r-section theorem of Hirsch, Pugh and Shub [1977].)

Thus, if we construct the map

$$
\begin{aligned}
\phi: B &\to \mathcal{N}^s \oplus \mathcal{N}^u \oplus TV \\
(x, y, \theta) &\mapsto (x - G^u(y, \theta), y - G^s(x, \theta), \theta)
\end{aligned}
\qquad (3.4.3)
$$

where $\phi(0,0,\theta) = (0,0,\theta)$ and $D\phi(0,0,\theta) =$ identity, and use (3.4.3) to define a new coordinate system, we see that locally $W^s(V)$ is given by $\mathcal{N}^s \oplus TV$ and $W^u(V)$ is given by $\mathcal{N}^u \oplus TV$ (i.e., (3.4.3) "straightens out" $W^s(V)$ and $W^u(V)$ in B). Hereafter, we shall assume that we are in this coordinate system and that $g_1(0,y,\theta) = g_2(x,0,\theta) = 0$.

In this case we can represent the tubular neighborhood B of V as follows

$$B = B^s \times B^u \qquad (3.4.4)$$

where

$$B^s = \{ (x,y,\theta) \in \mathbb{R}^n \times \mathbb{R}^m \times T^l \mid |x| \le \delta_s, \quad y = 0 \}$$
$$B^u = \{ (x,y,\theta) \in \mathbb{R}^n \times \mathbb{R}^m \times T^l \mid x = 0, \quad |y| \le \delta_u \}$$

for some $\delta_s, \delta_u > 0$.

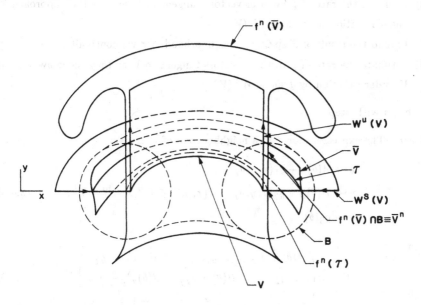

Figure 3.4.9. Geometry of $\bar{V}$ and $\bar{V}^n$, Cut Away View.

Next we give a preliminary result which describes the dynamics of f near V. This result is a generalization of the λ-lemma (see Palis and deMelo [1982]) for normally hyperbolic invariant tori. Let $\bar{V}$ be a μ_v-vertical slice intersecting B^s transversely in an l-dimensional torus τ. Let $\bar{V}^n$ denote the connected component of $f^n(\bar{V}) \cap B$ to which $f^n(\tau)$ belongs, see Figure 3.4.9. Then we have the following lemma.

Lemma 3.4.2 (Toral λ-lemma, Wiggins [1986a]). *Let $\epsilon > 0$ be given. Then, for B sufficiently small, there exists a positive integer n_0 such that for $n \geq n_0$ $\bar{V}^n$ is C^1 ϵ-close to $W^u(V)$.*

PROOF: The proof is very similar to the usual λ-lemma for hyperbolic periodic points (see Palis and deMelo [1982]). The only difference is that we must take account of the "θ dynamics." However, by normal hyperbolicity, we will see that the θ dynamics is dominated by the x-y dynamics. The proof proceeds in several steps.

1) Estimate the size of various partial derivatives in B.
2) Estimate the rate of growth of vectors tangent to $\bar{V}$ at τ as they approach V under iteration by f along $W^s(V)$.
3) Extend the result of Step 2 to a neighborhood of τ via continuity.
4) Estimate the rate of growth of vectors tangent to $\bar{V}$ as they move away from V under iteration by f along $W^u(V)$.

We begin with Step 1.

Step 1. The expression for f in a neighborhood of V is given by

$$f: \quad \begin{aligned} x &\mapsto A(\theta)x + g_1(x,y,\theta) \\ y &\mapsto B(\theta)y + g_2(x,y,\theta) \\ \theta &\mapsto g_3(\theta) \end{aligned} \qquad (x,y,\theta) \in \mathbb{R}^n \times \mathbb{R}^m \times T^l, \qquad (3.4.5)$$

with

$$Df = \begin{pmatrix} A(\theta) + g_{1x} & g_{1y} & (A(\theta)x)_\theta + g_{1\theta} \\ g_{2x} & B(\theta) + g_{2y} & (B(\theta)y)_\theta + g_{2\theta} \\ 0 & 0 & g_{3\theta} \end{pmatrix} \qquad (3.4.6)$$

and by the normal hyperbolicity assumptions given in (3.4.1) for δ_s, δ_u sufficiently small we have

$$\|A(\theta)\| \leq \lambda < 1, \qquad \|B^{-1}(\theta)\| \leq \lambda < 1,$$
$$\|A(\theta)\| < \|g_{3\theta}\| < \|B(\theta)\| \quad \forall \theta \in T^l. \qquad (3.4.7)$$

Let v_0 be a unit vector in the tangent bundle of $\bar{V}$ over τ denoted $T_\tau \bar{V}$. With respect to the splitting over V, v_0 can be written as $(v_0^x, v_0^y, v_0^\theta)$ where $v_0^x \in \mathcal{N}^s$, $v_0^y \in \mathcal{N}^u$, and $v_0^\theta \in TV$. Now, since $g_1 = g_{1x} = g_{1y} = g_{1\theta} = g_2 = g_{2x} = g_{2y} = g_{2\theta} = 0$ at $(0, 0, \theta)$ and $g_1(0, y, \theta) = g_2(x, 0, \theta) = 0$ in B, by continuity we can choose δ_s, δ_u small and $0 < k < 1$ such that

$$\lambda + k < 1,$$

$$b = \frac{1}{\lambda} - kL > 1,$$

$$\frac{\|g_{3\theta}\|}{\|B\|} \frac{1}{1 - \frac{kL}{\|B\|}} \le \alpha < 1, \quad \forall \theta \in T^l, \tag{3.4.8}$$

$$kL < \frac{(b-1)^2}{4},$$

with

$$L = \max_{v_0^\theta, v_0^y \in T_\tau \bar{V}} \left(1 + \left| \frac{v_0^\theta}{v_0^y} \right| \right),$$

and

$$k \ge \max_B \{ \|g_{1x}\|, \|g_{1y}\|, \|g_{2x}\|, \|g_{2y}\|, \|(Ax)_\theta + g_{1\theta}\|, \|(By)_\theta + g_{2\theta}\| \}.$$

We note that v_0^y is nonzero, since $\bar{V}$ is transverse to $W^s(V)$ at τ and, for notation, we let $Df^n(v_0^x, v_0^y, v_0^\theta) = (v_n^x, v_n^y, v_n^\theta)$.

Step 2. Now we can consider $\tau \in B$, and we want to study $\bar{V}$ and its tangent vectors at τ as they approach V under iteration. Note that $g_{2x} = 0$ since $g_2(x, 0, \theta) = 0$ in B. First we look at the ratio $\left| \frac{v_0^\theta}{v_0^y} \right|$. By (3.4.6), (3.4.7), and (3.4.8) we have

$$\left| \frac{v_1^\theta}{v_1^y} \right| \le \frac{\left| g_{3\theta} v_0^\theta \right|}{\left| Bv_0^y \right| - k \left| v_0^y \right| \left(1 + \left| \frac{v_0^\theta}{v_0^y} \right| \right)} \le \frac{\|g_{3\theta}\|}{\|B\|} \frac{1}{1 - \frac{kL}{\|B\|}} \left| \frac{v_0^\theta}{v_0^y} \right| \le \alpha \left| \frac{v_0^\theta}{v_0^y} \right|$$

$$\left| \frac{v_2^\theta}{v_2^y} \right| \le \frac{\left| g_{3\theta} v_1^\theta \right|}{\left| Bv_1^y \right| - k \left| v_1^y \right| \left(1 + \left| \frac{v_1^\theta}{v_1^y} \right| \right)} \le \frac{\|g_{3\theta}\|}{\|B\|} \frac{1}{1 - \frac{kL}{\|B\|}} \left| \frac{v_1^\theta}{v_1^y} \right| \le \alpha^2 \left| \frac{v_0^\theta}{v_0^y} \right|$$

$$\vdots \qquad\qquad\qquad \vdots$$

$$\left| \frac{v_n^\theta}{v_n^y} \right| \le \frac{\left| g_{3\theta} v_{n-1}^\theta \right|}{\left| Bv_{n-1}^y \right| - k \left| v_{n-1}^y \right| \left(1 + \left| \frac{v_{n-1}^\theta}{v_{n-1}^y} \right| \right)} \le \frac{\|g_{3\theta}\|}{\|B\|} \frac{1}{1 - \frac{kL}{\|B\|}} \left| \frac{v_{n-1}^\theta}{v_{n-1}^y} \right| \le \alpha^n \left| \frac{v_0^\theta}{v_0^y} \right|.$$

$$\tag{3.4.9}$$

(Note: this result is not surprising: it just tells us that vectors in $W^u(V)$ grow much faster than vectors tangent to V, with $\lim\limits_{n\to\infty}\left|\dfrac{v_n^\theta}{v_n}\right| = 0$.)

Next we look at the ratio $\chi_0 \equiv \left|\dfrac{v_0^x}{v_0^y}\right|$ and determine how it behaves under iteration by f. From (3.4.6) we have

$$\chi_1 \equiv \left|\frac{v_1^x}{v_1^y}\right| = \frac{\left|(A+g_{1x})v_0^x + g_{1y}v_0^y + ((Ax)_\theta + g_{1\theta})v_0^\theta\right|}{\left|(B+g_{2y})v_0^y + ((By)_\theta + g_{2\theta})v_0^\theta\right|} . \tag{3.4.10}$$

The numerator is bounded above by

$$(\lambda + k)\left|v_0^x\right| + k\left(\left|v_0^y\right| + \left|v_0^\theta\right|\right) \tag{3.4.11}$$

and the denominator is bounded below by

$$\left(\frac{1}{\lambda} - k\right)\left|v_0^y\right| - k\left|v_0^\theta\right| . \tag{3.4.12}$$

So we get

$$\chi_1 \equiv \left|\frac{v_1^x}{v_1^y}\right| \le \frac{(\lambda + k)\left|v_0^x\right| + k\left(\left|v_0^y\right| + \left|v_0^\theta\right|\right)}{\left(\frac{1}{\lambda} - k\right)\left|v_0^y\right| - k\left|v_0^\theta\right|} \le \frac{\chi_0 + kL}{\frac{1}{\lambda} - kL} . \tag{3.4.13}$$

From (3.4.6) and (3.4.8) it is easy to see that

$$\chi_2 \equiv \left|\frac{v_2^x}{v_2^y}\right| \le \frac{\chi_1 + k\left(1 + \frac{\left|v_1^\theta\right|}{\left|v_1^y\right|}\right)}{\frac{1}{\lambda} - k\left(1 + \frac{\left|v_1^\theta\right|}{\left|v_1^y\right|}\right)} , \tag{3.4.14}$$

and, using the estimate of $\left|\dfrac{v_1^\theta}{v_1^y}\right|$ given in (3.4.9), we get (recall from (3.4.8) that $b = 1/\lambda - kL$)

$$\chi_2 \le \frac{\chi_1 + kL}{\frac{1}{\lambda} - kL} \le \frac{\chi_0}{b^2} + \frac{kL}{b^2} + \frac{kL}{b} . \tag{3.4.15}$$

Continuing on in this manner, we find that

$$\chi_n \equiv \left|\frac{v_n^x}{v_n^y}\right| \le \frac{\chi_0}{b^n} + kL\sum_{i=1}^{n}\frac{1}{b^i} \le \frac{\chi_0}{b^n} + \frac{kL}{b-1} , \tag{3.4.16}$$

and, since $\chi_0/b^n \to 0$ as $n \uparrow \infty$ and $\frac{kL}{b-1} < \frac{(b-1)}{4}$, there exists an integer $\tilde{n}$ such that for any $n > \tilde{n}$ we have

$$\chi_n \le \frac{b-1}{4} . \qquad (3.4.17)$$

Now, originally v_0 could have been chosen so that $\left|\frac{v_0^x}{v_0^y}\right|$, $\left|\frac{v_0^\theta}{v_0^y}\right|$, were as large as possible in $T_\tau \bar{V}$, so there exists $\tilde{n}$ such that for all $n \ge \tilde{n}$, the nonzero vectors of $T_{f^n(\tau)}\big(f^n(\bar{V})\big)$ satisfy

$$\left|\frac{v_n^x}{v_n^y}\right| \le \frac{b-1}{4} \,, \qquad \left|\frac{v_n^\theta}{v_n^y}\right| \le \alpha^n \left|\frac{v_0^\theta}{v_0^y}\right| . \qquad (3.4.18)$$

Step 3. By continuity of the tangent spaces of $T_{f^n(\tau)}\big(f^n(\bar{V})\big)$, we can find a μ_v-vertical slice $\bar{\bar{V}} \subset f^n(\bar{V})$ with $f^n(\tau) \subset \bar{\bar{V}}$ such that the slopes of any vectors tangent to $\bar{\bar{V}}$ satisfy

$$\left|\frac{v_n^x}{v_n^y}\right| \le \frac{b-1}{2} \,, \qquad \left|\frac{v_n^\theta}{v_n^y}\right| \le \bar{\alpha}^n \left|\frac{v_0^\theta}{v_0^y}\right| \,, \qquad 0 < \alpha < \bar{\alpha} < 1 . \qquad (3.4.19)$$

Step 4. Let $v = (v^x, v^y, v^\theta) \in T_{f^n(\tau)}\bar{\bar{V}}$. We want to estimate the rate of growth of vectors in $T_{f^n(\tau)}\bar{\bar{V}}$. First we note that, if necessary, choosing δ_s, δ_u smaller we can assume there exists $k_1 > 0$ such that

$$0 < k_1 L < \min(\epsilon, kL) \,, \qquad (3.4.20)$$

and, since $g_1(0, y, \theta) = 0$ in B,

$$\max_B \{ \|g_{1y}\| , \|(Ax)_\theta + g_{1\theta}\| \} \le k_1 . \qquad (3.4.21)$$

Note that by (3.4.9) we have

$$\max_{v^\theta, v^y \in T_{f^n(\tau)}\bar{\bar{V}}} \left(1 + \left|\frac{v^\theta}{v^y}\right| \right) < L . \qquad (3.4.22)$$

We have

$$Df(p)v = \begin{pmatrix} (A + g_{1x})v^x + g_{1y}v^y + \big((Ax)_\theta + g_{1\theta}\big)v^\theta \\ g_{2x}v^x + (B + g_{2y})v^y + \big((By)_\theta + g_{2\theta}\big)v^\theta \\ g_{3\theta}v^\theta \end{pmatrix} \qquad (3.4.23)$$

and we have

$$\chi_{\tilde{n}+1} = \left| \frac{v^x_{\tilde{n}+1}}{v^y_{\tilde{n}+1}} \right| = \frac{\left| (A+g_{1x})v^x + g_{1y}v^y + ((Ax)_\theta + g1\theta)v^\theta \right|}{\left| g_{2x}v^x + (B+g_{2y})v^y + ((By)_\theta + g_{2\theta})v^\theta \right|} \tag{3.4.24}$$

where the numerator is bounded above by

$$(\lambda + k)\,|v^x| + k_1\,|v^y| + k_1\,\left| v^\theta \right| , \tag{3.4.25}$$

and the denominator is bounded below by

$$\left(\frac{1}{\lambda} - k \right) |v^y| - k\,|v^x| - k\,\left| v^\theta \right| . \tag{3.4.26}$$

$$\chi_{\tilde{n}+1} = \left| \frac{v^x_{\tilde{n}+1}}{v^y_{\tilde{n}+1}} \right| \le \frac{(\lambda+k)\,|v^x| + k_1\,|v^y| + k_1\,\left| v^\theta \right|}{\left(\frac{1}{\lambda} - k \right) |v^y| - k\,|v^x| - k\,\left| v^\theta \right|} \le \frac{(\lambda+k)\chi_{\tilde{n}} + k_1 L}{\frac{1}{\lambda} - kL - k\chi_{\tilde{n}}}$$

$$\le \frac{\chi_{\tilde{n}} + k_1 L}{b - k\chi_{\tilde{n}}}$$

$$\le \frac{\chi_{\tilde{n}} + k_1 L}{b - k\left(\frac{b-1}{2} \right)}$$

$$\le \frac{\chi_{\tilde{n}} + k_1 L}{b - \frac{1}{2}(b-1)} = \frac{\chi_{\tilde{n}} + k_1 L}{\frac{1}{2}(b+1)} . \tag{3.4.27}$$

Let $b_1 = \frac{1}{2}(b+1) > 1$. Carrying out similar calculations we find that

$$\chi_{\tilde{n}+n} \le \frac{\chi_{\tilde{n}}}{b_1^n} + \frac{k_1 L}{b_1 - 1} . \tag{3.4.28}$$

So there exists an $\bar{n}$ such that for $n \ge \bar{n}$

$$\chi_{\tilde{n}+n} \le \epsilon \left(1 + \frac{1}{b_1 - 1} \right) . \tag{3.4.29}$$

Since v could have been such that $\chi_{\tilde{n}}$ was as large as possible we see that for $n \ge \bar{n}$ any nonzero vector tangent to $f^n(\bar{\bar{V}}) \cap B$ satisfies $\left| \frac{v^x}{v^y} \right| \le \epsilon \left(1 + \frac{1}{b_1 - 1} \right)$. Thus, given $\epsilon > 0$, there exists n_0 such that for $n \ge n_0$, all nonzero vectors tangent to $f^n(\bar{\bar{V}}) \cap B$ satisfy

$$\left| \frac{v^x_n}{v^y_n} \right| < \epsilon .$$

Next we want to show that $f^n(\bar{\bar{V}}) \cap B$ is stretched in the direction of $W^u(V)$. We do this by examining ratios of tangent vectors perpendicular to V under iteration by f.

$$\sqrt{\frac{\left|v^x_{n+1}\right|^2 + \left|v^y_{n+1}\right|^2}{\left|v^x_n\right|^2 + \left|v^y_n\right|^2}} = \frac{\left|v^y_{n+1}\right|}{\left|v^y_n\right|} \sqrt{\frac{1 + \chi^2_{n+1}}{1 + \chi^2_n}}. \tag{3.4.30}$$

From (3.4.27) we see that

$$\frac{\left|v^y_{n+1}\right|}{\left|v^y_n\right|} \geq \frac{1}{\lambda} - kL - k\chi_n. \tag{3.4.31}$$

Now, since χ_{n+1} and χ_n are arbitrarily small, we see that the norms of iterates of tangent vectors normal to V are growing at a rate approaching $\frac{1}{\lambda} - kL > 1$. Thus, $f^n(\bar{\bar{V}}) \cap B$ is expanding along $W^u(V)$. This, together with the fact that vectors tangent to $f^n(\bar{\bar{V}}) \cap B$ satisfy $|v^x/v^y| < \epsilon$, shows that there exists an n_0 such that for $n \geq n_0$, $f^n(\bar{\bar{V}}) \cap B$ is C^1 ϵ-close to $W^u(V)$. $\qquad\square$

We are now ready to give the proof of Theorem 3.4.1.

Proof of Theorem 3.4.1.

Now $W^s(V)$ and $W^u(V)$ intersect transversely at τ; hence, there exist integers p_1, p_2 such that $f^{p_1}(\tau), f^{-p_2}(\tau) \in B$, with $W^s(V)$ and $W^u(V)$ intersecting transversely at $f^{p_1}(\tau)$ and $f^{-p_2}(\tau)$. We denote $f^{p_1}(\tau)$ and $f^{-p_2}(\tau)$ by $(x_1, 0, \theta)$ and $(0, y_2, \theta)$, respectively. Consider the following sets:

$$\begin{aligned} U_1 &= \{ (x, y, \theta) \mid |x - x_1| \leq \epsilon_1, \ |y| \leq \epsilon_1, \ \theta \in T^l \} \\ U_2 &= \{ (x, y, \theta) \mid |x| \leq \epsilon_2, \ |y - y_2| \leq \epsilon_2, \ \theta \in T^l \} \end{aligned} \tag{3.4.32}$$

for some $\epsilon_1, \epsilon_2 > 0$. Now the strategy is to show that there exists some set $\tilde{U}_1 \subset U_1$ which is mapped onto itself under some iterate of f and on which conditions A1 and A3 of Section 2.4 hold.

By Lemma 3.4.2 we know there exists a positive integer n_0 such that, for every $n > n_0$, $f^n(U_1) \cap U_2 \neq \emptyset$.

Let $\tilde{U}_2$ be the connected component of $f^n(U_1) \cap U_2$ which contains $f^n(x_1, 0, \theta)$. Then $f^{-n}(\tilde{U}_2) \equiv \tilde{U}_1$ is a subset of U_1 containing $(x_1, 0, \theta)$. We choose a foliation of $\tilde{U}_1$ (i.e., an n-parameter family) of μ_v-vertical slices $\bar{V}_\alpha$, $\alpha \in I$ where I is some n dimensional index set, such that the $\bar{V}_\alpha$ are parallel to the tangent bundle over

$(x_1, 0, \theta)$, $\theta \in T^l$, of the component of $W^u(V) \cap \tilde{U}_1$ containing $(x_1, 0, \theta)$, $\theta \in T^l$. Then, by Lemma 3.4.2, for $\tilde{U}_1$ sufficiently small, $f^n(\bar{V}_\alpha)$, $\alpha \in I$, are C^1 ϵ-close to $W^u(V)$ for $n \geq n_0$.

Now $f^{p_1 + p_2}(0, y_2, \theta) = (x_1, 0, \theta)$, so a neighborhood of $(0, y_2, \theta)$ is mapped onto a neighborhood of $(x_1, 0, \theta)$ and, by Taylor's theorem, $f^{p_1 + p_2}$ can be approximated C^1 ϵ-close by a rigid motion (i.e., translation plus rotation) whose derivative at $(0, y_2, \theta)$ we write as

$$\begin{pmatrix} a_{11} & a_{12} & a_{13} \\ a_{21} & a_{22} & a_{23} \\ a_{31} & a_{32} & a_{33} \end{pmatrix} \tag{3.4.33}$$

where $\det a_{22} \neq 0$ and $\det \begin{pmatrix} a_{22} & a_{23} \\ a_{32} & a_{33} \end{pmatrix} \neq 0$ since $W^s(V)$ and $W^u(V)$ intersect transversely at $(0, y_2, \theta)$. A direct consequence of this is that the $f^n(\bar{V}_\alpha) \subset \tilde{U}_2$, $\alpha \in I$, are mapped to an n-parameter family of μ_v-vertical slices which are C^1 ϵ-close to the component of $W^u(V) \cap \tilde{U}_1$ containing $(x_1, 0, \theta)$, $\theta \in T^l$. See Figure 3.4.10 for an illustration of the geometry.

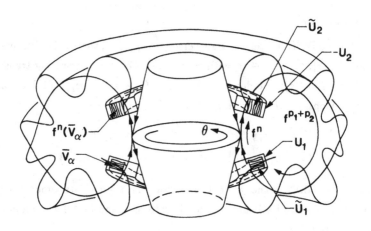

Figure 3.4.10. Geometry Near the Homoclinic Tangle.

Now we want to show that A1 and A3 of Section 2.4 hold for the map

$$F \equiv f^{p_1 + p_2} \circ f^n \colon \tilde{U}_1 \to \tilde{U}_1 . \tag{3.4.34}$$

A1. We can construct a countable set of μ_h-horizontal slabs H_i, $i = 1,\ldots$, in $\tilde{U}_1$ whose vertical boundaries are constructed from the $\bar{V}_\alpha$, $\alpha \in I$, such that $0 < \mu_v \mu_h < 1$ by Lemma 3.4.2. Then, from the previous discussion, $F(H_i) \subset \tilde{U}_1$ with the vertical boundaries of $F(H_i)$ C^1 ϵ-close to the vertical boundaries of the H_i. Moreover, since we can insure that the y direction is expanded by an arbitrary amount by choosing $\tilde{U}_1$ sufficiently small, then it follows that $F(H_i)$ intersects each of the H_j, $j = 1,\ldots$, properly; hence A1 holds.

A3. We give the argument for the stable sectors. The argument for the unstable sectors is virtually identical. We must show three things (see Section 2.4c).

1) $DF(S_X^u) \subset S_Y^u$,
2) $\left| \eta_{f(p_0)} \right| > \frac{1}{\mu} \left| \eta_{p_0} \right|$,
3) $1 < \left| \eta_{f(p_0)} \right| / \left| \chi_{f(p_0)} \right|$,

for all $p_0 \in X$, $0 < \mu < 1 - \mu_v \mu_h - \mu_h \bar{\mu}_v - \bar{\mu}_h$.

First some notation. We have

$$DF = \begin{pmatrix} a_{11} & a_{12} & a_{13} \\ a_{21} & a_{22} & a_{23} \\ a_{31} & a_{32} & a_{33} \end{pmatrix} Df^n. \tag{3.4.35}$$

For $v_{p_0} = (v_{p_0}^x, v_{p_0}^y, v_{p_0}^\theta) \in S_{p_0}^u$, $p_0 \in X$, let

$$Df^n(p_0)v_{p_0} = (v_{f^n(p_0)}^x, v_{f^n(p_0)}^y, v_{f^n(p_0)}^\theta).$$

Then

$$DF(p_0)v_{p_0} = \begin{pmatrix} a_{11}v_{f^n(p_0)}^x + a_{12}v_{f^n(p_0)}^y + a_{x13}v_{f^n(p_0)}^\theta \\ a_{21}v_{f^n(p_0)}^x + a_{22}v_{f^n(p_0)}^y + a_{x23}v_{f^n(p_0)}^\theta \\ a_{31}v_{f^n(p_0)}^x + a_{32}v_{f^n(p_0)}^y + a_{x33}v_{f^n(p_0)}^\theta \end{pmatrix} \equiv v_{F(p_0)}. \tag{3.4.36}$$

Also note that $\bar{\mu}_h$ and $\bar{\mu}_v$ can be chosen arbitrarily small by Lemma 3.4.2. We now begin the argument.

$\underline{DF(S_X^u) \subset S_Y^u}$. We choose the sectors $S_{p_0}^u$, $p_0 \in X$, to be centered along the $\bar{V}_\alpha$, $\alpha \in I$, in $\tilde{U}_1$. Let v_{p_0} be any vector in $S_{p_0}^u$. By Lemma 3.4.2, for n sufficiently large under $Df^n(p_0)$, we have

$$\left| \frac{v_{f^n(p_0)}^x}{v_{f^n(p_0)}^y} \right| < \epsilon \tag{3.4.37}$$

$$\left| \frac{v^x_{f^n(p_0)}}{v^\theta_{f^n(p_0)}} \right| < \epsilon.$$

(3.4.38)

Now under $Df^{p_1+p_2}$ the images of the tangent spaces of the $f^n(\bar{V}_\alpha)$ are C^1 ϵ-close to the tangent spaces of $\bar{V}_\alpha$. Hence, $DF(S^u_\mathcal{H}) \subset S^u_\mathcal{Y}$.

$\left| \eta_{f(p_0)} \right| > \frac{1}{\mu} \left| \eta_{p_0} \right|$. From (3.4.36) we have

$$\left| \frac{v^y_{p_0}}{v^y_{F(p_0)}} \right| = \frac{\left| v^y_{p_0} \right|}{\left| a_{21} v^x_{f^n(p_0)} + a_{22} v^y_{f^n(p_0)} + a_{23} v^\theta_{f^n(p_0)} \right|}.$$

(3.4.39)

By transversality a_{22} is invertible; hence, by Lemma 3.4.2 the denominator of (3.4.39) can be made arbitrarily large by taking n sufficiently large.

$1 < \left| \eta_{f(p_0)} \right| / \left| \chi_{f(p_0)} \right|$. This follows from normal hyperbolicity and estimates like those obtained in Lemma 3.4.2.   □

We make the following remarks concerning Theorem 3.4.1.

1) For the case $l = 0$ (i.e., the invariant torus reduces to a fixed point), Theorem 3.4.1 becomes the familiar Smale-Birkhoff homoclinic theorem, see Smale [1963].

2) Theorem 3.4.1 was proved by Wiggins [1986a], but a similar earlier result was obtained by Silnikov [1968b]. Meyer and Sell [1986] have studied orbits homoclinic to almost periodic orbits using very different techniques and have obtained a characterization of the dynamics similar to that given in Theorem 3.4.1.

Final Remarks.

1) The complicated dynamics associated with an orbit homoclinic to a hyperbolic fixed point were first noticed by Poincaré [1899] in his studies of the restricted three body problem. In fact, the term "homoclinic" is due to him. Birkhoff [1927] continued Poincaré's studies.

2) Most of the work done for orbits homoclinic to hyperbolic fixed points of maps has been for the two-dimensional case. Gavrilov and Silnikov [1972], [1973] have studied parametrized systems and have shown that infinite sequences of saddle-node and period doubling bifurcations accumulate on the bifurcation values for

the homoclinic orbits. Newhouse [1974], [1979] independently discovered many of the results of Gavrilov and Silnikov and went much further. In fact, the presence of "Newhouse sinks" is the main difficulty in proving the existence of a strange attractor in many systems, e.g., the forced Duffing oscillator (see Guckenheimer and Holmes [1983] for a discussion). See also Robinson [1985].

3) Orbits homoclinic to non-hyperbolic fixed points and non-transverse homoclinic orbits in dimensions ≥ 3 have not received much study.

4) For applications of knot theory to the study of the bifurcations associated with the formation of horseshoes in two-dimensional parametrized systems, see Holmes and Williams [1985].

5) Heteroclinic tangles formed via transverse heteroclinic orbits often yield the same type of chaotic dynamics as described in Theorem 3.4.1. The same techniques should apply, and we leave the details to the reader.

6) Orbits homoclinic to normally hyperbolic invariant tori have only recently been studied. There are many open questions. For example, how do the dynamics along the tori "couple" to the chaotic dynamics normal to the tori? Is it possible to "entrain" the Cantor set of tori? If so, then there may be "ordinary" horseshoes within the Cantor set of tori.

CHAPTER 4
Global Perturbation Methods
for Detecting Chaotic Dynamics

In Chapter 3 we saw that orbits homoclinic or heteroclinic to hyperbolic fixed points, hyperbolic periodic orbits, or normally hyperbolic invariant tori could often be mechanisms for producing deterministic chaos. In this chapter we will develop a variety of perturbation techniques which will allow us to detect such homoclinic and heteroclinic orbits.

The term "global perturbation" refers to the fact that we are concerned with perturbing a structure which exists throughout an extended region of the phase space. The main idea behind the methods can be found in the work of Melnikov [1963]. Melnikov considered an "unperturbed" system consisting of a planar ordinary differential equation having a hyperbolic fixed point connected to itself by a homoclinic orbit. As we saw in Section 3.2b, there are no complicated dynamical phenomena associated with such a system. He then perturbed this system with a time periodic perturbation. In this case the hyperbolic fixed point becomes a hyperbolic periodic orbit whose stable and unstable manifolds may intersect transversely, yielding Smale horseshoes and their attendant chaotic dynamics (see Section 3.3). Using a clever perturbation technique, he developed a computable formula for the distance between the stable and unstable manifolds of the hyperbolic periodic orbit, thus allowing him to explicitly determine the presence of chaotic dynamics in specific systems. The following year, Arnold [1964] generalized Melnikov's method to a specific example of a time periodic Hamiltonian perturbation of a two degree of freedom completely integrable Hamiltonian system. The method enabled Arnold to demonstrate the existence of a global type of instability for Hamiltonian systems which has come to be known as *Arnold Diffusion*. Following these developments of Melnikov and Arnold, the technique appears to have gone unused (at least in the west) until being rediscovered and applied by Holmes [1979] in his studies of the periodically forced Duffing oscillator. Since that time a variety of generalizations of the method have been developed by various workers, and we will describe these shortly.

This chapter is organized as follows. In Section 4.1 we will describe, in general terms, the structure of the three different types of systems we are considering and how they fit into a general theoretical framework. We will also comment on how our methods are generalizations of previous work. In Section 4.2 we will discuss a variety of examples which illustrate the theory, and in Section 4.3 we will make some comments regarding generalizations of our methods as well as some additional applications.

4.1. The Three Basic Systems and Their Geometrical Structure

In this section we describe the structure of the three types of systems under consideration and put them in the context of previous work.

Our goal is to develop perturbation techniques which will allow us to detect the presence of orbits which are homoclinic and heteroclinic to different types of invariant sets. As with most perturbation theories, we will begin with an unperturbed system of which we have considerable knowledge of the global dynamics. In our case, the unperturbed systems will be completely integrable Hamiltonian systems or parametrized families of completely integrable Hamiltonian systems which have a degenerate homoclinic or heteroclinic structure (more specifically, they will contain manifolds of nontransverse orbits homoclinic or heteroclinic to parametrized families of invariant tori). We will consider arbitrary perturbations of such systems (i.e., the perturbed systems need not be Hamiltonian), and determine the nature of any invariant sets which might remain in the perturbed systems. We will then use our knowledge of the nontransverse homoclinic structure of the unperturbed system to develop a measurement of the distance between the stable and unstable manifolds of certain invariant sets which are preserved in the perturbed system. Thus, we will be able to assert the existence (or nonexistence) of homoclinic and heteroclinic orbits in the perturbed systems.

The three types of systems which we will study have the forms:

System I. $\dot{x} = JD_xH(x,I) + \epsilon g^x(x,I,\theta,\mu;\epsilon)$

$\dot{I} = \epsilon g^I(x,I,\theta,\mu;\epsilon)$ $\qquad\qquad\qquad (x,I,\theta) \in \mathbf{R}^{2n} \times \mathbf{R}^m \times T^l$

$\dot{\theta} = \Omega(x,I) + \epsilon g^\theta(x,I,\theta,\mu;\epsilon)$

System II. $\dot{x} = JD_xH(x,I) + \epsilon g^x(x,I,\theta,\mu;\epsilon)$

$\dot{I} = \epsilon g^I(x,I,\theta,\mu;\epsilon)$ $\qquad\qquad\qquad (x,I,\theta) \in \mathbf{R}^{2n} \times T^m \times T^l$

$\dot{\theta} = \Omega(x,I) + \epsilon g^\theta(x,I,\theta,\mu;\epsilon)$

System III. $\dot{x} = JD_xH(x,I) + \epsilon JD_x\tilde{H}(x,I,\theta,\mu;\epsilon)$

$\dot{I} = -\epsilon D_\theta\tilde{H}(x,I,\theta,\mu;\epsilon)$ $\qquad\qquad (x,I,\theta) \in \mathbf{R}^{2n} \times \mathbf{R}^m \times T^m$

$\dot{\theta} = D_IH(x,I) + \epsilon D_I\tilde{H}(x,I,\theta,\mu;\epsilon)$

where $0 < \epsilon \ll 1$, $\mu \in \mathbf{R}^p$ is a vector of parameters, and J is the $2n \times 2n$ matrix given by

$$J = \begin{pmatrix} 0 & \text{Id} \\ -\text{Id} & 0 \end{pmatrix}$$

where "Id" denotes the $n \times n$ identity matrix and "0" denotes the $n \times n$ zero matrix. More detailed information concerning the structure of Systems I, II, and III will be given shortly; however, at this point we wish to make some general comments concerning the differences between the systems.

1) The systems obtained by setting $\epsilon = 0$ will be referred to as the *unperturbed systems*. Each of the three unperturbed systems has a very similar structure. However, the unperturbed System III is more special in that the entire vector field is derived from a Hamiltonian $H_\epsilon(x,I,\theta,\mu;\epsilon) = H(x,I) + \epsilon\tilde{H}(x,I,\theta,\mu;\epsilon)$. This need not be the case for the unperturbed Systems I and II.

2) At first glance it appears that Systems II and III are special cases of System I. This is true; however, there are vast differences in the underlying geometry of the vector fields, and what we prove in each context necessitates a separate discussion for each case.

3) In all three systems, the main difference in describing their general structures involves the behavior of the I component of the vector field. Specifically, we will need to have some type of control over the I variables for the perturbed system.

  In System I we will require the perturbed vector field to have a dissipative nature that results in the existence of a stationary point in the I component of the vector field in some averaged sense. This will result in certain nonresonance requirements among the frequencies Ω.

System II is periodic in each component of the I variable.

In System III the perturbations are Hamiltonian. For this case control over the I variables is obtained using KAM type arguments. As for System I, this will result in certain nonresonance conditions among the frequencies Ω.

Before proceeding with a general discussion of Systems I, II, and III we want to comment on previous generalizations of Melnikov's idea and how they fit into our general framework.

Melnikov's [1963] original work is a special case of System I. Setting $n = 1$, $m = 0$, and $l = 1$ with $\dot{\theta} = \omega$ =constant gives

$$\begin{aligned}
\dot{x} &= JD_xH(x) + \epsilon g^x(x, \theta; \mu; \epsilon) \\
\dot{\theta} &= \omega
\end{aligned} \qquad (x, \theta) \in \mathbb{R}^2 \times T^1 .$$

This equation has the form of a periodically perturbed oscillator and is the type of equation originally studied by Melnikov (note: Melnikov's work was actually more general in that he did not require the unperturbed system to be Hamiltonian; see Melnikov [1963] and Salam [1987]). Melnikov's results in a more abstract setting were later rediscovered by Chow, Hale, and Mallet-Paret [1980].

Holmes and Marsden [1982a] studied homoclinic orbits in dissipative and Hamiltonian perturbations of weakly coupled oscillators. For dissipative perturbations their work is a special case of System I with $n = 1$, $m = l = 1$, $\Omega(x, I) = D_IH(x, I)$ and with the resulting equations having the form

$$\begin{aligned}
\dot{x} &= JD_xH(x, I) + \epsilon g^x(x, I, \theta, \mu; \epsilon) \\
\dot{I} &= \epsilon g^I(x, I, \theta, \mu; \epsilon) \\
\dot{\theta} &= D_IH(x, I) + \epsilon g^\theta(x, I, \theta, \mu; \epsilon)
\end{aligned} \qquad (x, I, \theta) \in \mathbb{R}^2 \times \mathbb{R}^1 \times T^1 .$$

For Hamiltonian perturbations their work is a special case of System III with $n = 1$, $m = 1$, $l = 1$, and with the equations having the form

$$\begin{aligned}
\dot{x} &= JD_xH(x, I) + \epsilon JD_x\tilde{H}(x, I, \theta, \mu; \epsilon) \\
\dot{I} &= -\epsilon D_\theta\tilde{H}(x, I, \theta, \mu; \epsilon) \\
\dot{\theta} &= D_IH(x, I) + \epsilon D_I\tilde{H}(x, I, \theta, \mu; \epsilon)
\end{aligned} \qquad (x, I, \theta) \in \mathbb{R}^2 \times \mathbb{R}^1 \times T^1 .$$

Lerman and Umanski [1984] studied homoclinic orbits in strongly coupled oscillators subjected to both dissipative and Hamiltonian perturbations. For dissipative perturbations their work is a special case of System I with $n = 2$, $m = 0$, $l = 0$ and with the equations having the form

$$\dot{x} = JD_x H(x) + \epsilon g^x(x, \mu; \epsilon), \qquad x \in \mathbf{R}^4.$$

For Hamiltonian perturbations their work is a special case of System III with $n = 2$, $m = 0$, $l = 0$, and with the equations having the form

$$\dot{x} = JD_x H(x) + \epsilon JD_x \tilde{H}(x, \mu; \epsilon), \qquad x \in \mathbf{R}^4.$$

Holmes and Marsden [1982b], [1983] gave sufficient conditions for the existence of Arnold diffusion in a general class of systems. The systems they considered are a special case of System III with $n = 1$, m arbitrary, and with the equations having the form

$$
\begin{aligned}
\dot{x} &= JD_x H(x, I) + \epsilon JD_x \tilde{H}(x, I, \theta, \mu; \epsilon) \\
\dot{I} &= -\epsilon D_\theta \tilde{H}(x, I, \theta, \mu; \epsilon) \qquad\qquad (x, I, \theta) \in \mathbf{R}^2 \times \mathbf{R}^m \times T^m. \\
\dot{\theta} &= D_I H(I) + \epsilon D_I \tilde{H}(x, I, \theta, \mu; \epsilon)
\end{aligned}
$$

In the context of a problem concerning passage through resonance, Robinson [1983] studied homoclinic motions in a class of equations which are a special case of System I for $n = 1$, m arbitrary, $l = 0$, and with the equations having the form

$$
\begin{aligned}
\dot{x} &= JD_x H(x, I) + \epsilon g^x(x, I, \mu; \epsilon) \\
\dot{I} &= \epsilon g^I(x, I, \mu; \epsilon)
\end{aligned}
\qquad (x, I) \in \mathbf{R}^2 \times \mathbf{R}^m.
$$

Wiggins and Holmes [1987] later studied such systems with $n = 1$, $m = 1$, $l = 1$ with $\dot{\theta} = \omega =$ constant and with the equations having the form

$$
\begin{aligned}
\dot{x} &= JD_x H(x, I) + \epsilon g^x(x, I, \theta, \mu; \epsilon) \\
\dot{I} &= \epsilon g^I(x, I, \theta, \mu; \epsilon) \qquad\qquad (x, I, \theta) \in \mathbf{R}^2 \times \mathbf{R}^1 \times T^1. \\
\dot{\theta} &= \omega
\end{aligned}
$$

The general theory for System I was first given by Wiggins [1986b].

System II is a generalization of the work of Wiggins [1988] concerning homoclinic orbits in systems forced at low frequency and large amplitude.

Homoclinic orbits in quasiperiodically forced oscillators were studied by Wiggins [1986], [1987]. His work is a special case of System I with $n = 1$, $m = 0$, l arbitrary with $\dot{\theta} = \omega = $ constant and with the equations having the form

$$\begin{aligned} \dot{x} &= JD_xH(x) + \epsilon g^x(x,\theta,\mu;\epsilon) \\ \dot{\theta} &= \omega \end{aligned} \qquad (x,\theta) \in \mathbf{R}^2 \times T^l.$$

Techniques for studying homoclinic orbits in almost periodically forced oscillators have been developed by Meyer and Sell [1986] and Scheurle [1985].

Our methods rely heavily on the geometry of complete integrability associated with the unperturbed systems. However, Melnikov type techniques have been developed for time periodic perturbations of n dimensional systems possessing a hyperbolic fixed point connected to itself by a homoclinic orbit by Greundler [1985] and Palmer [1984]. Their methods are less geometrical and more functional analytic in nature and will not be covered in this book.

We will comment on additional applications of these techniques as we go along and mention further generalizations of the ideas at the end of this chapter. We next turn to a discussion of the structure of Systems I, II, and III.

4.1a. System I

The first type of system which we will consider has the following form

$$\begin{aligned} \dot{x} &= JD_xH(x,I) + \epsilon g^x(x,I,\theta,\mu;\epsilon) \\ \dot{I} &= \epsilon g^I(x,I,\theta,\mu;\epsilon) \\ \dot{\theta} &= \Omega(x,I) + \epsilon g^\theta(x,I,\theta,\mu;\epsilon) \end{aligned} \qquad (4.1.1)_\epsilon$$

with $0 < \epsilon \ll 1$, $(x,I,\theta) \in \mathbf{R}^{2n} \times \mathbf{R}^m \times T^l$, and $\mu \in \mathbf{R}^p$ is a vector of parameters. Additionally, we will assume the following.

I1. Let $V \subset \mathbf{R}^{2n} \times \mathbf{R}^m$ and $W \subset \mathbf{R}^p \times \mathbf{R}$ be open sets; then the functions

$$JD_xH : V \mapsto \mathbf{R}^{2n}$$
$$g^x : V \times T^l \times W \mapsto \mathbf{R}^{2n}$$
$$g^I : V \times T^l \times W \mapsto \mathbf{R}^m$$
$$\Omega : V \mapsto \mathbf{R}^l$$
$$g^\theta : V \times T^l \times W \mapsto \mathbf{R}^l$$

are defined and "sufficiently differentiable" on their respective domains of definition. By the phrase "sufficiently differentiable" we will mean C^r with $r \geq 6$ (the reason for this will be explained later on) and, in many cases, $r \geq 2$ will be sufficient. In any event, specifying the exact degree of differentiability is usually just a technical nuisance since all of our examples will be analytic. Finally, let us recall from Section 1.1i what it means for part of the phase space of $(4.1.1)_\epsilon$ to be the l-dimensional torus. Regarding x, I, μ, and ϵ as fixed, this means that g^x, g^I, and g^θ are 2π periodic *in each component* of their l-dimensional θ arguments, e.g., $g^x(x, I, \theta_1, \ldots, \theta_i, \ldots, \theta_l, \mu; \epsilon) = g^x(x, I, \theta_1, \ldots, \theta_i + 2\pi, \ldots, \theta_l, \mu; \epsilon)$ for any $1 \leq i \leq l$.

I2. $H = H(x, I)$ is a scalar valued function which can be thought of as an m-parameter family of Hamiltonians, and J is the $2n \times 2n$ "symplectic" matrix defined by

$$J = \begin{pmatrix} 0 & \text{Id} \\ -\text{Id} & 0 \end{pmatrix}$$

where Id denotes the $n \times n$ identity matrix and 0 represents the $n \times n$ zero matrix.

We will refer to $(4.1.1)_\epsilon$ as the *perturbed system*.

i) The Geometric Structure of the Unperturbed Phase Space

The system obtained by setting $\epsilon = 0$ in $(4.1.1)_\epsilon$ will be referred to as the unperturbed system.

$$\begin{aligned} \dot{x} &= JD_xH(x, I) \\ \dot{I} &= 0 \\ \dot{\theta} &= \Omega(x, I) . \end{aligned} \qquad (4.1.1)_0$$

Notice that since $\dot{I} = 0$, the x component of the unperturbed system has the form of an m-parameter family of Hamiltonian systems. Also, the x component of $(4.1.1)_0$ is independent of θ and, therefore, we can discuss the structure of the x component of $(4.1.1)_0$ independently of θ. We have the following two "structural assumptions" on the x component of $(4.1.1)_0$.

I3. There exists an open set $U \subset \mathbb{R}^m$ such that for each $I \in U$ the system

$$\dot{x} = JD_xH(x, I) \qquad (4.1.1)_{0,x}$$

is a completely integrable Hamiltonian system. By "completely integrable" we mean that there exist n scalar valued functions of (x, I), $H \equiv K_1, K_2, \ldots, K_n$, (the K_i are called the "integrals") which satisfy the following two conditions:

1) The set of vectors $D_x K_1, D_x K_2, \ldots, D_x K_n$ is pointwise linearly independent $\forall I \in U$ at all points of $\mathbf{R}^{2n}$ which are not fixed points of $(4.1.1)_{0,x}$.

2) We define the Poisson bracket of K_i, K_j (denoted $\{K_i, K_j\}$) as follows

$$\{K_i, K_j\} = \langle J D_x K_i, D_x K_j \rangle \tag{4.1.2}$$

where $\langle \cdot, \cdot \rangle$ denotes the usual Euclidean inner product, and we require that the pairwise Poisson brackets of the K_i vanish, i.e.,

$$\langle J D_x K_i, D_x K_j \rangle = 0 \qquad \forall i, j , \quad I \in U . \tag{4.1.3}$$

Furthermore, we assume that the K_i are at least C^{r+1}, $r \geq 6$.

We remark that our definition of complete integrability does not quite agree with the classical definition. The classical definition of integrability requires the integrals to be analytic functions and might also relax our requirement of independence of the integrals. For our purposes a finite degree of differentiability is sufficient.

(Note: for a more complete discussion of complete integrability (which we will not need for our purposes), we refer the reader to Abraham and Marsden [1978] or Arnold [1978].)

We want to emphasize that a background in Hamiltonian systems is not a prerequisite for the following material; rather, the geometrical consequences of completely integrable Hamiltonian systems will be important, and we will comment on those shortly.

I4. For every $I \in U$, $(4.1.1)_{0,x}$ possesses a hyperbolic fixed point which varies smoothly with I and has an n dimensional homoclinic manifold connecting the fixed point to itself. We will assume that trajectories along the homoclinic orbit can be represented in the form $x^I(t, \alpha)$ where $t \in \mathbf{R}^1$, $\alpha \in \mathbf{R}^{n-1}$. The reason we assume that the homoclinic manifold is n dimensional is related to the independence of the integrals and will be discussed shortly.

Let us now make the following remarks concerning the geometrical consequences of I3 and I4.

Consequences of I3

Let us consider the system $(4.1.1)_{0,x}$ for fixed $I = I_0 \in U$ as a vector field on $\mathbf{R}^{2n}$. Let x_0 be a hyperbolic fixed point of $(4.1.1)_{0,x}$ and denote the n dimensional stable and unstable manifolds of x_0 by $W^s_{I_0}(x_0)$ and $W^u_{I_0}(x_0)$, respectively (note: the dimensionality of the stable and unstable manifolds of x_0 is discussed more fully under the consequences of I4, below. At this point, we ask that the reader accept the above statement regarding the dimensions). We have the following preliminary lemma.

Lemma 4.1.1. *Suppose* $K_1(x_0, I_0) = c_1, \ldots, K_n(x_0, I_0) = c_n$, *then* $K_1(x, I_0) = c_1, \ldots, K_n(x, I_0) = c_n$ *for all* $x \in W^s_{I_0}(x_0) \cup W^u_{I_0}(x_0)$.

PROOF: This is an immediate consequence of the continuity of the K_i.   □

For our purposes, the important geometrical consequence associated with I3 is contained in the following two propositions (note: $T_x W^{s,u}_{I_0}(x_0)$ denotes the tangent space of $W^{s,u}_{I_0}(x_0)$ at x).

Proposition 4.1.2. *For any* $x \in W^s_{I_0}(x_0)$ *(resp.* $W^u_{I_0}(x_0)$*)* $T_x W^s_{I_0}(x_0)$ *(resp.* $T_x W^u_{I_0}(x_0)$*)* $= span\ \{JD_x K_1(x, I_0), \ldots, JD_x K_n(x, I_0)\}$. *Moreover,*

$$N_x = span\ \{D_x K_1(x, I_0), \ldots, D_x K_n(x, I_0)\}$$

is orthogonal to $T_x W^s_{I_0}(x)$ *(resp.* $T_x W^u_{I_0}(x)$*) and* $\mathbf{R}^{2n} = T_x W^s_{I_0}(x_0) + N_x$ *(resp.* $T_x W^u_{I_0}(x_0) + N_x$*).*

PROOF: From Lemma 4.1.1, for any $x \in W^s_{I_0}(x_0) \cup W^u_{I_0}(x_0)$ we have

$$K_1(x, I_0) - c_1 = 0$$

$$\vdots \qquad \vdots \qquad \vdots \qquad\qquad (4.1.4)$$

$$K_n(x, I_0) - c_n = 0.$$

Henceforth, for definiteness, we will give the argument for $W^s_{I_0}(x_0)$; however, the same argument applies to $W^u_{I_0}(x_0)$. Let $x \in W^s_{I_0}(x_0)$ and let $\beta(t)$ be a differentiable curve in $W^s_{I_0}(x_0)$ satisfying $\beta(0) = x$ for t contained in some open interval about the origin. Then $\beta(t)$ satisfies (4.1.4), i.e.,

$$K_1\big(\beta(t), I_0\big) - c_1 = 0$$

$$\vdots \qquad \vdots \qquad \vdots \qquad\qquad (4.1.5)$$

$$K_n\big(\beta(t), I_0\big) - c_n = 0.$$

Differentiating (4.1.5) with respect to t gives

$$\langle D_x K_1(x, I_0), \dot\beta(0)\rangle = 0$$

$$\vdots \qquad\qquad \vdots \qquad\qquad\qquad (4.1.6)$$

$$\langle D_x K_n(x, I_0), \dot\beta(0)\rangle = 0.$$

Geometrically, $\dot\beta(0)$ is a vector tangent to $W_{I_0}^s(x_0)$ at x, or, in other words, $\dot\beta(0) \in T_x W_{I_0}^s(x_0)$, and analytically $\dot\beta(0)$ can be viewed as a solution of (4.1.6). So, since $\beta(t)$ is an arbitrary curve in $W_{I_0}^s(x_0)$, any solution of (4.1.6) is an element of $T_x W_{I_0}^s(x_0)$. By I3 the n linearly independent vectors $J D_x K_1(x, I_0), \ldots, J D_x K_n(x, I_0)$ each solve (4.1.6). Therefore, $T_x W_{I_0}^s(x_0) = \text{span}\{J D_x K_1(x, I_0), \ldots, J D_x K_n(x, I_0)\}$.

The fact that $N_x = \text{span}\{D_x K_1(x, I_0), \ldots, D_x K_n(x, I_0)\}$ is orthogonal to $T_x W_{I_0}^s(x_0)$ is an immediate consequence of (4.1.3). $\qquad\square$

Regarding the dimension of $W_{I_0}^s(x_0)$ and $W_{I_0}^u(x_0)$, it should be noted that the fact that they are each n dimensional follows from (4.1.4), as does the fact that the $D_x K_i(x, I_0)$, $i = 1, \ldots, n$ are linearly independent.

Proposition 4.1.2 will be extremely useful later on when we construct "homoclinic coordinates." The next proposition indicates that if a homoclinic orbit exists it must necessarily be n dimensional.

Proposition 4.1.3. *Suppose $W_{I_0}^s(x_0)$ and $W_{I_0}^u(x_0)$ intersect. Then $W_{I_0}^s(x_0)$ and $W_{I_0}^u(x_0)$ coincide along the n-dimensional components of $W_{I_0}^s(x_0) - \{x_0\}$ and $W_{I_0}^u(x_0) - \{x_0\}$ which contain $W_{I_0}^s(x_0) \cap W_{I_0}^u(x_0) - \{x_0\}$. Hence, there exists an n-dimensional manifold of orbits homoclinic to x_0.*

PROOF: Consider the map

$$K(x, I_0) = \big(K_1(x, I_0), \ldots, K_n(x, I_0)\big) : \mathbb{R}^{2n} \to \mathbb{R}^n.$$

By Lemma 4.1.1 $K(x, I_0) = (c_1, \ldots, c_n) \equiv c$ for all $x \in W_{I_0}^s(x_0) \cup W_{I_0}^u(x_0)$. So $K^{-1}(c)$ is an invariant set containing $W_{I_0}^s(x_0) \cup W_{I_0}^u(x_0)$. Now by I3 the

$D_x K_i(x, I_0)$, $1 \leq i \leq n$, are linearly independent at each $x \in W^s_{I_0}(x_0) \cup W^u_{I_0}(x_0) - \{x_0\}$. Therefore, K is onto (i.e., $D_x K$ has maximal rank) for each $x \in W^s_{I_0}(x_0) \cup W^u_{I_0}(x_0) - \{x_0\}$. Hence, by the implicit function theorem (see also the "submersion theorem" in Guillemin and Pollack [1974]), $K^{-1}(c)$ has the structure of an n dimensional manifold near each $x \in W^s_{I_0}(x_0) \cup W^u_{I_0}(x_0) - \{x_0\}$. So, if $W^s_{I_0}(x_0)$ and $W^u_{I_0}(x_0)$ intersect, then they must coincide along the n dimensional components of $W^s_{I_0}(x_0) - \{x_0\}$ and $W^u_{I_0}(x_0) - \{x_0\}$, which contain $W^s_{I_0}(x_0) \cap W^u_{I_0}(x_0) - \{x_0\}$. $\square$

This proposition is not true if the integrals are not independent, as can be seen by the following example. Consider the system

$$\dot{x}_1 = x_1$$
$$\dot{x}_2 = x_4$$
$$\dot{x}_3 = -x_3 \qquad (x_1, x_2, x_3, x_4) \in \mathbf{R} \times \mathbf{R} \times \mathbf{R} \times \mathbf{R}.$$
$$\dot{x}_4 = x_2 - x_2^3$$

This system is just the Cartesian product of two integrable systems; the $x_1 - x_3$ components represent a linear system with a saddle point at the origin, and the $x_2 - x_4$ components are just the unforced, undamped Duffing equation. The Hamiltonian is given by

$$H(x_1, x_2, x_3, x_4) = x_1 x_3 + \frac{x_4^2}{2} - \frac{x_2^2}{2} + \frac{x_2^4}{4}$$

with an additional integral given by either

$$K_2(x_1, x_2, x_3, x_4) = x_1 x_3$$

or

$$K_2(x_1, x_2, x_3, x_4) = \frac{x_4^2}{2} - \frac{x_2^2}{2} + \frac{x_2^4}{4}.$$

It is an easy calculation to verify (4.1.3) for H and either choice of K_2.

Now the system has a hyperbolic fixed point at $(x_1, x_2, x_3, x_4) = (0, 0, 0, 0)$ having two dimensional stable and unstable manifolds. These manifolds intersect only along a one dimensional homoclinic orbit given by

$$\Gamma = \{ (x_1, x_2, x_3, x_4) \in \mathbb{R}^4 \,|\, x_1 = x_3 = 0, x_2 = \sqrt{2}\,\text{sech}\, t, x_4 = -\sqrt{2}\,\text{sech}\, t \tanh t \}.$$

A simple calculation shows that $D_x H$ and $D_x K_2$ (for either choice of K_2) are linearly *dependent* on Γ.

Consequences of I4

1. First we will show that the assumption that the fixed points of $(4.1.1)_{0,x}$ are hyperbolic $\forall I \in U$ leads to the fact that they can be represented as a C^r smooth function of the I variables (note: this will be useful for computations).

Recall that the condition for the existence of a fixed point of $(4.1.1)_{0,x}$ is that for some $x_0 \in \mathbb{R}^{2n}$, $I_0 \in U$, we have

$$JD_x H(x_0, I_0) = 0 \tag{4.1.7}$$

or, since J is nonsingular,

$$D_x H(x_0, I_0) = 0. \tag{4.1.8}$$

Now the assumption that the fixed points are hyperbolic implies

$$\det\left[JD_x^2 H(x_0, I_0)\right] \neq 0, \tag{4.1.9}$$

or, using the fact that J is nonsingular and the determinant of the product is the product of the determinants, (4.1.9) is equivalent to

$$\det\left[D_x^2 H(x_0, I_0)\right] \neq 0. \tag{4.1.10}$$

By the implicit function theorem, (4.1.10) is a sufficient condition for there to exist an m parameter family of solutions of (4.1.8), $\gamma(I)$, for I in some neighborhood of I_0. By I4 each of these new solutions $\gamma(I)$ of (4.1.8) is also a hyperbolic fixed point of $(4.1.1)_{0,x}$ so by the global implicit function theorem (see Chow and Hale [1982]) the function $\gamma(I)$ exists and is C^r for each $I \in U$. (Note: for fixed I, the solution of (4.1.8) may actually have many disconnected components. In that case our theory may be applied separately to each component.)

2. The symmetry properties of Hamiltonian systems require that, if λ is an eigenvalue of $JD_x^2 H(x_0, I_0)$, then so is $-\lambda$ (see Abraham and Marsden [1978] or Arnold [1978]). This implies that the stable and unstable manifolds of hyperbolic fixed points of Hamiltonian systems have equal dimensions.

3. We want to make some comments concerning the analytical expression for the trajectory along the n dimensional homoclinic manifold which we have assumed

could be written in the form $x^I(t, \alpha)$, $I \in U$, $t \in \mathbb{R}^1$, $\alpha \in \mathbb{R}^{n-1}$. The question we want to address is why did we choose this particular form for the homoclinic orbit?

The meaning of the superscript I should be clear. It just indicates the parametric dependence of the homoclinic trajectory on I. We remark that by Theorem 1.1.4 $x^I(t, \alpha)$ depends on I in a C^r manner.

The meaning of the variable α may be a bit more mysterious. However, it happens that, for some systems possessing certain symmetries, the existence of one homoclinic trajectory implies the existence of an entire surface or manifold of homoclinic trajectories. In this case, varying the α in the homoclinic trajectory $x^I(t, \alpha)$ acts to take us from solution to solution of the homoclinic manifold. We will give an explicit example of a system exhibiting this behavior in Section 4.2c.

4. We now want to discuss the distinction between the terms "trajectory" and "orbit" (see Section 1.1c). Consider the system $(4.1.1)_{0,x}$ for fixed I. Then, by I4, this system has a homoclinic trajectory $x^I(t, \alpha)$ connecting a hyperbolic fixed point to itself. The set of points through which this homoclinic trajectory passes as t varies between $+\infty$ and $-\infty$ is called a homoclinic orbit. We will be interested in the behavior of the perturbed system $(4.1.1)_\epsilon$ near this homoclinic orbit and, therefore, we would like a parametrization of the homoclinic orbit so that we can describe points along it. This can be accomplished by utilizing the fact that $(4.1.1)_{0,x}$ is autonomous. By Lemma 1.1.7, since $(4.1.1)_{0,x}$ is autonomous, then $x^I(t - t_0, \alpha)$ is also a homoclinic trajectory of $(4.1.1)_{0,x}$ for any $t_0 \in \mathbb{R}$. Thus, t_0 is defined to be the time that it takes for the point $x^I(-t_0, \alpha)$ on the homoclinic orbit to flow to the point $x^I(0, \alpha)$. Now, by uniqueness of solutions, there is only one solution passing through any given point $x^I(-t_0, \alpha)$. So, for fixed I, each point on the homoclinic orbit is uniquely specified by the coordinates $(t_0, \alpha) \in \mathbb{R}^1 \times \mathbb{R}^{n-1}$. Therefore, $x^I(-t_0, \alpha)$, $(t_0, \alpha) \in \mathbb{R}^1 \times \mathbb{R}^{n-1}$, is a parametrization of the homoclinic manifold.

This completes our discussion of the consequences of I3 and I4 for the system $(4.1.1)_{0,x}$. We now want to utilize these results to describe the phase space of the unperturbed system $(4.1.1)_0$ in the full (x, I, θ) phase space.

Consider the set of points M in $\mathbb{R}^{2n} \times \mathbb{R}^m \times T^l$ defined by

$$M = \left\{ (x, I, \theta) \in \mathbb{R}^{2n} \times \mathbb{R}^m \times T^l \mid x = \gamma(I) \text{ where } \gamma(I) \text{ solves } D_x H(\gamma(I), I) = 0 \right.$$
$$\left. \text{subject to } \det\left[D_x^2 H(\gamma(I), I) \right] \neq 0, \quad \forall I \in U, \quad \theta \in T^l \right\} . \qquad (4.1.11)$$

Then we have the following proposition.

Proposition 4.1.4. M *is a* C^r $m + l$ *dimensional normally hyperbolic invariant manifold of* $(4.1.1)_0$. *Moreover,* M *has* C^r $n+m+l$ *dimensional stable and unstable manifolds denoted* $W^s(M)$ *and* $W^u(M)$, *respectively, which intersect in the* $n+m+l$ *dimensional homoclinic manifold*

$$\Gamma \equiv \left\{ \left(x^I(-t_0, \alpha), I, \theta_0 \right) \in \mathbb{R}^{2n} \times \mathbb{R}^m \times T^l \mid (t_0, \alpha, I, \theta_0) \in \mathbb{R}^1 \times \mathbb{R}^{n-1} \times U \times T^l \right\} .$$

PROOF: Using the expression for M in (4.1.11), the vector field $(4.1.1)_0$ restricted to M is given by

$$\dot{x} = 0$$
$$\dot{I} = 0 \qquad\qquad I \in U \qquad\qquad (4.1.12)$$
$$\dot{\theta} = \Omega\big(\gamma(I), I\big)$$

with the flow on M given by

$$x(t) = \gamma(I) = \text{constant}$$
$$I(t) = I = \text{constant} \qquad I \subset U . \qquad\qquad (4.1.13)$$
$$\theta(t) = \Omega\big(\gamma(I), I\big)t + \theta_0$$

From (4.1.12) and (4.1.13) it should be clear that M is an invariant manifold (with boundary) having the structure of an m parameter family of l dimensional tori. Moreover, the flow on the tori is quite simple. Trajectories either close up (i.e., are periodic) or wind densely around the torus depending on whether the equation $m_1\Omega_1\big(\gamma(I), I\big) + \cdots + m_l\Omega_l\big(\gamma(I), I\big) = 0$ does or does not have solutions for integers $m_1, \ldots, m_l$ which are not all zero. The fact that $W^s(M)$ and $W^u(M)$ have dimension $n + m + l$ is an immediate consequence of I4, and the fact that Γ is $n + m + l$ dimensional follows from Proposition 4.1.3. We now want to discuss the hyperbolicity properties of M and compute the generalized Lyapunov type numbers.

In order to simplify certain calculations let us make the coordinate change

$$u = x - \gamma(I) . \qquad\qquad (4.1.14)$$

So in the (u, I, θ) coordinate system the invariant manifold M is given by

$$M = \left\{ (u, I, \theta) \in \mathbb{R}^{2n} \times \mathbb{R}^m \times T^l \mid u = 0, \quad I \in U \right\} . \qquad\qquad (4.1.15)$$

The vector field $(4.1.1)_0$ linearized about M is given by

$$\begin{pmatrix} \delta \dot{u} \\ \delta \dot{I} \\ \delta \dot{\theta} \end{pmatrix} = \begin{pmatrix} JD_u^2 H(0,I) & 0 & 0 \\ 0 & 0 & 0 \\ D_u \Omega(0,I) & D_I \Omega(0,I) & 0 \end{pmatrix} \begin{pmatrix} \delta u \\ \delta I \\ \delta \theta \end{pmatrix} \qquad (4.1.16)$$

where δu, δI, and $\delta \theta$ represent variations about orbits on $\mathcal{M}$. From (4.1.16) we can obtain the flow generated by $(4.1.1)_0$ linearized about $\mathcal{M}$. This is given by

$$D\phi_t(0,I,\theta) = D\phi_t(0,I)$$

$$= \begin{pmatrix} \exp\left[(JD_u^2 H(0,I))t\right] & 0 & 0 \\ 0 & \mathrm{id}_m & 0 \\ D_u \Omega(0,I)\left[JD_u^2 H(0,I)\right]^{-1} \exp\left[(JD_u^2 H(0,I))t\right] & D_I \Omega(0,I)t & \mathrm{id}_l \end{pmatrix}$$

$$(4.1.17)$$

where id_m and id_l denote the $m \times m$ and $l \times l$ dimensional identity matrices, respectively.

Note that in the u, I, θ coordinates, vectors tangent to $\mathcal{M}$ have zero u component. With this in mind notice the second and third columns of (4.1.17). These two columns span $T_p\mathcal{M}$ for any $p \in \mathcal{M}$. Thus, the projection onto $T_p\mathcal{M}$ is trivial and is given by

$$D\phi_t(p)\Pi^T = \begin{pmatrix} 0 & 0 & 0 \\ 0 & \mathrm{id}_m & 0 \\ 0 & D_I \Omega(0,I)t & \mathrm{id}_l \end{pmatrix} . \qquad (4.1.18)$$

We want to decompose $T\mathbb{R}^{2n+m+l}\big|_{\mathcal{M}}$ into three subbundles. First consider the linearized equation $(4.1.1)_{0,x}$ regarding I as fixed. We have

$$\delta \dot{u} = JD_u^2 H(0,I)\delta u, \quad I \in U \qquad (4.1.19)$$

where δu is the variation from $u = 0$. Now by I4, for each $I \in U$, $\mathbb{R}^{2n}$ splits into two n dimensional subspaces $E^s(I)$, $E^u(I)$, corresponding to the stable and unstable subspaces of (4.1.19). Let "0" denote the zero vector in $\mathbb{R}^{m+l}$ and consider the following two disjoint unions

$$E^s \equiv \bigcup_{I \subset U} \left(E^s(I),0\right)$$

$$E^u \equiv \bigcup_{I \subset U} \left(E^u(I),0\right) . \qquad (4.1.20)$$

Then we have

$$TR^{2n+m+l}\big|_M = TM \oplus E^s \oplus E^u \qquad (4.1.21)$$

and if we define

$$
\begin{aligned}
N^s &= TM \oplus E^s \\
N^u &= TM \oplus E^u
\end{aligned}
\qquad (4.1.22)
$$

then it should be clear that N^s is a positively invariant subbundle under (4.1.17) and N^u is a negatively invariant subbundle under (4.1.17).

We now want to compute the generalized Lyapunov type numbers associated with N^u in the context of Theorem 1.3.6. We let E^u and E^s play the roles of the subbundles I and J in the geometrical set-up for the theorem. Using (4.1.18) and the properties of (4.1.19) given in I4 we obtain for any $p = (u, I, \theta) \in M$

$$
\begin{aligned}
\lambda(p) &= \varlimsup_{t\to -\infty} \left\| \Pi^{E^u} D\phi_t(p) \right\|^{-1/t} = e^{-\lambda_u(I)} < 1 \\
\gamma(p) &= \varlimsup_{t\to -\infty} \left\| \Pi^{E^s} D\phi_t(p) \right\|^{1/t} = e^{\lambda_s(I)} < 1 \\
\sigma(p) &= \varlimsup_{t\to -\infty} \frac{\log \left\| D\phi_t(p)\Pi^T \right\|}{\log \left\| \Pi^{E^s} D\phi_t(p) \right\|} = 0
\end{aligned}
\qquad (4.1.23)
$$

where $\lambda_u(I)$ is the smallest real part of any of the n eigenvalues of $JD_u^2 H(0, I)$ which have positive real parts, and $\lambda_s(I)$ is the largest real part of any of the n eigenvalues of $JD_u^2 H(0, I)$ which have negative real parts. Recall that by I4 we have

$$-\lambda_u(I), \lambda_s(I) < 0 \quad \forall I \in U \qquad (4.1.24)$$

and, therefore,

$$\lambda(p) < 1, \quad \gamma(p) < 1, \quad \sigma(p) = 0 \quad \forall p \in M. \qquad (4.1.25)$$

A similar calculation follows for N^s under the time reversed vector field. In this case E^s and E^u are interchanged and we obtain

$$
\begin{aligned}
\lambda(p) &= e^{\lambda_s(I)} < 1 \\
\gamma(p) &= e^{-\lambda_u(I)} < 1 \\
\sigma(p) &= 0
\end{aligned}
\qquad (4.1.26)
$$

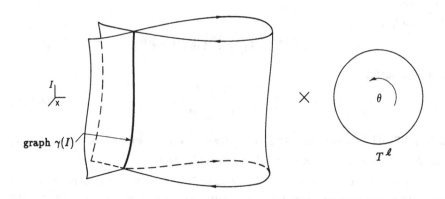

Figure 4.1.1. Unperturbed Phase Space of $(4.1.1)_0$.

and, therefore,

$$\lambda(p) < 1, \quad \gamma(p) < 1, \quad \sigma(p) = 0 \quad \forall \, p \in \mathcal{M}. \tag{4.1.27}$$

$\square$

So $\mathcal{M}$ satisfies the asymptotic stability properties for normally hyperbolic invariant manifolds described in Section 1.3. However, the perturbation theorems do not immediately apply since $\mathcal{M}$ is neither overflowing nor inflowing invariant. We will deal with this technical nuisance when we discuss the geometry of the perturbed phase space. See Figure 4.1.1 for an illustration of the unperturbed phase space of $(4.1.1)_0$.

ii) Homoclinic Coordinates

We now want to define a moving coordinate system along the homoclinic manifold Γ of the unperturbed system which will be useful for determining the splitting of the manifolds in the perturbed system.

For each $I \in U$, consider the following set of n linearly independent vectors in $\mathbb{R}^{2n} \times \mathbb{R}^m \times T^l$

$$\{(D_x H = D_x K_1, 0), (D_x K_2, 0), \ldots, (D_x K_n, 0)\} \tag{4.1.28}$$

where "0" denotes the $m + l$ dimensional zero vector. Also, we define a set of m linearly independent constant unit vectors in $\mathbb{R}^{2n} \times \mathbb{R}^m \times T^l$

$$\left\{ \hat{I}_1, \ldots, \hat{I}_m \right\} \tag{4.1.29}$$

where the $\hat{I}_i$ represent unit vectors in the I_i directions. As a convenient notation, for a given $(t_0, \alpha, I, \theta_0) \in \mathbb{R}^1 \times \mathbb{R}^{n-1} \times \mathbb{R}^m \times T^l$, let $p = \big(x^I(-t_0, \alpha), I, \theta_0\big)$ denote the corresponding point on $\Gamma = W^s(\mathcal{M}) \cap W^u(\mathcal{M}) - \mathcal{M}$. For any point $p \in \Gamma$, let Π_p denote the $m + n$ dimensional plane spanned by the vectors in (4.1.28) and (4.1.29) where the $D_x K_i$ are evaluated at p. Thus, varying p serves to move the plane Π_p along the homoclinic orbit Γ. See Figure 4.1.2.

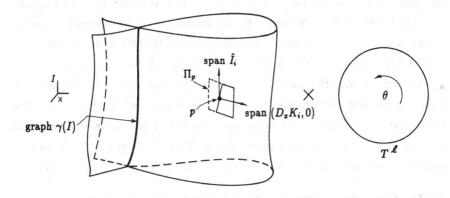

Figure 4.1.2. Geometry of Π_p.

We will be interested in how $W^s(\mathcal{M})$ and $W^u(\mathcal{M})$ intersect Π_p for each $p \in \Gamma$. In particular, we want to know the dimension of the intersection and whether or not the intersection is transversal (see Section 1.4).

Now $W^s(\mathcal{M})$ is $n + m + l$ dimensional. Therefore, for any point $p \in W^s(\mathcal{M})$, the tangent space of $W^s(\mathcal{M})$ at p, denoted $T_p W^s(\mathcal{M})$, is an $n + m + l$ dimensional linear vector space (see Section 1.3). By Proposition 4.1.2, an n dimensional vector space complementary to $T_p W^s(\mathcal{M})$ is given by

$$N_p = \operatorname{span}\big\{\big(D_x K_1(p), 0\big), \ldots, \big(D_x K_n(p), 0\big)\big\}. \tag{4.1.30}$$

So it should be evident that

$$T_p W^s(\mathcal{M}) + N_p = \mathbb{R}^{2n+m+l} \tag{4.1.31}$$

and, therefore, by Definition 1.4.1, $W^s(\mathcal{M})$ intersects Π_p transversely for all $p \in \Gamma$, since $N_p \subset \Pi_p$. Next we want to determine the dimension of the intersection.

Since the intersection is transverse, the formula for the dimension of the sum of two vector spaces gives

$$2n + m + l = \dim\left[T_p W^s(\mathcal{M})\right] + \dim T_p \Pi_p - \dim T_p\left(W^s(\mathcal{M}) \cap \Pi_p\right). \quad (4.1.32)$$

Now $\dim\left[T_p W^s(\mathcal{M})\right] = n+m+l$ and $\dim T_p \Pi_p = n+m$; therefore, $\dim T_p\left(W^s(\mathcal{M}) \cap \Pi_p\right) = m$, which implies that $W^s(\mathcal{M})$ intersects Π_p transversely in an m dimensional surface, which we refer to as S_p^s. A similar argument can be applied to conclude that $W^u(\mathcal{M})$ intersects Π_p transversely in an m dimensional surface S_p^u for each $p \in \Gamma$. Moreover, since $W^s(\mathcal{M})$ and $W^u(\mathcal{M})$ coincide along $\Gamma = W^s(\mathcal{M}) \cap W^u(\mathcal{M}) - \mathcal{M}$, we have $S_p^s = S_p^u$ for every $p \in \Gamma$.

We remark that the reason it is important to determine the dimensions of these intersections and whether or not they are transversal lies in the fact that transversal intersections persist under perturbations. This will be very important when we discuss the splitting of the manifolds. We refer the reader to Figure 4.1.3 for two possible scenarios for the intersection of $W^s(\mathcal{M})$ and $W^u(\mathcal{M})$ with Π_p.

iii) The Geometric Structure of the Perturbed Phase Space

We now describe some conclusions of a general nature that we can make concerning the structure of the phase space of the perturbed system. Recall that in the phase space of the unperturbed system we are concerned with three basic structures, the invariant manifold $\mathcal{M}$ and the stable and unstable manifolds of $\mathcal{M}$, $W^s(\mathcal{M})$ and $W^u(\mathcal{M})$, respectively. Let us now comment on the structure of each of these sets and how we might expect the structure to change under perturbation. Afterward we will give results describing the perturbed structures.

1) *The Invariant Manifold* $\mathcal{M}$. From Proposition 4.1.4, $\mathcal{M}$ is an $m+l$ dimensional normally hyperbolic invariant manifold which has the structure of an m parameter family of l dimensional tori. The flow on the tori is quite simple and is given by $\theta(t) = \Omega\left(\gamma(I), I\right)t + \theta_0$.

Now we would like to argue that most of this structure goes over for the perturbed system. However, there are two delicate points which deserve careful consideration. The first is that $\mathcal{M}$ is an invariant manifold in a very precarious sense due to the fact that it is a manifold with boundary, yet it is still invariant, since $\dot{I} = 0$, and therefore no orbits may cross the boundary. However, in the perturbed system, $\dot{I}$ need not be zero, and therefore we must consider the

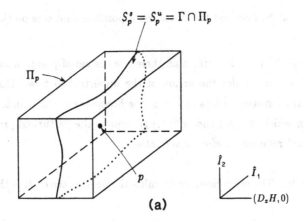

(a)

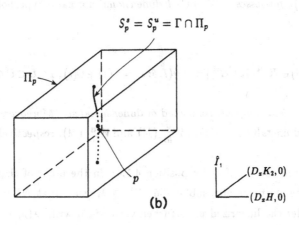

(b)

Figure 4.1.3. Geometry of the Intersection of $W^s(\mathcal{M})$, $W^u(\mathcal{M})$, and Γ with Π_p.

a) $n = 1$, $m = 2$.   b) $n = 2$, $m = 1$.

boundary more closely. Fortunately, Fenichel's invariant manifold theory will allow us to conclude that $\mathcal{M}$ persists for the perturbed system as a locally invariant manifold. The second point to be considered is the flow on $\mathcal{M}$ and how it goes over to the perturbed problem. Now an m parameter family of tori with rational or irrational flow is a very degenerate situation. Under perturbation we would expect "most" of these tori to be destroyed, and even complicated limit sets could result. We will need to develop some technique for determining the nature of the flow and the existence of possible limit sets on the perturbed manifold.

2) $W^s(M)$ and $W^u(M)$. By definition these are the set of points which approach M as $t \to \pm\infty$ under the action of the unperturbed flow. However if, in the perturbed system, orbits can cross the boundary of the manifold, then the manner in which $W^s(M)$ and $W^u(M)$ go over to the perturbed problem is not so clear and requires careful consideration.

Persistence of M. The main result concerning the persistence of M is the following.

Proposition 4.1.5. *There exists* $\epsilon_0 > 0$ *such that for* $0 < \epsilon \le \epsilon_0$ *the perturbed system* $(4.1.1)_\epsilon$ *possesses a* C^r $m + l$ *dimensional normally hyperbolic locally invariant manifold*

$$M_\epsilon = \{(x, I, \theta) \in \mathbb{R}^{2n} \times \mathbb{R}^m \times T^l \mid x = \tilde{\gamma}(I, \theta; \epsilon) = \gamma(I) + \mathcal{O}(\epsilon), I \subset \tilde{U} \subset U \subset \mathbb{R}^m, \theta \in T^l\}$$
(4.1.33)

where $\tilde{U} \subset U$ *is a compact, connected* m *dimensional set. Moreover,* M_ϵ *has local* C^r *stable and unstable manifolds,* $W^s_{loc}(M)$ *and* $W^u_{loc}(M)$, *respectively.*

PROOF: Recall the proof of Proposition 4.1.4. In the proof of that proposition we showed the existence of a subbundle $N^u \supset M$ such that N^u was negatively invariant under the linearized unperturbed vector field with $\lambda(p) < 1$, $\gamma(p) < 1$, and $\sigma(p) = 0$ for all $p \in M$. We also showed the existence of a subbundle $N^s \supset M$ such that N^s was negatively invariant under the linearized time reversed unperturbed vector field with $\lambda(p) < 1$, $\gamma(p) < 1$, and $\sigma(p) < 1$ for all $p \in M$. Moreover, $N^u \cap N^s = TM$. Now we would like to apply Theorem 1.3.6. However, there is a slight problem due to the fact that M is neither overflowing nor inflowing invariant, since the unperturbed vector field $(4.1.1)_0$ is identically zero on ∂M. This technical detail can be dealt with as follows.

Let $\tilde{U} \subset U$ be a compact m dimensional set. Choose open sets U_0 and U_i such that $\tilde{U} \subset U_0 \subset U_i \subset U$ with $\bar{U}_0 \subset U_i$. Next choose a C^∞ "bump" function

$$\omega : \mathbb{R}^m \to \mathbb{R}$$
(4.1.34)

such that

$$\omega(I) = 0 \qquad \text{for } I \in \tilde{U}$$
$$\omega(I) = 1 \qquad \text{for } I \in \partial U_0$$
$$\omega(I) = -1 \qquad \text{for } I \in \partial U_i \qquad\qquad (4.1.35)$$
$$\omega(I) = 0 \qquad \text{for } I \in \mathbb{R}^m - U$$

(see Spivak [1979]). Now consider the modified unperturbed vector field

$$\dot{x} = JD_x H(x, I)$$
$$\dot{I} = \delta\omega(I)I \qquad (x, I, \theta) \in \mathbb{R}^{2n} \times \mathbb{R}^m \times T^l \qquad (4.1.36)$$
$$\dot{\theta} = \Omega(x, I)$$

for some $\delta > 0$.

Let $\tilde{M}$, M_0, and M_i be subsets of M for which I is restricted to lie in $\tilde{U}$, U_0, and U_i, respectively. Then $\tilde{M} \subset M_0 \subset M_i \subset M$ and the following should be evident.

1) M_0 is an overflowing invariant manifold under (4.1.36) satisfying the hypotheses of Theorem 1.3.6.

2) M_i is an inflowing invariant manifold under (4.1.36) satisfying the hypotheses of Theorem 1.3.6 under the time reversed vector field.

Now, since (4.1.36) and $(4.1.1)_0$ are identical for $I \subset \tilde{U}$, it follows from Theorem 1.3.6 that the perturbed system $(4.1.1)_\epsilon$ possesses a C^r locally invariant manifold M_ϵ. Moreover, there exist locally invariant manifolds $W_{loc}^s(M_\epsilon)$ and $W_{loc}^u(M_\epsilon)$ which are C^r close to $W_{loc}^s(M)$ and $W_{loc}^u(M)$, respectively. $\qquad\square$

We make several remarks concerning Proposition 4.1.5.

1. *The Nature of* $W_{loc}^s(M_\epsilon)$ *and* $W_{loc}^u(M_\epsilon)$. M_ϵ is a locally invariant manifold, i.e., points may leave M_ϵ by crossing its boundary. We will refer to $W_{loc}^s(M_\epsilon)$ (resp. $W_{loc}^u(M_\epsilon)$) as the local stable (resp. unstable) manifold of M_ϵ. However, this terminology deserves some clarification. Normally, one defines the stable (resp. unstable) manifold of an invariant set as the set of points which are asymptotic to points on the invariant set as $t \to +\infty$ (resp. $-\infty$). This certainly need not be the case for points in $W_{loc}^s(M_\epsilon)$ (resp. $W_{loc}^u(M_\epsilon)$) since, although points in $W_{loc}^s(M_\epsilon)$ (resp. $W_{loc}^u(M_\epsilon)$) approach M_ϵ in forward (resp. backward) time, they need not actually limit on any points on M_ϵ as $t \to +\infty$ (resp. $-\infty$),

since all points on $\mathcal{M}_\epsilon$ may leave $\mathcal{M}_\epsilon$ in finite time. However, we will retain the terminology of stable (resp. unstable) manifolds when referring to $W^s_{\text{loc}}(\mathcal{M}_\epsilon)$ (resp. $W^u_{\text{loc}}(\mathcal{M}_\epsilon)$). We refer the reader to Figure 4.1.4 for an illustration of the geometry of the perturbed manifolds.

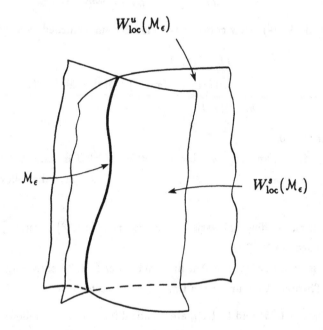

Figure 4.1.4. Geometry of the Perturbed Manifolds with the Angular Variables Suppressed for Clarity.

2. *Differentiability of the Manifolds with Respect to Parameters.* We want to show that a slight modification of the arguments given in Propositions 4.1.4 and 4.1.5 gives that $\mathcal{M}_\epsilon$, $W^s_{\text{loc}}(\mathcal{M}_\epsilon)$, and $W^u_{\text{loc}}(\mathcal{M}_\epsilon)$ are also C^r functions of ϵ and μ. Consider the vector field

$$\dot{x} = J D_x H(x, I)$$
$$\dot{I} = 0$$
$$\dot{\theta} = \Omega(x, I) \qquad (x, I, \theta, \epsilon, \mu) \in \mathbf{R}^{2n} \times \mathbf{R}^m \times T^l \times \mathbf{R} \times \mathbf{R}^p \qquad (4.1.37)$$
$$\dot{\epsilon} = 0$$
$$\dot{\mu} = 0$$

where the (x, I, θ) components of (4.1.37) satisfy the same hypotheses as before. Then the set

$$\bar{M} = \{ (x, I, \theta, \epsilon, \mu) \in \mathbb{R}^{2n} \times \mathbb{R}^m \times T^l \times \mathbb{R} \times \mathbb{R}^p \mid x = \gamma(I) \text{ where } \gamma(I) \text{ solves}$$
$$D_x H(\gamma(I), I) = 0 \quad \text{subject to } \det\left[D_x^2 H(\gamma(I), I) \right] \neq 0, \, \forall \, I \in U, \, \theta \in T^l \}$$

$$(4.1.38)$$

is an $m + l + 1 + p$ dimensional, normally hyperbolic invariant manifold having $n + m + l + 1 + p$ dimensional stable and unstable manifolds denoted $W^s(\bar{M})$ and $W^u(\bar{M})$, respectively. Generalized Lyapunov type numbers can be computed for $\bar{M}$ as in Proposition 4.1.4 and are identical to those given for $\bar{M}$. This is because the addition of ϵ and μ as new dependent variables only adds new *tangent* directions to M with rates of growth that are only linear in time. Thus, the argument given in Proposition 4.1.5 goes through identically the same in this case with the result that $\bar{M}_\epsilon$, $W^s(\bar{M}_\epsilon)$, and $W^u(\bar{M}_\epsilon)$ are C^r functions of x, I, θ, ϵ, and μ.

Dynamics on M_ϵ. Recall from our previous comments that it is possible there are no recurrent motions on M_ϵ, i.e., all orbits eventually leave M_ϵ by crossing its boundary. In this case, the dynamical consequences of intersections of $W^s(M_\epsilon)$ and $W^u(M_\epsilon)$ are not clear, but they deserve further study. In Chapter 3 we saw that there could be dramatic dynamical consequences associated with orbits homoclinic to fixed points, periodic orbits, or normally hyperbolic tori. With this in mind we would like to determine the existence of such motions on M_ϵ. Using the expression for M_ϵ given in Proposition 4.1.5, the perturbed vector field restricted to M_ϵ is given by

$$\dot{I} = \epsilon g^I(\gamma(I), I, \theta, \mu; 0) + \mathcal{O}(\epsilon^2)$$
$$\dot{\theta} = \Omega(\gamma(I), I) + \mathcal{O}(\epsilon) \qquad (I, \theta) \in \tilde{U} \times T^l. \qquad (4.1.39)$$

Let us consider the associated "averaged" equations

$$\dot{I} = \epsilon G(I) \qquad (4.1.40)$$

where $G(I) = \dfrac{1}{(2\pi)^l} \displaystyle\int_0^{2\pi} \cdots \int_0^{2\pi} g^I(\gamma(I), I, \theta, \mu; 0) \, d\theta_1 \ldots d\theta_l$.

We have the following result.

Proposition 4.1.6. *Suppose there exists* $I = \bar{I} \in \tilde{U}$ *such that*

1) *the equation*

$$m_1 \Omega_1\big(\gamma(\bar{I}), \bar{I}\big) + \cdots + m_l \Omega_l\big(\gamma(\bar{I}), \bar{I}\big) = 0 \qquad (4.1.41)$$

   has no solutions for any integers $m_1, \ldots, m_l$ *which are not all zero;*

2) *the averaged equations (4.1.40) have a hyperbolic fixed point at* $I = \bar{I}$ *with the linearized equation having* $m - j$ *eigenvalues having positive real parts and* j *eigenvalues having negative real parts.*

Then, for ϵ *sufficiently small, the vector field restricted to* $\mathcal{M}_\epsilon$, *i.e., equations (4.1.39), has a* C^r, $r \geq 3(s+1)$, l *dimensional normally hyperbolic invariant torus* $\tau_\epsilon(\bar{I})$ *having a* C^s $j + l$ *dimensional stable manifold and a* C^s $m - j + l$ *dimensional unstable manifold. (Note: by* C^r *we mean differentiability with respect to* $I, \theta, \epsilon,$ *and* μ.)

PROOF: See Arnold and Avez [1968] or Grebenikov and Ryabov [1983].   □

1) The question of smoothness of the stable and unstable manifolds in the averaged equations needs some clarification. For each fixed ϵ, $\mathcal{M}_\epsilon$ will have C^r stable and unstable manifolds. However, we will need to differentiate the manifolds with respect to ϵ at $\epsilon = 0$. In this case, the usual smoothness results do not go through. This problem was first studied by Schecter [1986], and the degree of differentiability of the stable and unstable manifolds with respect to ϵ at $\epsilon = 0$ based on the differentiability of the underlying vector field is due to him (note: his smoothness result is probably not optimal). This is the reason that, when averaging is necessary (i.e., when we have I variables in the problem with one or more frequencies $(T^l, l \geq 1)$), we must take $(4.1.1)_\epsilon$ to be at least C^6 in order to get C^1 stable and unstable manifolds for invariant sets on $\mathcal{M}_\epsilon$ found by averaging.

2) If $l = 0$ (i.e., there are no angular variables in the problem) then averaging is unnecessary and the flow on $\mathcal{M}_\epsilon$ is described by the equation

$$\dot{I} = \epsilon g^I\big(\gamma(I), I, \mu, 0\big) + \mathcal{O}(\epsilon^2). \qquad (4.1.42)$$

3) If $l = 1$ there is only one frequency and the nonresonance condition (4.1.41) is always satisfied.

4) The nonresonance requirement (4.1.41) implies that the flow on $\tau_\epsilon(\bar{I})$ is dense. Denseness of the flow on $\tau_\epsilon(\bar{I})$ will be necessary in order to show that certain improper integrals converge (see Lemma 4.1.27).

5) Suppose the perturbation is Hamiltonian as in System III. Then the vector field restricted to $\mathcal{M}_\epsilon$ and averaged over the angular variables becomes

$$G(I) = \frac{-1}{(2\pi)^l} \int_0^{2\pi} \cdots \int_0^{2\pi} D_\theta H(\gamma(I), I, \theta, \mu; 0) d\theta_1 \ldots d\theta_l = 0. \qquad (4.1.43)$$

Therefore, averaging gives no information in this case, and more sophisticated methods are needed. This is the reason why we give a separate discussion for System III.

Let us now view the l dimensional torus on $\mathcal{M}_\epsilon$ found via Proposition 4.1.6 in the context of the full $2n + m + l$ dimensional phase space.

Proposition 4.1.7. *Suppose we have* $I = \bar{I} \subset \tilde{U}$ *such that Proposition 4.1.6 is satisfied. Then* $\tau_\epsilon(\bar{I})$ *is a* C^r, $r \geq 3(s+1)$, l *dimensional normally hyperbolic invariant torus contained in* $\mathcal{M}_\epsilon$ *having a* C^s $n+j+l$ *dimensional stable manifold,* $W^s(\tau_\epsilon(\bar{I}))$, *and a* C^s $n+m-j+l$ *dimensional unstable manifold,* $W^u(\tau_\epsilon(\bar{I}))$. *Moreover,* $W^s(\tau_\epsilon(\bar{I})) \subset W^s(\mathcal{M}_\epsilon)$ *and* $W^u(\tau_\epsilon(\bar{I})) \subset W^u(\mathcal{M}_\epsilon)$.

PROOF: This is an immediate consequence of Propositions 4.1.5 and 4.1.6 and Theorem 6 in Fenichel [1974]. $\qquad\qquad\qquad\qquad\qquad\qquad\qquad\square$

Now our goal will be to determine whether or not $W^s(\tau_\epsilon(\bar{I}))$ and $W^u(\tau_\epsilon(\bar{I}))$ intersect, so it is to this that we turn our attention.

iv) The Splitting of the Manifolds

Suppose we have found an $\bar{I} \subset \tilde{U}$ such that $\tau_\epsilon(\bar{I}) \subset \mathcal{M}_\epsilon$ is an l dimensional normally hyperbolic invariant torus having an $n+j+l$ dimensional stable manifold, $W^s(\tau_\epsilon(\bar{I}))$, and an $n+m-j+l$ dimensional unstable manifold, $W^u(\tau_\epsilon(\bar{I}))$. We want to determine whether or not $W^s(\tau_\epsilon(\bar{I}))$ and $W^u(\tau_\epsilon(\bar{I}))$ intersect transversely. If this is the case, then depending on l, we can appeal to theorems from Chapter 3 and assert the existence of chaotic dynamics in the perturbed system $(4.1.1)_\epsilon$.

Let us first recall the geometry of the unperturbed system $(4.1.1)_0$.

Let $(t_0, \alpha, I, \theta_0) \in \mathbb{R}^1 \times \mathbb{R}^{n-1} \times U \times T^l$ be fixed and denote the corresponding point on $\Gamma = W^s(\mathcal{M}) \cap W^u(\mathcal{M}) - \mathcal{M}$ as $p \equiv \{x^I(-t_0, \alpha), I, \theta_0\}$. Let Π_p be the $m+n$

dimensional plane as previously defined with $W^s(M)$ and $W^u(M)$ intersecting Π_p
transversely in the m dimensional surfaces S_p^s and S_p^u, respectively, for each point
$p \in \Gamma$. Note also that Γ intersects Π_p transversely in an m dimensional surface with
$\Gamma \cap \Pi_p = S_p^s = S_p^u$ (see Figure 4.1.3).

Now we will consider the geometry of the perturbed system $(4.1.1)_\epsilon$ along Γ.
Since $W^s(M)$ and $W^u(M)$ intersect Π_p transversely for all $p \in \Gamma$, for ϵ sufficiently
small $W^s(M_\epsilon)$ and $W^u(M_\epsilon)$ intersect Π_p transversely for each $p \in \Gamma$ in the m
dimensional sets $W^s(M_\epsilon) \cap \Pi_p \equiv S_{p,\epsilon}^s$ and $W^u(M_\epsilon) \cap \Pi_p \equiv S_{p,\epsilon}^u$. However, in this
case, the sets $S_{p,\epsilon}^s$ and $S_{p,\epsilon}^u$ need not coincide. Also, it may be true that $W^s(M_\epsilon)$
and $W^u(M_\epsilon)$ intersect Π_p in a countable set of disconnected m dimensional sets
(see Figure 4.1.5). In this case, our choice of $S_{p,\epsilon}^s$ (resp. $S_{p,\epsilon}^u$) corresponds to the
set of points which is "closest" to M_ϵ in the sense of positive (resp. negative) time
of flow along $W^s(M_\epsilon)$ (resp. $W^u(M_\epsilon)$). We will elaborate more on this technical
point when we discuss the derivation of the Melnikov vector.

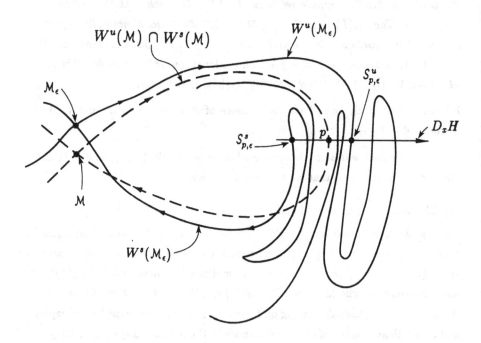

Figure 4.1.5. Sets $S_{p,\epsilon}^u$, $S_{p,\epsilon}^u$ "Closest" to M_ϵ (Note: the Figure
Is for the Case $n = 1$, $m = l = 0$).

Let us suppose we have used Proposition 4.1.6 and found an $I = \bar{I} \in \tilde{U} \subset \mathbf{R}^m$ such that $\tau_\epsilon(\bar{I}) \in \mathcal{M}_\epsilon$ is a normally hyperbolic invariant l-torus having an $n + j + l$ dimensional stable manifold, $W^s(\tau_\epsilon(\bar{I}))$ and an $n + m - j + l$ dimensional unstable manifold, $W^u(\tau_\epsilon(\bar{I}))$. Henceforth we will regard $I = \bar{I}$ as fixed. Then, since $W^s(\tau_\epsilon(\bar{I})) \subset W^s(\mathcal{M}_\epsilon)$ and $W^u(\tau_\epsilon(\bar{I})) \subset W^u(\mathcal{M}_\epsilon)$, we have $W^s(\tau_\epsilon(\bar{I})) \cap S^s_{p,\epsilon} \equiv W^s_p(\tau_\epsilon(\bar{I}))$ is a j dimensional set and $W^u(\tau_\epsilon(\bar{I})) \cap S^u_{p,\epsilon} \equiv W^u_p(\tau_\epsilon(\bar{I}))$ is an $m - j$ dimensional set. Let $p^s_\epsilon \equiv (x^s_\epsilon, I^s_\epsilon)$, and $p^u_\epsilon \equiv (x^u_\epsilon, I^u_\epsilon)$ be points in $W^s_p(\tau_\epsilon(\bar{I}))$ and $W^u_p(\tau_\epsilon(\bar{I}))$ which have the same I coordinate, i.e., $I^s_\epsilon = I^u_\epsilon$. It is always possible to choose two such points due to the fact that $\tau_\epsilon(\bar{I})$ is normally hyperbolic, implying that the angle between $W^s_{\text{loc}}(\tau_\epsilon(\bar{I}))$ and $W^u_{\text{loc}}(\tau_\epsilon(\bar{I}))$ is bounded away from zero independently of ϵ. Due to the importance of this assertion, we summarize the details of the argument in the following lemma.

Lemma 4.1.8. *For fixed* $p \in \Gamma$, *consider* $W^s_p(\tau_\epsilon(\bar{I}))$ *and* $W^u_p(\tau_\epsilon(\bar{I}))$ *as defined above. Then, for* ϵ *sufficiently small there exist two points* $p^s_\epsilon \equiv (x^s_\epsilon, I^s_\epsilon) \in W^s_p(\tau_\epsilon(\bar{I}))$ *and* $p^u_\epsilon \equiv (x^u_\epsilon, I^u_\epsilon) \in W^u_p(\tau_\epsilon(\bar{I}))$ *such that* $I^s_\epsilon = I^u_\epsilon$.

PROOF: Consider $\tau_\epsilon(\bar{I})$ restricted to $\mathcal{M}_\epsilon$. Since $\tau_\epsilon(\bar{I})$ is normally hyperbolic, the angle between the stable and unstable subspaces of the system restricted to $\mathcal{M}_\epsilon$ (equation (4.1.39)) and linearized about $\tau_\epsilon(\bar{I})$ is bounded away from zero independently of ϵ (note: the fact that the angle is independent of ϵ follows from the fact that the ϵ can be removed in the linearized system by a rescaling of time). Since the stable and unstable manifolds of $\tau_\epsilon(\bar{I})$ restricted to $\mathcal{M}_\epsilon$ are, for fixed ϵ, C^r-close to the stable and unstable subspaces of the linearized system, the stable and unstable manifolds of $\tau_\epsilon(\bar{I})$ intersect transversely at $\tau_\epsilon(\bar{I})$. Next, let us consider the stable and unstable manifolds of $\tau_\epsilon(\bar{I})$ in the full $2n + m + l$ dimensional phase space. We can view $W^s(\tau_\epsilon(\bar{I}))$ as the stable manifold of $\tau_\epsilon(\bar{I})$ restricted to $\mathcal{M}_\epsilon$ which has been carried into $W^s(\mathcal{M}_\epsilon)$ along trajectories in $W^s(\mathcal{M}_\epsilon)$ which are asymptotic to $\tau_\epsilon(\bar{I})$, similarly for $W^u(\tau_\epsilon(\bar{I}))$.

Choose $p \in \Gamma$ (with $I = \bar{I}$ fixed) to be in a sufficiently small neighborhood of $\mathcal{M}_\epsilon$ such that $W^s_{\text{loc}}(\mathcal{M}_\epsilon)$ and $W^u_{\text{loc}}(\mathcal{M}_\epsilon)$ intersect Π_p in the disjoint m dimensional sets $S^s_{p,\epsilon}$ and $S^u_{p,\epsilon}$, respectively, see Figure 4.1.6. Now $W^s_p(\tau_\epsilon(\bar{I}))$ is a j dimensional set contained in $S^s_{p,\epsilon}$, and $W^u_p(\tau_\epsilon(\bar{I}))$ is an $m - j$ dimensional set contained in $S^u_{p,\epsilon}$. Let Π^m_p denote the m dimensional subspace of Π_p spanned by the vectors $\hat{I}_i$, $i = 1, \ldots, m$, and let $W^s_{p,m}(\tau_\epsilon(\bar{I}))$ and $W^u_{p,m}(\tau_\epsilon(\bar{I}))$ denote the projections of

$W_p^s(\tau_\epsilon(\bar{I}))$ and $W_p^u(\tau_\epsilon(\bar{I}))$ onto Π_p^m. Since $(D_x K_i, 0)$, $i = 1, \ldots, n$, span an n dimensional space complementary to $T_{\bar{p}} W_{\bar{p}}^s(\tau_\epsilon(\bar{I}))$ for each $\bar{p} \in W_{\bar{p}}^s(\tau_\epsilon(\bar{I}))$, and to $T_{\bar{p}} W_{\bar{p}}^u(\tau_\epsilon(\bar{I}))$ for each $\bar{p} \in W_{\bar{p}}^u(\tau_\epsilon(\bar{I}))$, then the projections $W_{p,m}^s(\tau_\epsilon(\bar{I}))$ and $W_{p,m}^u(\tau_\epsilon(\bar{I}))$ are j and $m-j$ dimensional sets, respectively. Moreover, $W_{p,m}^s(\tau_\epsilon(\bar{I}))$ and $W_{p,m}^u(\tau_\epsilon(\bar{I}))$ must intersect transversely in at least one point, since the stable and unstable manifolds of $\tau_\epsilon(\bar{I})$ restricted to $\mathcal{M}_\epsilon$ intersect transversely at $\tau_\epsilon(\bar{I})$.

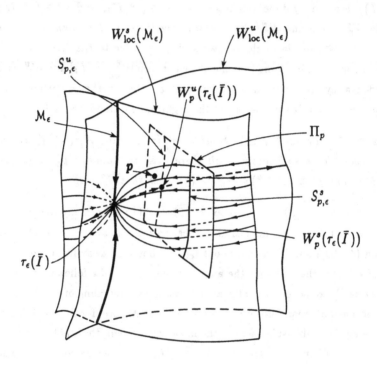

Figure 4.1.6. Intersection of $W_{\text{loc}}^s(\mathcal{M}_\epsilon)$ and $W_{\text{loc}}^u(\mathcal{M}_\epsilon)$
with Π_p, $n = 1$, $m = 1$, $l = 0$, $j = 1$.

This argument only proves the lemma for p near $\mathcal{M}_\epsilon$. Next, choose any point $p \in \Gamma$ outside of a neighborhood of $\mathcal{M}_\epsilon$. Then $S_{p,\epsilon}^s$ and $S_{p,\epsilon}^u$ are *finite time* images of $W_{\text{loc}}^s(\mathcal{M}_\epsilon)$ and $W_{\text{loc}}^u(\mathcal{M}_\epsilon)$, respectively, under the flow generated by $(4.1.1)_\epsilon$ for any $p \in \Gamma$. Then the result follows, since the angle between $W_{\text{loc}}^s(\tau_\epsilon(\bar{I}))$ and $W_{\text{loc}}^u(\tau_\epsilon(\bar{I}))$ will remain bounded away from zero under integration by finite time for ϵ sufficiently small by simple Gronwall type estimates. $\square$

We refer the reader to Figure 4.1.7 for two cases of the geometry of Lemma 4.1.8.

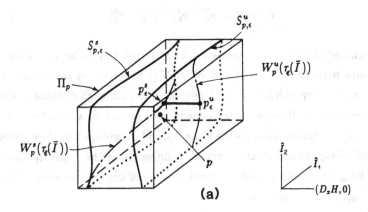

(a)

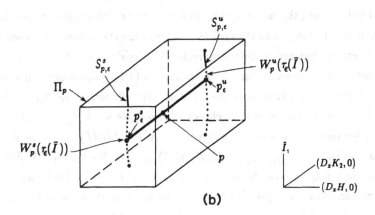

(b)

Figure 4.1.7. Geometry of Lemma 4.1.7. a) $n = 1$, $m = 2$, $j = 1$.

b) $m = 1$, $n = 2$, $j = 0$.

We remark that Lemma 4.1.8 need not be true in the case where the perturbation is Hamiltonian, since $\tau_\epsilon(\bar{I})$ would not be normally hyperbolic. This is one of the reasons why Hamiltonian systems are treated separately as System III.

We are now at the point where we can define the distance between $W^s\big(\tau_\epsilon(\bar{I})\big)$

and $W^u\big(\tau_\epsilon(\bar{I})\big)$. At any point $p \in \Gamma$, the distance between $W^s\big(\tau_\epsilon(\bar{I})\big)$ and $W^u\big(\tau_\epsilon(\bar{I})\big)$ might naively be defined as

$$d^{\bar{I}}(p,\epsilon) = |p_\epsilon^u - p_\epsilon^s| = |x_\epsilon^u - x_\epsilon^s| \tag{4.1.44}$$

with p_ϵ^u and p_ϵ^s chosen as in Lemma 4.1.8. Although this scalar measurement of the distance between $W^s\big(\tau_\epsilon(\bar{I})\big)$ and $W^u\big(\tau_\epsilon(\bar{I})\big)$ is correct, it fails to utilize the underlying geometry which we have developed, since, although distance is a scalar, the measurement of distance must be made with respect to a specific coordinate system. Our coordinate system is the plane Π_p, and we will see that the components of $p_\epsilon^u - p_\epsilon^s$ along the coordinate directions defining Π_p can be explicitly computed. The resulting vector will provide a signed measure of distance between $W^s\big(\tau_\epsilon(\bar{I})\big)$ and $W^u\big(\tau_\epsilon(\bar{I})\big)$ at the point p.

Before proceeding, let us make some additional comments concerning the geometry behind equation (4.1.44). To measure the distance between two surfaces at different points along the surfaces, the idea is to move around on the surfaces (i.e., to move in directions tangent to the surfaces) and measure the distances between them along directions that are complementary to directions tangent to the surfaces at each point. The arguments of the distance function (4.1.44) correspond to variables representing movement tangent to the manifolds (specifically, the t_0, θ_0, and α variables). Now, for $m-j$ appropriately chosen $\hat{I}_i$ vectors, denoted $\{\hat{I}_{i(1)}, \ldots, \hat{I}_{i(m-j)}\}$, an $m - j + n$ dimensional space complementary to $T_p W^u\big(\tau_\epsilon(\bar{I})\big)$ would be given by the span of $\{\hat{I}_{i(1)}, \ldots, \hat{I}_{i(m-j+1)}, (D_x K_1(p), 0), \ldots, (D_x K_n(p), 0)\}$. Similarly, for the remaining j $\hat{I}_i$ vectors, denoted $\{\hat{I}_{i(m-j+1)}, \ldots, \hat{I}_{i(m)}\}$, a $j + n$ dimensional space complementary to $T_p W^s\big(\tau_\epsilon(\bar{I})\big)$ would be given by the span of $\{\hat{I}_{i(m-j+1)}, \ldots, \hat{I}_{i(m)}, (D_x K_1(p), 0), \ldots, (D_x K_n(p), 0)\}$. Lemma 4.1.8 takes care of the necessity to measure along the I directions, since it assures us we can fix the components of distance equal to zero in these directions due to the normal hyperbolicity of $\tau_\epsilon(\bar{I})$. Therefore, in order to determine the distance between $W^s\big(\tau_\epsilon(\bar{I})\big)$ and $W^u\big(\tau_\epsilon(\bar{I})\big)$, we need only measure along the directions $(D_x K_1, 0), \ldots, (D_x K_n, 0)$.

Now our goal is to develop a computable expression for equation (4.1.44), and we do this by employing Melnikov's original trick (Melnikov [1963]). We define the signed component of the distance measurement along the directions $(D_x K_1, 0), \ldots, (D_x K_n, 0)$ as follows:

$$d_i^{\bar{I}}(p;\epsilon) = d_i^{\bar{I}}(t_0, \theta_0, \alpha, \mu; \epsilon)$$

$$\equiv \frac{\langle D_x K_i \left(x^{\bar{I}}(-t_0, \alpha), \bar{I} \right), x_\epsilon^u - x_\epsilon^s \rangle}{\left\| D_x K_i \left(x^{\bar{I}}(-t_0, \alpha), \bar{I} \right) \right\|}, \quad i = 1, \ldots, n \quad (4.1.45)$$

where we have replaced the symbol p by (t_0, θ_0, α), since for fixed $I = \bar{I}$, any point on Γ can be parametrized by $(t_0, \theta_0, \alpha) \in \mathbb{R}^1 \times T^l \times \mathbb{R}^{n-1}$, and we have also included the vector of parameters $\mu \in \mathbb{R}^p$ denoting possible parameter dependence of the perturbed vector field. It should be clear that, if for some $(t_0, \theta_0, \alpha, \mu; \epsilon)$ we have $d_i^{\bar{I}}(t_0, \theta_0, \alpha, \mu; \epsilon) = 0$ for $i = 1, \ldots, n$, then $W^s(\tau_\epsilon(\bar{I}))$ intersects $W^u(\tau_\epsilon(\bar{I}))$ at this point. At this stage it is still not clear that (4.1.45) gives us any computational advantage. However, if we Taylor expand (4.1.45) about $\epsilon = 0$ we obtain

$$d_i^{\bar{I}}(t_0, \theta_0, \alpha, \mu; \epsilon) = d_i^{\bar{I}}(t_0, \theta_0, \alpha, \mu; 0) + \epsilon \frac{\partial d_i^{\bar{I}}}{\partial \epsilon}(t_0, \theta_0, \alpha, \mu; 0) + \mathcal{O}(\epsilon^2), \quad i = 1, \ldots, n$$

$$(4.1.46)$$

where

$$d_i^{\bar{I}}(t_0, \theta_0, \alpha, \mu; 0) = 0, \quad i = 1, \ldots, n \quad \text{since } S_p^s = S_p^u$$

and

$$\frac{\partial d_i^{\bar{I}}}{\partial \epsilon}(t_0, \theta_0, \alpha, \mu; 0) = \frac{\langle D_x K_i \left(x^{\bar{I}}(-t_0, \alpha), \bar{I} \right), \frac{\partial x_\epsilon^u}{\partial \epsilon}\big|_{\epsilon=0} - \frac{\partial x_\epsilon^s}{\partial \epsilon}\big|_{\epsilon=0} \rangle}{\left\| D_x K_i \left(x^{\bar{I}}(-t_0, \alpha), \bar{I} \right) \right\|}, \quad i = 1, \ldots, n,$$

and we will shortly show that

$$\langle D_x K_i \left(x^{\bar{I}}(-t_0, \alpha), \bar{I} \right), \frac{\partial x_\epsilon^u}{\partial \epsilon}\bigg|_{\epsilon=0} - \frac{\partial x_\epsilon^s}{\partial \epsilon}\bigg|_{\epsilon=0} \rangle \equiv M_i^{\bar{I}}(\theta_0, \alpha; \mu)$$

$$= \int\limits_{-\infty}^{\infty} [\langle D_x K_i, g^x \rangle + \langle D_x K_i, (D_I J D_x H) \int^t g^I \rangle] \left(q_0^{\bar{I}}(t), \mu; 0 \right) dt,$$

$$(4.1.47a)$$

$$i = 1, \ldots, n,$$

or, equivalently,

$$M_i^{\bar{I}}(\theta_0, \alpha; \mu) = \int\limits_{-\infty}^{\infty} \left[\langle D_x K_i, g^x \rangle + \langle D_I K_i, g^I \rangle \right] \left(q_0^{\bar{I}}(t), \mu; 0 \right) dt$$

$$- \langle D_I K_i \left(\gamma(\bar{I}), \bar{I} \right), \int\limits_{-\infty}^{\infty} g^I (q_0^{\bar{I}}(t), \mu; 0) dt \rangle$$

$$(4.1.47b)$$

where $q_0^{\bar{I}}(t) = (x^I(t, \alpha), \bar{I}, \int_{}^{t} \Omega(x^I(s, \alpha) ds + \theta_0)$.

(Note: justification for the Taylor expansion of (4.1.45) follows from the fact that the manifolds vary with ϵ in a C^s $s \geq 1$ manner.)

In order to compute (4.1.47), we only need to know the unperturbed homoclinic manifold and the perturbed vector field. In particular, we do not need to compute expressions for solutions in $W^s(\tau_\epsilon(\bar{I}))$ and $W^u(\tau_\epsilon(\bar{I}))$. Since $\|D_x K_i(x^I(-t_0, \alpha), \bar{I})\| \neq 0$ on Γ, then $M_i = 0$ implies $\frac{\partial d_i}{\partial \epsilon} = 0$. So we see that M_i is essentially the leading order term in a Taylor series expansion for the distance between $W^s(\tau_\epsilon(\bar{I}))$ and $W^u(\tau_\epsilon(\bar{I}))$ along the direction $(D_x K_i(x^I(-t_0, \alpha), \bar{I}), 0)$. In honor of V. K. Melnikov we refer to the vector

$$M^{\bar{I}}(\theta_0, \alpha; \mu) \equiv \left(M_1^{\bar{I}}(\theta_0, \alpha; \mu), \dots, M_n^{\bar{I}}(\theta_0, \alpha; \mu) \right) \qquad (4.1.48)$$

as the *Melnikov vector*. We remark that the variable t_0 does not appear explicitly in the argument for the Melnikov vector. When we discuss the derivation of (4.1.47), we will show that it can be removed by a change of coordinates. This simply reflects the fact that, when invariant manifolds intersect, they must intersect along trajectories which are (at least) one dimensional. Therefore, to determine the distance between the manifolds at different points along the manifolds there is one direction in which we need not move. In reality, this fact would allow us to remove any one component of (t_0, θ_0); however, removal of the t_0 component is most convenient. More discussion of this point will appear in the section concerning the derivation of the Melnikov vector.

We are now at the point where we can state our main theorem.

Theorem 4.1.9. *Suppose there exists a point* $(\theta_0, \alpha, \mu) = (\bar{\theta}_0, \bar{\alpha}, \bar{\mu}) \in T^l \times \mathbf{R}^{n-1} \times \mathbf{R}^p$ *with* $l + n - 1 + p \geq n$ *such that*

1) $M^{\bar{I}}(\bar{\theta}_0, \bar{\alpha}; \bar{\mu}) = 0$,

2) $DM^{\bar{I}}(\bar{\theta}_0, \bar{\alpha}; \bar{\mu})$ is of rank n.

Then for ϵ sufficiently small $W^s(\tau_\epsilon(\bar{I}))$ and $W^u(\tau_\epsilon(\bar{I}))$ intersect near $(\bar{\theta}_0, \bar{\alpha}, \bar{\mu})$.

PROOF: By construction, $W^s(\tau_\epsilon(\bar{I}))$ and $W^u(\tau_\epsilon(\bar{I}))$ intersect if and only if

$$d^{\bar{I}}(t_0, \theta_0, \alpha, \mu; \epsilon) = \epsilon \left(\frac{M_1^{\bar{I}}(\theta_0, \alpha; \mu)}{\|D_x H\|}, \cdots, \frac{M_n^{\bar{I}}(\theta_0, \alpha; \mu)}{\|D_x K_n\|} \right) + \mathcal{O}(\epsilon^2) = 0 \quad (4.1.49)$$

where we leave out the arguments of the $D_x K_i$ for the sake of a more compact notation. Now consider the function

$$\tilde{d}^{\bar{I}}(\theta_0, \alpha, \mu; \epsilon) = \left(\frac{M_1^{\bar{I}}(\theta_0, \alpha; \mu)}{\|D_x H\|}, \cdots, \frac{M_n^{\bar{I}}(\theta_0, \alpha; \mu)}{\|D_x K_n\|} \right) + \mathcal{O}(\epsilon). \quad (4.1.50)$$

Then we have

$$\tilde{d}^{\bar{I}}(\bar{\theta}_0, \bar{\alpha}, \bar{\mu}; 0) = 0. \quad (4.1.51)$$

Now $DM^{\bar{I}}(\bar{\theta}_0, \bar{\alpha}; \bar{\mu})$ is of rank n. Therefore, we can find n of the variables (θ_0, α, μ), which we denote by u, such that $D_u M^{\bar{I}}(\bar{\theta}_0, \bar{\alpha}; \bar{\mu})$ is of rank n. Let v denote the remaining $l - 1 + p$ variables. Then we have

$$\det \left[D_u \tilde{d}^{\bar{I}}(\bar{\theta}_0, \bar{\alpha}, \bar{\mu}; 0) \right] = \frac{1}{\|D_x H\| \cdots \|D_x K_n\|} \det \left[D_u M^{\bar{I}}(\bar{\theta}_0, \bar{\alpha}, \bar{\mu}) \right] \neq 0. \quad (4.1.52)$$

So, by the implicit function theorem, we can find a C^r function $u = u(v, \epsilon)$ with $u_0 = u(v_0, 0)$ such that

$$\tilde{d}^{\bar{I}}(u(v, \epsilon), v, \epsilon) = 0 \quad (4.1.53)$$

for (v, ϵ) near $(v_0, 0)$, and the result follows, since $d^{\bar{I}} = \epsilon \tilde{d}^{\bar{I}}$. $\square$

From Chapter 3 we saw that often it is important to determine whether or not the intersection of $W^s(\tau_\epsilon(\bar{I}))$ and $W^u(\tau_\epsilon(\bar{I}))$ is transverse. For this we have the following theorem.

Theorem 4.1.10. *Suppose Theorem 4.1.9 holds at the point* $(\theta_0, \alpha, \mu) = (\bar{\theta}_0, \bar{\alpha}, \bar{\mu})$ $\in T^l \times \mathbf{R}^{n-1} \times \mathbf{R}^p$ *and that* $D_{(\theta_0, \alpha)} M^{\bar{I}}(\bar{\theta}_0, \bar{\alpha}, \bar{\mu})$ *is of rank* n. *Then, for* ϵ *sufficiently small,* $W^s(\tau_\epsilon(\bar{I}))$ *and* $W^u(\tau_\epsilon(\bar{I}))$ *intersect transversely near* $(\bar{\theta}_0, \bar{\alpha})$.

PROOF: Let p denote a point of intersection of $W^s(\tau_\epsilon(\bar{I}))$ and $W^u(\tau_\epsilon(\bar{I}))$. Then $T_p W^s(\tau_\epsilon(\bar{I}))$ is $n + j + l$ dimensional and $T_p W^u(\tau_\epsilon(\bar{I}))$ is $n + m - j + l$ dimensional. By Definition 1.4.1, $W^s(\tau_\epsilon(\bar{I}))$ and $W^u(\tau_\epsilon(\bar{I}))$ intersect transversely at p

if $T_p W^s(\tau_\epsilon(\bar{I})) + T_p W^u(\tau_\epsilon(\bar{I})) = \mathbb{R}^{2n+m+l}$. By the dimension formula for vector spaces we have

$$\dim\Big(T_p W^s(\tau_\epsilon(\bar{I})) + T_p W^u(\tau_\epsilon(\bar{I}))\Big) = \dim T_p W^s(\tau_\epsilon(\bar{I})) + \dim T_p W^u(\tau_\epsilon(\bar{I}))$$
$$- \dim T_p\Big(W^s(\tau_\epsilon(\bar{I})) \cap W^u(\tau_\epsilon(\bar{I}))\Big). \quad (4.1.54)$$

Thus, if $W^s(\tau_\epsilon(\bar{I}))$ intersects $W^u(\tau_\epsilon(\bar{I}))$ transversely at p, then $W^s(\tau_\epsilon(\bar{I}))$ intersects $W^u(\tau_\epsilon(\bar{I}))$ locally in an l dimensional set. Therefore, in order for $W^s(\tau_\epsilon(\bar{I}))$ and $W^u(\tau_\epsilon(\bar{I}))$ to intersect transversely at p, it is necessary and sufficient for $T_p W^s(\tau_\epsilon(\bar{I}))$ and $T_p W^u(\tau_\epsilon(\bar{I}))$ to contain $n+j$ and $n+m-j$ dimensional independent subspaces, respectively, which have no part contained in $T_p(W^s(\tau_\epsilon(\bar{I})) \cap W^u(\tau_\epsilon(\bar{I})))$.

Now let us recall the geometry of the unperturbed system. The invariant torus $\tau(\bar{I})$ has $n+l$ dimensional stable and unstable manifolds and an m dimensional center manifold. These manifolds coincide along an $n+m+l$ dimensional homoclinic manifold. Utilizing this information in the perturbed system, we need to show that the perturbation has created new independent $n+j$ and $n+m-j$ dimensional subspaces in $T_p W^s(\tau_\epsilon(\bar{I}))$ and $T_p W^u(\tau_\epsilon(\bar{I}))$, respectively, which are not contained in $T_p(W^s(\tau_\epsilon(\bar{I})) \cap W^u(\tau_\epsilon(\bar{I})))$.

Recall that, in the unperturbed system, each point along Γ can be parametrized by $(t_0, \alpha, I, \theta_0)$ (note: where I is fixed in Systems I and III), so for ϵ sufficiently small $S_{p,\epsilon}^s$ and $S_{p,\epsilon}^u$, $p \in \Gamma$, may also be parametrized by $(t_0, \alpha, I, \theta_0)$. Now let $q_\epsilon^u = (x_\epsilon^u, I_\epsilon^u, \theta_0) = q_\epsilon^s = (x_\epsilon^s, I_\epsilon^s, \theta_0) = p$ denote a point in $W^u(\tau_\epsilon(\bar{I})) \cap W^u(\tau_\epsilon(\bar{I}))$. Consider the $(2n+m+l) \times (l+n-1)$ matrices

$$A_\epsilon^u = \big(\begin{array}{cc} \frac{\partial q_\epsilon^u}{\partial \theta_0} & \frac{\partial q_\epsilon^u}{\partial \alpha} \end{array}\big)$$
$$A_\epsilon^s = \big(\begin{array}{cc} \frac{\partial q_\epsilon^s}{\partial \theta_0} & \frac{\partial q_\epsilon^s}{\partial \alpha} \end{array}\big). \quad (4.1.55)$$

The columns of these matrices represent vectors tangent to $T_p W^s(\tau_\epsilon(\bar{I}))$ and $T_p W^u(\tau_\epsilon(\bar{I}))$ along directions that were coincident in the unperturbed system, i.e., $A_0^u = A_0^s$. Consider the $n \times (2n+m+l)$ matrix

$$N = \begin{pmatrix} D_x K_1(p) & 0 \\ \vdots & \vdots \\ D_x K_n(p) & 0 \end{pmatrix}. \quad (4.1.56)$$

Since the $D_x K_i$ are independent along the homoclinic orbit, N has rank n. Consider the $n \times (l + n - 1)$ matrix

$$C_\epsilon = N(A_\epsilon^u - A_\epsilon^s). \tag{4.1.57}$$

Taylor expanding C_ϵ about $\epsilon = 0$, using the definition of the Melnikov vector (4.1.47), and using the fact that $M^I(\bar{\theta}_0, \bar{\alpha}, \bar{\mu}) = 0$, gives

$$C_\epsilon = \epsilon D_{(\theta_0, \alpha)} M^I(\bar{\theta}_0, \bar{\alpha}, \bar{\mu}) + \mathcal{O}(\epsilon^2) \tag{4.1.58}$$

where $(\bar{\theta}_0, \bar{\alpha})$ are parameters along the unperturbed homoclinic orbit corresponding to the point in $W^u(\tau_\epsilon(\bar{I})) \cap W^u(\tau_\epsilon(\bar{I}))$ given by Theorem 4.1.9. Now, by hypothesis, $D_{(\theta_0, \alpha)} M^I(\bar{\theta}_0, \bar{\alpha}, \bar{\mu})$ has rank n. So for ϵ sufficiently small C_ϵ also has rank n. Then, since N has rank n, we must have that $A_\epsilon^u - A_\epsilon^s$ is of rank n. This indicates that A_ϵ^u and A_ϵ^s each contain n linearly independent columns which correspond to n independent vectors in $T_p W^u(\tau_\epsilon(\bar{I}))$ and $T_p W^s(\tau_\epsilon(\bar{I}))$. Moreover, these vectors are not in $T_p(W^s(\tau_\epsilon(\bar{I})) \cap W^u(\tau_\epsilon(\bar{I})))$, since the rows of N span an n dimensional subspace complementary to $T_p(W^s(\tau_\epsilon(\bar{I})) \cap W^u(\tau_\epsilon(\bar{I})))$. The remaining j independent vectors in $T_p W^s(\tau_\epsilon(\bar{I}))$ and $m - j$ independent vectors in $T_p W^u(\tau_\epsilon(\bar{I}))$ which are not in $T_p(W^s(\tau_\epsilon(\bar{I})) \cap W^u(\tau_\epsilon(\bar{I})))$ come from the breakup of the m dimensional center manifold of the unperturbed system, cf. Lemma 4.1.8.   □

We make the following remarks regarding Theorem 4.1.10.

1. Theorem 4.1.10 provides only a sufficient condition for transversality. This can be seen from (4.1.57) and (4.1.58). The jacobian of the Melnikov vector is the leading order term in the projection onto a *particular* complementary n dimensional subspace of part of the difference of $T_p W^s(\tau_\epsilon(\bar{I}))$ and $T_p W^u(\tau_\epsilon(\bar{I}))$. Thus, the rank depends on the particular projection, since $A_\epsilon^u - A_\epsilon^s$ may be of rank n, but $N(A_\epsilon^u - A_\epsilon^s)$ may have rank less than n.

2. Note that if $l = 0$ then transversality is impossible. This is due to the fact that, in this case, the invariant set is a hyperbolic fixed point, and it is impossible for its stable and unstable manifolds to intersect transversely since they are constrained by uniqueness of solutions to intersect along a one dimensional trajectory.

4.1b. System II

The second type of system we will consider has the following form:

$$
\begin{aligned}
\dot{x} &= J D_x H(x, I) + \epsilon g^x(x, I, \theta, \mu; \epsilon) \\
\dot{I} &= \epsilon g^I(x, I, \theta, \mu; \epsilon) \\
\dot{\theta} &= \Omega(x, I) + \epsilon g^\theta(x, I, \theta, \mu; \epsilon)
\end{aligned}
\qquad (4.1.59)_\epsilon
$$

with $0 < \epsilon \ll 1$, $(x, I, \theta) \in \mathbb{R}^{2n} \times T^m \times T^l$, and $\mu \in \mathbb{R}^p$ is a vector of parameters. Additionally, we will assume:

II1. Let $V \subset \mathbb{R}^{2n}$ and $W \subset \mathbb{R}^p \times \mathbb{R}$ be open sets; then the functions

$$
\begin{aligned}
J D_x H &: V \times T^m \mapsto \mathbb{R}^{2n} \\
g^x &: V \times T^m \times T^l \times W \mapsto \mathbb{R}^{2n} \\
g^I &: V \times T^m \times T^l \times W \mapsto \mathbb{R}^m \\
\Omega &: V \times T^m \mapsto \mathbb{R}^l \\
g^\theta &: V \times T^m \times T^l \times W \mapsto \mathbb{R}^l
\end{aligned}
$$

are defined and C^r, $r \geq 2$.

II2. $H = H(x, I)$ is a scalar valued function which can be thought of as an m parameter family of Hamiltonians that is periodic of period 2π in each component of the I variable for x fixed. J is the $2n \times 2n$ symplectic matrix defined by

$$
J = \begin{pmatrix} 0 & \text{Id} \\ -\text{Id} & 0 \end{pmatrix}
$$

where Id denotes the $n \times n$ identity matrix and 0 denotes the $n \times n$ zero matrix.

We will refer to $(4.1.59)_\epsilon$ as the perturbed system.

i) The Geometric Structure of the Unperturbed Phase Space

The system obtained by setting $\epsilon = 0$ in $(4.1.59)_\epsilon$ will be referred to as the unperturbed system.

$$
\begin{aligned}
\dot{x} &= J D_x H(x, I) \\
\dot{I} &= 0 \\
\dot{\theta} &= \Omega(x, I) .
\end{aligned}
\qquad (4.1.59)_0
$$

We have the following two structural assumptions on the x-component of $(4.1.59)_0$.

II3. For each $I \in T^m$ the system

$$\dot{x} = JD_x H(x, I) \qquad (4.1.59)_{0,x}$$

is a completely integrable Hamiltonian system, i.e., there exist n scalar valued functions of (x, I), $H = K_1, \ldots, K_n$ which satisfy the following two conditions.

1) The set of vectors $D_x K_1, D_x K_2, \ldots, D_x K_n$ is pointwise linearly independent $\forall I \in T^m$ at all points of $\mathbb{R}^{2n}$ which are not fixed points of $(4.1.59)_{0,x}$.

2) $\langle JD_x K_i, K_j \rangle = 0 \quad \forall i, j,\ I \in T^m$ where $\langle \cdot, \cdot \rangle$ is the usual Euclidean inner product.

Furthermore, we assume that the K_i are C^r, $r \geq 2$.

The reader should compare II3 with I3 in our discussion of System I.

II4. For every $I \in T^m$ $(4.1.59)_{0,x}$ possesses a hyperbolic fixed point which varies smoothly with I and has an n dimensional homoclinic manifold connecting the fixed point to itself. We will assume that trajectories along the homoclinic manifold can be represented in the form $x^I(t, \alpha)$ where $t \in \mathbb{R}$, $\alpha \in \mathbb{R}^{n-1}$.

At this point, the reader should review the discussion of the geometrical consequences of I1–I4 in our discussion of System I. Much of the same follows in this case. In particular, consider the set of points $\mathcal{M} \subset \mathbb{R}^{2n} \times T^m \times T^l$ defined by

$$\mathcal{M} = \left\{ (x, I, \theta) \in \mathbb{R}^{2n} \times T^m \times T^l \mid x = \gamma(I) \text{ where } \gamma(I) \text{ solves } D_x H(\gamma(I), I) = 0 \right.$$
$$\left. \text{subject to } \det \left[D_x^2 H(\gamma(I), I) \right] \neq 0,\ I \in T^m,\ \theta \in T^l \right\}; \qquad (4.1.60)$$

then we have the following proposition.

Proposition 4.1.11. $\mathcal{M}$ is a C^r $m + l$ dimensional normally hyperbolic invariant manifold of $(4.1.59)_0$. Moreover, $\mathcal{M}$ has C^r $n + m + l$ dimensional stable and unstable manifolds denoted $W^s(\mathcal{M})$ and $W^u(\mathcal{M})$, respectively, which intersect in the $n + m + l$ dimensional homoclinic manifold

$$\Gamma = \left\{ (x^I(-t_0, \alpha), I, \theta_0) \in \mathbb{R}^{2n} \times T^m \times T^l \mid (t_0, \alpha, I, \theta_0) \in \mathbb{R}^1 \times \mathbb{R}^{n-1} \times T^m \times T^l \right\}.$$

PROOF: The proof is identical to the proof of Proposition 4.1.3.   □

Let us make several comments regarding the structure of M.

1. M is a boundaryless manifold. This eliminates the technicalities which were encountered in System I when proving that M persists under perturbation.

2. M has the structure of an $m + l$ dimensional torus with the flow on the torus given by

$$I(t) = I = \text{constant}$$
$$\theta(t) = \Omega(\gamma(I), I)t + \theta_0 \qquad (I, \theta) \in T^m \times T^l. \qquad (4.1.61)$$

See Figure 4.1.8 for an illustration of the geometry of the phase space of $(4.1.59)_0$.

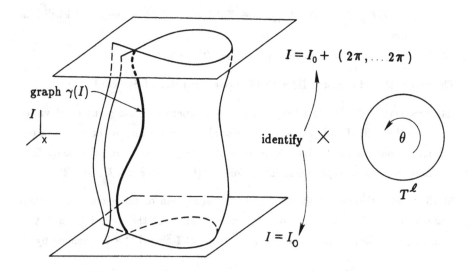

Figure 4.1.8. Unperturbed Phase Space of $(4.1.59)_0$.

ii) Homoclinic Coordinates

We define a moving system of homoclinic coordinates for System II in exactly the same way as we did for System I.

We consider the following $n + m$ linearly independent vectors

$$\{(D_x H = D_x K_1, 0), \cdots, (D_x K_n, 0)\} \qquad (4.1.62)$$

and

$$\left\{ \hat{I}_1, \ldots, \hat{I}_m \right\} \tag{4.1.63}$$

where the $D_x K_i$, $i = 1, \ldots, n$, are linearly independent for each $I \in T^m$ by II3 (except possibly at fixed points of $(4.1.59)_0$), "0" represents the $m + l$ dimensional zero vector, and $\hat{I}_i$, $i = 1, \ldots, m$, represent constant unit vectors in the I_i, $i = 1, \ldots, m$ directions. For a given $(t_0, \alpha, I, \theta_0) \in \mathbb{R}^1 \times \mathbb{R}^{n-1} \times T^m \times T^l$ we let $p = \left(x^I(-t_0, \alpha), I, \theta_0 \right)$ denote the corresponding point on $\Gamma = W^s(\mathcal{M}) \cap W^u(\mathcal{M}) - \mathcal{M}$. Then Π_p is defined to be the $m + n$ dimensional plane spanned by (4.1.62) and (4.1.63) where the $D_x K_i$ are evaluated at p. As in System I, we will be interested in the nature of the intersection of $W^s(\mathcal{M})$ and $W^u(\mathcal{M})$ with Π_p.

Using arguments identical to those given for System I, it is easy to see that $W^s(\mathcal{M})$ and $W^u(\mathcal{M})$ intersect Π_p transversely in an m dimensional manifold for each $p \in \Gamma$. We denote the intersection of $W^s(\mathcal{M})$ (resp. $W^u(\mathcal{M})$) with Π_p by S_p^s (resp. S_p^u). Moreover, we have $\Pi_p \cap \Gamma = S_p^s = S_p^u$.

See Figure 4.1.9 for an illustration of the geometry (note the similarity with Figure 4.1.2). Recall that the importance of determining whether or not the intersections are transversal lies in the fact that transversal intersections persist under small perturbation, and this fact is useful in determining the nature of the intersection of the manifolds in the perturbed system.

iii) The Geometric Structure of the Perturbed Phase Space

We now want to describe some general conclusions concerning the structure of the perturbed phase space which are due to the normal hyperbolicity of $\mathcal{M}$. There are fewer complications along these lines than in System I due to the fact that $\mathcal{M}$ is boundaryless. We will point out this fact as we go along.

The main result is the following.

Proposition 4.1.12. *There exists $\epsilon_0 > 0$ such that for $0 < \epsilon \leq \epsilon_0$ the perturbed system $(4.1.59)_\epsilon$ possesses a C^r $m + l$ dimensional normally hyperbolic invariant manifold*

$$\mathcal{M} = \left\{ (x, I, \theta) \in \mathbb{R}^{2n} \times T^m \times T^l \mid x = \tilde{\gamma}(I, \theta; \epsilon) = \gamma(I) + O(\epsilon) , \; I \in T^m, \; \theta \in T^l \right\} .$$

Moreover, $\mathcal{M}_\epsilon$ has local C^r stable and unstable manifolds, $W^s_{loc}(\mathcal{M}_\epsilon)$ and $W^u_{loc}(\mathcal{M}_\epsilon)$, which are of the same dimension and C^r close to $W^s_{loc}(\mathcal{M})$ and $W^u_{loc}(\mathcal{M})$, respectively.

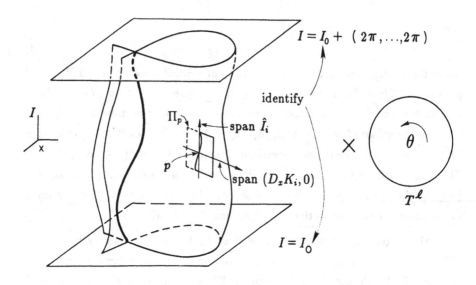

Figure 4.1.9. Homoclinic Coordinates.

PROOF: Due to the fact that M is boundaryless the technical modifications required in Proposition 4.1.4 are unnecessary. Thus the result follows immediately from Proposition 4.1.10 and Theorem 1.3.7.   □

Let us now make several remarks regarding Proposition 4.1.11 and its geometrical consequences.

1. *The Structure of M_ϵ and the Flow on M_ϵ*

M_ϵ has the structure of an $m+l$ dimensional torus. The vector field restricted to M_ϵ is given by

$$\dot{I} = \epsilon g^I(\gamma(I), I, \theta, \mu; 0) + \mathcal{O}(\epsilon^2)$$
$$\dot{\theta} = \Omega(\gamma(I), I) + \mathcal{O}(\epsilon). \tag{4.1.64}$$

In general, the flow on M_ϵ is unknown and could involve complicated limit sets such as Smale horseshoes or resonance phenomena amongst the different frequencies.

In System I it was necessary for us to first locate an invariant torus on M_ϵ, since M_ϵ was only a locally invariant manifold and not an invariant torus. This was accomplished via an averaging technique. This technique is unnecessary for System II, since M_ϵ is itself an invariant torus.

Our analysis is insensitive to the dynamics on M_ϵ in that there may be additional dynamical phenomena associated with limit sets or resonance phenomena on the torus. Such questions deserve further study.

2. *Differentiability of the Manifolds with Respect to Parameters*

Following the manner described in comment 2 after the proof of Proposition 4.1.5, ϵ and μ can be included explicitly as dependent variables in order to show that M_ϵ, $W^s_{\text{loc}}(M_\epsilon)$, and $W^u_{\text{loc}}(M_\epsilon)$ are C^r in ϵ and μ.

Our goal will be to determine whether or not $W^s(M_\epsilon)$ and $W^u(M_\epsilon)$ intersect. The motivation for this comes from Chapter 3, where we saw that orbits homoclinic to tori can often be the underlying mechanism for deterministic chaos. We emphasize that the term "torus" is used in a general sense. For $l = m = 0$, M_ϵ is a fixed point (0-torus), for $l = 1$, $m = 0$ or $m = 1$, $l = 0$, M_ϵ is a periodic orbit (1-torus), and for $m + l \geq 2$, M_ϵ is a torus with a nontrivial flow, and the dynamical consequences of homoclinic orbits are different in each case.

iv) The Splitting of the Manifolds

We now want to describe the geometry associated with our measure of the splitting of $W^s(M_\epsilon)$ and $W^u(M_\epsilon)$. This situation is less complicated than for System I, since in System I we were measuring the splitting of the stable and unstable manifolds of an invariant torus on M_ϵ which were contained in $W^s(M_\epsilon)$ and $W^u(M_\epsilon)$, respectively. In the present situation the whole of M_ϵ is the relevant invariant torus, and this simplifies the geometry.

Let us first recall the geometry of the unperturbed system $(4.1.59)_0$. Let $(t_0, \alpha, I, \theta_0) \in \mathbb{R}^1 \times \mathbb{R}^{n-1} \times T^m \times T^l$ be fixed and denote the corresponding point on Γ as $p \equiv \left(x^I(-t_0, \alpha), I, \theta_0\right)$. Let Π_p be the $m + n$ dimensional plane as previously defined, with $W^s(M)$ and $W^u(M)$ intersecting Π_p transversely in the m dimensional surfaces S^s_p and S^u_p, respectively, for each point $p \in \Gamma = W^s(M) \cap W^u(M) - M$ where $S^s_p = S^s_u$ (see Figure 4.1.10).

Now we will consider the geometry of the perturbed system $(4.1.59)_\epsilon$ along Γ. Since $W^s(M)$ and $W^u(M)$ intersect Π_p transversely for all $p \in \Gamma$, for ϵ sufficiently small $W^s(M_\epsilon)$ and $W^u(M_\epsilon)$ intersect Π_p transversely for each $p \in \Gamma$ in the m dimensional sets $W^s(M_\epsilon) \cap \Pi_p \equiv S^s_{p,\epsilon}$ and $W^u(M_\epsilon) \cap \Pi_p \equiv S^u_{p,\epsilon}$. However, in this case the sets $S^s_{p,\epsilon}$ and $S^u_{p,\epsilon}$ need not coincide. Also, it may be true that $W^s(M_\epsilon)$ and $W^u(M_\epsilon)$ intersect Π_p in a countable set of disconnected m dimensional sets as was discussed for System I (see Figure 4.1.5). In this case our choice of $S^s_{p,\epsilon}$ (resp. $S^u_{p,\epsilon}$)

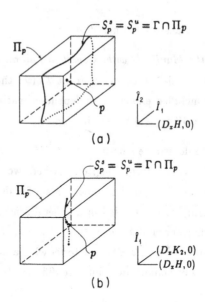

Figure 4.1.10. Intersection of $W^s(M)$ and $W^u(M)$ with Π_p.

a) $m = 2, n = 1$.   b) $m = 1, n = 2$.

corresponds to the set of points which is "closest" to M_ϵ in the sense of positive (resp. negative) time of flow along $W^s(M_\epsilon)$ (resp. $W^u(M_\epsilon)$). We will elaborate more on this technical point when we discuss the derivation of the Melnikov vector for System II.

We are now at the point where we can define the distance between $W^s(M_\epsilon)$ and $W^u(M_\epsilon)$. Let $p^s_\epsilon = (x^s_\epsilon, I^s_\epsilon)$ and $p^u_\epsilon = (x^u_\epsilon, I^u_\epsilon)$ be points in $S^s_{p,\epsilon}$ and $S^u_{p,\epsilon}$, respectively, *which have the same I coordinate*, i.e., $I^s_\epsilon = I^u_\epsilon$. We remark that, unlike the situation in System I, such a choice of points is no problem in this case. Then the distance between $W^s(M_\epsilon)$ and $W^u(M_\epsilon)$ at any point $p \in \Gamma$ may naively be defined as

$$d(p, \epsilon) = |x^u_\epsilon - x^s_\epsilon| . \qquad (4.1.65)$$

See Figure 4.1.11 for an illustration of the geometry.

However, we will develop a computable expression for (4.1.65) which utilizes the underlying geometry of the distance between $W^s(M_\epsilon)$ and $W^u(M_\epsilon)$ (cf. the discussion in 4.1a, iv). So, as in the case of System I, we define the signed component of the distance along the directions $(D_x K_1, 0), \cdots, (D_x K_n, 0)$ which are

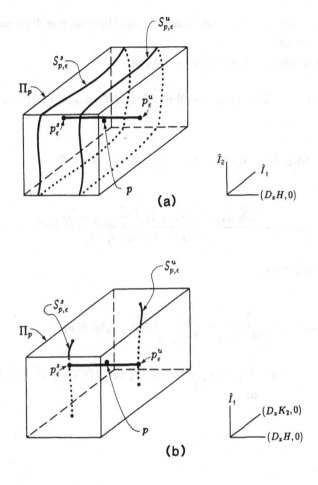

Figure 4.1.11. Intersection of $W^s(\mathcal{M}_\epsilon)$ and $W^u(\mathcal{M}_\epsilon)$ with Π_p.

a) $m = 2$, $n = 1$.   b) $m = 1$, $n = 2$.

complementary to the tangent spaces of the manifolds as follows.

$$d_i(p, \epsilon) = d_i(t_0, I, \theta_0, \alpha, \mu; \epsilon) = \frac{\left\langle D_x K_i\left(x^I(-t_0, \alpha), I\right), x_\epsilon^u - x_\epsilon^s\right\rangle}{\left\| D_x K_i\left(x^I(-t_0, \alpha), I\right)\right\|}, \quad i = 1, \ldots, n$$

$$(4.1.66)$$

where we have replaced the symbol p by $(t_0, I, \theta_0, \alpha) \in \mathbb{R}^1 \times T^m \times T^l \times \mathbb{R}^{n-1}$, since any point on Γ can be parametrized by $(t_0, I, \theta_0, \alpha)$, and we have also included the vector of parameters $\mu \in \mathbb{R}^p$, indicating the possible parameter dependence of the perturbed vector field.

Taylor expanding (4.1.66) about $\epsilon = 0$ gives

$$d_i(t_0, I, \theta_0, \alpha, \mu; \epsilon) = d_i(t_0, I, \theta_0, \alpha, \mu; 0) + \epsilon \frac{\partial d_i}{\partial \epsilon}(t_0, I, \theta_0, \alpha, \mu; 0) + \mathcal{O}(\epsilon^2), \quad i = 1, \ldots, n$$

$$(4.1.67)$$

where

$$d_i(t_0, I, \theta_0, \alpha, \mu; 0) = 0, \quad i = 1, \ldots, n \quad \text{since } S_p^s = S_p^u$$

and

$$\frac{\partial d_i}{\partial \epsilon}(t_0, I, \theta_0, \alpha, \mu; 0) = \frac{\langle D_x K_i(x^I(-t_0, \alpha), I), \frac{\partial x_\epsilon^u}{\partial \epsilon}\big|_{\epsilon=0} - \frac{\partial x_\epsilon^s}{\partial \epsilon}\big|_{\epsilon=0}\rangle}{\|D_x K_i(x^I(-t_0, \alpha), I)\|}, \quad i = 1, \ldots, n.$$

$$(4.1.68)$$

We will shortly show that

$$\langle D_x K_i(x^I(-t_0, \alpha), I), \frac{\partial x_\epsilon^u}{\partial \epsilon}\bigg|_{\epsilon=0} - \frac{\partial x_\epsilon^s}{\partial \epsilon}\bigg|_{\epsilon=0}\rangle \equiv M_i(I, \theta_0, \alpha; \mu)$$

$$= \int_{-\infty}^{\infty} [\langle D_x K_i, g^x\rangle + \langle D_x K_i, (D_I J D_x H)\int^t g^I\rangle](q_0^I(t), \mu; 0)\,dt ,$$

$$(4.1.69a)$$

$$i = 1, \ldots, n$$

or, equivalently,

$$M_i(I, \theta_0, \alpha; \mu) = \int_{-\infty}^{\infty} \left[\langle D_x K_i, g^x\rangle + \langle D_I K_i, g^I\rangle\right](q_0^I(t), \mu; 0)\,dt$$

$$-\langle D_I K_i(\gamma(I), I), \int_{-\infty}^{\infty} g^I(q_0^I(t), \mu; 0)\,dt\rangle$$

$$(4.1.69b)$$

where $q_0^I(t) \equiv (x^I(t, \alpha), I, \int^t \Omega(x^I(s, \alpha), I)\,ds + \theta_0)$,
and we define the n vector

$$M(I, \theta_0, \alpha; \mu) = (M_1(I, \theta_0, \alpha; \mu), \ldots, M_n(I, \theta_0, \alpha; \mu))$$

$$(4.1.70)$$

to be the *Melnikov vector*.

At this point we want to make several remarks concerning the geometry of our measurement of the distance between $W^s(\mathcal{M}_\epsilon)$ and $W^u(\mathcal{M}_\epsilon)$ and make some comparisons with System I.

1. As in System I, we do not explicitly show the variable t_0 in the argument of the Melnikov vector; this is because we will eliminate t_0 via a change of coordinates when we discuss the derivation of M. Our ability to do this arises from the fact that M measures the distance between manifolds of trajectories and, therefore, by uniqueness of solutions, if the manifolds intersect, they must intersect along (at least) one dimensional orbits. Hence, there is one direction along the manifolds in which we need not move in order to examine the distance between the manifolds. More details on this point will be given when we discuss the derivation of the Melnikov vector.

2. We do not need to solve the perturbed equations $(4.1.59)_\epsilon$ in order to compute M. We only need to know the unperturbed homoclinic manifold and the perturbed vector field.

3. $W^s(\mathcal{M}_\epsilon)$ and $W^u(\mathcal{M}_\epsilon)$ are codimension n manifolds, and the M_i, $i = 1, \ldots, n$, represent (to $\mathcal{O}(\epsilon^2)$) measurements along the n independent directions $(D_x K_i, 0)$, $i = 1, \ldots, n$, complementary to the manifolds.

4. In System II the I variables appear as explicit arguments of the Melnikov vector, as opposed to the situation for System I, where it was necessary to fix I in order to locate an invariant torus on $\mathcal{M}_\epsilon$. For System II, $\mathcal{M}_\epsilon$ is entirely an invariant torus and the I variables are part of the parametrization of the torus.

Now, by construction, if $d_i(p, \epsilon) = 0$, $i = 1, \ldots, n$, then $W^s(\mathcal{M}_\epsilon)$ and $W^u(\mathcal{M}_\epsilon)$ intersect near p. We now state our main theorem.

Theorem 4.1.13. *Suppose there exists a point* $(I, \theta_0, \alpha, \mu) = (\bar{I}, \bar{\theta}_0, \bar{\alpha}, \bar{\mu}) \in T^m \times T^l \times \mathbf{R}^{n-1} \times \mathbf{R}^p$ *with* $m + l + n - 1 + p \geq n$ *such that*

   1) $M(\bar{I}, \bar{\theta}_0, \bar{\alpha}, \bar{\mu}) = 0$.

   2) $DM(\bar{I}, \bar{\theta}_0, \bar{\alpha}, \bar{\mu})$ *is of rank* n.

Then for ϵ *sufficiently small* $W^s(\mathcal{M}_\epsilon)$ *and* $W^u(\mathcal{M}_\epsilon)$ *intersect near* $(\bar{I}, \bar{\theta}_0, \bar{\alpha}, \bar{\mu})$.

PROOF: The proof is identical to that of Theorem 4.1.9. $\square$

A sufficient condition for the intersection of $W^s(\mathcal{M}_\epsilon)$ and $W^u(\mathcal{M}_\epsilon)$ is given in the following theorem.

Theorem 4.1.14. *Suppose Theorem 4.1.13 holds at the point* $(I, \theta_0, \alpha, \mu) = (\bar{I}, \bar{\theta}_0, \bar{\alpha}, \bar{\mu}) \in T^m \times T^l \times \mathbb{R}^{n-1} \times \mathbb{R}^p$ *and that* $D_{(I,\theta_0,\alpha)} M(\bar{I}, \bar{\theta}_0, \bar{\alpha}, \bar{\mu})$ *is of rank* n. *Then for* ϵ *sufficiently small,* $W^s(\mathcal{M}_\epsilon)$ *and* $W^u(\mathcal{M}_\epsilon)$ *intersect transversely near* $(\bar{I}, \bar{\theta}_0, \bar{\alpha})$.

PROOF: The proof is very similar to the proof of Theorem 4.1.10, but the geometry of System II gives rise to some slight differences. Let $p \in W^s(\mathcal{M}_\epsilon) \cap W^u(\mathcal{M}_\epsilon)$. Then $T_p W^s(\mathcal{M}_\epsilon)$ and $T_p W^u(\mathcal{M}_\epsilon)$ are $n + m + l$ dimensional and, by Definition 1.4.1, $W^s(\mathcal{M}_\epsilon)$ and $W^u(\mathcal{M}_\epsilon)$ intersect transversely at p if $T_p W^s(\mathcal{M}_\epsilon) + T_p W^u(\mathcal{M}_\epsilon) = \mathbb{R}^{2n+m+l}$. So we need to show that $T_p W^s(\mathcal{M}_\epsilon)$ and $T_p W^u(\mathcal{M}_\epsilon)$ each contain n dimensional independent subspaces which have no part contained in $T_p(W^s(\mathcal{M}_\epsilon) \cap W^u(\mathcal{M}_\epsilon))$.

Let us recall the geometry of the unperturbed system. In this case, $\mathcal{M}$ is an $m + l$ dimensional invariant torus having $n + m + l$ dimensional stable and unstable manifolds which coincide along an $n + m + l$ dimensional homoclinic manifold. So we need to show that, in $T_p W^s(\mathcal{M}_\epsilon)$ and $T_p W^u(\mathcal{M}_\epsilon)$, new independent n dimensional subspaces are created which are not contained in $T_p(W^s(\mathcal{M}_\epsilon) \cap W^u(\mathcal{M}_\epsilon))$. The remaining part of the argument proceeds exactly as in the latter part of Theorem 4.1.10. □

4.1c. System III

We will now consider Hamiltonian perturbations of completely integrable Hamiltonian systems. These systems have the form

$$\dot{x} = J D_x H(x, I) + \epsilon J D_x \tilde{H}(x, I, \theta, \mu; \epsilon)$$
$$\dot{I} = -\epsilon D_\theta \tilde{H}(x, I, \theta, \mu; \epsilon) \qquad (4.1.71)_\epsilon$$
$$\dot{\theta} = D_I H(x, I) + \epsilon D_I \tilde{H}(x, I, \theta, \mu; \epsilon)$$

with $0 < \epsilon \ll 1$, $(x, I, \theta) \in \mathbb{R}^{2n} \times \mathbb{R}^m \times T^m$, and $\mu \in \mathbb{R}^p$ is a vector of parameters. Additionally, we will assume

III1. Let $V \subset \mathbb{R}^{2n} \times \mathbb{R}^m$ and $W \subset \mathbb{R}^{2n} \times \mathbb{R}^m \times \mathbb{R}^p \times \mathbb{R}$ be open sets; then the functions

$$H : V \to \mathbb{R}^1$$
$$\tilde{H} : W \times T^m \to \mathbb{R}^1$$

are defined and C^{r+1}, $r \geq 2m + 2$.

III2. J is the $2n \times 2n$ symplectic matrix defined by

$$J = \begin{pmatrix} 0 & \text{Id} \\ -\text{Id} & 0 \end{pmatrix}$$

where Id denotes the $n \times n$ zero matrix.

We will refer to $(4.1.71)_\epsilon$ as the perturbed system.

i) The Geometric Structure of the Unperturbed Phase Space

The system obtained by setting $\epsilon = 0$ in $(4.1.71)_\epsilon$ will be referred to as the unperturbed system.

$$\dot{x} = JD_x H(x, I)$$
$$\dot{I} = 0 \qquad\qquad (4.1.71)_0$$
$$\dot{\theta} = D_I H(x, I) .$$

We have the following two structural assumptions on the x-component of $(4.1.71)_0$.

III3. There exists an open set $U \subset \mathbb{R}^m$ such that for each $I \subset U$ the system

$$\dot{x} = JD_x H(x, I) \qquad\qquad (4.1.71)_{0,x}$$

is a completely integrable Hamiltonian system, i.e., there exist n scalar valued functions of (x, I), $H = K_1, K_2, \ldots, K_n$, which satisfy the following two conditions.

1) The set of vectors $D_x K_1, D_x K_2, \ldots, D_x K_n$ is pointwise linearly independent $\forall\, I \in U$ at all points of $\mathbb{R}^{2n}$ which are not fixed points of $(4.1.71)_{0,x}$.

2) $\langle JD_x K_i, K_j \rangle = 0 \ \forall i,j, \ I \in U \subset \mathbb{R}^m$ where $\langle \cdot, \cdot \rangle$ is the usual Euclidean inner product. Furthermore, we assume that the K_i are C^r, $r \geq 2m + 2$.

III4. For every $I \in U$, $(4.1.71)_{0,x}$ possesses a hyperbolic fixed point which varies smoothly with I and an n-dimensional homoclinic manifold connecting the fixed point to itself. We will assume that trajectories along the homoclinic manifold can be represented in the form $x^I(t, \alpha)$ where $t \in \mathbb{R}^1$ and $\alpha \in \mathbb{R}^{n-1}$.

At this point, the reader should note that the assumptions on the unperturbed structure of System III are identical to those for System I. The differences will occur when we consider the perturbed systems.

Analogous to System I and System II we consider the set of points $\mathcal{M}$ in $\mathbb{R}^{2n} \times \mathbb{R}^m \times T^m$ defined by

$$\mathcal{M} = \left\{ (x, I, \theta) \in \mathbb{R}^{2n} \times \mathbb{R}^m \times T^m \mid x = \gamma(I) \quad \text{where } \gamma(I) \text{ solves } D_x H(\gamma(I), I) = 0 \right.$$
$$\left. \text{subject to } \det \left[D_x^2 H(\gamma(I), I) \right] \neq 0, \quad \forall I \in U, \, \theta \in T^m \right\} \qquad (4.1.72)$$

and we have the following proposition.

Proposition 4.1.15. $\mathcal{M}$ *is a* C^r $2m$ *dimensional normally hyperbolic invariant manifold of* $(4.1.71)_0$. *Moreover,* $\mathcal{M}$ *has* C^r $2m+n$ *dimensional stable and unstable manifolds denoted* $W^s(\mathcal{M})$ *and* $W^u(\mathcal{M})$, *respectively, which intersect in the* $n + 2m$ *dimensional homoclinic manifold*

$$\Gamma = \left\{ (x^I(-t_0, \alpha), I, \theta_0) \in \mathbb{R}^{2n} \times \mathbb{R}^m \times T^m \mid (t_0, \alpha, I, \theta_0) \in \mathbb{R}^1 \times \mathbb{R}^{n-1} \times U \times T^m \right\}.$$

PROOF: The proof is identical to the proof of Proposition 4.1.4. $\square$

As was the case for System I, eventually we will be interested in the detailed dynamics on $\mathcal{M}$ in the perturbed system. For Hamiltonian perturbations we will see that much of the structure of the flow on $\mathcal{M}$ (in particular certain "nonresonant" motions) goes over for the perturbed system. For this reason we want to discuss in more detail the structure of the dynamics on $\mathcal{M}$.

The unperturbed vector field restricted to $\mathcal{M}$ is given by

$$\dot{I} = 0$$
$$\dot{\theta} = D_I H(\gamma(I), I) \qquad (I, \theta) \in U \times T^m \qquad (4.1.73)$$

with flow given by

$$I(t) = I = \text{constant}$$
$$\theta(t) = D_I H(\gamma(I), I)t + \theta_0. \qquad (4.1.74)$$

So $\mathcal{M}$ has the structure of an m parameter family of m tori with the flow on the tori being either rational or irrational. Let us denote these tori as follows: for a fixed $\bar{I} \in U$, the corresponding m-torus on $\mathcal{M}$ is

$$\tau(\bar{I}) \equiv \left\{ (x, I, \theta) \in \mathbb{R}^{2n} \times U \times T^m \mid x = \gamma(\bar{I}), \, I = \bar{I} \right\}. \qquad (4.1.75)$$

$\tau(\bar{I})$ has $n + m$ dimensional stable and unstable manifolds denoted $W^s(\tau(\bar{I}))$ and $W^u(\tau(\bar{I}))$, respectively, which intersect along the $n + m$ dimensional homoclinic manifold given by

$$\Gamma_{\bar{I}} = \left\{ (x^{\bar{I}}(-t_0, \alpha), \bar{I}, \theta_0) \in \mathbb{R}^{2n} \times \mathbb{R}^m \times T^m \mid (t_0, \alpha, \theta_0) \in \mathbb{R}^1 \times \mathbb{R}^{n-1} \times T^m \right\}.$$
$$(4.1.76)$$

Additionally, $\tau(\bar{I})$ has an m dimensional center manifold corresponding to the non-exponentially expanding or contracting directions tangent to $\mathcal{M}$. See Figure 4.1.12 for an illustration of the geometry of the unperturbed phase space.

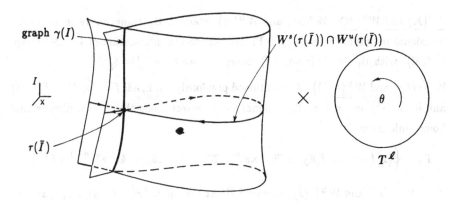

Figure 4.1.12. Geometry of the Unperturbed Phase Space of $(4.1.71)_0$.

ii) Homoclinic Coordinates

We define a moving system of homoclinic coordinates along Γ for System III in the same way as we did for System I and System II.

We consider the following $n+m$ linearly independent vectors in $\mathbb{R}^{2n} \times \mathbb{R}^m \times \mathbb{R}^m$:

$$\{(D_x H = D_x K_1, 0), \cdots, (D_x K_n, 0)\} \tag{4.1.77}$$

where $D_x K_i$ are linearly independent by III3 except at fixed points of $(4.1.71)_0$, and "0" denotes the $2m$ dimensional zero vector and

$$\left\{ \hat{I}_1, \dots, \hat{I}_m \right\}, \tag{4.1.78}$$

where the $\hat{I}_i$ are constant unit vectors in $\mathbb{R}^{2n} \times \mathbb{R}^m \times \mathbb{R}^m$ parallel to the I_i coordinate axes. For a given $(t_0, \alpha, I, \theta_0) \in \mathbb{R}^1 \times \mathbb{R}^{n-1} \times U \times T^m$, we let $p = \left(x^I(-t_0, \alpha), I, \theta_0 \right)$ denote the corresponding point on Γ. Then Π_p is defined to be the $m+n$ dimensional plane spanned by (4.1.77) and (4.1.78), where the $D_x K_i$ are evaluated at p and the $\hat{I}_i$ are viewed as emanating from p. So varying p serves to move Π_p along Γ.

As for Systems I and II, we will be interested in the nature of the intersection of $W^s(\mathcal{M}), W^u(\mathcal{M}), W^s(\tau(I))$, and $W^u(\tau(I))$ with Π_p. We will only state the results, since the details of the arguments are identical to those given in our discussion of homoclinic coordinates for System I.

$\underline{W^s(\mathcal{M}) \text{ and } W^u(\mathcal{M})}$. $W^s(\mathcal{M})$ and $W^u(\mathcal{M})$ intersect Π_p transversely in an m dimensional manifold for each $p \in \Gamma$. We denote the intersection of $W^s(\mathcal{M})$ (resp. $W^u(\mathcal{M})$) with Π_p by S_p^s (resp. S_p^u). Moreover, we have $S_p^s = S_p^u$.

$\underline{W^s(\tau(\bar{I})) \text{ and } W^u(\tau(\bar{I}))}$. As mentioned previously, for fixed $I = \bar{I} \in U, W^s(\tau(\bar{I}))$ and $W^u(\tau(\bar{I}))$ are $n + m$ dimensional and intersect along the $n + m$ dimensional homoclinic manifold

$$\Gamma_{\bar{I}} = \left\{ (x^{\bar{I}}(-t_0, \alpha), \bar{I}, \theta_0) \in \mathbb{R}^{2n} \times \mathbb{R}^m \times T^m \mid (t_0, \alpha, \theta_0) \in \mathbb{R}^1 \times \mathbb{R}^{n-1} \times T^m \right\}.$$

Now $W^s(\tau(\bar{I}))$ and $W^u(\tau(\bar{I}))$ intersect Π_p at the point $(x^{\bar{I}}(-t_0, \alpha), \bar{I}, \theta_0)$, and we want to argue that this intersection is transversal. We can take the tangent space of Π_p at p to be just Π_p, i.e., $T_p\Pi_p = \Pi_p$. Now the tangent space to $W^s(\tau(\bar{I}))$ (resp. $W^u(\tau(\bar{I}))$) at p is $n + m$ dimensional and can be viewed as being spanned by the n vectors $(JD_x K_i, 0)$, $i = 1, \ldots, n$ (where the $JD_x K_i$ are evaluated at p) and m vectors in the θ directions (see Proposition 4.1.2). Hence, $\Pi_p + T_p W^s(\tau(\bar{I}))$ (resp. $\Pi_p + T_p W^u(\tau(\bar{I}))) = \mathbb{R}^{2n} \times \mathbb{R}^{2m}$, and therefore $W^s(\tau(\bar{I}))$ (resp. $W^u(\tau(\bar{I}))$) intersects Π_p transversely in a point.

See Figure 4.1.13 for an illustration of the geometry (note the similarities with Figures 4.1.2 and 4.1.9). Recall that the importance of determining whether the intersections are transversal lies in the fact that transversal intersections persist under small perturbation; this fact is useful in determining the nature of the intersection of the manifolds in the perturbed system.

iii) The Geometric Structure of the Perturbed Phase Space

We now describe some general conclusions that can be made about the structure of the perturbed phase space. We will be concerned with $\mathcal{M}$, its local stable and unstable manifolds, and the flow on $\mathcal{M}$.

Persistence of $\mathcal{M}$.

The situation for System III regarding $\mathcal{M}$ is exactly the same as for System I: namely, $\mathcal{M}$ persists as a locally invariant manifold, $\mathcal{M}_\epsilon$.

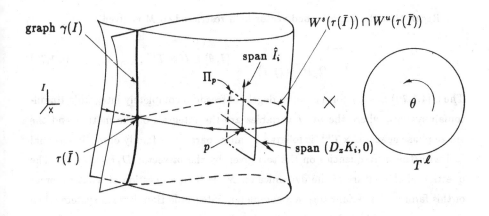

Figure 4.1.13. Homoclinic Coordinates.

Proposition 4.1.16. *There exists $\epsilon_0 > 0$ such that for $0 < \epsilon \leq \epsilon_0$ the perturbed system $(4.1.71)_\epsilon$ possesses a C^r $2m$ dimensional normally hyperbolic locally invariant manifold*

$$\mathcal{M}_\epsilon = \left\{ (x, I, \theta) \in \mathbb{R}^{2n} \times \mathbb{R}^m \times T^m \mid x = \tilde{\gamma}(I, \theta; \epsilon) = \gamma(I) + \mathcal{O}(\epsilon), \right.$$
$$\left. I \in \tilde{U} \subset U \subset \mathbb{R}^m, \theta \in T^m \right\}$$

where $\tilde{U} \subset U$ is a compact, connected m-dimensional set. Moreover, $\mathcal{M}_\epsilon$ has local C^r stable and unstable manifolds, $W^s_{loc}(\mathcal{M}_\epsilon)$ and $W^u_{loc}(\mathcal{M}_\epsilon)$, which are of the same dimension and C^r close to $W^s_{loc}(\mathcal{M}_\epsilon)$ and $W^u_{loc}(\mathcal{M}_\epsilon)$, respectively.

PROOF: The proof is similar to the proof of Proposition 4.1.5. $\square$

We remark that $\mathcal{M}$ is also C^r with respect to ϵ and μ (see the remark following the proof of Proposition 4.1.5).

Dynamics on $\mathcal{M}_\epsilon$.

We now want to address the question of whether there are any recurrent motions on $\mathcal{M}_\epsilon$. In particular, we want to know if any of the m parameter family of invariant m dimensional tori survive the perturbation. For System I we used the method of averaging and, as a result, only considered nonresonant motions. However, as discussed following Proposition 4.1.6, the method of averaging over angular variables does not work when the system is Hamiltonian, and a more sophisticated method is required.

Recall that the unperturbed vector field restricted to $\mathcal{M}$ is given by

$$\dot{I} = 0$$
$$\dot{\theta} = D_I H\big(\gamma(I), I\big) \qquad (I, \theta) \in \tilde{U} \times T^m. \qquad (4.1.79)$$

Thus, (4.1.79) has the form of an m degree of freedom completely integrable Hamiltonian system, where the m I variables are the integrals of the motion and the entire phase space $\tilde{U} \times T^m$ is foliated by an m parameter family of m-dimensional tori with the m frequencies on the tori given by the m vector $D_I H\big(\gamma(I), I\big)$. The question of the nature of the dynamics on $\mathcal{M}_\epsilon$ is then a question of what becomes of this family of invariant tori in a completely integrable Hamiltonian system when the system is subjected to a Hamiltonian perturbation. Some important results along these lines are provided by the Kolmogorov-Arnold-Moser (KAM) theorem (see Arnold [1978], Appendix 8), which we will now state in a form sufficient for our needs.

Theorem 4.1.17 (KAM). *Suppose*

$$\det \left[D_I^2 H\big(\gamma(I), I\big) \right] \neq 0, \qquad I \in \tilde{U} \subset \mathbf{R}^m.$$

Then "most" of the invariant tori persist in (4.1.79) for sufficiently small ϵ. The motion on these surviving tori is quasiperiodic, having m rationally incommensurate frequencies. The invariant tori form a majority in the sense that the Lebesque measure of the complement of their union is small when ϵ is small.

Let us now make several remarks concerning this important theorem.

1. An immediate question that arises concerning Theorem 4.1.17 is, what does the term "most" mean? Mathematically, "most" means a Cantor set of positive measure. For our purposes the important characteristic of the surviving "KAM" tori is that, since a Cantor set of positive measure persists, given a KAM torus there exists another KAM torus arbitrarily close. However, a more precise characterization of the surviving tori in terms of the unperturbed frequencies can be found in Arnold [1963] or Moser [1973].

2. Note the radical difference regarding the dynamics on $\mathcal{M}_\epsilon$ in the case of dissipative perturbations in System I and the Hamiltonian perturbations in System III. For System I, discrete nonresonant normally hyperbolic tori persisted on $\mathcal{M}_\epsilon$ and, for System III, most of the nonresonant non-normally hyperbolic tori

persisted. We thus might expect very different dynamical phenomena in the two cases. Note that our methods for determining the resulting motion on $\mathcal{M}_\epsilon$ allow us only to find certain nonresonant motions, and that more sophisticated techniques (as yet undeveloped) could reveal interesting dynamics that are missed with present techniques.

3. We now address the question of how differentiable (4.1.79) must be in order for the KAM theorem to hold. Originally, the theorem was announced by Kolmogorov [1954], with full details given by Arnold [1963] for the case of analytic Hamiltonians. The analogous theorem for vector fields with finitely many derivatives was first given by Moser [1966a,b]. Moser's result applies to C^r vector fields of the form of (4.1.79) with $r \geq 2m + 2$. For a recent review of KAM theory and related results see Bost [1986].

Recall the structure in the unperturbed system. For any $I = \bar{I} \in \tilde{U}$, $\tau(\bar{I})$ is an m dimensional torus on $\mathcal{M}$ having $m+n$ dimensional stable and unstable manifolds $W^s(\tau(\bar{I}))$ and $W^u(\tau(\bar{I}))$ which intersect along an $n + m$ dimensional homoclinic orbit $\Gamma_{\bar{I}} = \{ (x^{\bar{I}}(-t_0, \alpha), \bar{I}, \theta_0) \in \mathbb{R}^{2n} \times \mathbb{R}^m \times T^m \mid (t_0, \alpha, \theta_0) \in \mathbb{R}^1 \times \mathbb{R}^{n-1} \times T^m \}$. By Theorem 4.1.17, we know that most of these tori persist on $\mathcal{M}_\epsilon$. Let us denote the surviving tori by $\tau_\epsilon(I)$. Now, the standard KAM theorem tells us nothing about $W^s(\tau_\epsilon(I))$ and $W^u(\tau_\epsilon(I))$. However, a generalization of the KAM theorem due to Graff [1974] tells us that $\tau_\epsilon(I)$ has $m + n$ dimensional stable and unstable manifolds, which we denote as $W^s(\tau_\epsilon(I))$ and $W^u(\tau_\epsilon(I))$. By invariance of the manifolds, we have $W^s(\tau_\epsilon(I)) \subset W^s(\mathcal{M}_\epsilon)$ and $W^u(\tau_\epsilon(I)) \subset W^u(\mathcal{M}_\epsilon)$. Now Graff's theorem is proven only for analytic vector fields. However, he states that smoothing techniques developed by Moser [1966a,b] can be used to extend the result to the finite differentiable case. In any case, this is a technical difficulty that will cause us little concern, because all of our examples will be analytic.

iv) The Splitting of the Manifolds

We now develop a procedure for determining whether or not $W^s(\tau_\epsilon(I))$ and $W^u(\tau_\epsilon(I))$ intersect. The situation will be different than for Systems I and II, since $W^s(\tau_\epsilon(I))$ and $W^u(\tau_\epsilon(I))$ have larger codimension than the corresponding stable and unstable manifolds in Systems I and II, and the KAM tori are not normally hyperbolic. Hence, we might expect that a larger dimensional Melnikov vector is necessary in order to determine whether or not $W^s(\tau_\epsilon(I))$ and $W^u(\tau_\epsilon(I))$ intersect.

Recall the geometry of the unperturbed system. For any point $p \in \Gamma$, the $2m+n$ dimensional manifolds $W^s(M)$ and $W^u(M)$ intersect the $m+n$ dimensional plane Π_p transversely in the m dimensional surfaces S_p^s and S_p^u, respectively, with $S_p^s = S_p^u$. For any $I = \bar{I} \in \tilde{U}$, the corresponding invariant m torus $\tau(\bar{I}) \in M$ has $m+n$ dimensional stable and unstable manifolds $W^s(\tau(\bar{I}))$ and $W^u(\tau(\bar{I}))$, which intersect Π_p transversely in the point p. See Figure 4.1.14 for an illustration of the unperturbed geometry. Note that each point on $S_p^s = S_p^u = \Gamma \cap \Pi_p$ in the figure represents $W^s(\tau(I)) \cap W^u(\tau(I))$ for some torus $\tau(I)$.

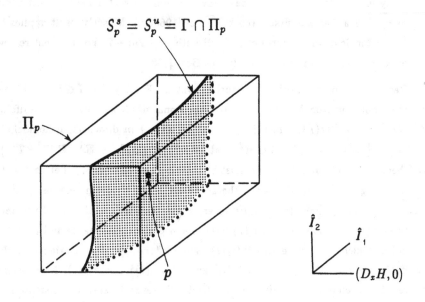

Figure 4.1.14. The Unperturbed Geometry of the Manifolds on Π_p, $n = 1$, $m = 2$.

Next we consider the geometry of the perturbed system along $\Gamma = W^s(M) - W^u(M) - M$. By transversality, for each point $p \in \Gamma$, $W^s(M_\epsilon)$ and $W^u(M_\epsilon)$ intersect Π_p in the m dimensional surfaces $S_{p,\epsilon}^s$ and $S_{p,\epsilon}^u$, respectively. As mentioned previously, $W^s(M_\epsilon)$ and $W^u(M_\epsilon)$ may intersect Π_p in a multiple number of m dimensional components; however, we choose $S_{p,\epsilon}^s$ (resp. $S_{p,\epsilon}^u$) to be the component which is closest to M_ϵ in the sense of positive (resp. negative) time of flow along $W^s(M_\epsilon)$ (resp. $W^u(M_\epsilon)$). More attention will be given to this technical detail when we discuss the derivation of the Melnikov vector. Also by transversality, for a given surviving KAM torus $\tau_\epsilon(\bar{I}) \in \mathcal{M}_\epsilon$, $W^s(\tau_\epsilon(\bar{I}))$ intersects Π_p transversely in a point $p_\epsilon^s = (x_\epsilon^s, I_\epsilon^s)$, and $W^u(\tau_\epsilon(\bar{I}))$ intersects Π_p transversely in a point $p_\epsilon^u = (x_\epsilon^u, I_\epsilon^u)$. Moreover, these points are contained in $S_{p,\epsilon}^s$ and $S_{p,\epsilon}^u$, respectively. See Figure 4.1.15 for an illustration of the perturbed geometry. Note that in Figure 4.1.15 the cloud of dots on $S_{p,\epsilon}^s$ (resp. $S_{p,\epsilon}^u$) represent the intersections of the stable (resp. unstable) manifolds of the surviving tori on $\mathcal{M}_\epsilon$ with Π_p.

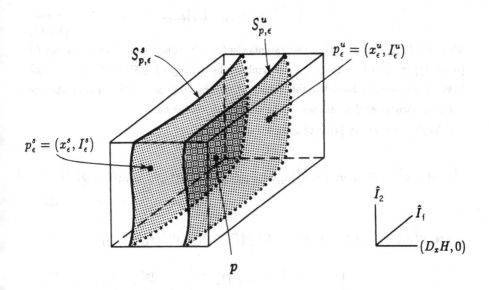

Figure 4.1.15. The Perturbed Geometry of the Manifolds on Π_p, $n = 1$, $m = 2$.

We might naively define the distance between $W^s(\tau_\epsilon(\bar{I}))$ and $W^u(\tau_\epsilon(\bar{I}))$ at the point p as

$$d(p, \epsilon) = |(x_\epsilon^u, I_\epsilon^u) - (x_\epsilon^s, I_\epsilon^s)| . \tag{4.1.80}$$

However, as for Systems I and II (cf. the discussion in 4.1a, iv), our goal is to develop a computable expression for (4.1.80) which is compatible with the underlying geometry. This will involve developing expressions for the components of the distance along the coordinates on Π_p. Notice an important difference in System III as opposed to Systems I and II. Since $W^s(\tau_\epsilon(\bar{I}))$ and $W^u(\tau_\epsilon(\bar{I}))$ intersect Π_p only in single points we cannot guarantee that $I_\epsilon^s = I_\epsilon^u$, as was the case for Systems I and II. Hence, for System III, we must also measure the distance along these directions.

We define the signed component of the distance between p_ϵ^s and p_ϵ^u along the $n + m$ directions $(D_x K_i, 0)$, $i = 1, \ldots, n$ and $\hat{I}_i$, $i = 1, \ldots, m$ as follows:

$$d_i^{\bar{I}}(p, \epsilon) = d_i^{\bar{I}}(t_0, \theta_0, \alpha, \mu; \epsilon) = \begin{cases} \dfrac{\langle D_x K_i(x^{\bar{I}}(-t_0, \alpha), \bar{I}), x_\epsilon^u - x_\epsilon^s \rangle}{\left\| D_x K_i(x^{\bar{I}}(-t_0, \alpha), \bar{I}) \right\|}, & i = 1, \ldots, n \\[2em] (I_\epsilon^u)_{i-n} - (I_\epsilon^s)_{i-n}, & i = n+1, \ldots, n+m, \end{cases}$$

(4.1.81)

where $(I_\epsilon^u)_{i-n}$ (resp. $(I_\epsilon^s)_{i-n}$) represents the $(i-n)^{\text{th}}$ component of the m vector I_ϵ^u (resp. I_ϵ^s), and we have replaced the symbol p by $(t_0, \theta_0, \alpha) \in \mathbb{R}^1 \times T^m \times \mathbb{R}^{n-1}$; we have also explicitly included the vector of parameters $\mu \in \mathbb{R}^p$ to indicate the possible parameter dependence of the perturbed vector field.

Taylor expanding (4.1.81) about $\epsilon = 0$ gives

$$d_i^{\bar{I}}(p, \epsilon) = d_i^{\bar{I}}(t_0, \theta_0, \alpha, \mu; \epsilon) = d_i^{\bar{I}}(t_0, \theta_0, \alpha, \mu; 0) + \epsilon \frac{\partial d_i^{\bar{I}}}{\partial \epsilon}(t_0, \theta_0, \alpha, \mu; 0) + O(\epsilon^2),$$

$$i = 1, \ldots, n+m \qquad (4.1.82)$$

where $d_i^{\bar{I}}(t_0, \theta_0, \alpha, \mu; 0) = 0$ since $W^s\big(\tau(\bar{I})\big) \cap \Pi_p = W^u\big(\tau(\bar{I})\big) \cap \Pi_p = p$ and

$$\frac{\partial d_i^{\bar{I}}}{\partial \epsilon}(t_0, \theta_0, \alpha, \mu; 0) = \begin{cases} \dfrac{\langle D_x K_i(x^{\bar{I}}(-t_0, \alpha), \bar{I}), \frac{\partial x_\epsilon^u}{\partial \epsilon}\big|_{\epsilon=0} - \frac{\partial x_\epsilon^s}{\partial \epsilon}\big|_{\epsilon=0} \rangle}{\left\| D_x K_i(x^{\bar{I}}(-t_0, \alpha), \bar{I}) \right\|}, & i = 1, \ldots, n \\[2em] \left(\frac{\partial I_\epsilon^u}{\partial \epsilon}\big|_{\epsilon=0}\right)_{i-n} - \left(\frac{\partial I_\epsilon^s}{\partial \epsilon}\big|_{\epsilon=0}\right)_{i-n}, & i = n+1, \ldots, n+m. \end{cases}$$

(4.1.83)

We will shortly show that

$$\langle D_x K_i(x^{\bar{I}}(-t_0, \alpha), \bar{I}), \frac{\partial x_\epsilon^u}{\partial \epsilon}\bigg|_{\epsilon=0} - \frac{\partial x_\epsilon^s}{\partial \epsilon}\bigg|_{\epsilon=0} \rangle \equiv M_i^{\bar{I}}(\theta_0, \alpha; \mu)$$

$$= \int_{-\infty}^{\infty} \left[\langle D_x K_i, J D_x \tilde{H} \rangle - \langle D_x K_i, (D_I J D_x H) \int^t D_\theta \tilde{H} \rangle \right](q_0^{\bar{I}}(t), \mu; 0)\, dt$$

(4.1.84a)

$$i = 1, \ldots, n$$

or, equivalently,

$$M_i(\theta_0, \alpha; \mu) = \int\limits_{-\infty}^{\infty} [\langle D_x K_i, J D_x \tilde{H} \rangle - \langle D_I K_i, D_\theta \tilde{H} \rangle](q_0^{\bar{I}}(t), \mu; 0) dt$$

$$+ \langle D_I K_i(\gamma(\bar{I}), \bar{I}), \int\limits_{-\infty}^{\infty} D_\theta \tilde{H}(q_0^{\bar{I}}(t), \mu; 0) dt \rangle$$

$$(4.1.84b)$$

and

$$\left(\frac{\partial I_\epsilon^u}{\partial \epsilon}\Big|_{\epsilon=0}\right)_{i-n} - \left(\frac{\partial I_\epsilon^s}{\partial \epsilon}\Big|_{\epsilon=0}\right)_{i-n} = - \int\limits_{-\infty}^{\infty} D_{\theta_{i-n}} \tilde{H}(q_0^{\bar{I}}(t), \mu; 0) dt$$

$$i = n+1, \ldots, n+m \qquad (4.1.85)$$

where $q_0^{\bar{I}}(t) = \left(x^{\bar{I}}(t, \alpha), \bar{I}, \int^t D_I H(x^{\bar{I}}(s, \alpha), \bar{I}) ds + \theta_0\right)$. Thus, (4.1.84) and (4.1.85) represent, to order ϵ, the $m+n$ components of the measurement of distance between $W^s(\tau_\epsilon(\bar{I}))$ and $W^u(\tau_\epsilon(\bar{I}))$ at the point p. However, there is a slight subtlety. Because the perturbed system is Hamiltonian in the $2n + 2m$ dimensional phase space, orbits are restricted to lie in $2n + 2m - 1$ dimensional "energy" surfaces given by the level sets of the Hamiltonian $H_\epsilon = H + \epsilon \tilde{H}$. Thus, we would expect to need only $n + m - 1$ independent measurements to determine whether or not $W^s(\tau_\epsilon(\bar{I}))$ and $W^u(\tau_\epsilon(\bar{I}))$ intersect rather than $n+m$. This problem is resolved in the following $2n+2m$ dimensional version of a lemma due to Lerman and Umanskii [1984].

Lemma 4.1.18. $p_\epsilon^u = p_\epsilon^s$ if and only if $d_i(p, \epsilon) = 0$, $i = 2, \ldots, n+m$.

PROOF: $p_\epsilon^u = p_\epsilon^s$ implies $d_i(p, \epsilon) = 0$, $i = 2, \ldots, n+m$ is obvious. We now show that $d_i(p, \epsilon) = 0$, $i = 2, \ldots, n+m$, implies $p_\epsilon^u = p_\epsilon^s$.

Let $p = (x^{\bar{I}}(-t_0, \alpha), \bar{I}, \theta_0)$; then any point $\bar{p} \in \Pi_p$ can be expressed as

$$\bar{p} = p + \xi_1(D_x H(p), 0) + \cdots + \xi_n(D_x K_n(p), 0) + \xi_{n+1} \hat{I}_1 + \cdots + \xi_{n+m} \hat{I}_m \quad (4.1.86)$$

where $(\xi_1, \ldots, \xi_{n+m})$ represent coordinates along the vectors $(D_x K_i(p), 0)$, $i = 1, \ldots, n$, and $\hat{I}_{i-n}$, $i = n+1, \ldots, n+m$ which define Π_p. Then, using (4.1.86), we have

$$p_\epsilon^u - p_\epsilon^s = (\xi_1^u - \xi_1^s)(D_x H(p), 0) + \cdots + (\xi_n^u - \xi_n^s)(D_x K_n(p), 0)$$
$$+ (\xi_{n+1}^u - \xi_{n+1}^s)\hat{I}_1 + \cdots + (\xi_{n+m}^u - \xi_{n+m}^s)\hat{I}_m .$$

$$(4.1.87)$$

Since p_ϵ^u and p_ϵ^s lie on $W^s(\tau_\epsilon(\bar{I}))$ and $W^u(\tau_\epsilon(\bar{I}))$, they must lie on the surface defined by the equation

$$
\begin{aligned}
R(\xi_1,\ldots,\xi_{n+m};\epsilon) = {} & H\Big(p + \xi_1(D_x H(p),0) + \cdots + \xi_n(D_x K_n(p),0) \\
& + \xi_{n+1}\hat{I}_1 + \cdots + \xi_{n+m}\hat{I}_m, \bar{I}\Big) \\
& + \epsilon\tilde{H}\Big(p + \xi_1(D_x H(p),0) + \cdots + \xi_n(D_x K_n(p),0) \quad \text{(4.1.88)} \\
& + \xi_{n+1}\hat{I}_1 + \cdots + \xi_{n+m}\hat{I}_m, \bar{I}, \theta_0, \epsilon\Big) \\
& - H(\tilde{\gamma}(\bar{I},\theta_0;\epsilon),\bar{I}) - \epsilon\tilde{H}(\tilde{\gamma}(\bar{I},\theta_0;\epsilon),\bar{I},\theta_0,\epsilon) = 0
\end{aligned}
$$

(note: we have suppressed the possible parameters μ since they will not affect the argument). We have

$$
R(0,\ldots,0;0) = H(x^{\bar{I}}(-t_0,\alpha),\bar{I}) - H(\gamma(\bar{I}),\bar{I}) = 0 \qquad \text{(4.1.89)}
$$

and $\ D_{\xi_1}R(0,\ldots,0;0) = \langle D_x H(x^{\bar{I}}(-t_0,\alpha),\bar{I}), D_x H(x^{\bar{I}}(-t_0,\alpha),\bar{I})\rangle \neq 0$. Hence, by the implicit function theorem, we have

$$
\xi_1 = \phi(\xi_2,\ldots,\xi_{n+m};\epsilon) \qquad \text{(4.1.90)}
$$

for $(\xi_1,\ldots,\xi_{n+m},\epsilon)$ sufficiently small and where ϕ is as smooth as $H + \epsilon\tilde{H}$. Using (4.1.87) and (4.1.90), we obtain

$$
\begin{aligned}
p_\epsilon^u - p_\epsilon^s = {} & \big(\phi(\xi_2^u,\ldots,\xi_{n+m}^u;\epsilon) - \phi(\xi_2^s,\ldots,\xi_{n+m}^s;\epsilon)\big)\big(D_x H(p),0\big) + \cdots \\
& + (\xi_n^u - \xi_n^s)\big(D_x K_n(p),0\big) + (\xi_{n+1}^u - \xi_{n+1}^s)\hat{I}_1 + \cdots + (\xi_{n+m}^u - \xi_{n+m}^s)\hat{I}_m \,.
\end{aligned}
$$
$$\text{(4.1.91)}$$

So from (4.1.90) and (4.1.91) it is clear that for $\ \xi_i^u = \xi_i^s, \ \ i = 2,\ldots,n+m\ $ we have $p_\epsilon^u = p_\epsilon^s$. This proves the lemma. $\qquad\qquad\square$

    This lemma tells us we do not need to measure along the direction $(D_x H, 0)$ in order to determine whether $W^s(\tau_\epsilon(\bar{I}))$ and $W^u(\tau_\epsilon(\bar{I}))$ intersect. Intuitively, this should be reasonable, since the energy manifold $H + \epsilon\tilde{H}$ is preserved and $(D_x H, 0)$ is a direction complementary to the energy manifold.

    We will define the Melnikov vector as

$$
M^{\bar{I}}(\theta_0,\alpha;\mu) = \big(M_2^{\bar{I}}(\theta_0,\alpha;\mu),\ldots,M_{n+m}^{\bar{I}}(\theta_0,\alpha;\mu)\big) \qquad \text{(4.1.92)}
$$

where we have left out the explicit dependence on t_0 for the same reasons discussed for Systems I and II (note: this will be elaborated on when we discuss our derivation of the Melnikov vector). We can now state our main theorem for System III.

Theorem 4.1.19. *Suppose there exists* $I = \bar{I} \in \tilde{U} \subset \mathbf{R}^m$ *such that* $\tau_\epsilon(\bar{I})$ *is a KAM torus on* M_ϵ. *Let* $(\theta_0, \alpha, \mu) = (\bar{\theta}_0, \bar{\alpha}, \bar{\mu}) \in T^m \times \mathbf{R}^{n-1} \times \mathbf{R}^p$ *be a point with* $m + n - 1 + p \geq m + n - 1$ *such that*

    1) $M^{\bar{I}}(\bar{\theta}_0, \bar{\alpha}, \bar{\mu}) = 0$

    2) $DM^{\bar{I}}(\bar{\theta}_0, \bar{\alpha}, \bar{\mu})$ *is of rank* $m + n - 1$;

then $W^s(\tau_\epsilon(\bar{I}))$ *and* $W^u(\tau_\epsilon(\bar{I}))$ *intersect near* $(\bar{\theta}_0, \bar{\alpha}, \bar{\mu})$.

PROOF: The proof is similar to the proof of Theorem 4.1.9.   □

A sufficient condition for the transversal intersection of $W^s(\tau_\epsilon(\bar{I}))$ and $W^u(\tau_\epsilon(\bar{I}))$ is given in the following theorem.

Theorem 4.1.20. *Suppose Theorem 4.1.19 holds at the point* $(\theta_0, \alpha, \mu) = (\bar{\theta}_0, \bar{\alpha}, \bar{\mu}) \in T^m \times \mathbf{R}^{n-1} \times \mathbf{R}^p$ *and that* $D_{(\theta_0, \alpha)} M^{\bar{I}}(\bar{\theta}_0, \bar{\alpha}, \bar{\mu})$ *is of rank* $m + n - 1$. *Then, for* ϵ *sufficiently small,* $W^s(\tau_\epsilon(\bar{I}))$ *and* $W^u(\tau_\epsilon(\bar{I}))$ *intersect transversely near* $(\bar{\theta}_0, \bar{\alpha})$ *in the* $2n + 2m - 1$ *dimensional energy surface.*

PROOF: The proof is similar to the proof of Theorem 4.1.10. Let $p \in W^s(\tau_\epsilon(\bar{I})) \cap W^u(\tau_\epsilon(\bar{I}))$. Then $T_p W^s(\tau_\epsilon(\bar{I}))$ and $T_p W^u(\tau_\epsilon(\bar{I}))$ are $m + n$ dimensional and, by Definition 1.4.1, $W^s(\tau_\epsilon(\bar{I}))$ intersects $W^u(\tau_\epsilon(\bar{I}))$ transversely at p in the $2n+2m-1$ dimensional energy manifold if $T_p W^s(\tau_\epsilon(\bar{I})) + T_p W^u(\tau_\epsilon(\bar{I})) = \mathbf{R}^{2n+2m-1}$. By the dimension formula for vector spaces we have

$$
\begin{aligned}
2n + 2m - 1 = {}& \dim T_p W^s(\tau_\epsilon(\bar{I})) + \dim T_p W^u(\tau_\epsilon(\bar{I})) \\
& - \dim T_p\big(W^s(\tau_\epsilon(\bar{I})) \cap W^u(\tau_\epsilon(\bar{I}))\big).
\end{aligned}
\tag{4.1.93}
$$

Thus, if $W^s(\tau_\epsilon(\bar{I}))$ intersects $W^u(\tau_\epsilon(\bar{I}))$ transversely in the $2n + 2m - 1$ dimensional energy manifold at p, then $W^s(\tau_\epsilon(\bar{I}))$ intersects $W^u(\tau_\epsilon(\bar{I}))$ in a one dimensional trajectory. Therefore, in order for $W^s(\tau_\epsilon(\bar{I}))$ to intersect $W^u(\tau_\epsilon(\bar{I}))$ transversely at p, it is necessary for $T_p W^s(\tau_\epsilon(\bar{I}))$ and $T_p W^u(\tau_\epsilon(\bar{I}))$ to each contain $n + m - 1$ dimensional independent subspaces which have no part contained in $T_p\big(W^s(\tau_\epsilon(\bar{I})) \cap W^u(\tau_\epsilon(\bar{I}))\big)$.

Let us recall the geometry of the unperturbed phase space. The m dimensional KAM torus $\tau(\bar{I})$ has $m+n$ dimensional stable and unstable manifolds which coincide along an $s + m$ dimensional homoclinic orbit. We need to show that independent $m + n - 1$ dimensional subspaces are created in $T_p W^s(\tau_\epsilon(\bar{I}))$ and $T_p W^u(\tau_\epsilon(\bar{I}))$ which are not contained in $T_p\big(W^s(\tau_\epsilon(\bar{I})) \cap W^u(\tau_\epsilon(\bar{I}))\big)$. The remainder of the proof is the same as the latter part of the proof of Theorem 4.1.10.   □

v) Horseshoes and Arnold Diffusion

For Systems I and II some dynamical consequences of the intersection of the stable and unstable manifolds of a normally hyperbolic invariant torus follow from the associated theorems given in Chapter 3 (note: the term torus is used in a general sense and also applies to the case of the 0-torus (fixed point) and the 1-torus (periodic orbit)). However, the somewhat more subtle geometry associated with the phase space of perturbed completely integrable Hamiltonian systems is responsible for exotic dynamics, which we will now discuss separately. There are two distinct cases corresponding to differences in the dimensions of the phase space.

$n \geq 1$, $m = 1$. In this case the phase space is $2n + 2$ dimensional and is foliated by invariant $2n + 1$ dimensional energy surfaces. In the unperturbed system, $\mathcal{M}$ is a 1-dimensional normally hyperbolic invariant manifold which has the structure of a 1 parameter family of 1-tori, $\tau(I)$, $I \subset U \subset \mathbb{R}^1$. Each torus has $n + 1$ dimensional stable and unstable manifolds coinciding along an $n + 1$ dimensional homoclinic orbit. In the perturbed system, $\mathcal{M}$ is preserved (denoted $\mathcal{M}_\epsilon$), and on $\mathcal{M}_\epsilon$ we have

$$H_\epsilon = H\big(\gamma(I), I\big) + \mathcal{O}(\epsilon) = \text{constant} . \tag{4.1.94}$$

Now, since $I \in U \subset \mathbb{R}^1$, from (4.1.94) we see that on a fixed energy manifold I is likewise fixed. So, in this case, the full results of the KAM theorem are not needed. On a fixed energy manifold an isolated 1-torus (i.e., periodic orbit) survives and is normally hyperbolic on the energy manifold. Thus, the Melnikov vector measures the distance between the stable and unstable manifolds of a normally hyperbolic periodic orbit, and the dynamical consequences associated with their intersection are ordinary Smale horseshoes.

$n \geq 1$, $m \geq 2$. In the unperturbed system on $\mathcal{M}$ we have

$$H\big(\gamma(I), I\big) = \text{constant} . \tag{4.1.95}$$

Thus, on a fixed $2n + 2m - 1$ dimensional energy manifold we have an $m - 1$ parameter family of m-tori. Each torus has $m + n$ dimensional stable and unstable manifolds which coincide along an $n + m$ dimensional homoclinic orbit. Note the important point that, since $m \geq 2$, the tori, along with their stable and unstable manifolds, are not isolated on the energy manifold.

In the perturbed system "most" of the $m - 1$ parameter family of tori survive on each energy manifold by the KAM theorem. In this case, it is possible to choose a set of KAM tori, $\tau_\epsilon(I_1)$, $\tau_\epsilon(I_2)$, ..., $\tau_\epsilon(I_N)$, with the property that $\tau_\epsilon(I_i)$ is arbitrarily close to $\tau_\epsilon(I_{i+1})$ for $i = 1,\ldots,N - 1$. Now, suppose that for some i with $1 \leq i \leq N - 1$, $W^u\big(\tau_\epsilon(I_i)\big)$ intersects $W^s\big(\tau_\epsilon(I_i)\big)$ transversely. Then, by arguments similar to those given in the proof of the Toral Lambda Lemma in Chapter 3, it can be shown that $W^u\big(\tau_\epsilon(I_i)\big)$ accumulates on $\tau_\epsilon(I_i)$, resulting in it also transversely intersecting $W^s\big(\tau_\epsilon(I_{i+1})\big)$ and $W^s\big(\tau_\epsilon(I_{i-1})\big)$ (for $i \geq 1$), which are arbitrarily close. This argument can be repeated, with the ultimate conclusion being that $W^u\big(\tau_\epsilon(I_i)\big)$ transversely intersects $W^s\big(\tau_\epsilon(I_j)\big)$ for any $1 \leq i, j \leq N$. The resulting tangle of manifolds provides a mechanism whereby orbits may wander in an apparently random fashion amongst the KAM tori. The sequence of tori $\tau_\epsilon(I_1)$, ..., $\tau_\epsilon(I_N)$ is referred to as a *transition chain*, and the resulting motion is called *Arnold diffusion*, see Figure 4.1.16 for a heuristic illustration of the geometry.

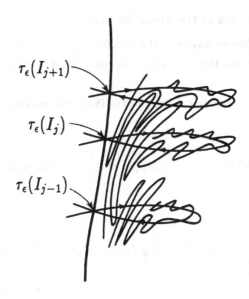

Figure 4.1.16. The Geometry of Arnold Diffusion.

Despite the ubiquity of Arnold diffusion in Hamiltonian systems having more than two degrees of freedom, there has been surprisingly little work done in the area since Arnold's original paper in 1964. Nehoroshev [1971], [1972] provided estimates on the rate of Arnold diffusion and, as mentioned earlier, Holmes and Marsden [1982b] developed the first general techniques to verify the existence of Arnold diffusion in specific systems. The previously described geometrical picture of the dense tangling of the manifolds of the KAM tori has yet to be put on a rigorous footing along the lines of Chapter 3, although results of Easton [1978], [1981] regarding certain model problems should go through in the general case and would provide a good starting point. Numerical simulations, which have yielded great insight into the global dynamics of one and two dimensional maps, have yet to be put to extensive use in the study of Arnold diffusion (but see Lichtenberg and Lieberman [1982]). This is probably due to the fact that at least a four dimensional volume preserving map would be needed in order to exhibit Arnold diffusion, and it is not immediately clear how to best display the dynamics of a four dimensional map on a two dimensional computer screen.

4.1d. The Derivation of the Melnikov Vector

We will now give the derivation of the Melnikov vectors for Systems I, II, and III. We will do this for the three systems simultaneously, discussing the differences as we go along.

Recall that the Melnikov vectors for the three systems are given by:

System I.

$$M^I(\theta_0, \alpha; \mu) = \left(M_1^{\bar{I}}(\theta_0, \alpha; \mu), \ldots, M_n^{\bar{I}}(\theta_0, \alpha; \mu)\right), \quad (\theta_0, \alpha; \mu) \in T^l \times \mathbb{R}^{n-1} \times \mathbb{R}^p$$

$$(4.1.96)$$

where we will show that

$$M_i^{\bar{I}}(\theta_0, \alpha; \mu) = \int_{-\infty}^{\infty} \left[\langle D_x K_i, g^x\rangle + \langle D_x K_i, (D_I J D_x H) \int^t g^I\rangle\right]\left(q_0^{\bar{I}}(t), \mu; 0\right) dt,$$

$$i = 1, \ldots, n \qquad (4.1.97a)$$

or, equivalently,

$$M_i^{\bar{I}}(\theta_0, \alpha; \mu) = \int_{-\infty}^{\infty} [\langle D_x K_i, g^x \rangle + \langle D_I K_i, g^I \rangle](q_0^{\bar{I}}(t), \mu; 0) dt$$

$$- \langle D_I K_i(\gamma(\bar{I}), \bar{I}), \int_{-\infty}^{\infty} g^I(q_0^{\bar{I}}(t), \mu; 0) dt \rangle$$

$$(4.1.97b)$$

and $q_0^{\bar{I}}(t) = (x^{\bar{I}}(t, \alpha), \bar{I}, \int^t \Omega(x^{\bar{I}}(s, \alpha), \bar{I}) ds + \theta_0)$, $\bar{I} \in \tilde{U} \subset U \subset \mathbf{R}^m$ being chosen such that $\tau_\epsilon(\bar{I})$ is a normally hyperbolic invariant torus on $\mathcal{M}_\epsilon$ (see Propositions 4.1.6 and 4.1.7).

System II.

$$M(I, \theta_0, \alpha; \mu) = \big(M_1(I, \theta_0, \alpha; \mu), \ldots, M_n(I, \theta_0, \alpha; \mu)\big),$$

$$(I, \theta_0, \alpha; \mu) \in T^m \times T^l \times \mathbf{R}^{n-1} \times \mathbf{R}^p \qquad (4.1.98)$$

where we will show that

$$M_i(I, \theta_0, \alpha; \mu) = \int_{-\infty}^{\infty} [\langle D_x K_i, g^x \rangle + \langle D_x K_i, (D_I J D_x H) \int^t g^I \rangle](q_0^I(t), \mu; 0) dt,$$

$$(4.1.99a)$$

$$i = 1, \ldots, n.$$

or, equivalently,

$$M_i(I, \theta_0, \alpha; \mu) = \int_{-\infty}^{\infty} [\langle D_x K_i, g^x \rangle + \langle D_I K_i, g^I \rangle](q_0^I(t), \mu; 0) dt$$

$$- \langle D_I K_i(\gamma(I), I), \int_{-\infty}^{\infty} g^I(q_0^I(t), \mu; 0) dt \rangle$$

$$(4.1.99b)$$

and $q_0^I(t) \equiv (x^I(t, \alpha), I, \int^t \Omega(x^I(s, \alpha), I) ds + \theta_0)$.

System III.

$$M^{\bar{I}}(\theta_0, \alpha; \mu) = \big(M_2^{\bar{I}}(\theta_0, \alpha; \mu), \ldots, M_n^{\bar{I}}(\theta_0, \alpha; \mu), M_{n+1}^{\bar{I}}(\theta_0, \alpha; \mu), \ldots, M_{n+m}^{\bar{I}}(\theta_0, \alpha; \mu)\big),$$

$$(\theta_0, \alpha; \mu) \in T^m \times \mathbf{R}^{n-1} \times \mathbf{R}^p \qquad (4.1.100)$$

where we will show that

$$M_i^{\bar{I}}(\theta_0, \alpha; \mu) =$$

$$\int\limits_{-\infty}^{\infty} \left[\langle D_x K_i, J D_x \tilde{H} \rangle - \langle D_x K_i, (D_I J D_x H) \int\limits^t D_\theta \tilde{H} \rangle \right] (q_0^{\bar{I}}(t), \mu; 0) dt,$$

$$(4.1.101a)$$

$$i = 2, \ldots, n$$

or, equivalently,

$$M_i^{\bar{I}}(\theta_0, \alpha; \mu) = \int\limits_{-\infty}^{\infty} \left[\langle D_x K_i J D_x \tilde{H} \rangle - \langle D_I K_i, D_\theta \tilde{H} \rangle \right] (q_0^{\bar{I}}(t), \mu; 0) dt$$

$$+ \langle D_I K_i(\gamma(\bar{I}), \bar{I}), \int\limits_{-\infty}^{\infty} D_\theta \tilde{H}(q_0^{\bar{I}}(t), \mu; 0) dt \rangle$$

$$(4.1.101b)$$

and

$$M_i^{\bar{I}}(\theta_0, \alpha; \mu) = - \int\limits_{-\infty}^{\infty} D_{\theta_{i-n}} \tilde{H}(q_0^{\bar{I}}(t), \mu; 0) dt,$$

$$i = n+1, \ldots, n+m \qquad (4.1.102)$$

and $q_0^{\bar{I}}(t) \equiv (x^{\bar{I}}(t, \alpha), \bar{I}, \int^t D_I H(x^{\bar{I}}(s, \alpha)\bar{I}) ds + \theta_0)$, $I = \bar{I} \in \tilde{U} \subset U \subset \mathbb{R}^m$ being chosen such that $\tau_\epsilon(\bar{I})$ is a KAM torus on M_ϵ (see Theorem 4.1.17).

The procedure for obtaining (4.1.97), (4.1.99), (4.1.101), and (4.1.102) will involve deriving a first order, linear ordinary differential equation that a time dependent Melnikov vector must satisfy, solving the equation for the time dependent Melnikov vector and, finally, evaluating the solution at the appropriate time so as to obtain (4.1.97), (4.1.99), (4.1.101), and (4.1.102). At the appropriate point we will discuss the convergence properties of the improper integrals.

As a preliminary to our derivation of the Melnikov vectors, let us recall the geometry associated with the splitting of the manifolds. We are interested in the perturbed systems in a neighborhood of the $n + m + l$ dimensional unperturbed homoclinic manifold $\Gamma = W^s(M) \cap W^u(M) - M$, which is parametrized as follows:

$$\Gamma = \left\{ (x^I(-t_0, \alpha), I, \theta_0) \in \mathbb{R}^{2n} \times \mathbb{R}^m \times T^l \mid (t_0, \alpha, I, \theta_0) \in \mathbb{R}^1 \times \mathbb{R}^{n-1} \times U \times T^l \right\}$$

(note: for System II $U = T^m$ and for System III $l = m$). At each point $p \in \Gamma$ we constructed the $n + m$ dimensional plane Π_p spanned by the n linearly independent vectors $(D_x K_i(p), 0)$, $i = 1, \ldots, n$, where "0" represents the $m + l$ dimensional zero vector, and the m linearly independent unit vectors $\hat{I}_i$, $i = 1, \ldots, m$, where the $\hat{I}_i$ represent constant vectors in the I_i directions. We then argued that $W^s(\mathcal{M})$ and $W^u(\mathcal{M})$ intersected Π_p transversely in the m dimensional coincident surfaces S_p^s and S_p^u, respectively, for each $p \in \Gamma$. This geometrical structure of the unperturbed system was the backbone on which we derived our measurement of the splitting of the manifolds of certain invariant tori in the perturbed system. In the perturbed system, $\mathcal{M}$ persisted (denoted by $\mathcal{M}_\epsilon$) and, by transversality, for ϵ sufficiently small, $W^s(\mathcal{M}_\epsilon)$ and $W^u(\mathcal{M}_\epsilon)$ intersected Π_p transversely in the m dimensional surfaces $S_{p,\epsilon}^s$ and $S_{p,\epsilon}^u$, respectively, for each $p \in \Gamma$. Unlike in the unperturbed systems, it was possible for $W^s(\mathcal{M}_\epsilon)$ and $W^u(\mathcal{M}_\epsilon)$ to intersect Π_p in multiple disconnected components. With this possibility in mind, we chose $S_{p,\epsilon}^s$ and $S_{p,\epsilon}^u$ to be the m dimensional components that were closest to $\mathcal{M}_\epsilon$ in the sense of elapsed negative and positive flow time along $W^s(\mathcal{M}_\epsilon)$ and $W^u(\mathcal{M}_\epsilon)$, respectively. The reasoning behind this choice will be explained shortly. Now, our interest was not necessarily in the splitting of the stable and unstable manifolds of $\mathcal{M}_\epsilon$, but rather in the splitting of the stable and unstable manifolds of invariant tori that were contained in $\mathcal{M}_\epsilon$. There were three distinct situations, as follows:

System I. An l dimensional normally hyperbolic invariant torus, $\tau_\epsilon(\bar{I})$, having an $n + j + l$ dimensional stable manifold, $W^s(\tau_\epsilon(\bar{I}))$, and an $n + m - j + l$ dimensional unstable manifold, $W^u(\tau_\epsilon(\bar{I}))$, was located on $\mathcal{M}_\epsilon$ using averaging. The averaging method required nonresonance conditions resulting in the flow on $\tau_\epsilon(\bar{I})$ being irrational with orbits densely filling the torus. We then argued that $W^s(\tau_\epsilon(\bar{I})) \cap S_{p,\epsilon}^s \equiv W_p^s(\tau_\epsilon(\bar{I}))$ was a j dimensional set, and $W^u(\tau_\epsilon(\bar{I})) \cap S_{p,\epsilon}^u \equiv W_p^u(\tau_\epsilon(\bar{I}))$ was an $m - j$ dimensional set for each $p \in \Gamma$. We chose points $p_\epsilon^s = (x_\epsilon^s, I_\epsilon^s) \in W_p^s(\tau_\epsilon(\bar{I}))$ and $p_\epsilon^u = (x_\epsilon^u, I_\epsilon^u) \in W_p^u(\tau_\epsilon(\bar{I}))$ such that $I_\epsilon^u = I_\epsilon^s$. This was possible due to the normal hyperbolicity of $\tau_\epsilon(\bar{I})$ (see Lemma 4.1.8). Then the signed distance between $W^s(\tau_\epsilon(\bar{I}))$ and $W^u(\tau_\epsilon(\bar{I}))$ along the remaining n independent directions on Π_p was given by

$$d_i^{\bar{I}}(p; \epsilon) = d_i^{\bar{I}}(t_0, \theta_0, \alpha; \mu; \epsilon) = \frac{\langle D_x K_i(x^{\bar{I}}(-t_0, \alpha), \bar{I}), x_\epsilon^u - x_\epsilon^s \rangle}{\left\| D_x K_i(x^{\bar{I}}(-t_0, \alpha), \bar{I}) \right\|}, \quad i = 1, \ldots, n.$$

$$(4.1.103)$$

Now, since $W^s(\mathcal{M}_\epsilon)$ and $W^u(\mathcal{M}_\epsilon)$ are differentiable in ϵ, we Taylor expanded (4.1.103) about $\epsilon = 0$ and obtained

$$d_i^{\bar{I}}(t_0,\theta_0,\alpha,\mu;\epsilon) = \epsilon \frac{\langle D_x K_i(x^{\bar{I}}(-t_0,\alpha),\bar{I}), \frac{\partial x_\epsilon^u}{\partial \epsilon}\big|_{\epsilon=0} - \frac{\partial x_\epsilon^s}{\partial \epsilon}\big|_{\epsilon=0}\rangle}{\|D_x K_i(x^{\bar{I}}(-t_0,\alpha),\bar{I})\|} + \mathcal{O}(\epsilon^2),$$

$$i = 1,\ldots,n \qquad (4.1.104)$$

where $d_i^{\bar{I}}(t_0,\theta_0,\alpha,\mu;0) = 0$ since $S_p^s = S_p^u$.

The Melnikov vector was defined to be

$$M^{\bar{I}}(\theta_0,\alpha,\mu) = \left(M_1^{\bar{I}}(\theta_0,\alpha;\mu),\ldots,M_n^{\bar{I}}(\theta_0,\alpha;\mu)\right) \qquad (4.1.105)$$

where

$$M_i^{\bar{I}}(\theta_0,\alpha;\mu) = \langle D_x K_i(x^{\bar{I}}(-t_0,\alpha),\bar{I}), \frac{\partial x_\epsilon^u}{\partial \epsilon}\big|_{\epsilon=0} - \frac{\partial x_\epsilon^s}{\partial \epsilon}\big|_{\epsilon=0}\rangle, \quad i = 1,\ldots,n.$$

$$(4.1.106)$$

System II. In this case $\mathcal{M}_\epsilon$ was itself an $m+l$ dimensional normally hyperbolic invariant torus having $n+m+l$ dimensional stable and unstable manifolds denoted $W^s(\mathcal{M}_\epsilon)$ and $W^u(\mathcal{M}_\epsilon)$, respectively. At each $p \in \Gamma$ we choose points $p_\epsilon^s = (x_\epsilon^s, I_\epsilon^s) \in S_{p,\epsilon}^s$ and $p_\epsilon^u = (x_\epsilon^u, I_\epsilon^u) \in S_{p,\epsilon}^u$ such that $I_\epsilon^s = I_\epsilon^u$. Then the signed distance between $W^s(\mathcal{M}_\epsilon)$ and $W^u(\mathcal{M}_\epsilon)$ at the point p along the n independent directions on Π_p was given by

$$d_i(p,\epsilon) = d_i(t_0,I,\theta_0,\alpha,\mu;\epsilon) = \frac{\langle D_x K_i(x^I(-t_0,\alpha),I), x_\epsilon^u - x_\epsilon^s\rangle}{\|D_x K_i(x^I(-t_0,\alpha),I)\|}, \quad i = 1,\ldots,n.$$

$$(4.1.107)$$

Taylor expanding (4.1.107) about $\epsilon = 0$ gave

$$d_i(t_0,I,\theta_0,\alpha,\mu;\epsilon) = \epsilon \frac{\langle D_x K_i(x^I(-t_0,\alpha),I), \frac{\partial x_\epsilon^u}{\partial \epsilon}\big|_{\epsilon=0} - \frac{\partial x_\epsilon^s}{\partial \epsilon}\big|_{\epsilon=0}\rangle}{\|D_x K_i(x^I(-t_0,\alpha),I)\|} + \mathcal{O}(\epsilon^2),$$

$$i = 1,\ldots,n \qquad (4.1.108)$$

where $d_i(t_0,I,\theta_0,\alpha,\mu;0) = 0$ since $S_p^s = S_p^u$.

The Melnikov vector was defined to be

$$M(I,\theta_0,\alpha,\mu) = \left(M_1(I,\theta_0,\alpha;\mu),\ldots,M_n(I,\theta_0,\alpha;\mu)\right) \qquad (4.1.109)$$

where

$$M_i(I,\theta_0,\alpha,\mu) = \langle D_x K_i(x^I(-t_0,\alpha),I), \frac{\partial x^u_\epsilon}{\partial \epsilon}\Big|_{\epsilon=0} - \frac{\partial x^s_\epsilon}{\partial \epsilon}\Big|_{\epsilon=0} \rangle, \quad i=1,\ldots,n.$$

(4.1.110)

Note that for System II, I is a variable of the Melnikov vector since it is an angular variable along the torus $\mathcal{M}_\epsilon$.

System III. In this case, we located an m dimensional invariant torus on $\mathcal{M}_\epsilon$, $\tau_\epsilon(\bar{I})$, using the KAM theorem. This invariant torus had $n+m$ dimensional stable and unstable manifolds denoted $W^s(\tau_\epsilon(\bar{I}))$ and $W^u(\tau_\epsilon(\bar{I}))$, respectively. For each $p \in \Gamma$, $W^s(\tau_\epsilon(\bar{I}))$ and $W^u(\tau_\epsilon(\bar{I}))$ intersected Π_p transversely in the points $p^s_\epsilon = (x^s_\epsilon, I^s_\epsilon) \in S^s_{p,\epsilon}$ and $p^u_\epsilon = (x^u_\epsilon, I^u_\epsilon) \in S^u_{p,\epsilon}$, respectively. The signed distance between $W^s(\tau_\epsilon(\bar{I}))$ and $W^u(\tau_\epsilon(\bar{I}))$ at the point p along the $n+m$ directions on Π_p was defined to be

$$d^{\bar{I}}_i(p,\epsilon) = d^{\bar{I}}_i(t_0,\theta_0,\alpha,\mu;\epsilon) = \begin{cases} \dfrac{\langle D_x K_i(x^{\bar{I}}(-t_0,\alpha),\bar{I}), x^u_\epsilon - x^s_\epsilon \rangle}{\|D_x K_i(x^{\bar{I}}(-t_0,\alpha),\bar{I})\|}, & i=2,\ldots,n \\[2mm] (I^u_\epsilon)_{i-n} - (I^s_\epsilon)_{i-n}, & i=n+1,\ldots,n+m. \end{cases}$$

(4.1.111)

Taylor expanding (4.1.111) about $\epsilon = 0$ gave

$$d^{\bar{I}}_i(t_0,\theta_0,\alpha,\mu;\epsilon) = \begin{cases} \epsilon\dfrac{\langle D_x K_i(x^{\bar{I}}(-t_0,\alpha),\bar{I}), \frac{\partial x^u_\epsilon}{\partial \epsilon}\big|_{\epsilon=0} - \frac{\partial x^s_\epsilon}{\partial \epsilon}\big|_{\epsilon=0} \rangle}{\|D_x K_i(x^{\bar{I}}(-t_0,\alpha),\bar{I})\|} + O(\epsilon^2), \\[2mm] \hspace{5cm} i=2,\ldots,n \\[4mm] \epsilon\left[\left(\frac{\partial I^u_\epsilon}{\partial \epsilon}\Big|_{\epsilon=0}\right)_{i-n} - \left(\frac{\partial I^s_\epsilon}{\partial \epsilon}\Big|_{\epsilon=0}\right)_{i-n}\right] + O(\epsilon^2), \\[2mm] \hspace{5cm} i=n+1,\ldots,n+m \end{cases}$$

(4.1.112)

where $d^{\bar{I}}_i(t_0,\theta_0,\alpha,\mu;0) = 0$ since $W^s(\tau(\bar{I})) \cap \Pi_p = W^u(\tau(\bar{I})) \cap \Pi_p = p$.

The Melnikov vector was defined to be

$$M^{\bar{I}}(\theta_0,\alpha;\mu) = \left(M^{\bar{I}}_2(\theta_0,\alpha;\mu),\ldots,M^{\bar{I}}_{n+m}(\theta_0,\alpha;\mu)\right)$$

(4.1.113)

where

$$M^{\bar{I}}_i(\theta_0,\alpha;\mu) = \langle D_x K_i(x^{\bar{I}}(-t_0,\alpha),\bar{I}), \frac{\partial x^u_\epsilon}{\partial \epsilon}\Big|_{\epsilon=0} - \frac{\partial x^s_\epsilon}{\partial \epsilon}\Big|_{\epsilon=0} \rangle, \quad i=2,\ldots,n$$

(4.1.114)

$$M_i^{\bar{I}}(\theta_0, \alpha; \mu) = \left(\frac{\partial I_\epsilon^u}{\partial \epsilon}\bigg|_{\epsilon=0}\right)_{i-n} - \left(\frac{\partial I_\epsilon^s}{\partial \epsilon}\bigg|_{\epsilon=0}\right)_{i-n}, \quad i = n+1,\ldots,n+m.$$

$$(4.1.115)$$

We remark that we have not measured along the direction $(D_x K_1, 0) = (D_x H, 0)$, since, for System III, the level surfaces of $H_\epsilon = H + \epsilon \tilde{H}$ are preserved under the perturbation and the direction $(D_x K_1, 0)$ is complementary to these surfaces (see Lemma 4.1.18).

Our goal now is to show that (4.1.106), (4.1.110), (4.1.114), and (4.1.115) are given by (4.1.97), (4.1.99), (4.1.101), and (4.1.102), respectively. However, first we want to establish some shorthand notation that will make the formulas more manageable.

a) We will denote the perturbed vector fields for Systems I, II, and III by

$$\dot{q} = f(q) + \epsilon g(q; \mu, \epsilon) \tag{4.1.116}$$

where $f = (JD_x H, 0, \Omega)$ for Systems I and II; $f = (JD_x H, 0, D_I H)$ for System III; $g = (g^x, g^I, g^\theta)$ for Systems I and II; and $g = (JD_x \tilde{H}, -D_\theta \tilde{H}, D_I \tilde{H})$ for System III.

b) We denote trajectories of the unperturbed system along the homoclinic manifold Γ by

$$q_0^I(t - t_0) = \left(x^I(t - t_0, \alpha), I, \int^t \Omega(x^I(s, \alpha)I)\,ds + \theta_0\right) \tag{4.1.117}$$

where $\Omega \equiv D_I H$ for System III, and we denote trajectories of the perturbed system in $W^{s,u}(M_\epsilon)$ by

$$q_\epsilon^{s,u}(t) = \left(x_\epsilon^{s,u}(t), I_\epsilon^{s,u}(t), \theta_\epsilon^{s,u}(t)\right). \tag{4.1.118}$$

i) The Time Dependent Melnikov Vector

We define time dependent Melnikov vectors for Systems I, II, and III as follows:

System I.

$$M_i^{\bar{I}}(t) = \langle D_x K_i(x^{\bar{I}}(t - t_0, \alpha), \bar{I}), \frac{\partial x_\epsilon^u(t)}{\partial \epsilon}\bigg|_{\epsilon=0} - \frac{\partial x_\epsilon^s(t)}{\partial \epsilon}\bigg|_{\epsilon=0}\rangle, \quad i = 1,\ldots,n.$$

$$(4.1.119)$$

System II.

$$M_i(t) = \langle D_x K_i\big(x^I(t - t_0, \alpha), I\big), \frac{\partial x_\epsilon^u(t)}{\partial \epsilon}\bigg|_{\epsilon=0} - \frac{\partial x_\epsilon^s(t)}{\partial \epsilon}\bigg|_{\epsilon=0} \rangle, \quad i = 1, \ldots, n.$$

$$(4.1.120)$$

System III.

$$M_i(t) = \begin{cases} \langle D_x K_i\big(x^{\bar I}(t - t_0, \alpha), \bar I\big), \frac{\partial x_\epsilon^u(t)}{\partial \epsilon}\bigg|_{\epsilon=0} - \frac{\partial x_\epsilon^s(t)}{\partial \epsilon}\bigg|_{\epsilon=0} \rangle, \quad i = 2, \ldots, n \\ \qquad\qquad\qquad\qquad\qquad\qquad\qquad\qquad\qquad\qquad (4.1.121) \\ \\ \left(\frac{\partial I_\epsilon^u(t)}{\partial \epsilon}\bigg|_{\epsilon=0}\right)_{i-n} - \left(\frac{\partial I_\epsilon^s(t)}{\partial \epsilon}\bigg|_{\epsilon=0}\right)_{i-n}, \quad i = n+1, \ldots, n+m \\ \qquad\qquad\qquad\qquad\qquad\qquad\qquad\qquad\qquad\qquad (4.1.122) \end{cases}$$

where the trajectories $q_\epsilon^s(t) = \big(x_\epsilon^s(t), I_\epsilon^s(t), \theta_\epsilon^s(t)\big)$ and $q_\epsilon^u(t) = \big(x_\epsilon^u(t), I_\epsilon^u(t), \theta_\epsilon^u(t)\big)$ lie in the stable and unstable manifolds of the invariant torus on M_ϵ and satisfy $q_\epsilon^s(0) = \big(x_\epsilon^s(0), I_\epsilon^s(0), \theta_\epsilon^s(0)\big) = (x_\epsilon^s, I_\epsilon^s, \theta_0)$ and $q_\epsilon^u(0) = \big(x_\epsilon^u(0), I_\epsilon^u(0), \theta_\epsilon^u(0)\big) = (x_\epsilon^u, I_\epsilon^u, \theta_0)$.

The trajectories $q_\epsilon^s(t)$ and $q_\epsilon^u(t)$ satisfy the equations

$$\dot q_\epsilon^s = f(q_\epsilon^s) + \epsilon g(q_\epsilon^s, \mu; \epsilon) \tag{4.1.123}$$

$$\dot q_\epsilon^u = f(q_\epsilon^u) + \epsilon g(q_\epsilon^u, \mu; \epsilon). \tag{4.1.124}$$

We will be interested in the length of the time interval on which these solutions are valid. We have the following lemma.

Lemma 4.1.21. *For ϵ sufficiently small, $q_\epsilon^s(t)$ and $q_\epsilon^u(t)$ are solutions of (4.1.123) and (4.1.124), respectively, which exist on the semi-infinite time intervals $[0, \infty)$ and $(-\infty, 0]$ for all initial conditions $q_\epsilon^s(0)$ and $q_\epsilon^u(0)$ contained in the stable and unstable manifolds of the invariant torus.*

PROOF: The proof of this is obvious since, by definition of the stable (resp. unstable) manifold, given any point in the stable (resp. unstable) manifold of the invariant torus, the trajectory through this point exists for all positive (resp. negative) time and is asymptotic to the invariant torus. It is necessary for ϵ to be taken small in order for M_ϵ, along with its stable and unstable manifolds, to exist. $\quad\square$

We remark that Lemma 4.1.21 does not imply that the trajectories $q_\epsilon^{u,s}(t)$ approximate $q_0^I(t)$ to within $\mathcal{O}(\epsilon)$ on the appropriate semi-infinite time intervals. This fact is not needed and, in general, is not true since, on the semi-infinite time intervals, the angular variables of perturbed and unperturbed trajectories may separate by an $\mathcal{O}(1)$ amount.

We will be interested in the time evolution of the quantities $\left.\dfrac{\partial x_\epsilon^{u,s}(t)}{\partial \epsilon}\right|_{\epsilon=0}$ and $\left.\dfrac{\partial I_\epsilon^{u,s}(t)}{\partial \epsilon}\right|_{\epsilon=0}$. As a shorthand notation, we define

$$
\begin{aligned}
x_1^{u,s}(t) &= \left.\frac{\partial x_\epsilon^{u,s}(t)}{\partial \epsilon}\right|_{\epsilon=0} \\
I_1^{u,s}(t) &= \left.\frac{\partial I_\epsilon^{u,s}(t)}{\partial \epsilon}\right|_{\epsilon=0} \\
\theta_1^{u,s}(t) &= \left.\frac{\partial \theta_\epsilon^{u,s}(t)}{\partial \epsilon}\right|_{\epsilon=0} .
\end{aligned}
\tag{4.1.125}
$$

Now, by Theorem 1.1.4, solutions of (4.1.123) and (4.1.124) are differentiable with respect to ϵ. By Theorem 1.1.5, the solutions $\left(x_1^{u,s}(t), I_1^{u,s}(t), \theta_1^{u,s}(t)\right)$ satisfy the first variational equation given by

$$
\begin{pmatrix} \dot{x}_1^{u,s} \\ \dot{I}_1^{u,s} \\ \dot{\theta}_1^{u,s} \end{pmatrix} = \begin{pmatrix} JD_x^2 H & D_I JD_x H & 0 \\ 0 & 0 & 0 \\ D_x \Omega & D_I \Omega & 0 \end{pmatrix} \begin{pmatrix} x_1^{u,s} \\ I_1^{u,s} \\ \theta_1^{u,s} \end{pmatrix} + \begin{pmatrix} g^x(q_0^I(t-t_0),\mu;0) \\ g^I(q_0^I(t-t_0),\mu;0) \\ g^\theta(q_0^I(t-t_0),\mu;0) \end{pmatrix}
\tag{4.1.126}
$$

where the entries of the matrix are evaluated on $\left(x^I(t-t_0,\alpha),I\right)$, $q_0^I(t-t_0) = \left(x^I(t-t_0,\alpha), I, \int^t \Omega(x^I(s,\alpha),I)ds + \theta_0\right)$ is an unperturbed homoclinic trajectory, and the vector (g^x,g^I,g^θ) is modified appropriately for System III (i.e., recall in System III we have $(g^x,g^I,g^\theta) = (JD_x\tilde{H}, -D_\theta\tilde{H}, D_I\tilde{H})$).

ii) An Ordinary Differential Equation for the Melnikov Vector

Consider the expression

$$
M_i(t) = \begin{cases} \langle D_x K_i(x^I(t-t_0,\alpha),I), x_1^u(t) - x_1^s(t)\rangle, & i = 1,\ldots,n \\ \left(I_1^u(t)\right)_{i-n} - \left(I_1^s(t)\right)_{i-n}, & i = n+1,\ldots,n+m . \end{cases}
\tag{4.1.127}
$$

We will derive a linear ordinary differential equation which (4.1.127) must satisfy. The solution of this equation evaluated at $t = 0$ will yield the Melnikov vectors for Systems I, II, and III. However, it will be necessary to impose conditions at $\pm\infty$ on the solutions of the equation, and these conditions will be dictated by dynamical phenomena that are specific to Systems I, II, and III individually.

As a shorthand notation we have

$$\Delta_i^{u,s}(t) \equiv \begin{cases} \langle D_x K_i(x^I(t-t_0,\alpha),I), x_1^{u,s}(t) \rangle, & i=1,\ldots,n \\ (I_1^{u,s}(t))_{i-n}, & i=n+1,\ldots,n+m \end{cases} \tag{4.1.128}$$

where now

$$M_i(t) = \Delta_i^u(t) - \Delta_i^s(t), \quad i=1,\ldots,n+m. \tag{4.1.129}$$

Differentiating (4.1.129) with respect to t gives

$$\dot{M}_i(t) = \dot{\Delta}_i^u(t) - \dot{\Delta}_i^s(t), \quad i=1,\ldots,n+m \tag{4.1.130}$$

where

$$\dot{\Delta}_i^{u,s}(t) = \begin{cases} \langle \frac{d}{dt}(D_x K_i(x^I(t-t_0,\alpha),I)), x_1^{u,s}(t) \rangle + \langle D_x K_i(x^I(t-t_0,\alpha),I), \dot{x}_1^{u,s}(t) \rangle, \\ \hspace{9cm} i=1,\ldots,n \\ (\dot{I}_1^{u,s}(t))_{i-n}, \hspace{6cm} i=n+1,\ldots,n+m. \end{cases}$$
$$\tag{4.1.131}$$

Using the chain rule, and the fact that $\dot{I}=0$ in the unperturbed system, we obtain

$$\frac{d}{dt}(D_x K_i(x^I(t-t_0,\alpha),I)) = D_x^2 K_i(x^I(t-t_0,\alpha),I)\dot{x}^I(t-t_0,\alpha). \tag{4.1.132}$$

Using the fact that $\dot{x}^I(t-t_0,\alpha) = J D_x H(x^I(t-t_0,\alpha),I)$, (4.1.132) becomes

$$\frac{d}{dt}(D_x K_i(x^I(t-t_0,\alpha),I)) = D_x^2 K_i(x^I(t-t_0,\alpha),I) J D_x H(x^I(t-t_0,\alpha),I). \tag{4.1.133}$$

From the first variational equation (4.1.126), we have

$$\dot{x}_1^{u,s} = J D_x^2 H(x^I(t-t_0,\alpha),I)x_1^{u,s} + D_I J D_x H(x^I(t-t_0,\alpha),I)I_1^{u,s} + g^x(q_0^I(t-t_0),\mu;0)$$
$$\dot{I}_1^{u,s} = g^I(q_0^I(t-t_0),\mu;0). \tag{4.1.134}$$

Substituting (4.1.134), (4.1.131) gives (note: henceforth we will leave out the arguments of the functions for the sake of a less cumbersome notation)

$$\dot{\Delta}_i^{u,s}(t) = \begin{cases} \langle D_x K_i, (J D_x^2 H)x_1^{u,s} \rangle + \langle D_x K_i, (D_I J D_x H)I_1^{u,s} \rangle \\ \quad + \langle D_x K_i, g^x \rangle + \langle x_1^{u,s}, (D_x^2 K_i)(J D_x H) \rangle, & i=1,\ldots,n \\ (g^I)_{i-n}, & i=n+1,\ldots,n+m. \end{cases}$$
$$\tag{4.1.135}$$

The first n components of (4.1.135) simplify considerably with the following lemma.

Lemma 4.1.22. $\langle D_x K_i, (JD_x^2 H)x_1^{u,s}\rangle + \langle x_1^{u,s}, (D_x^2 K_i)(JD_x H)\rangle = 0, i = 1, \ldots, n.$

PROOF: By I3, II3, or III3 we have

$$\langle JD_x H, D_x K_i\rangle = 0, \qquad i = 1, \ldots, n. \tag{4.1.136}$$

Differentiating (4.1.136) with respect to x gives

$$D_x\langle JD_x H, D_x K_i\rangle = (JD_x^2 H)^T D_x K_i + (D_x^2 K_i)JD_x H = 0, \quad i = 1, \ldots, n \tag{4.1.137}$$

where "T" denotes the matrix transpose. Taking the inner product of (4.1.137) with $x_1^{u,s}$ gives

$$\langle (JD_x^2 H)^T D_x K_i, x_1^{u,s}\rangle + \langle (D_x^2 K_i)(JD_x H), x_1^{u,s}\rangle = 0, \quad i = 1, \ldots, n \tag{4.1.138}$$

or

$$\langle D_x K_i, (JD_x^2 H)x_1^{u,s}\rangle + \langle x_1^{u,s}, (D_x^2 K_i)(JD_x H)\rangle = 0, \quad i = 1, \ldots, n. \tag{4.1.139}$$

$\square$

Using Lemma 4.1.22, (4.1.135) reduces to

$$\dot{\Delta}_i^{u,s}(t) = \begin{cases} \langle D_x K_i, g^x\rangle + \langle D_x K_i, (D_I JD_x H)I_1^{u,s}\rangle, & i = 1, \ldots, n \\ (g^I)_{i-n}, & i = n+1, \ldots, n+m. \end{cases} \tag{4.1.140}$$

iii) Solution of the Ordinary Differential Equation

Integrating $\dot{\Delta}_i^u(t)$ from $-T^u$ to 0 and $\dot{\Delta}_i^s(t)$ from 0 to T^s for some $T^s, T^u > 0$ gives

$$\Delta_i^u(0) - \Delta_i^u(-T^u) =$$

$$\begin{cases} \displaystyle\int_{-T^u}^0 [\langle D_x K_i, g^x\rangle + \langle D_x K_i, (D_I JD_x H)I_1^u\rangle] \, (q_0^I(t-t_0), \mu; 0)dt, \\ \hspace{6cm} i = 1, \ldots, n \\[4mm] \displaystyle\int_{-T^u}^0 (g^I)_{i-n}(q_0^I(t-t_0), \mu; 0)dt, \quad i = n+1, \ldots, n+m \end{cases} \tag{4.1.141}$$

$$\Delta_i^s(T^s) - \Delta_i^s(0) =$$

$$\begin{cases} \displaystyle\int_0^{T^s} \left[\langle D_x K_i, g^x \rangle + \langle D_x K_i, (D_I J D_x H) I_1^s \rangle \right] (q_0^I(t - t_0), \mu; 0) dt, \\ \qquad\qquad\qquad\qquad\qquad\qquad i = 1, \dots, n \\ \displaystyle\int_0^{T^s} (g^I)_{i-n} (q_0^I(t - t_0), \mu; 0) dt, \quad i = n + 1, \dots, n + m. \end{cases} \tag{4.1.142}$$

We will want to consider the limit of (4.1.142) as $T^s \to +\infty$ and (4.1.141) as $-T^u \to -\infty$. The following lemma will be useful.

Lemma 4.1.23. $D_x K_i (\gamma(I), I) = 0, \ i = 1, \dots, n.$

PROOF: $\gamma(I)$ is the surface of hyperbolic fixed points of the x components of the unperturbed vector fields. Therefore,

$$J D_x H (\gamma(I), I) = J D_x K_1 (\gamma(I), I) = 0 \tag{4.1.143}$$

and, since J is nondegenerate,

$$D_x H (\gamma(I), I) = D_x K_1 (\gamma(I), I) = 0 . \tag{4.1.144}$$

Now from (4.1.137) we have

$$D_x \langle J D_x H, D_x K_i \rangle = (J D_x^2 H)^T D_x K_i + (D_x^2 K_i) J D_x H = 0 , \qquad i = 1, \dots, n . \tag{4.1.145}$$

Evaluating (4.1.145) on $(\gamma(I), I)$ and using (4.1.143) gives

$$(J D_x^2 H(\gamma(I), I))^T D_x K_i (\gamma(I), I) = 0 , \quad i = 1, \dots, n . \tag{4.1.146}$$

Now, since $(\gamma(I), I)$ is a hyperbolic fixed point, $\det \left[J D_x^2 H(\gamma(I), I) \right] \neq 0$, and therefore $D_x K_i (\gamma(I), I) = 0, i = 1, \dots, n.$ $\qquad\square$

Using (4.1.129) the components of the Melnikov vector are given by

$$M_i^I(\theta_0, \alpha, \mu) = \Delta_i^u(0) - \Delta_i^s(0) . \tag{4.1.147}$$

Using (4.1.141) and (4.1.142), we will evaluate (4.1.147) for the three systems individually.

System I. From the first variational equation (4.1.126) we have

$$I_1^s(t) = I_1^u(t) = \int^{t-t_0} g^I .$$

(4.1.148)

Substituting (4.1.148) into (4.1.141) into (4.1.142) gives

$$M_i^{\bar{I}}(\theta_0, \alpha; \mu) = \Delta_i^u(0) - \Delta_i^s(0)$$

$$= \int_{-T^u}^{T^s} [\langle D_x K_i, g^x \rangle + \langle D_x K_i, (D_I J D_x H) \int^{t-t_0} g^I \rangle](q_0^I(t-t_0), \mu; 0)dt$$

$$+\Delta_i^u(-T^u) - \Delta_i^s(T^s) , \qquad i = 1, \dots, n .$$

(4.1.149)

Now we want to consider the limit of (4.1.149) as $-T^u \to -\infty$ and $T^s \to +\infty$.

Lemma 4.1.24. $\lim\limits_{-T^u \to -\infty} \Delta_i^u(-T^u) = \lim\limits_{T^s \to \infty} \Delta_i^s(T^s) = 0.$

PROOF: We will give the argument for Δ_i^u; the argument for Δ_i^s is similar.

From (4.1.129) we have

$$\Delta_i^u(t) = \langle D_x K_i(x^{\bar{I}}(t-t_0, \alpha), \bar{I}), x_1^u(t) \rangle .$$

(4.1.150)

Now, $x_i^u(t)$ can grow at best linearly in time and, by Lemma 4.1.23, $D_x K_i(x^{\bar{I}}(t - t_0, \alpha), \bar{I})$ goes to zero exponentially fast as $t \to -\infty$. Therefore, $\lim\limits_{t \to -\infty} \Delta_i^u(t) = 0.$ □

So now we have obtained (4.1.97a)

$$M_i^{\bar{I}}(\theta_0, \alpha; \mu) =$$

$$\int_{-\infty}^{\infty} [\langle D_x K_i, g^x \rangle + \langle D_x K_i, (D_I J D_x H) \int^{t-t_0} g^I \rangle](q_0^{\bar{I}}(t-t_0), \mu; 0)dt ,$$

$$i = 1, \dots, n .$$

(4.1.151)

Proposition 4.1.25. *The improper integrals in (4.1.151) converge absolutely.*

PROOF: $x^{\bar{I}}(t - t_0, \alpha) \to \gamma(\bar{I})$ exponentially fast as $t \to \pm\infty$, since $\gamma(\bar{I})$ is a hyperbolic fixed point of the x component of the unperturbed vector field.

Therefore, by Lemma 4.1.23, $D_x K_i(x^I(t - t_0, \alpha), I) \to 0$ exponentially fast as $t \to \pm\infty$. Now, since g^x and $(D_I J D_x H) \int^{t-t_0} g^I$ are bounded on bounded subsets of their respective domains of definition, we can conclude that the integral in (4.1.151) converges absolutely as $T^s \to +\infty$, $-T^u \to -\infty$. $\qquad\square$

We remark that convergence properties of Melnikov type integrals were first studied in detail by Robinson [1985].

We now show how to obtain the form of equation (4.1.97b).

Lemma 4.1.26. $\langle D_x K_i, (D_I J D_x H) I_1^{u,s} \rangle = -\langle \frac{d}{dt}(D_I K_i), I_1^{u,s} \rangle$ *evaluated on the unperturbed homoclinic orbit* $(x^I(t - t_0, \alpha), I)$.

PROOF: On the unperturbed homoclinic orbit we have

$$\frac{d}{dt}(D_I K_i) = (D_x D_I K_i)(J D_x H) \qquad (4.1.152)$$

since $\dot{I} = 0$. Differentiating the Poisson Bracket we have

$$D_I \langle J D_x H, D_x K_i \rangle = (D_I J D_x H)^T D_x K_i + (D_I D_x K_i)^T J D_x H = 0 \qquad (4.1.153)$$

where "T" denotes the matrix transpose. Combining (4.1.152) and (4.1.153), and using the fact that $D_x D_I K_i = (D_I D_x K_i)^T$, we get the following identity on the unperturbed homoclinic orbit

$$(D_I J D_x H)^T D_x K_i = -\frac{d}{dt}(D_I K_i). \qquad (4.1.154)$$

Taking the inner product of (4.1.154) with $I_1^{u,s}$ gives

$$\langle (D_I J D_x H)^T D_x K_i, I_1^{u,s} \rangle = \langle -\frac{d}{dt}(D_I K_i), I_1^{u,s} \rangle \qquad (4.1.155)$$

but

$$\langle (D_I J D_x H)^T D_x K_i, I_1^{u,s} \rangle = \langle D_x K_i, (D_I J D_x H) I_1^{u,s} \rangle \qquad (4.1.156)$$

which gives the result. $\qquad\square$

Now, from (4.1.149), we have

$$M_i^I(\theta_0, \alpha, \mu) = \Delta_i^u(0) - \Delta_i^s(0)$$

$$= \int_{-T^u}^{0} \left[\langle D_x K_i, g^x \rangle - \langle \frac{d}{dt}(D_I K_i), I_1^u \rangle \right] dt + \int_{0}^{T^s} \left[\langle D_x K_i, g^x \rangle - \langle \frac{d}{dt}(D_I K_i), I_1^s \rangle \right] dt$$

$$+ \Delta_i^u(-T^u) - \Delta_i^s(T^s), \qquad i = 1, \ldots, n \qquad (4.1.157)$$

where we have left out the argument of the integrand for the sake of a less cumbersome notation. Integrating the second term in each integrand once by parts, and using the fact that $\dot{I}_1^u = \dot{I}_1^s = g^I$, gives

$$M_i^{\bar{I}}(\theta_0,\alpha,\mu) = \int_{-T^u}^{T^s} \Big[\langle D_x K_i, g^x\rangle + \langle D_I K_i, g^I\rangle\Big]dt - \langle D_I K_i, I_1^u\rangle\Big|_{-T^u}^0 - \langle D_I K_i, I_1^s\rangle\Big|_0^{T^s}$$

$$+\Delta_i^u(-T^u) - \Delta_i^s(T^s), \qquad i = 1,\ldots,n. \qquad (4.1.158)$$

As before, we want to consider the limit of (4.1.158) as T^s, $T^u \to \infty$. First we give two preliminary lemmas.

Lemma 4.1.27. *For each ϵ sufficiently small there exists monotonely increasing sequences of real numbers $\{T_j^s\}$, $\{T_j^u\}$, $j = 1,2,\ldots$, with $\lim_{j\to\infty} T_j^{s,u} = \infty$ such that*

1) $\lim_{j\to\infty} \Big|q_\epsilon^s(T_j^s) - q_\epsilon^u(-T_j^u)\Big| = 0;$

2) $\lim_{j\to\infty} \Big|g^I\big(q_\epsilon^s(T_j^s),\mu;0\big)\Big| = \lim_{j\to\infty} \Big|g^I\big(q_\epsilon^u(-T_j^u),\mu;0\big)\Big| = 0.$

PROOF: 1) This follows from the fact that orbits in the stable and unstable manifolds of the torus approach the torus with asymptotic phase (see Fenichel [1974], [1979]). Then we can choose sequences of times so that $q_\epsilon^s(t)$ and $q_\epsilon^u(t)$ approach the same point on the torus along these sequences of times. 2) Recall that g^I has zero average on the torus (see Proposition 4.1.6). Thus, rather than choosing sequences of times such that $q_\epsilon^s(t)$ and $q_\epsilon^u(t)$ approach any arbitrary point on the torus, we can choose the sequences so that $q_\epsilon^s(t)$ and $q_\epsilon^u(t)$ approach a point on the torus such that g^I vanishes at that point. This uses the fact that trajectories on the torus are dense, see proposition 4.16. $\qquad \square$

Lemma 4.1.28. $\lim_{j\to\infty} \left[-\langle D_I K_i, I_1^u\rangle\Big|_{-T_j^u}^0 - \langle D_I K_i, I_1^s\rangle\Big|_0^{T_j^s}\right] = -\langle D_I K_i(\gamma(I),I),$

$\int_{-\infty}^{\infty} g^I(q_0^{\bar{I}}(t-t_0),\mu;0)dt\rangle$, *where $\{T_j^s\}$, $\{T_j^u\}$ are chosen as in Lemma 4.1.27.*

PROOF: Writing out this expression in detail we obtain

$$-\langle D_I K_i\big(x^{\bar{I}}(-t_0,\alpha),\bar{I}\big),I_1^u(0)\rangle + \langle D_I K_i\big(x^{\bar{I}}(-T_j^u-t_0,\alpha),\bar{I}\big),I_1^u(-T_j^u)\rangle$$

$$-\langle D_I K_i\big(x^{\bar{I}}(T_j^s-t_0,\alpha),\bar{I}\big),I_1^s(T_j^s)\rangle + \langle D_I K_i\big(x^{\bar{I}}(-t_0,\alpha),\bar{I}\big),I_1^s(0)\rangle.$$

$$(4.1.159)$$

Since $I_1^u(0) = I_1^s(0)$, (4.1.159) reduces to

$$\langle D_I K_i(x^{\bar{I}}(-T_j^u - t_0, \alpha), \bar{I}) I_1^u(-T_j^u) \rangle - \langle D_I K_i(x^{\bar{I}}(T_j^s - t_0, \alpha), \bar{I}) I_1^s(T_j^s) \rangle. \quad (4.1.160)$$

Now as $-T_j^u \to -\infty$ we have

$$x^{\bar{I}}(-T_j^u - t_0, \alpha) \to \gamma(\bar{I}), \quad (4.1.161)$$

and as $T_j^s \to \infty$ we have

$$x^{\bar{I}}(T_j^s - t_0, \alpha) \to \gamma(\bar{I}). \quad (4.1.162)$$

Also, from the first Variational equation (4.1.126) we obtain

$$\lim_{j \to \infty} I_1^s(T_j^s) - I_1^u(-T_j^u) = \int_{-\infty}^{\infty} g^I(q_0^{\bar{I}}(t - t_0), \mu; 0) dt. \quad (4.1.163)$$

Therefore, using (4.1.160), (4.1.161), and (4.1.163), we have

$$\lim_{j \to \infty} \left[\langle D_I K_i(x^{\bar{I}}(-T_j^u - t_0, \alpha), \bar{I}) I_1^u(-T_j^u) \rangle - \langle D_I K_i(x^{\bar{I}}(T_j^s - t_0, \alpha), \bar{I}) I_1^s(T_j^s) \rangle \right]$$

$$= -\langle D_I K_i(\gamma(I), I), \int_{-\infty}^{\infty} g^I(q_0^{\bar{I}}(t - t_0), \mu; 0) dt \rangle. \quad (4.1.164)$$

$$\square$$

So, using Lemmas 4.1.24 and 4.1.28, we obtain

$$M_i^{\bar{I}}(\theta_0, \alpha; \mu) = \int_{-\infty}^{\infty} \left[\langle D_x K_i, g^x \rangle + \langle D_I K_i, g^I \rangle \right] (q_0^{\bar{I}}(t - t_0), \mu; 0) dt,$$

$$-\langle D_I K_i(\gamma(\bar{I}), \bar{I}), \int_{-\infty}^{\infty} g^I(q_0^{\bar{I}}(t - t_0), \mu; 0) dt \rangle$$

$$i = 1, \ldots, n. \quad (4.1.165)$$

Now (4.1.165) converges absolutely, since it is just another way of writing (4.1.151) (note: $(D_I K_i(x^{\bar{I}}(t - t_0, \alpha), \bar{I}) - D_I K_i(\gamma(\bar{I}), \bar{I})) \to 0$ exponentially fast as $t \to \pm\infty$

by Lemma 4.1.23, and by the fact that $x^{\bar{I}}(t - t_0, \alpha) \to \gamma(\bar{I})$ exponentially fast as $t \to \pm\infty$). However, the two terms in the integral

$$\int_{-\infty}^{\infty} \langle D_I K_i(x^{\bar{I}}(t - t_0, \alpha), \bar{I}), g^I(q_0^{\bar{I}}(t - t_0), \mu; 0)\rangle dt \qquad (4.2.166)$$

$$\langle D_I K_i(\gamma(\bar{I}), \bar{I}), \int_{-\infty}^{\infty} g^I(q_0^{\bar{I}}(t - t_0), \mu; 0)dt \rangle \qquad (4.2.167)$$

each individually only converge conditionally. This is expressed in the following proposition.

Proposition 4.1.29. *Let* $\{T_j^s\}$, $\{T_j^u\}$, $j = 1, 2, \ldots$, *be chosen as in Lemma 4.1.27. Then (4.2.166) and (4.2.167) converge conditionally when the limits of integration are allowed to approach* $+\infty$ *and* $-\infty$ *along the sequences* $\{T_j^s\}$ *and* $\{-T_j^u\}$, *respectively.*

As $j \to \infty$, the homoclinic trajectory approaches the invariant torus $\tau_\epsilon(\bar{I})$ exponentially fast along the sequences of times $\{T_j^s\}$, $\{-T_j^u\}$. So, by Lemma 4.1.27, along these sequences of times g^I goes to zero exponentially fast along the homoclinic trajectory. Recall that $D_I K_i$ is assumed to be bounded on bounded subsets of its domain of definition. Therefore, the term $\langle D_I K_i, g^I \rangle$ goes to zero exponentially fast on the homoclinic trajectory with the choice of sequences of times satisfying Lemma 4.1.27. $\qquad\qquad\qquad\square$

System II. The components of the Melnikov vector for System II are also given by (4.1.151):

$$M_i(I, \theta_0, \alpha; \mu) =$$
$$\int_{-\infty}^{\infty} [\langle D_x K_i, g^x \rangle + \langle D_x K_i, (D_I J D_x H) \int^{t-t_0} g^I \rangle](q_0^I(t - t_0), \mu; 0)dt, \qquad (4.1.168a)$$

or, using Lemma 4.1.26 and 4.1.28,

$$M_i(I, \theta_0, \alpha; \mu) = \int\limits_{-\infty}^{\infty} \left[\langle D_x K_i, g^x \rangle + \langle D_I K_i, g^I \rangle \right] (q_0^I(t - t_0), \mu; 0) \, dt$$

$$-\langle D_I K_i(\gamma(I), I), \int\limits_{-\infty}^{\infty} g^I(q_0^I(t - t_0), \mu; 0) \, dt \rangle$$

$$i = 1, \ldots, n \quad . \qquad (4.1.168b)$$

The absolute convergence of (4.1.168) is established using an argument identical to that given in Proposition 4.1.25 (note: recall that I is an m vector of angular variables for System II).

System III. Substituting $g^x = J D_x \tilde{H}$ and $g^I = -D_\theta \tilde{H}$ into (4.1.151) gives

$$M_i^{\bar{I}}(\theta_0, \alpha; \mu) =$$

$$\int\limits_{-\infty}^{\infty} \left[\langle D_x K_i, J D_x \tilde{H} \rangle - \langle D_x K_i, (D_I J D_x H) \int\limits^{t-t_0} D_\theta \tilde{H} \rangle \right] (q_0^{\bar{I}}(t - t_0), \mu; 0) \, dt,$$

$$(4.1.169a)$$

or, using Lemma 4.1.26 and 4.1.28,

$$M_i^{\bar{I}}(\theta_0, \alpha; \mu) = \int\limits_{-\infty}^{\infty} \left[\langle D_x K_i, J D_x \tilde{H} \rangle - \langle D_I K_i, D_\theta \tilde{H} \rangle \right] (q_0^{\bar{I}}(t - t_0), \mu; 0) \, dt$$

$$+ \langle D_I K_i(\gamma(\bar{I}), \bar{I}), \int\limits_{-\infty}^{\infty} D_\theta \tilde{H}(q_0^{\bar{I}}(t - t_0), \mu; 0) \, dt \rangle$$

$$(4.1.169b)$$

$$i = 2, \ldots, n$$

for the first $n - 1$ components of the Melnikov vector. Absolute convergence of (4.1.169) is an immediate result of Proposition 4.1.25. The remaining m components of the Melnikov vector require a more careful consideration.

From (4.1.141) and (4.1.142) we have

$$M_i^{\bar{I}}(\theta_0, \alpha; \mu) = \Delta_i^u(0) - \Delta_i^s(0)$$

$$= -\int\limits_{-T^u}^{T^s} D_{\theta_{i-n}} \tilde{H}(q_0^{\bar{I}}(t - t_0), \mu; 0) \, dt + \Delta_i^u(-T^u) - \Delta_i^s(T^s),$$

$$i = n + 1, \ldots, n + m. \qquad (4.1.170)$$

Now we want to consider the limit of (4.1.170) as T^s, $T^u \to \infty$.

Lemma 4.1.30. *For* $\{T_j^s\}$, $\{T_j^u\}$ *chosen according to Lemma 4.1.27,*

$$\lim_{j \to \infty} \left| \Delta_i^u(-T_j^u) - \Delta_i^s(T_j^s) \right| = 0 , \qquad i = n+1, \ldots, n+m .$$

PROOF: From (4.1.128) we have

$$\Delta_i^u(-T_j^u) - \Delta_i^s(T^s) = \left(I_1^u(-T_j^u) \right)_{i-n} - \left(I_1^s(T^s) \right)_{i-n} , \quad i = n+1, \ldots, n+m .$$
$$(4.1.171)$$

Since the I variables were chosen to lie in the unstable and stable manifolds of the KAM torus $\tau_\epsilon(\bar{I})$ the lemma is an immediate consequence of Lemma 4.1.27. $\quad\square$

So we have

$$M_i^{\bar{I}}(\theta_0, \alpha; \mu) = - \int\limits_{-\infty}^{\infty} D_{\theta_{i-n}} \tilde{H}(q_0^{\bar{I}}(t - t_0), \mu; 0) dt ,$$

$$i = n+1, \ldots, n+m . \qquad (4.1.172)$$

Proposition 4.1.29 applies directly to (4.1.172) and allows us to conclude that (4.1.172) converges conditionally.

iv) The Choice of $S_{p,\epsilon}^s$ and $S_{p,\epsilon}^u$

Recall our discussions of the splitting of the manifolds for the three systems. In the unperturbed systems, for each $p \in \Gamma$, $W^s(M)$ and $W^u(M)$ intersect the plane Π_p transversely in the coincident m dimensional surfaces S_p^s and S_p^u, respectively. Therefore, in the perturbed system, for ϵ sufficiently small, $W^s(M_\epsilon)$ and $W^u(M_\epsilon)$ intersect the plane Π_p transversely in the m dimensional surfaces $S_{p,\epsilon}^s$ and $S_{p,\epsilon}^u$, respectively, for each $p \in \Gamma$. However, it is possible for $W^s(M_\epsilon)$ and $W^u(M_\epsilon)$ to intersect Π_p in countably many disconnected components, see Figure 4.1.5.

In our construction of the measurement of distance between the stable and unstable manifolds of the invariant torus we chose points in $S_{p,\epsilon}^s$ and $S_{p,\epsilon}^u$ that were defined to be the components of $W^s(M_\epsilon) \cap \Pi_p$ and $W^u(M_\epsilon) \cap \Pi_p$ closest to M_ϵ in terms of positive and negative time of flow along $W^s(M_\epsilon)$ and $W^u(M_\epsilon)$, respectively. Before proceeding, let us define these sets more precisely.

Definition 4.1.1. Let $q_\epsilon^s(t)$ be any trajectory in $W^s(\mathcal{M}_\epsilon)$ with $q_\epsilon^s(0) \in S_{p,\epsilon}^s \subset W^s(\mathcal{M}_\epsilon) \cap \Pi_p$. Then $S_{p,\epsilon}^s$ is said to be the component of $W^s(\mathcal{M}_\epsilon) \cap \Pi_p$ closest to $\mathcal{M}_\epsilon$ in the sense of positive time of flow along $W^s(\mathcal{M}_\epsilon)$ if for all $t > 0$ $q_\epsilon^s(t) \cap \Pi_p = \emptyset$. A similar definition holds for $S_{p,\epsilon}^u$ in negative time with the obvious modifications.

Now we want to argue that the procedure utilized in deriving the computable form of the Melnikov vector given in (4.1.97), (4.1.99), (4.1.101), and (4.1.102) results in the Melnikov vector being a measure of the distance between points in the stable and unstable manifolds of the invariant torus which satisfy Definition 4.1.1.

For a fixed $p = \left(x^I(-t_0, \alpha), I, \theta_0\right) \in \Gamma$, let $S_{p,\epsilon}^s$ and $S_{p,\epsilon}^u$ be the components of $W^s(\mathcal{M}_\epsilon) \cap \Pi_p$ and $W^u(\mathcal{M}_\epsilon) \cap \Pi_p$ which are closest to $\mathcal{M}_\epsilon$ in the sense of positive and negative time of flow along $W^s(\mathcal{M}_\epsilon)$ and $W^u(\mathcal{M}_\epsilon)$, respectively. Let $\hat{S}_{p,\epsilon}^s$ and $\hat{S}_{p,\epsilon}^u$ denote additional components of $W^s(\mathcal{M}_\epsilon) \cap \Pi_p$ and $W^u(\mathcal{M}_\epsilon) \cap \Pi_p$. Recall that Π_p is defined to be the span of $\left\{\left(D_x K_i(x^I(-t_0, \alpha), I), 0\right)\right\}$, $i = 1, \dots, n$, and $\{\hat{I}_i\}$, $i = 1, \dots, m$. We will denote the time varying plane used in the construction of the time dependent Melnikov vector as $\Pi_{p(t)}$, which is the span of the time varying vectors $\left\{\left(D_x K_i(x^I(t - t_0, \alpha), I), 0\right)\right\}$, $i = 1, \dots, n$, and the constant vectors $\{\hat{I}_i\}$, $i = 1, \dots, m$.

Now consider the expression $\Delta_i^s(t)$ defined in (4.1.128). $\Delta_i^s(0)$ represents the $\mathcal{O}(\epsilon)$ term of the projection of the point $p_\epsilon^s = (x_\epsilon^s, I_\epsilon^s)$ in the stable manifold of the invariant torus along the ith coordinate on Π_p. $\Delta_i^s(t)$ represents the evolution of $\Delta_i^s(0)$ on the time interval $[0, \infty)$ with the plane $\Pi_{p(t)}$ evolving along a trajectory in the unperturbed homoclinic orbit Γ and the point $p_\epsilon^s(t)$ evolving along a trajectory in the perturbed stable manifold of the invariant torus. Now suppose that at $t = 0$ $p_\epsilon^s(0) \equiv p_\epsilon^s$ is contained in $\hat{S}_{p,\epsilon}^s$ rather than $S_{p,\epsilon}^s$. Then, by Definition 4.1.1, there exists some $T > 0$ such that $p_\epsilon^s(T) \in S_{p,\epsilon}^s \subset \Pi_p$. But, in this case, the plane Π_p has moved to $\Pi_{p(T)} \neq \Pi_p$. Therefore, $\Delta_i^s(T)$ does not approximate to $\mathcal{O}(\epsilon)$ the projection onto the ith coordinate of $\Pi_{p(T)}$ of a point in the stable manifold of the invariant torus which is contained in $W^s(\mathcal{M}_\epsilon) \cap \Pi_{p(T)}$. So we see that the only way in which $\Delta_i^s(t)$ can be defined for all $t \in [0, \infty)$ is if the point in the stable manifold of the torus is contained in $S_{p(t),\epsilon}^s$ as defined in Definition 4.1.1. A similar argument follows for $\Delta_i^u(t)$ on the time interval $(-\infty, 0]$. See Figure 4.1.17 for an illustration of the geometry behind this argument.

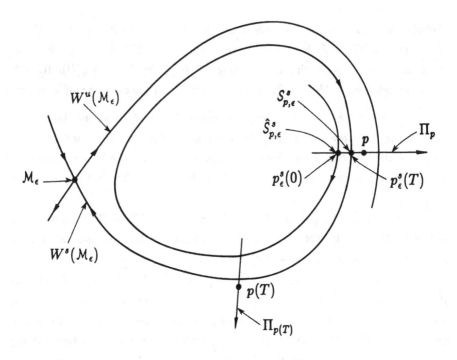

Figure 4.1.17. Geometry of $\Delta_i^s(t)$.

v) Elimination of t_0

We now discuss how the parameter t_0 can be eliminated as an independent variable of the Melnikov vector. Recall that the unperturbed homoclinic orbit can be parametrized by $(t_0, \alpha, I, \theta_0)$ where I is fixed in Systems I and III. Varying $(t_0, \alpha, I, \theta_0)$ corresponds to moving along the unperturbed homoclinic orbit and measuring the distance between the perturbed stable and unstable manifolds of the invariant torus at each point. By uniqueness of solutions, if the stable and unstable manifolds intersect at a point, they must intersect along (at least) a one dimensional orbit (note: the intersection could be higher dimensional if there are symmetries in the system). Therefore, since the Melnikov vector measures the distance between trajectories in the stable and unstable manifolds, then one zero of the Melnikov vector should imply the existence of a one parameter family of zeros, with this parameter being redundant in order to determine whether the stable and unstable manifolds intersect. Now we discuss how this geometric fact is manifested mathematically in the Melnikov vector.

The arguments of the integrands defining the components of the Melnikov vector for Systems I, II, and III were

$$\left(x^I(t - t_0, \alpha), I, \int^t \Omega(x^I(s, \alpha), I)ds + \theta_0, \mu; 0\right). \tag{4.1.173}$$

If we make the change of variables $t \to t + t_0$, the limits of integration of the integrals defining the components of the Melnikov vector do not change; however, the arguments of the integrands become

$$\left(x^I(t, \alpha), I, \int^{t+t_0} \Omega(x^I(s, \alpha), I)ds + \theta_0, \mu; 0\right). \tag{4.1.174}$$

Now t_0 still appears explicitly in the argument of the integrands. However, note that all functions are periodic in each component of the θ variable, so for any fixed component of the θ variable, say the ith component, varying θ_{i0} is equivalent to varying t_0. Therefore we may consider t_0 as fixed, and, for convenience, we take $t_0 = 0$. With this choice we arrive at the form of the components of the Melnikov vector given in (4.1.97), (4.1.99), (4.1.101), and (4.1.102). Note that it would be equivalent to let t_0 vary and fix any one component of θ_0. We will interpret this geometrically in terms of a Poincaré map in Section 4.1e.

4.1e. Reduction to a Poincaré Map

The theorems regarding the dynamical consequences of the intersection of the stable and unstable manifolds of normally hyperbolic l-tori ($l \geq 1$) were stated in the context of maps; for $l \geq 1$ we had Theorem 3.4.1. The techniques of this chapter were developed in terms of vector fields; however, it is a simple matter to reduce the study of Systems I, II, and III to the study of a local Poincaré map defined in a neighborhood of Γ.

A local cross-section Σ of the phase space is constructed by fixing any one component of the angular variables whose time derivative is nonzero (in the perturbed system) in a neighborhood of Γ. Then the Poincaré map associates points on the cross-section Σ with their first return to Σ under the action of the flow generated by the perturbed vector field. We remark that, in applying the theorems of Chapter 3 to specific problems, it is not so important that the Poincaré map is actually constructed, but merely that it can be constructed.

Now we want to describe the elimination of the parameter t_0 or any one component of θ_0 in the argument of the Melnikov vector in the context of the Poincaré map.

Recall that the homoclinic orbit Γ can be parametrized by the $n + m + l$ parameters $(t_0, \alpha, I, \theta_0)$ (note: for System III, $l = m$). Hence, the intersection of Γ with the cross-section can be described by $n + m + l - 1$ parameters where one angular variable has been fixed corresponding to the one which defines Σ. So, in this case, the elimination of the angular variable defining the cross-section from the argument of the Melnikov vector leads us to the interpretation that the Melnikov vector is restricted to the cross-section and measures the distance between the stable and unstable manifolds of an invariant torus of the Poincaré map. The elimination of t_0 from the argument of the Melnikov vector (i.e., setting $t_0 = 0$) could then be thought of as measuring the distance between the stable and unstable manifolds of the invariant torus of the Poincaré map only along the α, I, and $l - 1$ of the θ_0 directions, and then varying the cross-section Σ. Mathematically, both points of view are equivalent.

4.2. Examples

We now give a variety of examples which will serve to illustrate the theory developed in Section 4.1.

4.2a. Periodically Forced Single Degree of Freedom Systems

We give two examples of the simple pendulum subjected to time periodic external forcing. The first example involves a forcing function having $\mathcal{O}(\epsilon)$ amplitude and $\mathcal{O}(1)$ frequency and is an example of System I. The second example involves a forcing function having $\mathcal{O}(1)$ amplitude but $\mathcal{O}(\epsilon)$ frequency and is an example of System II. More details on these examples can be found in Wiggins [1988].

i) The Pendulum: Parametrically Forced at $O(\epsilon)$ Amplitude, $O(1)$ Frequency

We consider a simple planar pendulum whose base is subjected to a vertical, periodic excitation given by $\epsilon\gamma\sin\Omega t$, where ϵ is regarded as small and fixed. The equation of motion for the system is given by

$$\ddot{x}_1 + \epsilon\delta\dot{x}_1 + (1 - \epsilon\gamma\sin\Omega t)\sin x_1 = 0 \tag{4.2.1}$$

where x_1 represents the angular displacement from the vertical and δ represents damping. See Figure 4.2.1 for an illustration of the geometry.

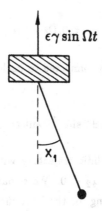

Figure 4.2.1. The Simple Pendulum.

Writing (4.2.1) as a first order system of equations gives

$$\dot{x}_1 = x_2$$
$$\dot{x}_2 = -\sin x_1 + \epsilon[\gamma\sin\theta\sin x_1 - \delta x_2] \qquad (x_1, x_2, \theta) \in T^1 \times \mathbb{R}^1 \times T^1. \tag{4.2.2}$$
$$\dot{\theta} = \Omega$$

The unperturbed system is given by

$$\dot{x}_1 = x_2$$
$$\dot{x}_2 = -\sin x_1 \tag{4.2.3}$$
$$\dot{\theta} = \Omega$$

and it should be clear that the (x_1, x_2) component of (4.2.3) is Hamiltonian with a Hamiltonian function given by

$$H = \frac{x_2^2}{2} - \cos x_1 \,. \tag{4.2.4}$$

The (x_1, x_2) component of (4.2.3) has a hyperbolic fixed point given by

$$(\bar{x}_1, \bar{x}_2) = (\pi, 0) = (-\pi, 0) \,. \tag{4.2.5}$$

Thus, when viewed in the full $x_1 - x_2 - \theta$ phase space, (4.2.3) has a hyperbolic periodic orbit given by

$$\mathcal{M} = \big(\bar{x}_1, \bar{x}_2, \theta(t)\big) = (\pi, 0, \Omega t + \theta_0) = (-\pi, 0, \Omega t + \theta_0) \,. \tag{4.2.6}$$

This hyperbolic periodic orbit is connected to itself by a pair of homoclinic trajectories given by

$$\big(x_{1h}^{\pm}(t), x_{2h}^{\pm}(t), \theta(t)\big) = (\pm 2\sin^{-1}(\tanh t), \pm 2\operatorname{sech} t, \Omega t + \theta_0) \tag{4.2.7}$$

where "$+$" refers to the homoclinic trajectory with $x_2 > 0$ and "$-$" refers to the homoclinic trajectory with $x_2 < 0$. We remark that the (x_1, x_2) component of (4.2.7) can be found by solving for the level curve of the Hamiltonian given by $H = 1$. So the periodic orbit is connected to itself by a pair of two dimensional homoclinic orbits, and from 4.1a, i) these homoclinic orbits, denoted $\Gamma^{\pm}$, can be parametrized as

$$\Gamma^{\pm} = \big\{ \big(x_{1h}^{\pm}(-t_0), x_{2h}^{\pm}(-t_0), \theta_0\big) \in T^1 \times \mathbb{R}^1 \times T^1 \mid (t_0, \theta_0) \in \mathbb{R}^1 \times T^1 \big\} \,. \tag{4.2.8}$$

So, for any fixed $(t_0, \theta_0) \in \mathbb{R}^1 \times T^1$, $p^{\pm} \equiv \big(x_{1h}^{\pm}(-t_0), x_{2h}^{\pm}(-t_0), \theta_0\big)$ denotes unique points on $\Gamma^{\pm}$ and, in this case, the plane $\Pi_{p^{\pm}}$ is one dimensional and is the span of the vector

$$\big(D_x H(x_{1h}^{\pm}(-t_0), x_{2h}^{\pm}(-t_0)), 0\big) = (\sin x_{1h}^{\pm}(-t_0), x_{2h}^{\pm}(-t_0), 0) \tag{4.2.9}$$

which intersects $\Gamma^{\pm}$ transversely at each $p^{\pm} \in \Gamma^{\pm}$, i.e., for all $(t_0, \theta_0) \in \mathbb{R}^1 \times T^1$. See Figure 4.2.2 for an illustration of the geometry of the unperturbed phase space.

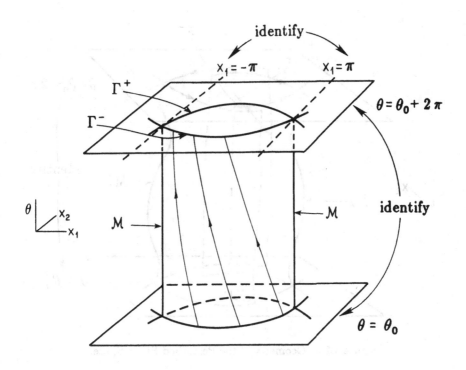

Figure 4.2.2. Geometry of the Unperturbed Phase Space.

It should be clear that $\Gamma^{\pm} = W^{s}(\mathcal{M}) \cap W^{u}(\mathcal{M}) - \mathcal{M}$.

With the geometry of the unperturbed phase space described, we now ask what becomes of this degenerate homoclinic structure for $\epsilon \neq 0$. By Proposition 4.1.5 we know that the hyperbolic periodic orbit persists, which we denote as $\mathcal{M}_{\epsilon}$, and its local stable and unstable manifolds, which are denoted $W^{s}_{loc}(\mathcal{M}_{\epsilon})$ and $W^{u}_{loc}(\mathcal{M}_{\epsilon})$, respectively, are C^{r} close to $W^{u}_{loc}(\mathcal{M})$ and $W^{u}_{loc}(\mathcal{M})$, respectively. We would like to determine if $W^{s}(\mathcal{M}_{\epsilon})$ and $W^{u}(\mathcal{M}_{\epsilon})$ intersect transversely for, if this is the case, then we can appeal to the Smale-Birkhoff homoclinic theorem to assert the existence of horseshoes and their attendant chaotic dynamics in our system.

Recall that for ϵ sufficiently small, for each point $p^{\pm} \in \Gamma^{\pm}$, $W^{s}(\mathcal{M}_{\epsilon})$ and $W^{u}(\mathcal{M}_{\epsilon})$ intersect $\Pi_{p^{\pm}}$ transversely in the points $S^{s}_{p^{\pm},\epsilon}$ and $S^{u}_{p^{\pm},\epsilon}$, respectively. This is because $\Pi_{p^{\pm}}$ intersects $\Gamma^{\pm}$ transversely for $\epsilon = 0$, and the manifolds vary smoothly with ϵ. See Figure 4.2.3 for an illustration of the geometry of the perturbed phase space.

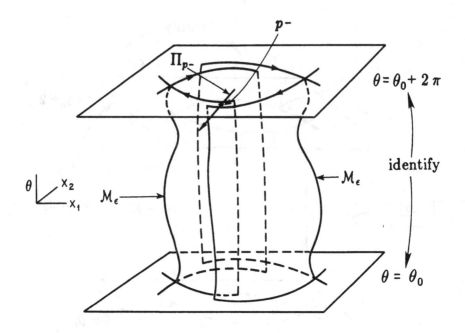

Figure 4.2.3. Geometry of the Perturbed Phase Space.

Now the distance between $W^s(\mathcal{M}_\epsilon)$ and $W^u(\mathcal{M}_\epsilon)$ at the point $p^\pm \in \Gamma^\pm$ has been shown to be

$$d^\pm(t_0,\theta_0,\delta,\gamma,\Omega) = \epsilon \frac{M^\pm(t_0,\theta_0,\delta,\gamma,\Omega)}{\left\| D_x H(x_{1h}^\pm(-t_0), x_{2h}^\pm(-t_0)) \right\|} + O(\epsilon^2) \qquad (4.2.10)$$

where

$$\left\| D_x H(x_{1h}^\pm(-t_0), x_{2h}^\pm(-t_0)) \right\| =$$
$$\sqrt{[D_{x_1} H(x_{1h}^\pm(-t_0), x_{2h}^\pm(-t_0))]^2 + [D_{x_2} H(x_{1h}^\pm(-t_0), x_{2h}^\pm(-t_0))]^2} \quad (4.2.11)$$

and by (4.1.47)

$$M^\pm(t_0,\theta_0,\delta,\gamma,\Omega) = \int_{-\infty}^{\infty} \left\{ -\delta \left[x_{2h}^\pm(t-t_0) \right]^2 + \gamma x_{2h}^\pm(t-t_0) \sin x_{1h}^\pm(t-t_0) \sin(\Omega t + \theta_0) \right\} dt.$$

$$(4.2.12)$$

Substituting (4.2.7) into the integral (4.2.12) gives

$$M^+(t_0,\theta_0,\delta,\gamma,\Omega) = M^-(t_0,\theta_0,\delta,\gamma,\Omega) \equiv M(t_0,\theta_0,\delta,\gamma,\Omega)$$
$$= -8\delta + \frac{2\gamma\pi\Omega^2}{\sinh\frac{\pi\Omega}{2}} \cos(\Omega t_0 + \theta_0). \qquad (4.2.13)$$

Before proceeding with an analysis of the Melnikov function and a discussion of its dynamical implications, let us make some remarks concerning t_0 and θ_0 in (4.2.13) (cf. 4.1d and 4.1e). Notice that varying either t_0 or θ_0 in (4.2.13) has the same effect; thus, we can view one or the other as fixed. Geometrically, fixing θ_0 corresponds to fixing a cross-section Σ^{θ_0} of the phase space (cf. Section 1.6) and considering the associated Poincaré map. Then, varying t_0 corresponds to moving along the unperturbed homoclinic orbit Γ and measuring the distance between the perturbed stable and unstable manifolds of the hyperbolic fixed point of the Poincaré map. Alternatively, fixing t_0 and varying θ_0 corresponds to fixing a point on Σ^{θ_0} and then varying the cross-section Σ^{θ_0}. Either point of view is mathematically equivalent as we showed in 4.1.d and 4.1.e.

Using (4.2.13), along with Theorems 4.1.9 and 4.1.10, we can show that for ϵ sufficiently small there exists a surface in the $\Gamma - \delta - \Omega$ parameter space given by

$$\epsilon\gamma = \frac{4\epsilon\delta}{\pi\Omega^2} \sinh\frac{\pi\Omega}{2} + \mathcal{O}(\epsilon^2) \qquad (4.2.14)$$

above which transverse intersections of the stable and unstable manifolds of the hyperbolic periodic orbit occur.

In order to more easily present the information obtained in equation (4.2.14), we give two graphs that give the shape of the curves defined by (4.2.14) when one of the parameters is fixed, one in $\gamma - \delta$ space with $\Omega \neq 0$ fixed and the other in $\gamma - \delta$ space with $\delta \neq 0$ fixed. In each case, the curves in Figures 4.2.4a and 4.2.4b are such that quadratic homoclinic tangencies occur on the curves, and transverse homoclinic orbits occur above the curves (note: see Guckenheimer and Holmes [1983] for bifurcation theorems concerning quadratic homoclinic tangencies). Notice in Figure 4.2.4b that, along the bifurcation curve as $\Omega \to 0$, it appears that $\gamma \to \infty$. Of course γ cannot become too large, in which case we would be outside the range of validity of the theory. Thus, we have no information about the low frequency limit; however, from these results we might expect that the amplitude of the excitation must become large in order for transverse homoclinic orbits to exist. Our next example will verify this conjecture.

Now our results show that (4.2.2) contains transverse homoclinic orbits to a hyperbolic periodic orbit. So by the Smale-Birkhoff homoclinic theorem, (4.2.2) contains an invariant Cantor set on which the dynamics can be described symbolically via the techniques in Section 2.2. However, we want to go a bit further and

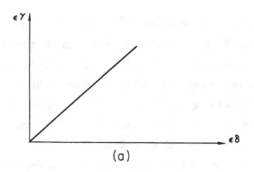

(a)

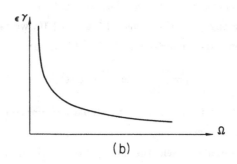

(b)

Figure 4.2.4. a) Graph of (4.2.14), $\Omega \neq 0$ fixed.

b) Graph of (4.2.14), $\delta \neq 0$ fixed.

describe the dynamical implications of the transverse homoclinic orbits in terms of the oscillatory motions of the pendulum. For this purpose it will be useful to deform the pair of homoclinic orbits into a shape which is more amenable to our geometric arguments. Consider Figure 4.2.5; in 4.2.5a we show the pair of homoclinic orbits in the unperturbed system on the cylinder. The $+$ sign refers to the upper homoclinic orbit on which the corresponding motion of the pendulum is clockwise, and the $-$ sign refers to the lower homoclinic orbit on which the corresponding motion of the pendulum is counterclockwise. In 4.2.5b imagine that the pair of homoclinic orbits have been slipped off the cylinder and flattened out in the plane in 4.2.5c. Figure 4.2.5d just represents a convenient rotation and deformation of 4.2.5c.

Now let us consider the time $\frac{2\pi}{\Omega}$ Poincaré map of the perturbed system denoted

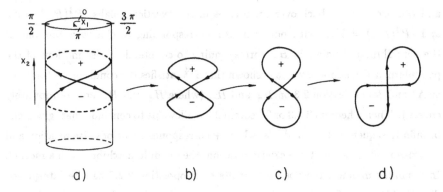

Figure 4.2.5. Geometry of the Unperturbed Homoclinic Orbits.

by P. In this case, the map has a hyperbolic fixed point whose stable and unstable manifolds may intersect transversely to give the familiar homoclinic tangle shown in Figure 4.2.6.

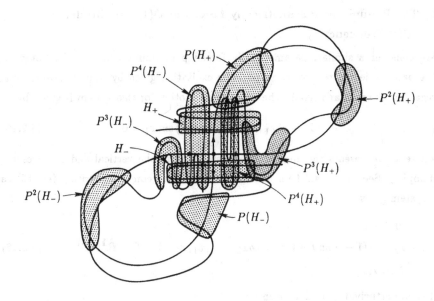

Figure 4.2.6. The Formation of the Horseshoe.

Notice the "horizontal" slabs H_+ and H_- in Figure 4.2.6. Under P^4 H_+

and H_- are mapped back over themselves in the "vertical" slabs $P^4(H_+) = V_+$ and $P^4(H_-) = V_-$, with points in H_+ corresponding to clockwise motions of the pendulum and points in H_- corresponding to counterclockwise motions of the pendulum. Now, suppose we have shown that P^4 satisfies the conditions A1 and A2 or A1 and A3 of Section 2.3 on H_+ and H_- (where H_+ and H_- are appropriately chosen). Then Theorem 2.3.3 or Theorem 2.3.5 allows us to conclude that, given *any* bi-infinite sequence of $+$'s and $-$'s, where $+$ corresponds to a clockwise rotation and $-$ corresponds to a counterclockwise rotation, the pendulum exhibits such a motion. In a similar manner, the abstract results of Proposition 2.2.7 can be interpreted directly in terms of clockwise or counterclockwise rotations of the pendulum. The verification of the conditions of Theorem 2.3.3 and 2.3.5 we leave as an exercise to the reader since they are similar to examples in Chapter 3. Also, see Holmes and Marsden [1982a] for an estimate of the number of iterates of the Poincaré map which are necessary to form the horseshoe in terms of the perturbation parameter ϵ and the Melnikov function.

ii) The Pendulum: Parametrically Forced at $O(1)$ Amplitude, $O(\epsilon)$ Frequency

We consider a similar, parametrically forced pendulum as in 4.2a, i), but with the base subjected to a vertical, periodic excitation given by $\gamma \sin \epsilon \Omega t$, where ϵ is regarded as small and fixed. The equation of motion for this system is given by

$$\ddot{x}_1 + \epsilon \delta \dot{x}_1 + (1 - \gamma \sin \epsilon \Omega t) \sin x_1 = 0 \qquad (4.2.15)$$

where x_1 represents the angular displacement from the vertical and δ represents damping. See Figure 4.2.1 for an illustration of the geometry. Writing (4.2.15) as a system gives

$$\begin{aligned}
\dot{x}_1 &= x_2 \\
\dot{x}_2 &= -(1 - \gamma \sin I) \sin x_1 - \epsilon \delta x_2 \qquad (x_1, x_2, I) \in T^1 \times \mathbb{R}^1 \times T^1 . \qquad (4.2.16) \\
\dot{I} &= \epsilon \Omega
\end{aligned}$$

The unperturbed system is given by

$$\begin{aligned}
\dot{x}_1 &= x_2 \\
\dot{x}_2 &= -(1 - \gamma \sin I) \sin x_1 \qquad\qquad\qquad (4.2.17) \\
\dot{I} &= 0
\end{aligned}$$

and it is easily seen that (4.2.17) has the form of a 1-parameter family of Hamiltonian systems with Hamiltonian given by

$$H(x_1, x_2, I) = \frac{x_2^2}{2} - (1 - \gamma \sin I) \cos x_1 . \tag{4.2.18}$$

The unperturbed system has a fixed point at

$$(\bar{x}_1, \bar{x}_2) = (\pi, 0) = (-\pi, 0) \tag{4.2.19}$$

for each $I \in [0, 2\pi)$ which is hyperbolic provided

$$0 \leq \gamma < 1 . \tag{4.2.20}$$

Henceforth, we will always assume that (4.2.20) is satisfied. In the full $x_1 - x_2 - I$ phase space we can view

$$\mathcal{M} = (\bar{x}_1, \bar{x}_2, I) = (\pi, 0, I) = (-\pi, 0, I) , \quad I \in [0, 2\pi) \tag{4.2.21}$$

as a periodic orbit. Two homoclinic trajectories which connect $\mathcal{M}$ to itself are given by

$$(x_{1h}^{\pm}(t), x_{2h}^{\pm}(t), I) =$$
$$(\pm 2 \sin^{-1}[\tanh \sqrt{1 - \gamma \sin I}\, t], \pm 2\sqrt{1 - \gamma \sin I}\, \text{sech}\sqrt{1 - \gamma \sin I}\, t, I) \tag{4.2.22}$$

where "+" refers to the homoclinic trajectory with $x_2 > 0$ and "−" refers to the homoclinic trajectory with $x_2 < 0$. Thus $\mathcal{M}$ is connected to itself by a pair of two dimensional homoclinic orbits, denoted $\Gamma^{\pm}$, which can be parametrized by

$$\Gamma^{\pm} = \left\{ (x_{1h}^{\pm}(-t_0), x_{2h}^{\pm}(-t_0), I) \in T^1 \times \mathbb{R}^1 \times T^1 \mid (t_0, I) \in \mathbb{R}^1 \times T^1 \right\} . \tag{4.2.23}$$

This system is therefore an example of System II with $n = 1$, $m = 1$, $l = 0$. We remark that the unperturbed phase space of (4.2.17) is much the same as (4.2.3), which is sketched in Figure 4.2.2. The difference lies in the fact that $x_1 - x_2$ coordinates of $\Gamma^{\pm}$ do not depend on the angular variable θ_0 in (4.2.3), but they do depend on the angular variable I in (4.2.17).

We now consider the perturbed system (4.2.16). By Proposition 4.1.12, $\mathcal{M}$ persists (denoted by $\mathcal{M}_\epsilon$) as a periodic orbit of period $\frac{2\pi}{\epsilon\Omega}$. We want to determine the behavior of the stable and unstable manifolds of $\mathcal{M}_\epsilon$. From 4.1b, v), the distance between the manifolds is given by the scalar function

$$d^\pm(t_0, I, \delta, \gamma, \Omega) = \epsilon \frac{M^\pm(I, \delta, \gamma, \Omega)}{\left\| D_x H\left(x_{1h}^\pm(-t_0), x_{2h}^\pm(-t_0)\right) \right\|} + \mathcal{O}(\epsilon^2) \qquad (4.2.24)$$

where

$$\left\| D_x H\left(x_{1h}^\pm(-t_0), x_{2h}^\pm(-t_0)\right) \right\| =$$
$$\sqrt{\left[D_{x_1} H\left(x_{1h}^\pm(-t_0), x_{2h}^\pm(-t_0)\right)\right]^2 + \left[D_{x_2} H\left(x_{1h}^\pm(-t_0), x_{2h}^\pm(-t_0)\right)\right]^2}. \qquad (4.2.25)$$

From (4.1.69), the Melnikov function is given by

$$M^\pm(I; \delta, \gamma, \Omega) = \int_{-\infty}^{\infty} \left[-\delta\left(x_{2h}^\pm(t)\right)^2 + \gamma \Omega t (\cos I) x_{2h}^\pm(t) \sin x_{1h}^\pm(t)\right] dt. \qquad (4.2.26)$$

Substituting (4.2.22) into (4.2.26) we obtain

$$M^+(I; \delta, \gamma, \Omega) = M^-(I; \delta, \gamma, \Omega) \equiv M(I; \delta, \gamma, \Omega) = -8\delta\sqrt{1 - \gamma \sin I} + \frac{4\gamma\Omega \cos I}{\sqrt{1 - \gamma \sin I}}. \qquad (4.2.27)$$

Using (4.2.27) and Theorems 4.1.13 and 4.1.14 we obtain, after some algebra, an equation whose graph in (δ, γ, Ω) space is a surface above which transverse homoclinic orbits occur. This equation is given by

$$\gamma = \frac{\frac{2\delta}{\Omega}}{\sqrt{1 + \left(\frac{2\delta}{\Omega}\right)^2}} + \mathcal{O}(\epsilon). \qquad (4.2.28)$$

As in the previous example, we will present two graphs representing the shapes of the curves obtained from (4.2.28) when one of the parameters is viewed as fixed, one in $\gamma - \delta$ space with $\Omega \neq 0$ fixed and the other in $\gamma - \Omega$ space with $\delta \neq 0$ fixed. In each case, the curves in Figure 4.2.7 are such that quadratic homoclinic tangencies occur on the curves, and transverse homoclinic orbits occur above the curves. Note that our theory is not valid for $\gamma = 1$, since in this case some of the fixed points in the unperturbed system are nonhyperbolic, and that would violate II4.

The dynamical consequences of the transverse homoclinic orbits can be interpreted in terms of symbolic dynamics in a manner virtually identical to that described at the end of the previous example.

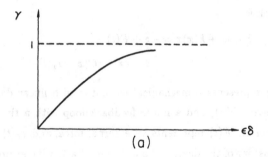

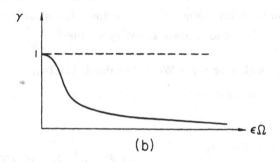

Figure 4.2.7. a) Graph of (4.2.28), $\Omega \neq 0$ Fixed.
              b) Graph of (4.2.28), $\delta \neq 0$ Fixed.

4.2b. Slowly Varying Oscillators

We will now give two examples of System I that have the structure of a periodically forced, single degree of freedom, nonlinear oscillator containing a parameter that obeys a slow $(\mathcal{O}(\epsilon))$ first order ordinary differential equation.

The first example we shall consider arises from a class of third order nonautonomous systems proposed by Holmes and Moon [1983] to model certain feedback-controlled mechanical devices. Imagine a mechanical device with multiple equilibrium positions in the absence of feedback. A controller is added to move the system from one equilibrium position to another. A possible model for such systems with first order feedback is

$$\ddot{x} + \delta\dot{x} + k(x)x = -z + F(t)$$
$$\dot{z} + \epsilon z = \epsilon G[x - x_r(t)] . \tag{4.2.29}$$

Equation (4.2.29) represents a mechanical oscillator with linear damping δ, non-linear spring constant $k(x)$, and a linear feedback loop with a time constant $1/\epsilon$ and gain parameter G. $F(t)$ represents an external force, and $x_r(t)$ represents the desired position history of the device. We will present a specific example of (4.2.29), which consists of the Duffing oscillator with a first order linear feedback loop. More details on this example can be found in Wiggins and Holmes [1987].

The second example we will consider is that of a pendulum attached to a rotating frame. This is an example of a class of systems which frequently arise in rotational dynamics, and it exhibits a very rich homoclinic structure. More details of this example can be found in Shaw and Wiggins [1988].

i) The Duffing Oscillator with Weak Feedback Control

We consider the following system

$$\begin{aligned}
\dot{x}_1 &= x_2 \\
\dot{x}_2 &= x_1 - x_1^3 - I - \epsilon\delta x_2 \\
\dot{I} &= \epsilon(\gamma x_1 - \alpha I + \beta\cos\theta) \\
\dot{\theta} &= 1
\end{aligned} \qquad (x_1, x_2, I, \theta) \in \mathbb{R}^1 \times \mathbb{R}^1 \times \mathbb{R}^1 \times T^1 \tag{4.2.30}$$

where α, β, γ, and δ are parameters, and ϵ is small and fixed. The unperturbed system is given by

$$\begin{aligned}
\dot{x}_1 &= x_2 \\
\dot{x}_2 &= x_1 - x_1^3 - I \\
\dot{I} &= 0 \\
\dot{\theta} &= 1
\end{aligned} \tag{4.2.31}$$

and the $x_1 - x_2$ component of (4.2.31) has the form of a 1-parameter family of Hamiltonian systems with Hamiltonian function given by

$$H(x_1, x_2; I) = \frac{x_2^2}{2} - \frac{x_1^2}{2} + \frac{x_1^4}{4} + Ix_1 . \tag{4.2.32}$$

We now want to describe the geometrical structure of the unperturbed phase space.

Fixed Points. Fixed points of the $x_1 - x_2 - I$ component of (4.2.31) are given by

$$(x_1(I), 0, I) \qquad (4.2.33)$$

where $x_1(I)$ is a solution of

$$x_1^3 - x_1 + I = 0. \qquad (4.2.34)$$

For $I \in \left(-\frac{2}{3\sqrt{3}}, \frac{2}{3\sqrt{3}}\right)$, (4.2.34) has three solutions, with the intermediate root corresponding to a hyperbolic fixed point. For $I > \frac{2}{3\sqrt{3}}$ and $I < -\frac{2}{3\sqrt{3}}$, there exists only one solution of (4.2.34) corresponding to an elliptic fixed point and, for $I = \pm\frac{2}{3\sqrt{3}}$, (4.2.34) has two solutions corresponding to an elliptic fixed point and a saddle-node fixed point. See Figure 4.2.8 for an illustration of the graph of (4.2.34). We will only be interested in the hyperbolic fixed points.

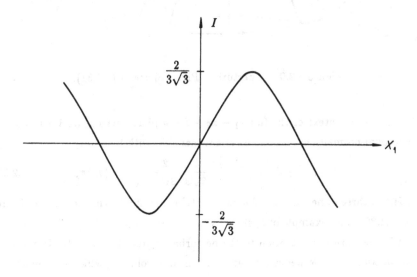

Figure 4.2.8. Graph of (4.2.34).

Homoclinic Orbits. We denote the 1-manifold of hyperbolic fixed points of the $x_1 - x_2 - I$ component of (4.2.31) by

$$\gamma(I) = (\bar{x}_1(I), 0, I) \qquad (4.2.35)$$

where $\bar{x}_1(I)$ is the intermediate solution of (4.2.34) for $I \in \left(-\frac{2}{3\sqrt{3}}, \frac{2}{3\sqrt{3}}\right)$. Each of these fixed points is connected to itself by a pair of homoclinic orbits which satisfy

$$\left[\frac{x_2^2}{2} - \frac{x_1^2}{2} + \frac{x_1^4}{4} + Ix_1\right] - \left[-\frac{\bar{x}_1(I)^2}{2} + \frac{\bar{x}_1(I)^4}{4} + I\bar{x}_1(I)\right] = 0. \qquad (4.2.36)$$

Thus, the phase space of (4.2.31) appears as in Figure 4.2.9.

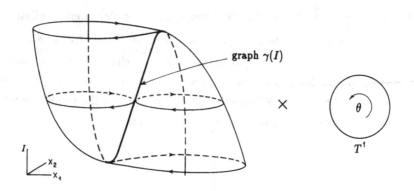

Figure 4.2.9. Unperturbed Phase Space of (4.2.31).

So, in the context of the full $x_1 - x_2 - I - \theta$ phase space, (4.2.31) has a two dimensional normally hyperbolic invariant manifold with boundary

$$\mathcal{M} = (\gamma(I), \theta_0), \quad I \in \left(-\frac{2}{3\sqrt{3}}, \frac{2}{3\sqrt{3}}\right), \ \theta_0 \in [0, 2\pi) \qquad (4.2.37)$$

and $\mathcal{M}$ has three dimensional stable and unstable manifolds which coincide. Therefore, (4.2.30) is an example of System I with $n = m = l = 1$.

We now turn our attention to the perturbed system (4.2.30). By Proposition 4.1.5, we know that $\mathcal{M}$ persists an an invariant manifold $\mathcal{M}_\epsilon$, which we denote

$$\mathcal{M}_\epsilon = (\gamma(I) + \mathcal{O}(\epsilon), \theta_0), \quad I \in \left(-\frac{2}{3\sqrt{3}}, \frac{2}{3\sqrt{3}}\right), \ \theta_0 \in [0, 2\pi). \qquad (4.2.38)$$

The procedure is to determine if $\mathcal{M}_\epsilon$ contains any periodic orbits by using Proposition 4.1.6, and then to determine whether or not the stable and unstable manifolds of these periodic orbits intersect by computing the appropriate Melnikov integral.

The Flow on $\mathcal{M}_\epsilon$. The perturbed vector field restricted to $\mathcal{M}_\epsilon$ is given by

$$\begin{aligned} \dot{I} &= \epsilon[\gamma x_1(I) - \alpha I + \cos\theta] + \mathcal{O}(\epsilon^2) \\ \dot{\theta} &= 1 \end{aligned} \quad I \in \left(-\frac{2}{3\sqrt{3}}, \frac{2}{3\sqrt{3}}\right). \qquad (4.2.39)$$

The averaged vector field is

$$\dot{I} = \frac{\epsilon}{2\pi} \int\limits_{0}^{2\pi} \left[\gamma x_1(I) - \alpha I + \beta \cos\theta\right] d\theta \tag{4.2.40}$$

$$= \epsilon\left[\gamma x_1(I) - \alpha I\right].$$

Fixed points of (4.2.40) must satisfy

$$I = \frac{\gamma}{\alpha} x_1(I). \tag{4.2.41}$$

Using (4.2.41), along with the fact that $x_1(I)$ must satisfy (4.2.34), gives the following expression for fixed points of (4.2.40):

$$I = 0, \ \pm\frac{\gamma}{\alpha}\sqrt{1 - \frac{\gamma}{\alpha}}. \tag{4.2.42}$$

So, for $\frac{\gamma}{\alpha} < 1$, (4.2.40) has three fixed points and, for $\frac{\gamma}{\alpha} > 1$, (4.2.40) has one fixed point (note: a pitchfork bifurcation is said to have occurred at $\frac{\gamma}{\alpha} = 1$, see Guckenheimer and Holmes [1983]). See figure 4.2.10.

We next want to calculate the nature of the stability of these fixed points of the averaged equations. This is given by the sign of

$$\frac{d}{dI}\epsilon\left(\gamma x_1(I) - \alpha I\right). \tag{4.2.43}$$

The fixed point is unstable (on $\mathcal{M}_\epsilon$) if (4.2.43) is positive and stable if (4.2.43) is negative. A simple calculation shows that

$$\frac{d}{dI}\left[\epsilon\left(\gamma x_1(I) - \alpha I\right)\right] = \begin{cases} \epsilon(\gamma - \alpha) & \text{for } I = 0 \\ \dfrac{-2\epsilon(\gamma - \alpha)}{3\frac{\gamma}{\alpha} - 2} & \text{for } I = \pm\dfrac{\gamma}{\alpha}\sqrt{1 - \dfrac{\gamma}{\alpha}}. \end{cases} \tag{4.2.44}$$

Thus, $I = 0$ is stable for $\gamma < \alpha$, unstable for $\gamma > \alpha$, and $I = \pm\frac{\gamma}{\alpha}\sqrt{1 - \frac{\gamma}{\alpha}}$ are unstable for $\frac{2\alpha}{3} < \gamma < \alpha$.

By Proposition 4.1.6, these fixed points of the averaged equations (4.2.40) correspond to periodic orbits of (4.2.39) of period 2π having the same stability type as the fixed points of the averaged equations. If we consider a three dimensional Poincaré map formed from (4.2.30) by fixing $\theta = \bar{\theta}$ with the map taking points

(x_1, x_2, I) to their image under the perturbed flow after a time 2π (see Section 1.6), then the periodic orbits on $\mathcal{M}_\epsilon$ become fixed points of the Poincaré map.

The Structure of the Poincaré Map. Using our knowledge of the flow on $\mathcal{M}_\epsilon$ as well as the structure of the unperturbed phase space, we see that the Poincaré map has the following structure.

$\underline{\gamma > \alpha}$. Hyperbolic fixed point at $(x_1, x_2, I) = (0, 0, 0)$ having a two dimensional unstable manifold and a one dimensional stable manifold.

$\underline{\frac{2\alpha}{3} < \gamma < \alpha}$. Three hyperbolic fixed points at $(x_1, x_2, I) = (0, 0, 0)$ and $\left(\pm\sqrt{1 - \frac{\gamma}{\alpha}}, 0, \pm\frac{\gamma}{\alpha}\sqrt{1 - \frac{\gamma}{\alpha}}\right)$, where $(0, 0, 0)$ has a two dimensional stable manifold and a one dimensional unstable manifold, and $\left(\pm\sqrt{1 - \frac{\gamma}{\alpha}}, 0, \pm\frac{\gamma}{\alpha}\sqrt{1 - \frac{\gamma}{\alpha}}\right)$ have two dimensional unstable manifolds and one dimensional stable manifolds.

$\underline{\gamma < \frac{2\alpha}{3}}$. Hyperbolic fixed point at $(x_1, x_2, I) = (0, 0, 0)$ having a two dimensional stable manifold and a one dimensional unstable manifold.

See Figure 4.2.10 for an illustration of the geometry of the Poincaré map.

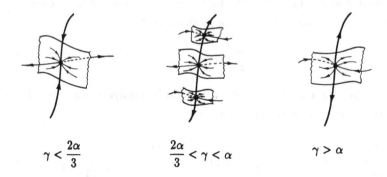

$$\gamma < \frac{2\alpha}{3} \qquad\qquad \frac{2\alpha}{3} < \gamma < \alpha \qquad\qquad \gamma > \alpha$$

Figure 4.2.10. Geometry of the Poincaré Map.

Calculation of the Melnikov Integrals. Next we calculate the Melnikov integrals. These will give us sufficient conditions for the stable and unstable manifolds of the hyperbolic fixed points of the Poincaré map to intersect transversely.

From (4.1.47), the Melnikov integral is given by

$$M^{\bar{I}}(\theta_0, \alpha, \beta, \gamma, \delta) = \int_{-\infty}^{\infty} [\langle D_x H, g^x \rangle + \langle D_I H, g^I \rangle] (x^{\bar{I}}(t), \bar{I}, t + t_0) dt$$

$$- \langle D_I H(x(\bar{I}), 0, \bar{I}), \int_{-\infty}^{\infty} g^I(x^{\bar{I}}(t), \bar{I}, t + t_0 dt \rangle$$

$$= \int_{-\infty}^{\infty} [-\delta(x_2^{\bar{I}}(t))^2 + \gamma(x_1^{\bar{I}}(t))^2 - \alpha \bar{I} x_1^{\bar{I}}(t) + \beta x_1^{\bar{I}}(t) \cos(t + t_0)] dt$$

$$- x_1(\bar{I}) \int_{-\infty}^{\infty} [\gamma x_1^{\bar{I}}(t) - \alpha \bar{I} + \beta \cos(t + t_0)] dt$$

$$(4.2.45)$$

where $x^{\bar{I}}(t) = (x_1^{\bar{I}}(t), x_2^{\bar{I}}(t))$ is a homoclinic trajectory of the unperturbed system on the $I = \bar{I}$ level corresponding to the hyperbolic fixed point of the averaged vector field on $\mathcal{M}_\epsilon$. From (4.2.36) these homoclinic trajectories are found to be

$\underline{\bar{I} = 0}.$ $x^{\pm}(t) = (x_1^{\pm}(t), x_2^{\pm}(t)) = (\pm\sqrt{2} \operatorname{sech} t, \mp\sqrt{2} \operatorname{sech} t \tanh t)$ (4.2.46)

$\underline{\bar{I} = -\frac{\gamma}{\alpha}\sqrt{1 - \gamma/\alpha}}.$

$$x^+(t) = \left(\frac{2cS + ab}{2bS - a}, \frac{-2ad^3 ST}{(2bS - a)^2} \right)$$

$$x^-(t) = \left(\frac{2cS - ab}{2bS + a}, \frac{2ad^3 ST}{(2bS + a)^2} \right)$$

$$(4.2.46)_l$$

$\underline{\bar{I} = +\frac{\gamma}{\alpha}\sqrt{1 - \gamma/\alpha}}.$

$$x^+(t) = \left(\frac{-2cS - ab}{2bS - a}, \frac{2ad^3 ST}{(2bS - a)^2} \right)$$

$$x^-(t) = \left(\frac{-2cS + ab}{2bS + a}, \frac{-2ad^3 ST}{(2bS + a)^2} \right)$$

$$(4.2.46)_u$$

where

$$a = \sqrt{\frac{2\gamma}{\alpha}}, \quad b = \sqrt{1 - \frac{\gamma}{\alpha}}, \quad c = 1 - \frac{2\gamma}{\alpha}, \quad d = \sqrt{\frac{3\gamma}{\alpha} - 2}$$

$$S = \operatorname{sech}(dt), \quad T = \tanh(dt)$$

where the subscripts "u" and "l" refer to the "upper" homoclinic orbits on the $I = (\gamma/\alpha)\sqrt{1-\gamma/\alpha}$ level and the "lower" homoclinic orbits on the $I=-(\gamma/\alpha)\sqrt{1-\gamma/\alpha}$ level.

Substituting (4.2.26), (4.2.26)$_u$, and (4.2.26)$_l$ into (4.2.45) and computing the integrals gives

$\underline{I = 0.}$
$$M^{\pm}(t_0, \alpha, \beta, \gamma, \delta) = \frac{-4\delta}{3} + 4\gamma \pm \sqrt{2}\beta\pi\,\text{sech}\,\frac{\pi}{2}\cos t_0 \qquad (4.2.47)^{\pm}$$

$\underline{I = -\frac{\gamma}{\alpha}\sqrt{1-\gamma/\alpha}.}$

$$
\begin{aligned}
M_l^{\pm}(t_0, \alpha, \beta, \gamma, \delta) = &-4\delta\left[\frac{d}{3} + \frac{\gamma b}{\sqrt{2}\alpha}\left(\frac{\pi}{2} \pm \sin^{-1}\sqrt{\frac{2\alpha}{\gamma}b}\right)\right] \\
&+ 2\gamma\left[2d - 2\sqrt{2}b\left(\frac{\pi}{2} \pm \sin^{-1}\sqrt{\frac{2\alpha}{\gamma}b}\right)\right] \qquad (4.2.47)_l^{\pm} \\
&\mp 2\sqrt{2}\pi\beta\frac{\sinh(\frac{1}{d}\sin^{-1}\sqrt{\frac{\alpha}{\gamma}d})}{\sinh\frac{\pi}{d}}\cos t_0
\end{aligned}
$$

$\underline{I = +\frac{\gamma}{\alpha}\sqrt{1-\gamma/\alpha}.}$

$$
\begin{aligned}
M_u^{\pm}(t_0, \alpha, \beta, \gamma, \delta) = &-4\delta\left[\frac{d}{3} + \frac{\gamma b}{\sqrt{2}\alpha}\left(\frac{\pi}{2} \pm \sin^{-1}\sqrt{\frac{2\alpha}{\gamma}b}\right)\right] \\
&+ 2\gamma\left[2d - \sqrt{2}b\left(\frac{\pi}{2} \pm \sin^{-1}\sqrt{\frac{2\alpha}{\gamma}b}\right)\right] \qquad (4.2.47)_u^{\pm} \\
&\pm 2\sqrt{2}\pi\beta\frac{\sinh(\frac{1}{d}\sin^{-1}\sqrt{\frac{\alpha}{\gamma}d})}{\sinh\frac{\pi}{d}}\cos t_0
\end{aligned}
$$

where, on the $I = \pm\frac{\gamma}{\alpha}\sqrt{1-\gamma/\alpha}$ levels, the superscript "+" refers to the larger homoclinic loop.

We present graphs of $(4.2.47)^{\pm}$, $(4.2.47)_u^{\pm}$, and $(4.2.47)_l^{\pm}$ in Figure 4.2.11 for $\alpha = 1$, $\beta = 1$. In the region bounded by $(4.2.47)_u^-$ and $(4.2.47)_l^-$, the stable and unstable manifolds of the hyperbolic fixed point on the $I = \pm\frac{\gamma}{\alpha}\sqrt{1-\gamma/\alpha}$ corresponding to the small homoclinic orbit intersect transversely; in the region bounded by $(4.2.47)_u^+$ and $(4.2.47)_l^+$, the stable and unstable manifolds of the hyperbolic fixed point on the $I = \pm\frac{\gamma}{\alpha}\sqrt{1-\gamma/\alpha}$ levels corresponding to the larger homoclinic orbit intersect transversely; and in the region bounded by $(4.2.47)^+$ and $(4.2.47)^-$, both

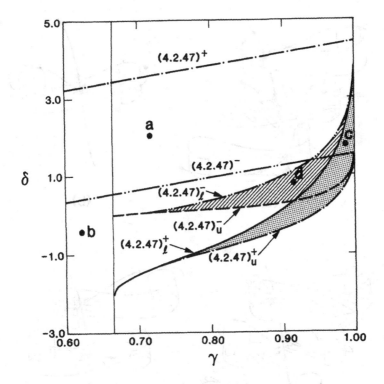

Figure 4.2.11. Regions where Horseshoes Exist.

branches of the stable and unstable manifolds of the hyperbolic fixed point on the
$I = 0$ level intersect transversely.

In Figure 4.2.12, we illustrate the behavior of the stable and unstable manifolds
of the Poincaré map for the four different parameter values indicated in Figure
4.2.11.

Now let us interpret our results in the context of feedback control systems.
There are two aspects which we want to consider: 1) modification of the region of
chaos in parameter space by the feedback loop, and 2) introduction of chaos by the
feedback loop. We emphasize that, in this example, chaos means Smale horseshoes.
We consider each aspect individually.

1. *Modification of Chaos Via Feedback.* Let us consider the situation of the gain,
γ, going to zero. In this case, the $x_1 - x_2$ component of the vector field decouples
from the I component. Thus, I can be solved for as an explicit function of time

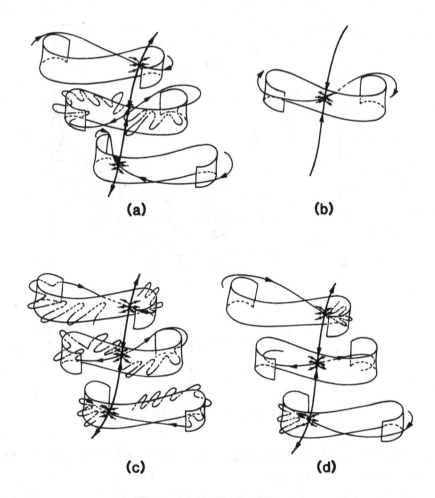

Figure 4.2.12. Poincaré Maps.

which is asymptotically periodic $(I \sim \epsilon\beta \sin t + O(\epsilon^2)$ as $t \to \infty)$, and the solution can be substituted into the $x_1 - x_2$ components of the equation. The result is an equation for a periodically forced Duffing oscillator

$$\dot{x}_1 = x_2$$
$$\dot{x}_2 = x_1 - x_1^3 - \epsilon[\delta x_2 + \beta \sin t] + O(\epsilon^2). \quad (4.2.48)$$

Equation (4.2.48) has been studied in great detail by Holmes [1979] and Greenspan and Holmes [1983], and the original Melnikov [1963] method gives a curve in $\beta - \delta$ space above which transverse homoclinic orbits to a hyperbolic periodic orbit exist.

This curve is given by

$$\beta = \frac{4\delta}{3\sqrt{2\pi}} \cosh(\pi/2) . \qquad (4.2.49)$$

From (4.2.47) we see that a similar curve, above which there exist transverse homoclinic orbits to a hyperbolic periodic orbit on the $I = 0$ level in the presence of nonzero gain γ, is given by

$$\beta = \frac{\left(\frac{4\delta}{3} - 4\gamma\right)}{\sqrt{2\pi}} \cosh\frac{\pi}{2} . \qquad (4.2.50)$$

These curves are shown in Figure 4.2.13.

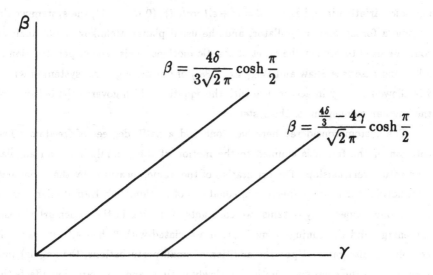

Figure 4.2.13. Graphs of (4.2.49) and (4.2.50).

Thus, we see from Figure 4.2.13 that the effect of the gain is to lower the boundary, and hence increase the area of the region in $\beta - \delta$ space in which Smale horseshoes are present in the dynamics of (4.2.50).

2. *Introduction of Chaos Via Feedback.* The fact that the feedback loop has introduced chaos into the system is evident from Figure 4.2.11. The fixed points and horseshoes on the $I = \pm\frac{\gamma}{\alpha}\sqrt{1 - \gamma/\alpha}$ levels are there solely as a result of the feedback.

ii) The Whirling Pendulum

The whirling pendulum is shown in Figure 4.2.14. It consists of a rigid frame, that freely rotates about a vertical axis and to which a planar pendulum is attached, the pivot being on the vertical axis. The behavior of this system is well known if the frame rotation rate, $\dot{\theta}$, is held at a constant value, say Ω. Below a critical Ω, the pendulum behaves essentially like a nonrotating pendulum; it has a stable equilibrium at $\phi = 0$ and an unstable one at $\phi = \pi$. Above the critical Ω value, $\phi = 0$ becomes unstable, and two new equilibria appear at $\phi = \bar{\phi} = \pm \cos^{-1}\left(\frac{g}{l\Omega^2}\right)$. As $\Omega \to \infty$, $\bar{\phi} \to \pm\pi/2$ as expected.

If one were to add small dissipation at the pendulum pivot and allow a small periodic variation in $\dot{\theta}$, i.e., set $\dot{\theta} = \Omega + \epsilon\tilde{\Omega}\cos(\omega t)$ $(0 < \epsilon \ll 1)$, the system would become a forced planar oscillator, and the usual planar Melnikov [1963] analysis could be used to predict the onset of chaotic motions. This type of perturbation is a limiting case (see Shaw and Wiggins [1988]) of our more general system in which $\dot{\theta}$ is allowed to vary in accordance with the equation which governs the behavior of the angular momentum of the system.

The system considered here has "one and a half" degrees of freedom. The rotation of the frame is coupled to the motion of the pendulum via an angular momentum relationship. The orientation of the frame, measured by the variable θ, does not appear in the unperturbed equations of motion. In a Hamiltonian formulation, one immediately obtains two constants of motion in the unperturbed case: the energy and the conjugate momentum associated with θ; hence, this system is completely integrable. Upon the addition of small perturbations, the angular momentum and the energy will both vary slowly in time, and this variation affects the occurrence of chaotic motions. These results should be of interest to experimentalists, since often in rotating systems one can specify the applied torques but not necessarily the rotation speed itself.

In dimensionless form (see Shaw and Wiggins [1988] for details), the equations of motion are given by

$$\dot{\phi} = p_\phi \tag{4.2.51a}$$

$$\dot{p}_\phi = \sin\phi\left[-1 + p_\theta^2\cos\phi/(\mu + \sin^2\phi)^2\right] + \epsilon Q_\phi(p_\phi) \tag{4.2.51b}$$

$$\dot{\theta} = \frac{p_\theta}{\mu + \sin^2\phi} \tag{4.2.51c}$$

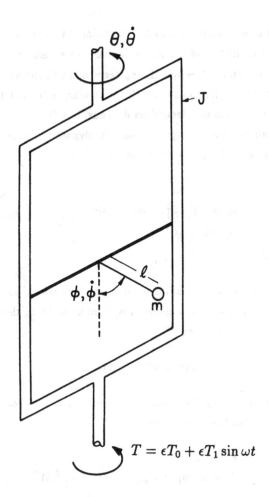

Figure 4.2.14. The Whirling Pendulum.

$$\dot{p}_\theta = \epsilon Q_\theta(\phi, p_\theta, t) \qquad\qquad (4.2.51d)$$

$$(\phi, p_\phi, \theta, p_\theta) \in T^1 \times \mathbb{R}^1 \times T^1 \times \mathbb{R}^1$$

where $\mu = J/ml^2$, $Q_\phi = -c_\phi p_\phi$, and $Q_\theta = -c_\theta p_\theta/(\mu + \sin^2 \phi) + T_0 + T_1 \sin(\omega t)$. Physically, $c_\theta > 0$ is the damping constant representing viscous damping in the bearings of the frame, $c_\phi > 0$ is the damping constant associated with viscous damping in the pendulum pivot, T_0 represents a constant torque applied to the frame about the vertical axis, and $T_1 \sin(\omega t)$ represents an oscillating torque applied to the frame also about the vertical axis.

The form of these equations is quite interesting; (4.2.51a, b) are of the form of a weakly damped oscillator with a particular form of parametric excitation. This small excitation is applied through the p_θ term in (4.2.51b) and is governed by its own differential equation, (4.2.51d). In particular, note that the $\phi - p_\phi - p_\theta$ components of (4.2.51) do not depend on θ. Thus, it suffices to analyze this three dimensional nonautonomous subsystem, since its dynamics determine $\theta(t)$.

Thus, the system which we will analyze is

$$
\begin{aligned}
\dot\phi &= p_\phi \\
\dot p_\phi &= \sin\phi\left[-1 + p_\theta^2 \cos\phi/(\mu + \sin^2\phi)^2\right] + \epsilon Q_\phi(p_\phi) \\
\dot p_\theta &= \epsilon Q_\theta(\phi, p_\theta, \psi) \\
\dot\psi &= \omega
\end{aligned}
\tag{4.2.52}
$$

where we have rewritten the nonautonomous three dimensional subsystem (4.2.51a, b, d) as a four dimensional autonomous subsystem by utilizing the time periodicity of the perturbation and defining

$$
\psi(t) = \omega t, \qquad \mod 2\pi. \tag{4.2.53}
$$

We begin by describing the unperturbed system and the geometry of its phase space. The unperturbed system is given by

$$
\begin{aligned}
\dot\phi &= p_\phi \\
\dot p_\phi &= \sin\phi\left[-1 + p_\theta^2 \cos\phi/(\mu + \sin^2\phi)^2\right] \\
\dot p_\theta &= 0 \\
\dot\psi &= \omega.
\end{aligned}
\tag{4.2.54}
$$

It is easily seen that the $\phi - p_\phi$ components of (4.2.54) have the form of a 1-parameter family of Hamiltonian systems with Hamiltonian function given by

$$
H(\phi, p_\phi; p_\theta) = \frac{1}{2}\left[\frac{p_\theta^2}{\mu + \sin^2\phi}\right] + \frac{1}{2}p_\phi^2 + (1 - \cos\phi). \tag{4.2.55}
$$

Fixed Points. Fixed points of the $\phi - p_\phi - p_\theta$ component of (4.2.54) are given by

$$
(\phi(p_\theta), 0, p_\theta) \tag{4.2.56}
$$

where $\phi(p_\theta)$ is a solution of

$$\sin\phi\left[-1 + p_\theta^2\cos\phi/(\mu + \sin^2\phi)^2\right] = 0. \tag{4.2.57}$$

For $\mu < p_\theta$, (4.2.57) has two solutions corresponding to $\phi = 0$ and $\phi = \pi$. These solutions exist for all p_θ, but a change of stability occurs at $\mu = p_\theta$ for $\phi = 0$. The solution $\phi = 0$ is a center for $\mu < p_\theta$ and a saddle for $\mu > p_\theta$. At $\mu = p_\theta$, a pitchfork bifurcation occurs at which two centers bifurcate from $\phi = 0$ and approach $\pm\pi/2$, respectively, as $p_\theta \to \infty$. The fixed points are shown in the $\phi - p_\theta$ plane in Figure 4.2.15.

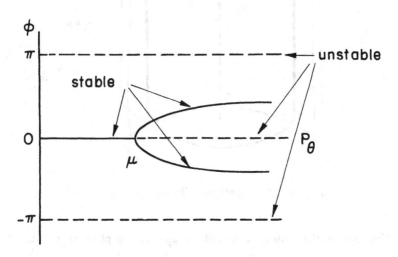

Figure 4.2.15. Solutions of (4.2.57).

Homoclinic Orbits. Using the Hamiltonian function (4.2.55), it can be shown that the saddle point

$$\gamma_P(p_\theta) = (\pi, 0, p_\theta) \tag{4.2.58}$$

is connected to itself by a pair of homoclinic orbits for all values of p_θ, and the saddle point

$$\gamma_D(p_\theta) = (0, 0, p_\theta) \tag{4.2.59}$$

is connected to itself by a pair of homoclinic orbits for all $p_\theta > \mu$. The phase space of (4.2.54) appears as in Figure 4.2.16. We use the subscripts P in (4.2.58), since the homoclinic orbits in that case are reminiscent of those in the simple pendulum, and D in (4.2.59), since the homoclinic orbits in that case are reminiscent of those in the Duffing oscillator (see Guckenheimer and Holmes [1983]).

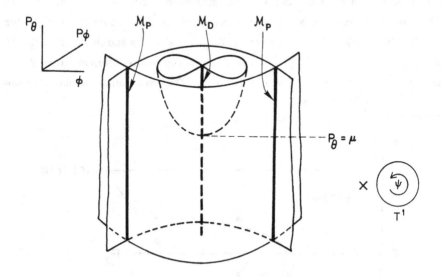

Figure 4.2.16. Unperturbed Phase Space of (4.2.54).

Therefore, in the context of the full $\phi - p_\phi - p_\theta - \psi$ phase space, (4.2.54) has two two-dimensional normally hyperbolic invariant manifolds with boundary

$$\mathcal{M}_P = \left(\gamma_P(p_\theta), \psi_0\right), \quad \psi_0 \in [0, 2\pi) \tag{4.2.60}$$

$$\mathcal{M}_D = \left(\gamma_D(p_\theta), \psi_0\right), \quad p_\theta > \mu, \quad \psi_0 \in [0, 2\pi) \tag{4.2.61}$$

and $\mathcal{M}_P$ and $\mathcal{M}_D$ have three dimensional stable and unstable manifolds which coincide. Thus, (4.2.51) is an example of System I with $n = m = l = 1$.

We now turn our attention to the perturbed system. By Proposition 4.1.5, we know that $\mathcal{M}_P$ and $\mathcal{M}_D$ persist as invariant manifolds, which we denote as $\mathcal{M}_{P,\epsilon}$ and $\mathcal{M}_{D,\epsilon}$, respectively. We then use Proposition 4.1.6 to determine if there are any periodic orbits on these manifolds and, if so, we compute the appropriate Melnikov integral to determine whether or not the stable and unstable manifolds of the periodic orbit intersect.

The Flow on $\mathcal{M}_{D,\epsilon}$ *and* $\mathcal{M}_{P,\epsilon}$.

The perturbed vector field restricted to $\mathcal{M}_{D,\epsilon}$ and $\mathcal{M}_{P,\epsilon}$ is identical and is given by

$$\dot{p}_\theta = -\epsilon \left[\frac{c_\theta p_\theta}{\mu} + T_0 + T_1 \sin \psi \right] + \mathcal{O}(\epsilon^2)$$

$$\dot{\psi} = \omega \, .$$

(4.2.62)

The averaged vector field is given by

$$\dot{p}_\theta = -\epsilon \left(\frac{c_\theta p_\theta}{\mu} + T_0 \right) \, .$$

(4.2.63

This equation has a unique fixed point given by

$$p_\theta = \frac{\mu T_0}{c_\theta} \, .$$

(4.2.64)

The fixed point is stable on $\mathcal{M}_{D,\epsilon}$ and $\mathcal{M}_{P,\epsilon}$, since

$$\frac{d}{dp_\theta} \left(-\epsilon \left[\frac{c_\theta p_\theta}{\mu} + T_0 \right] \right) = -\epsilon \frac{c_\theta}{\mu} < 0 \, .$$

(4.2.65)

Now, by Proposition 4.1.6, these fixed points of the averaged equations correspond to periodic orbits on $\mathcal{M}_{D,\epsilon}$ and $\mathcal{M}_{P,\epsilon}$, each having period $\frac{2\pi}{\omega}$ and the same stability type as the fixed points of the averaged equations.

In order to better understand the geometry of the phase space, we will consider a three dimensional Poincaré map constructed in the usual way of sampling the variables (ϕ, p_ϕ, p_θ) at discrete times corresponding to the period of the external forcing (i.e., at time intervals of $\frac{2\pi}{\omega}$); cf., the previous example and Section 1.6. In this case the periodic orbits become fixed points of the Poincaré map.

The Structure of the Poincaré Map. Using our knowledge of the flow on $\mathcal{M}_{P,\epsilon}$ and $\mathcal{M}_{D,\epsilon}$ as well as our knowledge of the structure of the unperturbed phase space, it is easy to see that the Poincaré map has one hyperbolic fixed point for $p_\theta < \mu$ which has a two dimensional stable manifold and a one dimensional unstable manifold. For $p_\theta > \mu$ the Poincaré map has two hyperbolic fixed points, each having two dimensional stable manifolds and one dimensional unstable manifolds. See Figure 4.2.17 for an illustration of the geometry of the three dimensional Poincaré map.

Calculation of the Melnikov Integrals. We now compute the Melnikov integrals for the corresponding fixed point of the Poincaré map so that we can determine the existence of orbits homoclinic to the fixed points of the Poincaré map.

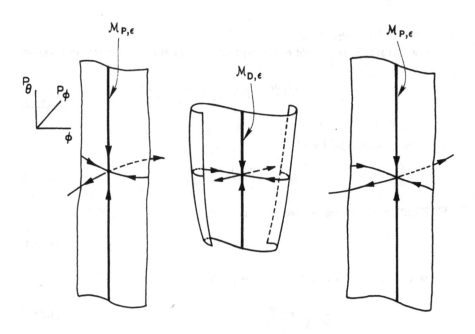

Figure 4.2.17. Geometry of the Poincaré Map, $p_\theta > \mu$.

From (4.1.47) the Melnikov vector is a scalar and is given by

$$M_{P,D}(t_0; \mu, c_\theta, c_\phi, T_0, T_1, \omega)$$

$$= \int_{-\infty}^{\infty} [(D_{p_\phi}H)(Q_\phi) + (D_{p_\theta}H)(Q_\theta)] (\phi^{\bar{p}_\theta}(t), p_\phi^{\bar{p}_\theta}(t), \bar{p}_\theta, t + t_0) dt$$

$$- D_{p_\theta}H(\gamma_{P,D}(\bar{p}_\theta)) \int_{-\infty}^{\infty} Q_\theta(\phi^{\bar{p}_\theta}(t), p_\phi^{\bar{p}_\theta}(t), \bar{p}_\theta, t + t_0) dt$$

$$= \int_{-\infty}^{\infty} \left[-c_\phi \left(p_\phi^{\bar{p}_\theta}(t) \right)^2 + \frac{\bar{p}_\theta}{\mu + \sin^2 \phi^{\bar{p}_\theta}(t)} \left(\frac{-c_\theta \bar{p}_\theta}{\mu + \sin^2 \phi^{\bar{p}_\theta}(t)} + T_0 + T_1 \sin \omega(t+t_0) \right) \right] dt$$

$$\frac{-\bar{p}_\theta}{\mu} \int_{-\infty}^{\infty} \left[\frac{-c_\theta \bar{p}_\theta}{\mu + \sin^2 \phi^{\bar{p}_\theta}(t)} + T_0 + T_1 \sin \omega(t + t_0) \right] dt \quad (4.2.66)$$

where $\left(\phi^{\bar{p}_\theta}(t), p_\phi^{\bar{p}_\theta}(t) \right)$ is the homoclinic trajectory of the unperturbed system on the $p_\theta = \bar{p}_\theta = \frac{\mu T_0}{c_\theta}$ level corresponding to the hyperbolic fixed point of the averaged vector field on $M_{D,\epsilon}$ or $M_{P,\epsilon}$. Unlike the previous example, we will not explicitly compute the homoclinic orbits; however, with qualitative arguments we will be able to go quite far.

The Melnikov integral is more conveniently written as

$$M_{P,D}(t_0, c_\phi, T_1, \omega) = -c_\phi J_1 - c_\theta J_2 + T_1 J_3(\omega, t_0) \tag{4.2.67}$$

where

$$J_1 = \int_{-\infty}^{\infty} \left(p_\phi^{\bar{p}_\theta}(t) \right)^2 dt \tag{4.2.68}$$

$$J_2 = \int_{-\infty}^{\infty} \left(\frac{\bar{p}_\theta}{\mu + \sin^2 \phi^{\bar{p}_\theta}(t)} - \frac{\bar{p}_\theta}{\mu} \right)^2 dt \tag{4.2.69}$$

$$J_3(\omega, t_0) = \int_{-\infty}^{\infty} \left(\frac{\bar{p}_\theta}{\mu + \sin^2 \phi^{\bar{p}_\theta}(t)} - \frac{\bar{p}_\theta}{\mu} \right) \sin \omega(t + t_0) dt . \tag{4.2.70}$$

We will consider μ, T_0, and c_θ as fixed, and c_ϕ, T_1, and ω as parameters.

Lemma 4.2.1. 1) $J_1 > 0$

     2) $J_2 > 0$

     3) $J_3(\omega, t_0) = \overline{J_3(\omega)} \sin \omega t_0$ where $\overline{J_3(\omega)} = -\frac{1}{\omega} \int_{-\infty}^{\infty} \theta^{\bar{p}_\theta}(t) \sin \omega t\, dt$.

PROOF: 1) and 2) are obvious from (4.2.68) and (4.2.64), respectively. 3) First we consider the term $-\frac{\bar{p}_\theta}{\mu} \int_{-\infty}^{\infty} \sin \omega(t + t_0) dt$ in (4.2.70). Expanding $\sin \omega(t + t_0)$ and considering the improper integral as a limit of a sequence, we obtain

$$-\frac{\bar{p}_\theta}{\mu} \int_{-\infty}^{\infty} \sin \omega(t + t_0) dt = \lim_{n \to \infty} \left[\frac{-\bar{p}_\theta}{\mu} \cos \omega t_0 \int_{-\frac{2\pi n}{\omega}}^{\frac{2\pi n}{\omega}} \sin \omega t\, dt - \frac{\bar{p}_\theta}{\mu} \sin \omega t_0 \int_{-\frac{2\pi n}{\omega}}^{\frac{2\pi n}{\omega}} \cos \omega t\, dt \right] = 0.$$

$$\tag{4.2.71}$$

We next treat the remaining part of (4.2.70). First, one notes that $\bar{p}_\theta / (\mu + \sin^2 \phi^{\bar{p}_\theta}(t))$ can be taken to be an even function of t if we choose the unperturbed homoclinic trajectory such that $\phi^{\bar{p}_\theta}(0) = 0$. Then, by expanding $\sin(\omega(t + t_0)) = \sin(\omega t) \cos(\omega t_0) + \cos(\omega t) \sin(\omega t_0)$, the odd part of the integrand, $\left(\bar{p}_\theta / (\mu + \sin^2 \phi^{\bar{p}_\theta}(t)) \right) \sin \omega t \cos(\omega t_0)$, can be eliminated, since it integrates to zero.

The remaining term, $\left(\bar{p}_\theta/(\mu + \sin^2 \phi^{\bar{p}_\theta}(t))\right)\cos(\omega t)\sin(\omega t_0)$, is still only conditionally convergent, see Section 4.1d. It is computed by considering the integral as a limit and integrating once by parts as follows:

$$J_3(\omega, t_0) = \sin \omega t_0 \left(\lim_{n \to \infty} \left[\frac{\bar{p}_\theta \sin \omega t}{\mu + \sin^2 \phi^{\bar{p}_\theta}(t)} \Big|_{\frac{-2\pi n}{\omega}}^{\frac{2\pi n}{\omega}} - \frac{1}{\omega} \int_{\frac{-2\pi n}{\omega}}^{\frac{2\pi n}{\omega}} \theta^{\bar{p}_\theta}(t) \sin \omega t \, dt \right] \right)$$

$$(4.2.72)$$

where $\theta^{\bar{p}_\theta}(t)$ is the solution of

$$\dot{\theta}^{\bar{p}_\theta}(t) = \frac{\bar{p}_\theta}{\mu + \sin^2 \phi^{\bar{p}_\theta}(t)} \, . \qquad (4.2.73)$$

The result follows from (4.2.72). $\square$

Now for fixed c_θ, ω, and T_0 the integrals J_1, J_2, and $\overline{J_3}$ are constant, and the Melnikov function can be written in the form

$$M(t_0; c_\phi, T_1) = -c_\phi J_1 - c_\theta J_2 + T_1 \overline{J_3}(\omega) \sin \omega t_0 \qquad (4.2.74)$$

where J_1, J_2, and $\overline{J_3}(\omega)$ do depend on whether or not they are calculated along the homoclinic orbits connecting the unperturbed fixed point on $\mathcal{M}_D$ or $\mathcal{M}_P$. There are two distinct possibilities; either

$$(-J_2/J_1)_D > (-J_2/J_1)_P \qquad (4.2.75)$$

or

$$(-J_2/J_1)_P > (-J_2/J_1)_D \qquad (4.2.76)$$

where the subscript D represents the integrals computed along the unperturbed homoclinic orbit connecting the fixed point on $\mathcal{M}_D$, and the subscript P represents the integrals computed along the unperturbed homoclinic trajectory connecting the fixed point on $\mathcal{M}_P$. We will assume that (4.2.76) holds. In this case, we can appeal to Theorems 4.1.9 and 4.1.10 and use (4.2.74) to plot regions in $c_\phi - T_1$ space where transverse homoclinic orbits to hyperbolic fixed points of the Poincaré map occur. These are shown in Figure 4.2.18 as two wedge-shaped regions. Inside the wedge labeled D, transverse homoclinic orbits to the hyperbolic periodic on $\mathcal{M}_{D,\epsilon}$ exist, and, inside the wedge labeled P, transverse homoclinic orbits to the hyperbolic fixed

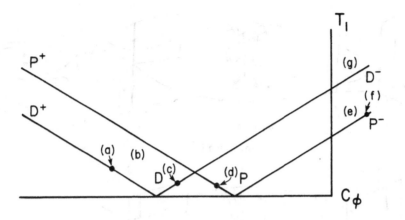

Figure 4.2.18. Regions where Horseshoes Exist.

point on $\mathcal{M}_{P,\epsilon}$ exist. In Figure 4.2.19, we show pictures of the homoclinic tangles in the Poincaré maps in the different regions.

Several interesting dynamic behaviors are possible. For instance, inside the wedge D, chaotic motions in which the pendulum erratically swings back and forth past $\phi = 0$, but not over the top, are possible; this is very much like the chaos observed in Duffing's equation, see Holmes [1979]. Inside the wedge P, chaotic motions exist in which the pendulum undergoes arbitrary sequences of clockwise and counterclockwise rotations about $\phi = 0$, see 4.2a, i. This example provides two distinct types of chaotic motions that are commonly studied. In fact, in the region where the interiors of both wedges intersect, both types of chaos are simultaneously possible. The dynamics in that region have the potential for the system "hopping" from one type of chaos to another; this cannot be proved using the present methods, since the invariant manifolds for the two types of chaos remain bounded away from one another. However, for $\bar{p}_\theta = O(\frac{1}{\epsilon^2})$, using methods similar to the present ones, it may be possible to predict when these manifolds mingle, thus proving the existence of such "hopping."

The types of chaos that exist here involve arbitrary sequences of physically different events. For the pendulum type of chaos (P), there exist motions in which the pendulum swings through approximately full 2π revolutions in arbitrary clockwise and counterclockwise sequences. For the Duffing type of chaos (D), there exist motions which swing back and forth through $\phi = 0$ toward $\phi > 0$ and $\phi < 0$

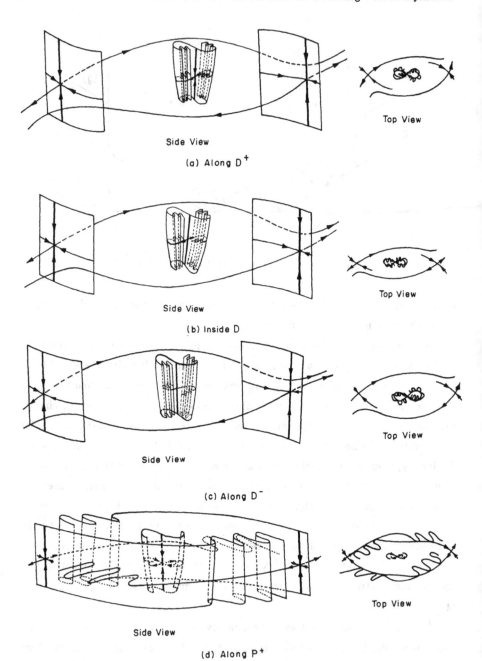

Figure 4.2.19. Poincaré Maps.

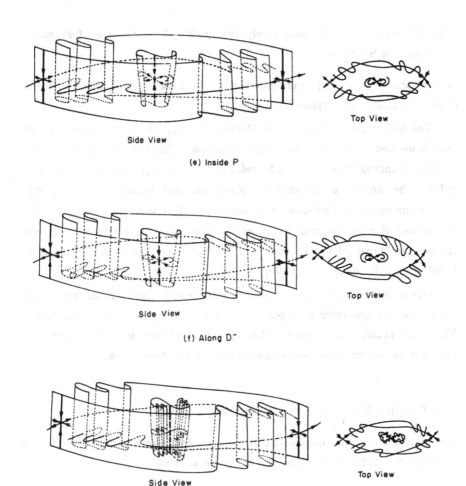

Side View

(e) Inside P

Top View

Side View

(f) Along D⁻

Top View

Side View

(g) Inside D∩P

Top View

Figure 4.2.19 Continued. Poincaré Maps.

in arbitrary orders. The proofs of these statements involve the use of symbol sequences and symbolic dynamics and arguments similar to those given at the end of Section 4.2a, i.

It must be pointed out that a different bifurcation diagram must be considered if $(-J_2/J_1)_D > (J_2/J_1)_D$. Essentially, the order of wedges P and D is switched and the sequences of bifurcations are changed. The details of that case can easily be worked out in the manner presented above.

4.2c. Perturbations of Completely Integrable, Two Degree of Freedom Hamiltonian Systems

We now present two examples of perturbations of completely integrable, two degree of freedom Hamiltonian systems.

The first is due to Holmes and Marsden [1982a], who first generalized Melnikov's method to such systems. Their technique utilized the "method of reduction" for Hamiltonian systems (see Marsden and Weinstein [1974]), which essentially reduced the problem to the standard planar Melnikov theory (Melnikov [1963]) when Hamiltonian perturbations were considered. Our techniques do not require the method of reduction, and the reader should compare our results with those of Holmes and Marsden [1982a] (note: for an example where the method of reduction is not applicable, see Holmes [1985]).

Our second example is due to Lerman and Umanski [1984]. It considers orbits homoclinic to a hyperbolic fixed point in two strongly coupled nonlinear oscillators. This is an example of an unperturbed system which contains a symmetry that results in the unperturbed homoclinic orbit being two dimensional.

i) A Coupled Pendulum and Harmonic Oscillator

We consider a linearly coupled simple pendulum and harmonic oscillator (see Holmes and Marsden [1982a]). The equations of motion are given by

$$
\begin{aligned}
\dot{x}_1 &= x_2 \\
\dot{x}_2 &= -\sin x_1 + \epsilon(x_1 - x_3) \\
\dot{x}_3 &= x_4 \\
\dot{x}_4 &= -\omega^2 x_3 + \epsilon(x_3 - x_1)
\end{aligned}
\qquad (x_1, x_2, x_3, x_4) \in T^1 \times \mathbb{R} \times \mathbb{R} \times \mathbb{R} . \qquad (4.2.77)
$$

As a convenience, we will transform the harmonic oscillator to action-angle coordinates via the transformation

$$
\begin{aligned}
x_3 &= \sqrt{\tfrac{2I}{\omega}} \sin \theta \\
x_4 &= \sqrt{2I\omega} \cos \theta .
\end{aligned}
\qquad (4.2.78)
$$

Using this transformation, (4.2.77) becomes

$$\dot{x}_1 = x_2$$
$$\dot{x}_2 = -\sin x_1 + \epsilon\left(\sqrt{\tfrac{2I}{\omega}}\sin\theta - x_1\right)$$
$$\dot{I} = -\epsilon\left(\sqrt{\tfrac{2I}{\omega}}\sin\theta - x_1\right)\sqrt{\tfrac{2I}{\omega}}\cos\theta \qquad (x_1, x_2, I, \theta) \in T^1 \times \mathbb{R} \times \mathbb{R}^+ \times T^1 \qquad (4.2.79)$$
$$\dot{\theta} = \omega + \epsilon\left(\sqrt{\tfrac{2I}{\omega}}\sin\theta - x_1\right)\frac{\sin\theta}{\sqrt{2I\omega}}$$

where $\mathbb{R}^+$ denotes the nonnegative real numbers. This system is Hamiltonian with Hamiltonian function given by

$$H_\epsilon = H(x_1, x_2, I) + \epsilon\tilde{H}(x_1, x_2, I, \theta) = \frac{x_2^2}{2} - \cos x_1 + I\omega + \frac{\epsilon}{2}\left(\sqrt{\tfrac{2I}{\omega}}\sin\theta - x_1\right)^2.$$
$$(4.2.80)$$

The unperturbed system is given by

$$\dot{x}_1 = x_2$$
$$\dot{x}_2 = -\sin x_1$$
$$\dot{I} = 0 \qquad\qquad (4.2.81)$$
$$\dot{\theta} = \omega.$$

It should be clear that the $x_1 - x_2$ component of (4.2.81) has a hyperbolic fixed point at $(\bar{x}_1, \bar{x}_2) = (\pi, 0) = (-\pi, 0)$ for all $I \in \mathbb{R}^+$. This fixed point is connected to itself by a pair of homoclinic trajectories

$$\left(x_{1h}^{\pm}(t), x_{2h}^{\pm}(t)\right) = \left(\pm 2\sin^{-1}(\tanh t), \pm 2\operatorname{sech} t\right). \qquad (4.2.82)$$

Thus, in the full $x_1 - x_2 - I - \theta$ phase space, (4.2.81) has a two dimensional normally hyperbolic invariant manifold with boundary

$$\mathcal{M} = \left\{ (x_1, x_2, I, \theta \mid (x_1, x_2) = (\pi, 0), \quad I \in \mathbb{R}^+, \quad \theta \in [0, 2\pi) \right\}. \qquad (4.2.83)$$

$\mathcal{M}$ has three dimensional stable and unstable manifolds which coincide along the three dimensional homoclinic orbit Γ parametrized by

$$\Gamma = \left\{ (\pm 2\sin^{-1}(\tanh(-t_0)), \pm 2\operatorname{sech}(-t_0), I, \theta_0) \in T^1 \times \mathbb{R} \times \mathbb{R}^+ \times T^1 \mid \right.$$
$$\left. (t_0, I, \theta_0) \in \mathbb{R} \times \mathbb{R}^+ \times T^1 \right\}. \qquad (4.2.84)$$

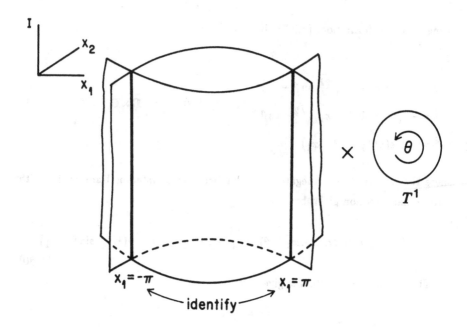

Figure 4.2.20. Geometry of the Phase Space of (4.2.81).

The geometry of the unperturbed phase space is shown in Figure 4.2.20. Thus, (4.2.77) has the form of System III with $n = m = 1$.

We now turn our attention to the perturbed system. By Proposition 4.1.16, $\mathcal{M}$ persists as well as the collection of three dimensional energy manifolds given by the level sets of (4.2.80). These energy manifolds intersect $\mathcal{M}$ in a periodic orbit which can be parametrized by I (cf., Section 4.1c,v). We now compute the Melnikov integrals in order to determine if the two dimensional stable and unstable manifolds of the periodic orbits intersect in the three dimensional energy surfaces.

By (4.1.92) and (4.1.85) the Melnikov vector is a scalar and is given by

$$M^{I\pm}(t_0;\omega) = -\int_{-\infty}^{\infty} D_\theta \tilde{H}\big(x_{1h}^{\pm}(t), x_{2h}^{\pm}(t), I, \omega(t + t_0)\big)\,dt$$

$$= -\int_{-\infty}^{\infty} \Big(\sqrt{\tfrac{2I}{\omega}}\sin(\omega(t + t_0)) - x_{1h}^{\pm}(t)\Big)\sqrt{\tfrac{2I}{\omega}}\cos(\omega(t + t_0))\,dt \,.$$

$$(4.2.85)$$

Evidently (4.2.85) converges at best conditionally. By Lemma 4.1.29 the improper integral makes sense if we approach the limits $\pm\infty$ along the sequences of times $\{T_j^s\}, \{-T_j^u\}, j = 1, 2, \ldots$, where for this problem we choose $T_j^s = T_j^u = \frac{\pi}{2\omega}(2j + 1)$.

Using (4.2.82), the integral becomes

$$M^{I^{\pm}}(t_0; \omega) = \mp \frac{2\pi}{\omega} \sqrt{\frac{2I}{\omega}} \operatorname{sech} \frac{\pi\omega}{2} \sin \omega t_0 . \tag{4.2.86}$$

Thus, (4.2.86) has zeros at $t_0 = \frac{\pi j}{\omega}$, $j = 0, \pm 1, \pm 2, \ldots$, and $D_{t_0} M^{I^{\pm}}(\bar{t}_0; \omega) \neq 0$ for all $I > 0$. Thus, by Theorems 4.1.19 and 4.1.20 for each $I > 0$ the stable and unstable manifolds of the corresponding hyperbolic periodic orbit intersect transversely yielding Smale horseshoes on the appropriate energy manifold. (Note: the question of what happens to these horseshoes for $I \to 0$ is an open problem.)

ii) A Strongly Coupled Two Degree of Freedom System

The following example is due to Lerman and Umanski [1984]. They consider a completely integrable two degree of freedom Hamiltonian system of the form

$$\begin{pmatrix} \dot{x}_1 \\ \dot{x}_2 \\ \dot{x}_3 \\ \dot{x}_4 \end{pmatrix} = \begin{pmatrix} \lambda & -\omega & 0 & 0 \\ \omega & \lambda & 0 & 0 \\ 0 & 0 & -\lambda & \omega \\ 0 & 0 & -\omega & -\lambda \end{pmatrix} \begin{pmatrix} x_1 \\ x_2 \\ x_3 \\ x_4 \end{pmatrix} - 2\sqrt{2}\lambda \begin{pmatrix} x_3(x_3^2 + x_4^2) \\ x_4(x_3^2 + x_4^2) \\ -x_1(x_1^2 + x_2^2) \\ -x_2(x_1^2 + x_2^2) \end{pmatrix} \tag{4.2.87}$$

where $\lambda < 0$ and $\omega > 0$. The two integrals are

$$H = K_1 = \lambda(x_1 x_3 + x_2 x_4) - \omega(x_2 x_3 - x_1 x_4) - \frac{\lambda}{\sqrt{2}}[(x_1^2 + x_2^2)^2 + (x_3^2 + x_4^2)^2]$$

$$K_2 = x_2 x_3 - x_1 x_4 . \tag{4.2.88}$$

The equation (4.2.87) has a hyperbolic fixed point at $(x_1, x_2, x_3, x_4) = (0, 0, 0, 0)$ where the eigenvalues of (4.2.87) linearized about the fixed point are easily seen to be $\pm\lambda \pm i\omega$. Thus, $(0, 0, 0, 0)$ has a two dimensional stable manifold and a two dimensional unstable manifold that coincide along the two dimensional homoclinic manifold Γ, which can be parametrized by

$$\Gamma = \{ (x_{1h}(t), x_{2h}(t), x_{3h}(t), x_{4h}(t)) =$$
$$2^{-1/4}[\cosh(-4\lambda t_0)]^{-1/2} (e^{\lambda t_0} \cos a, e^{\lambda t_0} \sin a, e^{-\lambda t_0} \cos a, e^{-\lambda t_0} \sin a) \in \mathbb{R}^4 \mid$$
$$a = (-\omega t_0 + \alpha), \quad (t_0, \alpha) \in \mathbb{R}^1 \times T^1 \} . \tag{4.2.89}$$

We remark that the reason the stable and unstable manifolds coincide is that the second integral, K_2, induces a rotational symmetry in the vector field (heuristically,

K_2 can be thought of as "angular momentum"). The variable "α" in (4.2.89) arises due to this rotational symmetry.

In this example, the origin plays the role of M in the general theory. So, for a sufficiently small perturbation of (4.2.87), the fixed point persists (note: in particular, for our perturbation, the origin actually remains a hyperbolic fixed point of the perturbed system). Whether or not any orbits homoclinic to the fixed point survive can be determined by computing the appropriate Melnikov vector. There are two distinct situations depending upon whether or not the perturbation is Hamiltonian. We present two examples illustrating the different cases for this problem.

Non-Hamiltonian Perturbation of (4.2.87). Consider the following perturbation of (4.2.87)

$$\epsilon g^x(x; \mu) = \epsilon \begin{pmatrix} (\mu_1 + \mu_2)x_2 \\ 0 \\ \mu_1 x_4 \\ -\mu_2 x_3 \end{pmatrix} \tag{4.2.90}$$

where $\mu = (\mu_1, \mu_2) \in \mathbf{R}^2$ is a parameter.

So, in this case, the perturbed system is an example of System I with $n = 2$, $m = l = 0$. For this problem, the Melnikov vector has two components, which are given by

$$M_1(\alpha; \mu) = \int_{-\infty}^{\infty} \langle D_x H, g^x \rangle (x_{1h}(t), x_{2h}(t), x_{3h}(t), x_{4h}(t); \mu_1, \mu_2) dt \tag{4.2.91a}$$

$$M_2(\alpha; \mu) = \int_{-\infty}^{\infty} \langle D_x K_2, g^x \rangle (x_{1h}(t), x_{2h}(t), x_{3h}(t), x_{4h}(t); \mu_1, \mu_2) dt . \tag{4.2.91b}$$

Using (4.2.89), the integrals are

$$M_1(\alpha; \mu) = \frac{-\pi \mu_1}{4\sqrt{2}} \operatorname{sech} \frac{\pi \omega}{4\lambda} \sin 2\alpha + \frac{\mu_2 \omega}{4\sqrt{2}\lambda} + \frac{\sqrt{2}\pi \mu_2 \omega}{8\lambda} \operatorname{sech} \frac{\pi \omega}{4\lambda} \cos 2\alpha \tag{4.2.92a}$$

$$M_2(\alpha; \mu) = \frac{-\pi \mu_2}{4\sqrt{2}\lambda} . \tag{4.2.92b}$$

It should be clear that $M(\alpha; \mu) \equiv (M_1(\alpha; \mu), M_2(\alpha; \mu))$ is zero at

$$\mu_2 = \bar{\mu}_2 = 0, \quad \alpha = \bar{\alpha} = 0, \tfrac{\pi}{2}, \pi, \tfrac{3\pi}{2} \tag{4.2.93}$$

and

$$
\det\left[D_{(\alpha_1,\mu_2)} M(\bar{\alpha},\mu_1,\bar{\mu}_2)\right] = \begin{cases} \dfrac{\pi^2\mu_1}{16\lambda}\operatorname{sech}\dfrac{\pi\omega}{4\lambda} & \alpha = 0, \pi \\[2mm] -\dfrac{\pi^2\mu_1}{16\lambda}\operatorname{sech}\dfrac{\pi\omega}{4\lambda} & \alpha = \dfrac{\pi}{2}, \dfrac{3\pi}{2} \end{cases} \tag{4.2.94}
$$

Thus, by Theorem 4.1.9, for μ_2 near zero and for any $\mu_1 \neq 0$ there exist four orbits homoclinic to the origin in the perturbed system. We cannot immediately appeal to the results of Section 3.2d, ii) to assert the existence of Smale horseshoe-like behavior in the perturbed system. This is because the eigenvalues of the unperturbed system do not satisfy the hypotheses of the example in Section 3.2d, ii) so, necessarily, we would have to compute the $O(\epsilon)$ correction to the eigenvalues due to the perturbation and then recheck the hypotheses of the example.

Hamiltonian Perturbation of (4.2.87).

We consider a Hamiltonian perturbation of (4.2.67) where we perturb H by $\epsilon \tilde{H}$ where

$$
\tilde{H} = x_1^2 + x_3^2 . \tag{4.2.95}
$$

So in this case the perturbed system is an example of System III with $n = 2$, $m = 0$.

Since the perturbation is Hamiltonian and, therefore, the three dimensional invariant energy manifolds are preserved, the Melnikov vector is just a scalar and, by (4.1.92) and (4.1.85), it is given by

$$
M_2(\alpha) = \int_{-\infty}^{\infty} \langle D_x K_2, JD\tilde{H}\rangle \big(x_{1h}(t), x_{2h}(t), x_{3h}(t), x_{4h}(t)\big) dt . \tag{4.2.96}
$$

Using (4.2.89), the integral can be computed and is found to be

$$
M_2(\alpha) = \frac{-\pi}{2\lambda} \frac{\cosh\frac{\pi\omega}{4\lambda}}{\cosh\frac{\pi\omega}{2\lambda}} \sin 2\alpha . \tag{4.2.97}
$$

Thus, $M_2(\alpha) = 0$ at $\alpha = \bar{\alpha} = 0, \frac{\pi}{2}, \pi, \frac{3\pi}{2}$, and at these points $D_\alpha M_2 \neq 0$. So, by Theorem 4.1.19, the perturbed system has four orbits homoclinic to the origin and, by Theorem 4.1.20, the two dimensional stable and unstable manifolds of the origin intersect transversely along these orbits in the three dimensional energy surface. Thus, by Devaney's theorem (Theorem 3.2.22), the perturbed system contains horseshoes near these orbits (note: the reader can verify directly by computation that $M_1 = 0$ for this problem for all α).

4.2d. Perturbation of a Completely Integrable Three Degree of Freedom System : Arnold Diffusion

We now give an example of a three degree of freedom Hamiltonian system due to Holmes and Marsden [1982b] which exhibits the phenomenon of Arnold diffusion, see Section 4.1d,v. Our methods differ slightly from those of Holmes and Marsden in that they utilize the method of reduction to first reduce the order of the system. Our methods are more direct in that we avoid this preliminary transformation of the system.

We consider the following system

$$
\begin{aligned}
\dot{x}_1 &= x_2 \\
\dot{x}_2 &= -\sin x_1 + \epsilon\big[\sqrt{2I_1}\sin\theta_1 + \sqrt{2I_2}\sin\theta_2 - 2x_1\big] \\
\dot{I}_1 &= -\epsilon\big[\sqrt{2I_1}\sin\theta_1 - x_1\big]\sqrt{2I_1}\cos\theta_1 \\
\dot{I}_2 &= -\epsilon\big[\sqrt{2I_2}\sin\theta_2 - x_1\big]\sqrt{2I_2}\cos\theta_2 \\
\dot{\theta}_1 &= D_{I_1}G_1(I_1) + \epsilon\big[\sqrt{2I_1}\sin\theta_1 - x_1\big]\frac{\sin\theta_1}{\sqrt{2I_1}} \\
\dot{\theta}_2 &= D_{I_2}G_2(I_2) + \epsilon\big[\sqrt{2I_2}\sin\theta_2 - x_2\big]\frac{\sin\theta_2}{\sqrt{2I_2}}
\end{aligned}
$$

$$(x_1, x_2, I_1, I_2, \theta_1, \theta_2) \in$$
$$T^1 \times \mathbb{R} \times \mathbb{R}^+ \times \mathbb{R}^+ \times T^1 \times T^1.$$

$$(4.2.98)$$

This system is Hamiltonian with Hamiltonian function given by

$$
\begin{aligned}
H_\epsilon &= H(x_1, x_2, I_1, I_2) + \epsilon\tilde{H}(x_1, x_2, I_1, I_2, \theta_1, \theta_2) \\
&= \frac{x_2^2}{2} - \cos x_1 + G_1(I_1) + G_2(I_2) + \frac{\epsilon}{2}\big[\big(\sqrt{2I_1}\sin\theta_1 - x_1\big)^2 \\
&\qquad\qquad + \big(\sqrt{2I_2}\sin\theta_2 - x_2\big)^2\big]
\end{aligned}
$$

$$(4.2.99)$$

where $G_1(I_1)$ and $G_2(I_2)$ are arbitrary C^2 functions which satisfy the nondegeneracy requirement for the KAM Theorem 4.1.17 given by

$$D_{I_1}^2 G_1(I_1)\, D_{I_2}^2 G_2(I_2) \neq 0. \qquad (4.2.100)$$

The unperturbed system is given by

$$\dot{x}_1 = x_2$$
$$\dot{x}_2 = -\sin x_1$$
$$\dot{I}_1 = 0$$
$$\dot{I}_2 = 0 \tag{4.2.101}$$
$$\dot{\theta}_1 = D_{I_1} G_1(I_1)$$
$$\dot{\theta}_2 = D_{I_2} G_2(I_2) .$$

We can think of the $I_1 - \theta_1$ and $I_2 - \theta_2$ components of (4.2.101) as being nonlinear oscillators expressed in action angle coordinates, with $D_{I_1} G_1(I_1)$ and $D_{I_2} G_2(I_2)$ being the frequencies of the oscillators. The $x_1 - x_2$ component has a hyperbolic fixed point at $(\bar{x}_1, \bar{x}_2) = (\pi, 0) = (-\pi, 0)$ which is connected to itself by a pair of homoclinic trajectories given by $(x_{1h}^{\pm}(t), x_{2h}^{\pm}(t)) = (\pm 2\sin^{-1}(\tanh t), \pm 2 \operatorname{sech} t)$. Thus, in the full $x_1 - x_2 - I_1 - I_2 - \theta_1 - \theta_2$ phase space, (4.2.101) has a 4 dimensional normally hyperbolic invariant manifold given by

$$\mathcal{M} = \{ (x_1, x_2, I_1, I_2, \theta_1, \theta_2) \in T^1 \times \mathbb{R} \times \mathbb{R}^+ \times \mathbb{R}^+ \times T^1 \times T^1 \mid (x_1, x_2) = (\pi, 0) = (-\pi, 0) \} \tag{4.2.102}$$

which has the structure of a two parameter family of 2-tori. $\mathcal{M}$ has a 5 dimensional stable manifold and a 5 dimensional unstable manifold that coincide along the 5 dimensional homoclinic orbits $\Gamma^{\pm}$, which can be parametrized by

$$\Gamma^{\pm} = [(\pm 2\sin^{-1}(\tanh(-t_0)), \pm 2\operatorname{sech}(-t_0)), I_1, I_2, \theta_{10}, \theta_{20})$$
$$\in T^1 \times \mathbb{R} \times \mathbb{R}^+ \times T^1 \times T^1 \mid t_0 \in \mathbb{R} \}. \tag{4.2.103}$$

Note that on $\mathcal{M}$, for each $I = (I_1, I_2)$, there corresponds a two torus $\tau(I)$ having three dimensional stable and unstable manifolds coinciding along the three dimensional homoclinic orbits $\Gamma_I^{\pm}$, where $\Gamma_I^{\pm}$ is obtained from (4.2.103) by fixing the I component. Thus, (4.2.98) is an example of System III with $n = 1$, $m = 2$.

We now turn our attention to the perturbed system where the perturbation corresponds to linear coupling of the nonlinear oscillators with the pendulum. By Proposition 4.1.16, $\mathcal{M}$ persists (denoted by $\mathcal{M}_\epsilon$) and intersects each 5 dimensional invariant energy surface given by the level sets of (4.2.99) in a three dimensional set of which, by the KAM Theorem 4.1.17, "most" of a one parameter family of invariant two tori persist (note: (4.2.100) is equivalent to the nondegeneracy hypothesis of the KAM theorem for our system). We then compute the Melnikov vector in order to determine if the stable and unstable manifolds of the KAM tori intersect transversely. If so, we can conclude that (4.2.98) exhibits Arnold diffusion.

From (4.1.92) and (4.1.85), the Melnikov vector has two components given by

$$M_2^\pm(\theta_{10}, \theta_{20}, I_1, I_2) = \int_{-\infty}^{\infty} D_{\theta_1}\tilde{H}\big(x_{1h}^\pm(t), x_{2h}^\pm(t), I_1, I_2, \omega_1 t + \theta_{10}, \omega_2 t + \theta_{20}\big)dt$$

(4.2.104a)

$$M_3^\pm(\theta_{10}, \theta_{20}, I_1, I_2) = \int_{-\infty}^{\infty} D_{\theta_2}\tilde{H}\big(x_{1h}^\pm(t), x_{2h}^\pm(t), I_1, I_2, \omega_1 t + \theta_{10}, \omega_2 t + \theta_{20}\big)dt \ .$$

(4.2.104b)

Using (4.2.103) (cf., equations (4.2.85) and (4.2.86)) the Melnikov integrals can be computed and are found to be

$$M_{i+1}^{I\pm}(\theta_{10}, \theta_{20}) = \mp\frac{2\pi}{\Omega_i}\sqrt{2I_i}\text{sech}\,\frac{\pi\Omega_i}{2}\sin\theta_{i0} \ , \quad i = 1, 2 \tag{4.2.105}$$

where $\Omega_i \equiv D_{I_i}G_i(I_i)$ with

$$\det\big[D_{(\theta_{10}, \theta_{20})}M^{I\pm}(\theta_{10}, \theta_{20})\big] = \frac{8\pi^2\sqrt{I_1 I_2}}{\Omega_1\Omega_2}\text{sech}\,\frac{\pi\Omega_1}{2}\text{sech}\,\frac{\pi\Omega_2}{2}\cos\theta_{10}\cos\theta_{20} \ .$$

(4.2.106)

Thus, $M^{I\pm}(\theta_{10}, \theta_{20})$ has zeros at $(\theta_{10}, \theta_{20}) = (\bar\theta_{10}, \bar\theta_{20}) = (k\pi, k\pi)$, $k = 0, \pm1, \pm2, \ldots$, and $D_{(\theta_{10}, \theta_{20})}M^{I\pm}(\theta_{10}, \theta_{20})$ has rank 2 at these points. So, by Theorems 4.1.19 and 4.1.20, the stable and unstable manifolds of the KAM tori intersect transversely, and hence Arnold diffusion occurs in (4.2.98).

4.2e. Quasiperiodically Forced Single Degree of Freedom Systems

We now consider two examples of single degree of freedom systems subjected to quasiperiodic excitation. In these examples the Melnikov vector will detect orbits homoclinic to normally hyperbolic invariant tori with the resulting chaotic dynamics being characterized by Theorem 3.4.1.

The first example is the quasiperiodically forced Duffing oscillator studied by Wiggins [1987]. In this example, the existence of transverse homoclinic tori is established, and the effect of the number of forcing frequencies on the chaotic region in parameter space is examined. Additionally, a relationship of these theoretical results with experimental work of Moon and Holmes [1985] on the quasiperiodically forced beam is discussed.

The second example is the parametrically excited pendulum whose base is vertically oscillated with a combination $\mathcal{O}(\epsilon)$ amplitude $\mathcal{O}(1)$ frequency and $\mathcal{O}(1)$ amplitude $\mathcal{O}(\epsilon)$ frequency excitation.

i) The Duffing Oscillator: Forced at $\mathcal{O}(\epsilon)$ Amplitude with N $\mathcal{O}(1)$ Frequencies

We first consider the quasiperiodically forced Duffing oscillator forced with two frequencies

$$
\begin{aligned}
\dot{x}_1 &= x_2 \\
\dot{x}_2 &= \frac{1}{2}x_1(1 - x_1^2) + \epsilon[f\cos\theta_1 + f\cos\theta_2 - \delta x_2] \\
\dot{\theta}_1 &= \omega_1 \qquad\qquad\qquad\qquad (x_1, x_2, \theta_1, \theta_2) \in \mathbb{R} \times \mathbb{R} \times T^1 \times T^1 \\
\dot{\theta}_2 &= \omega_2
\end{aligned}
$$

$$(4.2.107)$$

where f and δ are positive, and ω_1 and ω_2 are positive real numbers. We can reduce the study of (4.2.107) to the study of an associated three dimensional Poincaré map obtained by defining a three dimensional cross-section to the four dimensional phase space by fixing the phase of one of the angular variables and allowing the remaining three variables that start on the cross-section to evolve in time under the action of the flow generated by (4.2.107) until they return to the cross-section, see Section 1.6. The return occurs after one period of the angular variable whose phase was fixed in order to define the cross-section. To be more precise, the cross-section, $\Sigma^{\theta_{20}}$, is given by

$$\Sigma^{\theta_{20}} = \{ (x_1, x_2, \theta_1, \theta_2) \in \mathbb{R} \times \mathbb{R} \times T^1 \times T^1 \mid \theta_2 = \theta_{20} \} \qquad (4.2.108)$$

where, for definiteness, we fix the phase of θ_2. The Poincaré map is then defined to be

$$
\begin{aligned}
P_\epsilon &: \Sigma^{\theta_{20}} \to \Sigma^{\theta_{20}} \\
&(x_1(0), x_2(0), \theta_1(0) = \theta_{10}) \mapsto \left(x_1\left(\frac{2\pi}{\omega_2}\right), x_2\left(\frac{2\pi}{\omega_2}\right), \theta_1\left(\frac{2\pi}{\omega_2}\right) = \frac{2\pi\omega_1}{\omega_2} + \theta_{10}\right).
\end{aligned}
$$

$$(4.2.109)$$

For $\epsilon = 0$, the $x_1 - x_2$ component of (4.2.107) is a completely integrable Hamiltonian system with Hamiltonian function given by

$$H(x_1, x_2) = \frac{x_2^2}{2} - \frac{x_1^2}{4} + \frac{x_1^4}{8} . \qquad (4.2.110)$$

It also has a hyperbolic fixed point at $(x_1, x_2) = (0,0)$ which is connected to itself by a symmetric pair of homoclinic trajectories given by $(x_1(t), x_2(t)) = (\pm\sqrt{2}\,\text{sech}\,\frac{t}{\sqrt{2}}, \mp\,\text{sech}\,\frac{t}{\sqrt{2}}\tanh\frac{t}{\sqrt{2}})$. Thus, in the full $x_1 - x_2 - \theta_1 - \theta_2$ phase space, the unperturbed system has a two dimensional normally hyperbolic invariant torus given by

$$\mathcal{M} = \{\,(x_1, x_2, \theta_1, \theta_2) \in \mathbb{R}\times\mathbb{R}\times T^1\times T^1 \mid x_1 = x_2 = 0, \quad \theta_1, \theta_2 \in [0, 2\pi)\,\} \quad (4.2.111)$$

with trajectories on the torus given by $(x_1(t), x_2(t), \theta_1(t), \theta_2(t)) = (0, 0, \omega_1 t + \theta_{10}, \omega_2 t + \theta_{20})$. The torus has a symmetric pair of coincident three dimensional stable and unstable manifolds with trajectories in the respective branches given by

$$(x_{1h}^{\pm}(t), x_{2h}^{\pm}(t), \theta_1(t), \theta_2(t)) = (\pm\sqrt{2}\,\text{sech}\,\frac{t}{\sqrt{2}}, \mp\,\text{sech}\,\frac{t}{\sqrt{2}}\tanh\frac{t}{\sqrt{2}}, \omega_1 t + \theta_{10}, \omega_2 t + \theta_{20}).$$
$$(4.2.112)$$

Thus, (4.2.107) is an example of System I with $n = 1$, $m = 0$, $l = 2$.

Utilizing this information, we can obtain a complete picture of the global integrable dynamics of the unperturbed Poincaré map, P_0. In particular, P_0 has a one dimensional normally hyperbolic invariant torus, $T_0 = \mathcal{M} \cap \Sigma^{\theta_{20}}$, that has a symmetric pair of two dimensional stable and unstable manifolds, $W^s(T_0)$ and $W^u(T_0)$, that are coincident, see Figure 4.2.21.

By Proposition 4.1.5, for $\epsilon \neq 0$ and small, the perturbed Poincaré map, P_ϵ, still possesses an invariant one dimensional normally hyperbolic invariant torus $T_\epsilon = \mathcal{M}_\epsilon \cap \Sigma^{\theta_{20}}$ having two dimensional stable and unstable manifolds, $W^s(T_\epsilon)$ and $W^u(T_\epsilon)$, which may now intersect transversely yielding transverse homoclinic orbits to T_ϵ. Intersections of $W^s(T_\epsilon)$ and $W^u(T_\epsilon)$ can be determined from the Melnikov vector. From (4.1.47), the Melnikov vector has one component and is given by

$$M^{\pm}(\theta_{10}, \theta_{20}; f, \delta, \omega_1, \omega_2) = \int_{-\infty}^{\infty} \langle D_x H, g^x\rangle(x_{1h}^{\pm}(t), x_{2h}^{\pm}(t), \omega_1 t + \theta_{10}, \omega_2 t + \theta_{20})\,dt .$$
$$(4.2.113)$$

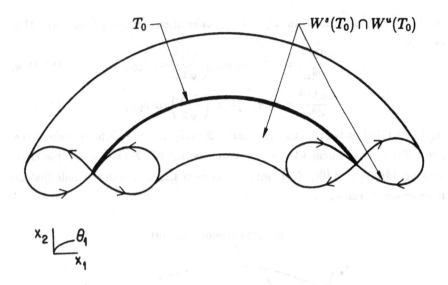

Figure 4.2.21. Homoclinic Geometry of the Phase Space of P_0, Cut Away Half View.

Using (4.2.110) and from (4.2.107), $g^x = (0, f\cos\theta_1 + f\cos\theta_2 - \delta x_2)$, (4.2.113) becomes

$$M^\pm(\theta_{10}, \theta_{20}; f, \delta, \omega_1, \omega_2)$$

$$= \int_{-\infty}^{\infty} \left[-\delta \left(x_{2h}^\pm(t)\right)^2 + f x_{2h}^\pm(t)\cos(\omega_1 t + \theta_{10}) + f x_{2h}^\pm(t)\cos(\omega_2 t + \theta_{20}) \right] dt$$

$$= \frac{-2\sqrt{2}}{3}\delta \pm 2\pi f \omega_1 \text{sech}\frac{\pi\omega_1}{\sqrt{2}}\sin\theta_{10} \pm 2\pi f \omega_2 \text{sech}\frac{\pi\omega_2}{\sqrt{2}}\sin\theta_{20} .$$

$$(4.2.114)$$

It should be clear that (4.2.114) has zeros provided

$$f > \frac{\sqrt{2}\delta}{3\pi\left(\omega_1 \text{sech}\frac{\pi\omega_1}{\sqrt{2}} + \omega_2 \text{sech}\frac{\pi\omega_2}{\sqrt{2}}\right)} . \qquad (4.2.115)$$

We have the following theorem.

Theorem 4.2.2. For all $f, \delta, \omega_1, \omega_2$ such that (4.2.115) is satisfied, $W^s(T_\epsilon)$ and $W^u(T_\epsilon)$ intersect transversely in a transverse homoclinic torus.

PROOF: From (4.2.114), $M^\pm = 0$ implies

$$f = \frac{\pm\sqrt{2}\delta}{3\pi\left[\omega_1 \text{sech}\frac{\pi\omega_1}{\sqrt{2}}\sin\theta_{10} + \omega_2 \text{sech}\frac{\pi\omega_2}{\sqrt{2}}\sin\theta_{20}\right]} . \qquad (4.2.116)$$

So, if (4.2.115) is satisfied, then (4.2.116) has solutions for some $(\theta_{10}, \theta_{20})$. Also, we have

$$\frac{\partial M^{\pm}}{\partial \theta_{10}} = \pm 2\pi f \omega_1 \operatorname{sech}\left(\frac{\pi \omega_1}{\sqrt{2}}\right) \cos \theta_{10}, \qquad (4.2.117a)$$

$$\frac{\partial M^{\pm}}{\partial \theta_{20}} = \pm 2\pi f \omega_2 \operatorname{sech}\left(\frac{\pi \omega_2}{\sqrt{2}}\right) \cos \theta_{20}, \qquad (4.2.117b)$$

so, if (4.2.115) is satisfied, (4.2.117a) and (4.2.117b) cannot each be zero simultaneously. Then, by Theorem 4.1.9, we can conclude that $W^s(T_\epsilon)$ and $W^u(T_\epsilon)$ intersect at some $(\theta_{10}, \theta_{20}) = (\bar{\theta}_{10}, \bar{\theta}_{20})$ and, by Theorem 4.1.10, we can conclude that the intersection is transversal.

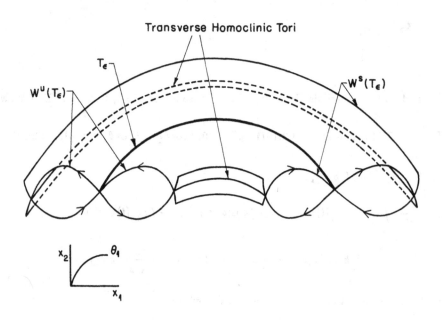

Figure 4.2.22. Transverse Homoclinic Torus for P_ϵ, Cut Away Half View.

This establishes that $W^s(T_\epsilon)$ and $W^u(T_\epsilon)$ intersect transversely at some point $(\theta_1, \theta_2) = (\bar{\theta}_1, \bar{\theta}_2)$; we now need to argue that (4.2.115) is actually a sufficient condition for $W^s(T_\epsilon)$ and $W^s(T_\epsilon)$ to intersect transversely in a 1-torus. The argument goes as follows: since, by (4.2.115), $\frac{\partial M^{\pm}}{\partial \theta_{10}}$ and $\frac{\partial M^{\pm}}{\partial \theta_{20}}$ cannot both vanish simultaneously, suppose for definiteness, $\frac{\partial M^{\pm}}{\partial \theta_{10}}(\bar{\theta}_{10}, \bar{\theta}_{20}) \neq 0$. Then, by the global implicit function theorem (Chow and Hale [1982]), there exists a function of θ_{20}, say $h^{\pm}(\theta_{20})$, whose graph is a zero of $M^{\pm}$, i.e., $M^{\pm}(h^{\pm}(\theta_{20}), \theta_{20})$ for all θ_{20} such that $\frac{\partial M^{\pm}}{\partial \theta_{10}}(h^{\pm}(\theta_{20}), \theta_{20}) \neq 0$. At points where $\frac{\partial M^{\pm}}{\partial \theta_{10}} = 0$ then, since $\frac{\partial M^{\pm}}{\partial \theta_{20}}$ is not also zero, there exists a function $g^{\pm}(\theta_{10})$ such that $\frac{\partial M^{\pm}}{\partial \theta_{20}}(\theta_{10}, g^{\pm}(\theta_{10})) \neq 0$. Then, since $\frac{\partial M^{\pm}}{\partial \theta_{10}}$ and $\frac{\partial M^{\pm}}{\partial \theta_{20}}$ are never both zero, graph $h^{\pm}(\theta_{20}) \cup$ graph $g^{\pm}(\theta_{10})$ forms a differentiable circle which is a zero of $M^{\pm}$. $\qquad\square$

Thus, Theorem 3.4.1 applies to this system and implies the existence of chaotic dynamics for parameter values satisfying (4.2.24).

We illustrate the geometry of the homoclinic orbits of P_{ϵ} in Figures 4.2.22 and 4.2.23. In Figure 4.2.22 we show a transverse homoclinic torus for P_{ϵ}. Using an argument similar to that given for concluding the existence of the homoclinic tangle for orbits homoclinic to fixed points of maps (see Section 3.4 or Abraham and Shaw [1984]), we can conclude that a *homoclinic torus tangle* results as shown in Figure 4.2.23.

The homoclinic torus tangle appears to form the backbone of the strange attractor experimentally observed by Moon and Holmes [1985] for this system. They studied the structure of the strange attractor by utilizing a technique due to Lorenz [1984], which involves constructing a double Poincaré section or Lorenz cross-section by fixing the phase of one of the angular variables and a small window about a fixed phase of the remaining angular variable. The map of this "section of a section" into itself revealed a fractal nature of the strange attractor similar to that found in the usual Duffing-Holmes strange attractor (Holmes [1979]) which was not apparent in the three dimensional Poincaré map. Our results give much insight into the nature of this phenomenon. In Figure 4.2.23, it is clear that the intersection of $W^s(T_{\epsilon})$ and $W^u(T_{\epsilon})$ with the double Poincaré section yields a geometric structure that is quite similar to the homoclinic tangle which occurs in the periodically forced Duffing equation and that is responsible for the fractal structure of the Duffing-Holmes strange attractor.

Next we want to consider the effect on the region where transverse homoclinic

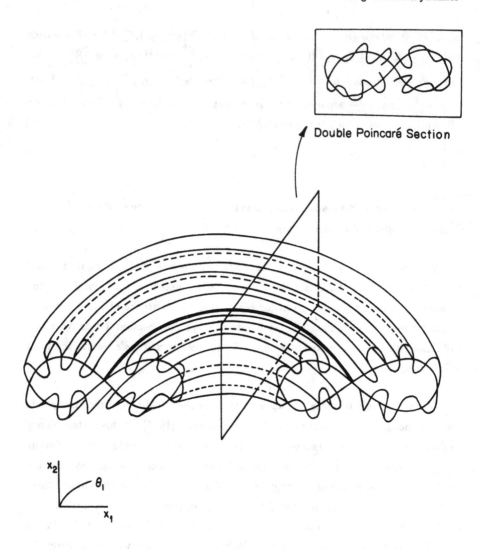

Figure 4.2.23. Homoclinic Torus Tangle for P_ϵ and the Double Poincaré Section, Cut Away Half View.

tori exist caused by adding additional forcing functions to (4.2.107), i.e., we consider

$$\dot{x}_1 = x_2$$

$$\dot{x}_2 = \frac{1}{2}x_1(1 - x_1^2) + \epsilon[f\cos\theta_1 + f\cos\theta_2 + \cdots + f\cos\theta_n - \delta y]$$

$$\dot{\theta}_1 = \omega_1 \tag{4.2.118}$$

$$\vdots$$

$$\dot{\theta}_2 = \omega_n .$$

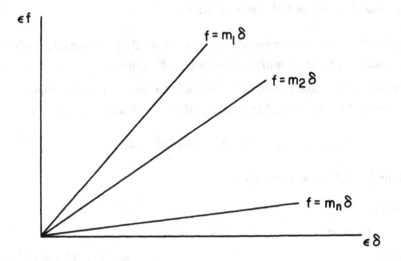

Figure 4.2.24. Regions of Chaos in $f - \delta$ Space as a Function of the Number of Forcing Frequencies.

We reduce the study of (4.2.118) to the study of an associated $n-1$ dimensional Poincaré map having an $n - 2$ dimensional normally hyperbolic invariant manifold with $n - 1$ dimensional stable and unstable manifolds. Intersection of the stable and unstable manifolds is determined by calculating the Melnikov function. In Figure 4.2.24, the lines $f = m_1\delta$, $f = m_2\delta$, and $f = m_n\delta$ represent lines above which transverse homoclinic tori occur for the Duffing oscillator forced at 1, 2, and n frequencies, respectively; m_1, m_2, and m_n are obtained from the Melnikov function and are given by

$$m_1 = \frac{\sqrt{2}}{3\pi\omega_1 \operatorname{sech} \frac{\pi\omega_1}{\sqrt{2}}}$$

$$m_2 = \frac{\sqrt{2}}{3\pi \left(\omega_1 \text{sech} \frac{\pi\omega_1}{\sqrt{2}} + \omega_2 \text{sech} \frac{\pi\omega_2}{\sqrt{2}} \right)}$$

and

$$m_n = \frac{\sqrt{2}}{3\pi \left(\omega_1 \text{sech} \frac{\pi\omega_1}{\sqrt{2}} + \omega_2 \text{sech} \frac{\pi\omega_2}{\sqrt{2}} + \cdots + \omega_n \text{sech} \frac{\pi\omega_n}{\sqrt{2}} \right)} .$$

Thus, we see that the effect of increasing the number of forcing frequencies is to increase the area in parameter space where chaotic behavior can occur, and hence to increase the likelihood of chaotic dynamics.

ii) The Pendulum: Parametrically Forced at $\mathcal{O}(\epsilon)$ Amplitude with $\mathcal{O}(1)$ Frequency and $\mathcal{O}(1)$ Amplitude with $\mathcal{O}(\epsilon)$ Frequency

We consider the same system as in Section 4.2a but with the combined forcing functions of Examples 4.2a, i and 4.2a, ii. More specifically, we have

$$\ddot{x}_1 + \epsilon\delta\dot{x}_1 + (1 - \epsilon\Gamma \sin \Omega t - \gamma \sin \epsilon\omega t) \sin x_1 = 0 . \tag{4.2.119}$$

Writing (4.2.119) as a system gives

$$\begin{aligned} \dot{x}_1 &= x_2 \\ \dot{x}_2 &= -(1 - \gamma \sin I) \sin x_1 + \epsilon[\Gamma \sin \theta \sin x_1 - \delta x_2] \\ \dot{I} &= \epsilon\omega \\ \dot{\theta} &= \Omega \end{aligned} \qquad (x_1, x_2, I, \theta) \in T^1 \times \mathbb{R} \times T^1 \times T^1 .$$

$$\tag{4.2.120}$$

The unperturbed system is given by

$$\begin{aligned} \dot{x}_1 &= x_2 \\ \dot{x}_2 &= -(1 - \gamma \sin I) \sin x_1 \\ \dot{I} &= 0 \\ \dot{\theta} &= \Omega \end{aligned} \tag{4.2.121}$$

and the $x_1 - x_2$ component of (4.2.121) is Hamiltonian with Hamiltonian function given by

$$H(x_1, x_2) = \frac{x_2^2}{2} - (1 - \gamma \sin I) \cos x_1 . \tag{4.2.122}$$

The unperturbed system has a hyperbolic fixed point at $(\bar{x}_1, \bar{x}_2) = (\pi, 0) = (-\pi, 0)$ for each $I \in (0, 2\pi]$, $\theta \in (0, 2\pi]$ provided $0 \leq \gamma < 1$. Each of these fixed points is connected to itself by a pair of homoclinic trajectories given by

$$\left(x_{1h}^{\pm}(t), x_{2h}^{\pm}(t)\right) =$$
$$\left(\pm 2 \sin^{-1}\left[\tanh\sqrt{1 - \gamma \sin I}\, t\right], \pm 2\sqrt{1 - \gamma \sin I}\,\operatorname{sech}\sqrt{1 - \gamma \sin I}\, t\right). \tag{4.2.123}$$

Thus, the unperturbed system has a normally hyperbolic invariant two torus whose stable and unstable manifolds coincide; see Figure 4.2.25 for an illustration of the unperturbed phase space of (4.2.121). Thus, (4.2.120) is an example of System II with $n = m = l = 1$.

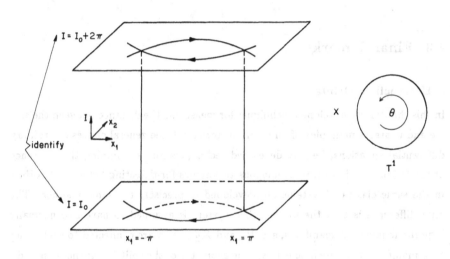

Figure 4.2.25. The Phase Space of (4.2.121).

The Melnikov function will give us information concerning the behavior of the stable and unstable manifolds of the torus in the perturbed system. From (4.1.89), the Melnikov function is given by

$$M^{\pm}(t_0, I; \delta, \gamma, \omega, \Gamma, \Omega) = \int_{-\infty}^{\infty} \left[-\delta\left(x_{2h}^{\pm}(t)\right)^2 + \gamma \omega t \cos I\, x_{2h}^{\pm}(t) \sin x_{1h}^{\pm}(t) \right.$$
$$\left. + \Gamma x_{2h}^{\pm}(t) \sin x_{1h}^{\pm}(t) \sin \Omega(t + t_0) \right] dt . \tag{4.2.124}$$

Using (4.2.31), we obtain

$$M^+(t_0, I; \delta, \gamma, \omega, \Gamma, \Omega) = M^-(t_0, I; \delta, \gamma, \omega, \Gamma, \Omega) \equiv M(t_0, I; \delta, \gamma, \omega, \Gamma, \Omega)$$

$$= -8\delta\sqrt{1 - \gamma\sin I} + \frac{4\gamma\omega\cos I}{\sqrt{1 - \gamma\sin I}} + \frac{2\pi\Omega^2\Gamma}{\sinh\left[\frac{\pi\Omega}{2\sqrt{1-\gamma\sin I}}\right]}\cos\Omega t_0 \,.$$

$$(4.2.125)$$

Using arguments similar to those given for the quasiperiodically forced Duffing oscillator, (4.2.125) can be used to prove the existence of transverse homoclinic tori in (4.2.120). However, the term $\sinh\frac{\pi\Omega}{2\sqrt{1-\gamma\sin I}}$ renders the necessary computations analytically intractable, and the details will probably need to be carried out numerically. We leave this to the interested reader.

4.3. Final Remarks

i) Heteroclinic Orbits

In this chapter, we developed techniques for measuring the distance between the stable and unstable manifolds of an invariant torus in three general classes of ordinary differential equations, i.e., we developed techniques for determining the existence of orbits homoclinic to tori. Analogous techniques for detecting heteroclinic orbits in the same classes of systems are developed in precisely the same manner. The only difference is that the unperturbed system is assumed to have two normally hyperbolic invariant manifolds, say M_1 and M_2, which are connected to each other by a manifold of heteroclinic orbits. The geometry of the splitting of the manifolds, the measure of the splitting of the manifolds, and the Melnikov vector are the same as in the homoclinic case for orbits heteroclinic to tori of the *same dimension*. The case of orbits heteroclinic to different types of invariant sets (e.g., a periodic orbit and a two torus) has not been worked out as yet.

ii) Additional Applications of Melnikov's Method

The following are references to additional applications of Melnikov's Method which are grouped according to the different fields.

Fluid Mechanics. Holmes [1985], Knobloch and Weiss [1986], Slemrod and Marsden [1985], Suresh [1985], Ziglin [1980], Rom-Kedar, Leonard, and Wiggins [1987].

The Josephson Junction. Holmes [1981], Salam and Sastry [1985], Hockett and Holmes [1986].

Power System Dynamics. Kopell and Washburn [1982], Salam, Marsden, and Varaiya [1984].

Condensed Matter Physics. Coullet and Elphick [1987], Koch [1986].

Rigid Body Dynamics. Holmes and Marsden [1981], [1983], Koiller [1984], Shaw and Wiggins [1988].

Yang-Mills Field Theory. Nikolaevskii and Shchur [1983].

Power Spectra of Strange Attractors. Brunsden and Holmes [1987].

iii) Exponentially Small Melnikov Functions

The methods of averaging and normal forms are often useful techniques for transforming analytically intractable problems into "almost" tractable problems, see Guckenheimer and Holmes [1983] or Sanders and Verhulst [1985]. In particular, these techniques can often be utilized to transform systems into "near integrable" systems which appear to be ideal candidates for Melnikov type analyses. However, formidable mathematical difficulties lie lurking in the background. We will briefly sketch the main problem.

Suppose we have a planar, time periodic ordinary differential equation in the standard form for application of the method of averaging, i.e.,

$$\dot{x} = \epsilon f(x,t) + \epsilon^2 g(x,t;\epsilon) \qquad (4.2.126)$$

with $0 < \epsilon \ll 1$, f and g $C^r, r \geq 2$, and bounded on bounded subsets of $\mathbb{R}^2$ for each $t \in [0,T)$ where T is the period, and $\lim_{\epsilon \to 0} g(x,t;\epsilon)$ exists uniformly. Application of the averaging transformation to (4.2.126) yields

$$\dot{x} = \epsilon \bar{f}(x) + \epsilon^2 \tilde{g}(x,t;\epsilon) \qquad (4.2.127)$$

where $\bar{f}(x) = \frac{1}{T} \int\limits_0^T f(x,t)dt$ (the exact form of $\tilde{g}$ is not important to us but it can be found in Guckenheimer and Holmes [1983] or Sanders and Verhulst [1985]). Rescaling time, $t \to t/\epsilon$, transforms (4.2.127) into

$$\dot{x} = \bar{f}(x) + \epsilon \tilde{g}(x,\frac{t}{\epsilon},\epsilon) . \qquad (4.2.128)$$

Now suppose that, for $\epsilon = 0$, (4.2.128) is Hamiltonian with Hamiltonian function $H(x)$ and, in particular, it has a homoclinic orbit $q_0(t)$ connecting a hyperbolic fixed point p_0 to itself. Then the Melnikov function for (4.2.128) is given by

$$M\left(\frac{t_0}{\epsilon}\right) = \int\limits_{-\infty}^{\infty} \langle D_x H(q_0(t)), \tilde{g}(q_0(t), \frac{t + t_0}{\epsilon}; 0)\rangle dt \quad . \tag{4.2.129}$$

From (4.2.129) the problem should be apparent; that is, the Melnikov function depends explicitly on ϵ. Moreover, the relatively rapid oscillation (period ϵT) in general results in the Melnikov function being exponentially small in ϵ. Thus, without a careful consideration of the errors in the expansion of the formula for the distance between the manifolds in powers of ϵ, we cannot claim that the $O(\epsilon)$ term (i.e. the Melnikov function) dominates the higher order terms for sufficiently small ϵ. In particular, Theorems 4.1.9, 4.1.13, and 4.1.19 are not valid. These problems were first pointed out by Sanders [1982]. Let us illustrate with a specific calculation.

Consider the simple pendulum

$$\ddot{x} + \sin x = \epsilon \sin \omega t . \tag{4.2.130}$$

The Melnikov function for (4.2.130) is given by

$$M(t_0) = 2\pi \text{sech}\left(\frac{\pi \omega}{2}\right) \cos \omega t_0 . \tag{4.2.131}$$

The *splitting distance* of the separatrices is proportional to

$$d_{\text{split}} \approx \epsilon \max_{t_0} |M(t_0)| + O(\epsilon^2) . \tag{4.2.132}$$

Next consider the rapidly forced pendulum

$$\ddot{x} + \sin x = \epsilon \sin \frac{t}{\epsilon} . \tag{4.2.133}$$

Using (4.2.131), for (4.2.133) we find

$$\max_{t_0} |M(t_0)| \approx 2\pi e^{\frac{-\pi}{2\epsilon}} . \tag{4.2.134}$$

Thus the Melnikov function is smaller than any power of ϵ.

A theoretical breakthrough on problems of this sort has recently been made by Holmes, Marsden and Scheurle [1987 a,b], and there are two results dealing with these problems which we now mention.

Upper Estimate *Consider*

$$\ddot{x} + \sin x = \delta \sin(t/\epsilon).$$

For any $\eta > 0$ there is a $\delta_0 > 0$ and a constant $C = C(\eta, \delta_0)$ such that, for all ϵ and δ satisfying $0 < \epsilon \leq 1$ and $0 < \delta \leq \delta_0$, we have

$$\text{splitting distance} \ \leq C\delta \exp\left[-\left(\frac{\pi}{2} - \eta\right)\frac{1}{\epsilon}\right].$$

Lower Estimate And Sharp Upper Estimate *Consider*

$$\ddot{x} + \sin x = \epsilon^p \delta \ \sin(t/\epsilon).$$

If $p \geq 8$, then there is a $\delta_0 > 0$ and (absolute) constants C_1 and C_2 such that, for all ϵ, δ satisfying $0 < \epsilon \leq 1$ and $0 < \delta \leq \delta_0$, we have

$$C_2 \epsilon^p \delta e^{-\pi/2\epsilon} \ \leq \text{splitting distance} \ \leq C_1 \epsilon^p \delta e^{-\pi/2\epsilon}.$$

These estimates are special cases of estimates for a planar system

$$\dot{u} = g(u, \epsilon) + \epsilon^p \delta h\left(u, \epsilon, \frac{t}{\epsilon}\right), \qquad u \in \mathbb{R}^2$$

where one assumes:

1) g and h are entire in u and ϵ;

2) h is of Sobolev class H^1 (for the splitting distance results) and T-periodic in the variable $\theta = t/\epsilon$;

3) $\dot{u} = g(u, \epsilon)$ has a homoclinic orbit $\bar{u}(\epsilon, t)$ which is analytic on a strip in the complex t-plane, with width r.

One needs to make additional assumptions on the fundamental solution of the first variation equation

$$\dot{v} = D_\mu g(\bar{u}, \epsilon) \cdot v$$

which can be checked to hold in the pendulum example. There are analogues of the upper and lower estimates above for this general situation, with $\pi/2$ replaced by a positive constant r; we refer to Holmes, Marsden, and Scheurle [1987a] for details. The proofs depend on detailed estimates of the terms in an iterative process in the complex strip that are used to define the invariant manifolds. It is important to extend these iterates to the complex strip in the proper way; for example, $\sin(\frac{t}{\epsilon})$

becomes very large for complex t, and naively extended iteration procedures for the stable and unstable manifolds will lead to unbounded sequences of functions.

The Holmes-Marsden-Scheurle techniques have immediate applications to the structure of the resonance bands in KAM theory and to the unfolding of degenerate singularities of vector fields. Specific examples can be found in Holmes, Marsden, and Scheurle [1987b].

REFERENCES AND INDEX

REFERENCES

Abraham, R. H., and Marsden, J. E. [1978]. *Foundations of Mechanics*. Benjamin/Cummings: Reading, MA.

Abraham, R. H., and Shaw, C. D. [1984]. *Dynamics – The Geometry of Behavior, Part Three: Global Behavior*. Aerial Press, Inc.: Santa Cruz.

Afraimovich, V. S., Bykov, V. V., and Silnikov, L. P. [1983]. On structurally unstable attracting limit sets of Lorenz attractor type. *Trans. Mosc. Math. Soc.*, **2**, 153–216.

Andronov, A. A., and Pontryagin, L. [1937]. Systèmes grossiers. *Dokl. Akad. Nauk. SSSR*, **14**, 247–251.

Andronov, A. A., Leontovich, E. A., Gordon, I. I., and Maier, A. G. [1971]. *Theory of Bifurcations of Dynamic Systems on a Plane*. Israel Program of Scientific Translations: Jerusalem.

Armbruster, D., Guckenheimer, J., and Holmes, P. [1988]. Heteroclinic cycles and modulated traveling waves in systems with $O(2)$ symmetry. To appear *Physica D*.

Arnéodo A., Coullet, P., and Tresser, C. [1981a]. A possible new mechanism for the onset of turbulence. *Phys. Lett.* 81A, 197–201.

Arnéodo A., Coullet, P., and Tresser, C. [1981b]. Possible new strange attractors with spiral structure. *Comm. Math. Phys.*, **79**, 573–579.

Arnéodo A., Coullet, P., and Tresser, C. [1982]. Oscillators with chaotic behavior: An illustration of a theorem by Shil'nikov. *J. Stat. Phys.*, **27**, 171–182.

Arnéodo A., Coullet, P., and Spiegel, E. [1982]. Chaos in a finite macroscopic system. *Phys. Lett.* 92A, 369–373.

Arnéodo A., Coullet, P., Spiegel, E., and Tresser, C. [1985]. Asymptotic chaos. *Physica* 14D, 327–347.

Arnold, V. I. [1964]. Instability of dynamical systems with many degrees of freedom. *Sov. Math. Dokl*, **5**, 581–585.

Arnold, V. I. [1973]. *Ordinary Differential Equations*. M.I.T. Press: Cambridge MA.

Arnold, V. I. [1978]. *Mathematical Methods of Classical Mechanics*. Springer-Verlag: New York, Heidelberg, Berlin.

Arnold, V. I. [1982]. *Geometrical Methods in the Theory of Ordinary Differential*

Equations. Springer-Verlag: New York, Heidelberg, Berlin..

Aronson, D. G., Chory, M. A., Hall, G. R., and McGeehee, R. P. [1982]. Bifurcations from an invariant circle for two parameter families of maps of the plane: A computer assisted study. *Comm. Math. Phys.*, **83**, 303–354.

Aubry, N., Holmes, P. J., Lumley, J. L., and Stone, E. [1986]. The dynamics of coherent structures in the wall region of a turbulent boundary layer. To appear *J. Fluid Mech.*

Bernoussou, J. [1977]. *Point Mapping Stability.* Pergamon: Oxford.

Birkhoff, G. D. [1927]. *Dynamical Systems.* A. M. S. Publications: Providence, reprinted 1966.

Birman, J. S., and Williams, R. F. [1983a]. Knotted periodic orbits in dynamical systems I: Lorenz's equations. *Topology*, **22**, 47–82.

Birman, J. S., and Williams, R. F. [1983b]. Knotted periodic orbits in dynamical systems II: Knot holders for fibred knots. *Contemp. Math.*, **20**, 1–60.

Bogoliobov, N., and Mitropolsky, Y. [1961]. *Asymptotic Methods in the Theory of Nonlinear Oscillations.* Gordon and Breach: New York.

Bost, J. [1986]. Tores Invariants des Systèmes Dynamiques Hamiltoniens. *Astérisque*, 133–134. 113–157.

Bowen, R. [1970]. Markov partitions for Axiom A diffeomorphisms. *Amer. J. Math.*, **92**, 725–747.

Bowen, R. [1978]. *On Axiom A Diffeomorphisms*, CBMS Regional Conference Series in Mathematics, Vol. 35, A.M.S. Publications: Providence.

Brunsden, V., and Holmes, P. J. [1987]. Power spectra of strange attractors near homoclinic orbits. *Phys. Rev. Lett.*, **58**, 1699-1702.

Carr, J. [1981]. *Applications of Center Manifold Theory.* Springer-Verlag: New York, Heidelberg, Berlin.

Carr, J. [1983]. Phase transitions via bifurcation from heteroclinic orbits. In *Proc. of the NATO Advanced Study Institute on Systems of Nonlinear Partial Differential Equations*, J. Ball (ed.), pp.333–342. D. Reidel: Holland.

Chillingworth, D. R. J. [1976]. *Differentiable Topology with a View to Applications.* Pitman: London.

Chow, S. N., Hale, J. K., and Mallet-Paret, J. [1980]. An example of bifurcation to homoclinic orbits. *J. Diff. Eqns.*, **37**, 351–373.

Chow, S. N., and Hale, J. K. [1982]. *Methods of Bifurcation Theory.* Springer-Verlag: New York, Heidelberg, Berlin..

Coddington, E. A., and Levinson, N. [1955]. *Theory of Ordinary Differential Equations.* McGraw-Hill: New York.

Conley, C. [1975]. On traveling wave solutions of nonlinear diffusion equations. In *Dynamical Systems and Applications.* Springer Lecture Notes in Physics, 38, pp. 498–510, Springer-Verlag: New York, Heidelberg, Berlin.

Conley, C. [1978]. *Isolated Invariant Sets and the Morse Index.* CBMS Regional Conference Series in Mathematics, Vol. 38, A.M.S. Publications: Providence.

Coppel, W. A. [1978]. *Dichotomies in Stability Theory.* Springer Lecture Notes in Mathematics, Vol. 629. Springer-Verlag: New York, Heidelberg, Berlin.

Coullet, P., and Elphick, C. [1987]. Topological defects dynamics and Melnikov's Theory. *Phys. Lett.*, 121A, 233–236.

Dangelmayr, G., and Guckenheimer, J. [1987]. On a four parameter family of planar vector fields. *Arch. Rat. Mech. Anal.*, 97, 321–352.

Devaney, R. [1976]. Homoclinic orbits in Hamiltonian systems. *J. Diff. Eqns.*, 21, 431–438.

Devaney, R. [1977]. Blue sky catastrophes in reversible and Hamiltonian systems. *Ind. Univ. J. Math.*, 26, 247–263.

Devaney, R. [1978]. Transversal homoclinic orbits in an integrable system. *Amer. J. Math.*, 100, 631–642.

Devaney, R. [1986]. *An Introduction to Chaotic Dynamical Systems.* Benjamin/ Cummings: Menlo Park, CA.

Devaney, R., and Nitecki, Z. [1979]. Shift automorphisms in the Hénon mapping. *Comm. Math. Phys.*, 67, 137–148.

Diliberto, S. P. [1960]. Perturbation theorems for periodic surfaces, I. *Rend. Circ. Mat. Palermo*, 9, 265–299.

Diliberto, S. P. [1961]. Perturbation theorems for periodic surfaces, II. *Rend. Circ. Mat. Palermo*, 10, 111.

Dugundji, J. [1966]. *Topology.* Allyn and Bacon: Boston.

Easton, R. W. [1978]. Homoclinic phenomena in Hamiltonian systems with several degrees of freedom. *J. Diff. Eqns.*, 29, 241–252.

Easton, R. W. [1981]. Orbit structure near trajectories biasymptotic to invariant tori. In *Classical Mechanics and Dynamical Systems*, R. Devaney and Z. Nitecki (eds.), pp. 55–67. Marcel Dekker, Inc: New York.

Evans, J. W., Fenichel, N., and Feroe, J. A. [1982]. Double impulse solutions in nerve axon equations. *SIAM J. Appl. Math.*, 42(2), 219–234.

Fenichel, N. [1971]. Persistence and smoothness of invariant manifolds for flows. *Ind. Univ. Math. J.*, 21, 193–225.

Fenichel, N. [1974]. Asymptotic stability with rate conditions. *Ind. Univ. Math. J.*, 23, 1109–1137.

Fenichel, N. [1977]. Asymptotic stability with rate conditions, II. *Ind. Univ. Math. J.*, 26, 81–93.

Fenichel, N. [1979]. Geometric singular perturbation theory for ordinary differential equations. *J. Diff. Eqns.*, 31, 53–98.

Feroe, J. A. [1982]. Existence and stability of multiple impulse solutions of a nerve equation. *SIAM J. Appl. Math.*, 42, 235–246.

Fowler, A. C., and Sparrow, C. [1984]. Bifocal homoclinic orbits in four dimensions. University of Oxford preprint.

Franks, J. M. [1982]. *Homology and Dynamical Systems.* CBMS Regional Conference Series in Mathematics, Vol. 49, A.M.S. Publications: Providence.

Gaspard, P. [1983]. Generation of a countable set of homoclinic flows through bifurcation. *Phys. Lett.* , 97A, 1–4.

Gaspard, P., and Nicolis, G. [1983]. What can we learn from homoclinic orbits in chaotic systems? *J. Stat. Phys.*, 31, 499–518.

Gaspard, P., Kapral, R., and Nicolis, G. [1984]. Bifurcation phenomena near homoclinic systems: A two parameter analysis. *J. Stat. Phys.*, **35**, 697–727.

Gavrilov, N. K., and Silnikov, L. P. [1972]. On three dimensional dynamical systems
close to systems with a structurally unstable homoclinic curve, I. *Math. USSR
Sb.*, **17**, 467–485.

Gavrilov, N. K., and Silnikov, L. P. [1973]. On three dimensional dynamical systems
close to systems with a structurally unstable homoclinic curve, II. *Math. USSR
Sb.*, **19**, 139–156.

Glendinning, P. [1984]. Bifurcations near homoclinic orbits with symmetry. *Phys.
Lett.*, **103A**, 163–166.

Glendinning, P. [1987]. Subsidiary bifurcations near bifocal homoclinic orbits. University of Warwick preprint.

Glendinning, P. [1987]. Homoclinic Bifurcations in Ordinary Differential Equations.
In NATO ASI Life Sciences Series, *Chaos in Biological Systems*, H. Degn, A.
V. Holden, and L. Olsen (eds.). Plenum: New York.

Glendinning, P., and Sparrow, C. [1984]. Local and global behavior near homoclinic
orbits. *J. Stat. Phys.*, **35**, 645–696.

Glendinning, P., and Tresser, C. [1985]. Heteroclinic loops leading to hyperchaos.
J. Physique Lett., **46**, L347–L352.

Goldstein, H. [1980]. *Classical Mechanics*, 2nd ed., Addison-Wesley: Reading.

Golubitsky, M., and Guillemin, V. [1973]. *Stable Mappings and Their Singularities*.
Springer-Verlag: New York, Heidelberg, Berlin..

Grebenikov, E. A., and Ryabov, Yu. A. [1983]. *Constructive Methods in the Analysis
of Nonlinear Systems*. Mir: Moscow.

Greenspan, B. D., and Holmes, P. J. [1983]. Homoclinic orbits, subharmonics, and
global bifurcations in forced oscillations. In *Nonlinear Dynamics and Turbulence*, G. Barenblatt, G. Iooss, and D.D. Joseph (eds.), pp. 172–214. Pitman,
London.

Greundler, J. [1985]. The existence of homoclinic orbits and the method of Melnikov
for systems in $\mathbf{R}^n$. *SIAM J. Math. Anal.*, **16**, 907–931.

Guckenheimer, J. [1981]. On a codimension two bifurcation. In *Dynamical Systems
and Turbulence*, D. A. Rand and L. S. Young (eds.), pp. 99–142. Springer Lecture Notes in Mathematics, Vol. 898. Springer-Verlag: New York, Heidelberg,
Berlin.

Guckenheimer, J., and Holmes, P. J. [1983]. *Nonlinear Oscillations, Dynamical Systems, and Bifurcations of Vector Fields*. Springer-Verlag: New York, Heidelberg, Berlin.

Guckenheimer, J., and Holmes, P. J. [1986]. Structurally stable heteroclinic cycles.
Cornell University preprint.

Guckenheimer, J., and Williams, R. F. [1980]. Structural stability of the Lorenz
attractor. *Publ. Math. IHES*, **50**, 73–100.

Hadamard, J. [1901]. Sur l'itération et les solutions asymptotiques des équations
différentielles. *Bull. Soc. Math. France*, **29**, 224–228.

Hale, J. [1961]. Integral manifolds of perturbed differential systems. *Ann. Math.*,
73, 496–531.

Hale, J. [1980]. *Ordinary Differential Equations.* Robert E. Krieger Publishing Co., Inc.: Malabar Florida.

Hale, J., Magalhães, L., and Oliva, W. [1984]. *An Introduction to Infinite Dimensional Dynamical Systems - Geometric Theory.* Springer-Verlag: New York, Heidelberg, Berlin..

Hartman, P. [1964]. *Ordinary Differential Equations.* Wiley: New York.

Hastings, S. [1982]. Single and multiple pulse waves for the Fitzhugh-Nagumo equations. *SIAM J. Appl. Math.,* 42, 247–260.

Hausdorff. [1962]. *Set Theory.* Chelsea: New York.

Henry, D. [1981]. *Geometric Theory of Semilinear Parabolic Equations.* Springer Lecture Notes in Mathematics, Vol. 840. Springer-Verlag: New York, Heidelberg, Berlin..

Hirsch, M. W. [1976]. *Differential Topology.* Springer-Verlag: New York, Heidelberg, Berlin..

Hirsch, M. W., and Pugh, C. C. [1970]. Stable manifolds and hyperbolic sets. *Proc. Symp. Pure Math.,* 14, 133–163.

Hirsch, M. W., Pugh, C. C., and Shub, M. [1977]. *Invariant Manifolds.* Springer Lecture Notes in Mathematics, Vol. 583. Springer-Verlag: New York, Heidelberg, Berlin..

Hirsch, M. W., and Smale, S. [1974]. *Differential Equations, Dynamical Systems, and Linear Algebra.* Academic Press: New York.

Hockett, K. and Holmes, P. [1986]. Josephson's junction, annulus maps, Birkhoff attractors, horseshoes, and rotation sets. *Ergod. Th. and Dynam. Sys.,* 6, 205–239.

Holmes, P. J. [1979]. A nonlinear oscillator with a strange attractor. *Phil. Trans. Roy. Soc. A,* 292, 419-448.

Holmes, P. J. [1980]. A strange family of three-dimensional vector fields near a degenerate singularity. *J. Diff. Eqns.,* 37, 382–404.

Holmes, P. J. [1980]. Periodic, nonperiodic, and irregular motions in a Hamiltonian system. *Rocky Mountain J. Math.,* 10, 679–693.

Holmes, P. J. [1981]. Space and time-periodic perturbations of the Sine-Gordon equation. *Proc. Univ. of Warwick Conference on Turbulence and Dynamical Systems,* Springer Lecture Series in Math., Vol. 898, D. A. Rand and L. S. Young (eds.), pp. 164–191. Springer-Verlag: New York, Heidelberg, Berlin.

Holmes, P. J. [1985]. Chaotic motions in a weakly nonlinear model for surface waves. *J. Fluid Mech.,* 162, 365–388.

Holmes, P. J. [1986]. Knotted periodic orbits in suspensions of Smale's horseshoe: Period multiplying and cabled knots. *Physica,* 21D, 7–41.

Holmes, P. J. [1987]. Knotted periodic orbits in suspensions of annulus maps. *Proc. R. Soc. Lond.,* A 411, 351–378.

Holmes, P. J., and Marsden, J. E. [1981]. A partial differential equation with infinitely many periodic orbits: Chaotic oscillations of a forced beam. *Arch. Rat. Mech. Analysis,* 76, 135–166.

Holmes, P. J., and Marsden, J. E. [1982a]. Horseshoes in perturbations of Hamiltonian systems with two degrees of freedom. *Comm. Math. Phys.,* 82, 523–544.

Holmes, P. J., and Marsden, J. E. [1982b]. Melnikov's method and Arnold diffusion for perturbations of integrable Hamiltonian systems. *J. Math. Phys.*, **23**, 669-675.

Holmes, P. J., and Marsden, J. E. [1983]. Horseshoes and Arnold diffusion for Hamiltonian systems on Lie groups, *Indiana Univ. Math J.*, **32**, 273-309.

Holmes, P. J., Marsden, J. E., and Scheurle, J. [1987a]. Exponentially small splitting of separatrices. Preprint.

Holmes, P. J., Marsden, J. E., and Scheurle, J. [1987b]. Exponentially small splittings of separatrices in KAM theory and degenerate bifurcations. Preprint.

Holmes, P., and Moon, F. [1983]. Strange attractors and chaos in nonlinear mechanics. *J. App. Mech.*, **50**, 1021-1032.

Holmes, P. J., and Williams, R. F. [1985]. Knotted periodic orbits in suspensions of Smale's horseshoe: Torus knots and bifurcation sequences. *Arch. Rat. Mech. Analysis*, **90**, 115-194.

Hufford, G. [1956]. Banach spaces and the perturbation of ordinary differential equations. In *Contributions to the Theory of Nonlinear Oscillations, III*. Annals of Mathematical Studies, Vol 36. Princeton University Press: Princeton.

Irwin, M. C. [1980]. *Smooth Dynamical Systems*. Academic Press: London, New York.

Katok, A. [1983]. Periodic and quasiperiodic orbits for twist maps. Springer Lecture Notes in Physics, Vol. 179, pp. 47-65. Springer-Verlag: New York, Heidelberg, Berlin..

Kelley, A. [1967]. The stable, center-stable, center, center-unstable, unstable manifolds. An appendix in *Transversal Mappings and Flows* by R. Abraham and J. Robbin. Benjamin: New York.

Knobloch, E., and Proctor, M. R. E. [1981]. Nonlinear periodic convection in double diffusive systems. *J. Fluid. Mech.*, **108**, 291-313.

Knobloch, E., and Weiss, J. B. [1986]. Chaotic advection by modulated travelling waves. U.C. Berkeley Physics preprint.

Koch, B. P. [1986]. Horseshoes in superfluid ^3He. *Phys. Lett.*, 117A, 302-306.

Koiller, J. [1984]. A mechanical system with a "wild" horseshoe. *J. Math. Phys.*, **25**, 1599-1604.

Kolmogorov, A. N. [1954]. On conservation of conditionally periodic motions for a small change in Hamilton's function. *Dokl. Akad. Nauk. SSSR*, 98:4, 525-530 (Russian).

Kopell, N. [1977]. Waves, shocks, and target patterns in an oscillating chemical reagent. *Proceedings of the 1976 Symposium on Nonlinear Equations at Houston*. Pitman Press: London.

Kopell, N., and Washburn, R. B. [1982]. Chaotic motion in the two-degree of freedom swing equation. *IEEE Trans. Circ.*, Vol. CAS-24, 738-46.

Krishnaprasad, P. S., and Marsden, J. E. [1987]. Hamiltonian structures and stability for rigid bodies with flexible attachments. *Arch. Rat. Mech. Anal.*, **98**, 71-93.

Kurzweil, J. [1968]. Invariant manifolds of differential systems. *Differencial'nye Uravnenija*, **4**, 785-797.

Kyner, W. T. [1956]. A fixed point theorem. In *Contributions to the Theory of Nonlinear Oscillations, III*. Annals of Mathematical Studies, Vol 36. Princeton University Press: Princeton.

Lamb, H. [1945]. *Hydrodynamics*. Dover: New York.

Langford, W. F. [1979]. Periodic and steady mode interactions lead to tori. *SIAM J. Appl. Math.*, **37** (1), 22–48.

LaSalle, J. P. [1976]. *The Stability of Dynamical Systems*. CBMS Regional Conference Series in Applied Mathematics, Vol. 25. SIAM: Philadelphia.

Lerman, L. M., Umanski, Ia. L. [1984]. On the existence of separatrix loops in four dimensional systems similar to integrable Hamiltonian systems. *PMM U.S.S.R.*, **47**, 335–340.

Levinson, N. [1950]. Small periodic perturbations of an autonomous system with a stable orbit. *Ann. Math.*, **52**, 727–738.

Lichtenberg, A. J., and Lieberman, M. A. [1982]. *Regular and Stochastic Motion*. Springer-Verlag: New York, Heidelberg, Berlin.

Lorenz, E. N. [1984]. The local structure of a chaotic attractor in four dimensions. *Physica* 13D, 90–104.

Marcus, M. [1956]. An invariant surface theorem for a non-degenerate system. In *Contributions to the Theory of Nonlinear Oscillations, III*. Annals of Mathematical Studies, Vol. 36. Princeton University Press: Princeton.

Marsden, J., and Weinstein, A. [1974]. Reduction of symplectic manifolds with symmetry. *Rep. Math. Phys.*, **5**, 121–130.

McCarthy, J. [1955]. Stability of invariant manifolds, abstract. *Bull. Amer. Math. Soc.*, **61**, 149–150.

Melnikov, V. K. [1963]. On the stability of the center for time periodic perturbations. *Trans. Moscow Math.*, **12**, 1–57.

Meyer, K. R., and Sell, G. R. [1986]. Homoclinic orbits and Bernoulli bundles in almost periodic systems. Univ. of Cincinnati preprint.

Milnor, J. W. [1965]. *Topology from the Differentiable Viewpoint*. University of Virginia Press: Charlottesville.

Moon, F.C. and Holmes, W.T. [1985]. Double Poincaré sections of a quasiperiodically forced, chaotic attractor. *Phys. Lett.* **111A**, 157–160.

Moon, F. C. [1980]. Experiments on chaotic motions of a forced nonlinear oscillator. *J. Appl. Mech.*, **47**, 638–644.

Moser, J. [1966a]. A rapidly convergent iteration method and nonlinear partial differential equations, I. *Ann. Scuola Norm. Sup. Pisa*, Ser. III, **20**, 265–315.

Moser, J. [1966b]. A rapidly convergent iteration method and nonlinear differential equations, II. *Ann. Scuola Norm. Sup. Pisa*, Ser. III, **20**, 499–535.

Moser, J. [1973]. *Stable and Random Motions in Dynamical Systems*. Princeton University Press: Princeton.

Nehoroshev, N. N. [1971]. On the behavior of Hamiltonian systems close to integrable ones. *Funct. Anal. Appl.*, **5**, 82–83.

Nehoroshev, N. N. [1972]. Exponential estimate of the time of stability of nearly integrable Hamiltonian systems. *Russ. Math. Surv.*, **32**(6), 1–65.

Newhouse, S. E. [1974]. Diffeomorphisms with infinitely many sinks. *Topology*, **13**, 9–18.

Newhouse, S. [1979]. The abundance of wild hyperbolic sets and non-smooth stable sets for diffeomorphisms. *Publ. Math. IHES*, **50**, 101–151.

Newhouse, S., and Palis, J. [1973]. Bifurcations of Morse-Smale dynamical systems. In *Dynamical Systems*, M. M. Peixoto (ed.). Academic Press: New York and London.

Nikolaevskii, E. S., and Shchur, L. N. [1983]. The nonintegrability of the classical Yang-Mills equations. *Soviet Physics JETP*, **58**, 1–7.

Nitecki. Z. [1971]. *Differentiable Dynamics*. M.I.T. Press: Cambridge.

Palis, J., and deMelo, W. [1982]. *Geometric Theory of Dynamical Systems: An Introduction*. Springer-Verlag: New York, Heidelberg, Berlin..

Palmer, K. J. [1984]. Exponential dichotomies and transversal homoclinic points. *J. Diff. Eqns.*, **55**, 225–256.

Peixoto, M. M. [1962]. Structural stability on two-dimensional manifolds. *Topology*, **1**, 101–120.

Perron, O. [1928]. Über Stabilität und asymptotisches verhalten der Integrale von Differentialgleichungssystem. *Math. Z.*, **29**, 129–160.

Perron, O. [1929]. Über Stabilität und asymptotisches verhalten der Losungen eines systems endlicher Differenzengleichungen. *J. Reine Angew. Math.*, **161**, 41–64.

Perron, O. [1930]. Die Stabilitäts Frage bei Differentialgleichungen. *Math. Z.*, **32**, 703–728.

Pesin, Ja. B. [1977]. Characteristic Lyapunov exponents and smooth ergodic theory. *Russ. Math. Surv.*, **32**, 55–114.

Pesin, Ja. B. [1976]. Families of invariant manifolds corresponding to nonzero characteristic exponents. *Math. USSR Izvestiji*, **10**, 1261–1305.

Pikovskii, A.S., Rabinovich, M.I., and Trakhtengerts, V.Yu. [1979]. Onset of stochasticity in decay confinement of parametric instability. *Sov. Phys. JETP*, **47**, 715–719.

Poincaré, H. [1899]. *Les Méthodes Nouvelles de la Mécanique Céleste*, 3 Vols. Gauthier-Villars: Paris.

Rabinovich, M.I. [1978]. Stochastic self-oscillations and turbulence. *Sov. Phys. Usp.*, **21**, 443–469.

Rabinovich, M.I., and Fabrikant, A.L. [1979]. Stochastic self-oscillation of waves in non-
equilibrium media. *Sov. Phys. JETP*, **50**, 311–323.

Robinson, C. [1983]. Sustained resonance for a nonlinear system with slowly varying coefficients. *SIAM J. Math. Anal.*, **14**, 847–860.

Robinson, C. [1985]. Bifurcation to infinitely many sinks. *Comm. Math. Phys.*, **90**, 433–459.

Robinson, C. [1985]. Horseshoes for autonomous Hamiltonian systems using the Melnikov integral. Northwestern University preprint.

Robinson, R. C. [1970]. Generic properties of conservative systems. *Am. J. Math.*, **92**, 562–603.

Robinson, R. C. [1970]. Generic properties of conservative systems II. *Am. J. Math.*, **92**, 897–906.

Rom-Kedar, V., Leonard, A., and Wiggins, S. [1988]. An analytical study of transport, mixing, and chaos in an unsteady vortical flow. Caltech preprint.

Rouche, N., Habets, P., and Laloy, M. [1977]. *Stability Theory by Liapunov's Direct Method*. Springer-Verlag: New York, Heidelberg, Berlin..

Roux, J.C., Rossi, A., Bachelart, S., and Vidal, C. [1981]. Experimental observations of complex dynamical behavior during a chemical reaction. *Physica*, 2D, 395–403.

Sacker, R. J. [1964] *On Invariant Surfaces and Bifurcation of Periodic Solutions of Ordinary Differential Equations*. New York University, IMM-NYU 333.

Sacker, R. J., and Sell G. R. [1974]. Existence of dichotomies and invariant splittings for linear differential systems. *J. Diff. Eqns.*, **15**, 429–458.

Sacker, R. J., and Sell G. R. [1978]. A spectral theory for linear differential systems. *J. Diff. Eqns.*, **27**, 320–358.

Salam, F. M. A. [1987]. The Melnikov technique for highly dissipative systems. *SIAM J. App. Math.*, **47**, 232–243.

Salam, F. M. A., and Sastry, S. S. [1985]. Dynamics of the forced Josephson junction: The regions of chaos. *IEEE Trans. Circ. Syst.*, Vol. CAS-32, No. 8, 784–796.

Salam, F. M. A., Marsden, J. E., and Varaiya, P. P. [1984]. Arnold diffusion in the swing equations of a power system. *IEEE Trans. Circ. Sys.*,, Vol CAS-31, no. 8, 673–688.

Samoilenko, A.M., [1972]. On the reduction to canonical form of a dynamical system in the neighborhood of a smooth invariant torus. *Math. USSR IZV.*, **6**, 211–234.

Sanders, J. A. [1982]. Melnikov's method and averaging. *Celestial Mechanics*, **28**, 171–181.

Sanders, J. A., and Verhulst, F. [1985]. *Averaging Methods in Nonlinear Dynamical Systems*. Springer-Verlag: New York, Heidelberg, Berlin., Tokyo.

Schecter, S. [1986]. Stable manifolds in the method of averaging. North Carolina State University preprint.

Scheurle, J. [1986]. Chaotic solutions of systems with almost periodic forcing. *ZAMP*, **37**, 12–26.

Sell, G. R. [1978]. The structure of a flow in the vicinity of an almost periodic motion. *J. Diff. Eqns.*, **27**, 359–393.

Sell, G. R. [1979]. Bifurcation of higher dimensional tori. *Arch. Rat. Mech. Anal.*, **69**, 199–230.

Sell, G. R. [1981]. Hopf-Landau bifurcation near strange attractors. *Chaos and Order in Nature*, Proc. Int. Symp. on Synergetics. Springer-Verlag: New York, Heidelberg, Berlin..

Sell, G. R. [1982]. Resonance and bifurcation in Hopf-Landau dynamical systems. In *Nonlinear Dynamics and Turbulence*, G. I. Barenblatt, G. Iooss, D. D. Joseph (eds.). Pitman: London.

Shaw, S.W. and Wiggins, S. [1988]. Chaotic dynamics of a whirling pendulum. *Physica* D, to appear.

Shub, M. [1987]. *Global Stability of Dynamical Systems.* Springer-Verlag: New York, Heidelberg, Berlin..

Sijbrand, J. [1985]. Properties of center manifolds. *Trans. Amer. Math. Soc.*, **289**, 431–469.

Silnikov, L. P. [1965]. A case of the existence of a denumerable set of periodic motions. *Sov. Math. Dokl.*, **6**, 163–166.

Silnikov, L. P. [1967]. On a Poincaré-Birkhoff problem. *Math. USSR Sb.*, **3**, 353–371.

Silnikov, L. P. [1967]. The existence of a denumerable set of periodic motions in four-dimensional space in an extended neighborhood of a saddle-focus. *Sov. Math. Dokl.*, **8**(1), 54–58.

Silnikov, L. P. [1968a]. On the generation of a periodic motion from a trajectory doubly asymptotic to an equilibrium state of saddle type. *Math. USSR Sb.*, **6**, 428–438.

Silnikov, L. P. [1968b]. Structure of the neighborhood of a homoclinic tube of an invariant torus. *Sov. Math. Dokl.*, **9**, 624–628.

Silnikov, L. P. [1970]. A contribution to the problem of the structure of an extended neighborhood of a rough equilibrium state of saddle-focus type. *Math. USSR Sb.*, **10**(1), 91–102.

Slemrod, M. [1983]. An admissibility criterion for fluids exhibiting phase transitions. In *Proc. of the NATO Advanced Study Institute on Systems of Nonlinear Partial Differential Equations*, J. Ball (ed.), pp. 423–432. D. Reidel: Holland.

Slemrod, M., and Marsden, J. E. [1985]. Temporal and spatial chaos in a Van der Waals fluid due to periodic thermal fluctuations. *Adv. Appl. Math.*, **6**, 135–158.

Smale, S. [1963]. Diffeomorphisms with many periodic points. In *Differential and Combinatorial Topology*, S. S. Cairns (ed.), pp. 63–80. Princeton University Press: Princeton.

Smale, S. [1966]. Structurally stable systems are not dense. *Amer. J. Math.*, **88**, 491–496.

Smale, S. [1967]. Differentiable dynamical systems. *Bull. Amer. Math. Soc.*, **73**, 747–817.

Smale, S. [1980]. *The Mathematics of Time: Essays on Dynamical Systems, Economic Processes and Related Topics.* Springer-Verlag: New York, Heidelberg, Berlin..

Smoller, J. [1983]. *Shock Waves and Reaction-Diffusion Equations.* Springer-Verlag: New York, Heidelberg, Berlin.

Sparrow, C. [1982]. *The Lorenz Equations.* Springer-Verlag: New York, Heidelberg, Berlin..

Spivak, M. [1979]. *Differential Geometry, Vol. I,.* second edition. Publish or Perish, Inc.: Wilmington.

Suresh, A. [1985]. Nonintegrability and chaos in unsteady ideal fluid flow. *AIAA Journal*, **23**, 1285-1287.

Takens, F. [1974]. Singularities of vector fields. *Publ. Math. IHES*, **43**, 47–100.

Tresser, C. [1984]. About some theorems by L. P. Silnikov. *Ann. Inst. Henri Poincaré*, **40**, 440–461.

Vyshkind, S.Ya. and Rabinovich, M.I. [1976]. The phase stochastization mechanism and the structure of wave turbulence in dissipative media. *Sov. Phys. JETP*, **44**, 292–299.

Wiggins, S. [1988]. On the detection and dynamical consequences of orbits homoclinic to hyperbolic periodic orbits and normally hyperbolic invariant tori in a class of ordinary differential equations. To appear in *SIAM J. App. Math.*

Wiggins, S. [1987]. Chaos in the quasiperiodically forced Duffing oscillator. *Phys. Lett.*, 124A, 138–142.

Wiggins, S. [1986a]. The orbit structure near a transverse homoclinic torus. Caltech preprint.

Wiggins, S. [1986b]. A generalization of the method of Melnikov for detecting chaotic invariant sets. Caltech preprint.

Wiggins, S. and Holmes, P.J. [1987]. Homoclinic orbits in slowly varying oscillators. *SIAM J. Math Anal.*, **18**, 612–629 with errata to appear.

Williams, R. F. [1980]. Structure of Lorenz attractors. *Publ. Math. IHES*, **50**, 59–72.

Yanagida, E. [1987]. Branching of double pulse solutions from single pulse solutions in nerve axon equations. *J. Diff. Eqn.*, **66**, 243–262.

Yoshizawa, T. [1966]. *Stability Theory by Liapunov's Second Method*. Math. Soc. of Japan: Tokyo.

Ziglin, S. L. [1980]. Nonintegrability of a problem on the motion of four point vortices. *Sov. Math. Dokl.*, **21**, 296–299.

INDEX

Applied Mathematical Sciences

cont. from page ii